MODE SELECTIVE CHEMISTRY

THE JERUSALEM SYMPOSIA ON
QUANTUM CHEMISTRY AND BIOCHEMISTRY

Published by the Israel Academy of Sciences and Humanities, distributed by Academic Press (N.Y.)

1. *The Physicochemical Aspects of Carcinogenesis* (October 1968)
2. *Quantum Aspects of Heterocyclic Compounds in Chemistry and Biochemistry* (April 1969)
3. *Aromaticity, Pseudo-Aromaticity, Antiaromaticity* (April 1970)
4. *The Purines: Theory and Experiment* (April 1971)
5. *The Conformation of Biological Molecules and Polymers* (April 1972)

Published by the Israel Academy of Sciences and Humanities, distributed by D. Reidel Publishing Company (Dordrecht, Boston, Lancaster, and Tokyo)

6. *Chemical and Biochemical Reactivity* (April 1973)

Published and distributed by D. Reidel Publishing Company (Dordrecht, Boston, Lancaster, and Tokyo)

7. *Molecular and Quantum Pharmacology* (March/April 1974)
8. *Environmental Effects on Molecular Structure and Properties* (April 1975)
9. *Metal-Ligand Interactions in Organic Chemistry and Biochemistry* (April 1976)
10. *Excited States in Organic Chemistry and Biochemistry* (March 1977)
11. *Nuclear Magnetic Resonance Spectroscopy in Molecular Biology* (April 1978)
12. *Catalysis in Chemistry and Biochemistry Theory and Experiment* (April 1979)
13. *Carcinogenesis: Fundamental Mechanisms and Environmental Effects* (April/May 1980)
14. *Intermolecular Forces* (April 1981)
15. *Intermolecular Dynamics* (Maart/April 1982)
16. *Nucleic Acids: The Vectors of Life* (May 1983)
17. *Dynamics on Surfaces* (April/May 1984)
18. *Interrelationship Among Aging, Cancer and Differentiation* (April/May 1985)
19. *Tunneling* (May 1986)
20. *Large Finite Systems* (May 1987)

Published and distributed by Kluwer Academic Publishers (Dordrecht, Boston, London)

21. *Transport through Membranes: Carriers, Channels and Pumps* (May 1988)
22. *Perspectives in Photosynthesis* (May 1989)
23. *Molecular Basis of Specificity in Nucleic Acid-Drug Interaction* (May 1990)
24. *Mode Selective Chemistry* (May 1991)

VOLUME 24

MODE SELECTIVE CHEMISTRY

PROCEEDINGS OF THE TWENTY-FOURTH JERUSALEM SYMPOSIUM ON QUANTUM CHEMISTRY AND BIOCHEMISTRY HELD IN JERUSALEM, ISRAEL, MAY 20–23, 1991

Edited by

J. JORTNER
The Israel Academy of Sciences of Humanities, Jerusalem, Israel

R.D. LEVINE
The Fritz Haber Center for Molecular Dynamics, The Hebrew University of Jerusalem, Israel

and

B. PULLMAN
Institut de Biologie Physico-Chimique (Fondation Edmond de Rothschild), Paris, France

KLUWER ACADEMIC PUBLISHERS
DORDRECHT / BOSTON / LONDON

Library of Congress Cataloging-in-Publication Data

Jerusalem Symposium on Quantum Chemistry and Biochemistry (24th : 1991)
Mode selective chemistry : proceedings of the Twenty Fourth Jerusalem Symposium on Quantum Chemistry and Biochemistry, Jerusalem, May 20-23, 1991 / edited by J. Jortner, R.D. Levine, B.Pullman.
p. cm. -- (The Jerusalem symposia on quantum chemistry and biochemistry ; v. 24)
Includes index.
ISBN 0-7923-1421-2 (HB : acid-free paper)
1. Molecular dynamics--Congresses. 2. Quantum chemistry--Congresses. 3. Photochemistry--Congresses. I. Jortner, Joshua. II. Levine, Raphael D. III. Pullman, Bernard, 1919- . IV. Title. V. Series.
QD461.J47 1991
541.2'8--dc20 91-31455

ISBN 0-7923-1421-2

Published by Kluwer Academic Publishers,
P.O. Box 17, 3300 AA Dordrecht, The Netherlands.

Kluwer Academic Publishers incorporates
the publishing programmes of
D. Reidel, Martinus Nijhoff, Dr W. Junk and MTP Press.

Sold and distributed in the U.S.A. and Canada
by Kluwer Academic Publishers,
101 Philip Drive, Norwell, MA 02061, U.S.A.

In all other countries, sold and distributed
by Kluwer Academic Publishers Group,
P.O. Box 322, 3300 AH Dordrecht, The Netherlands.

Printed on acid-free paper

Printed in the Netherlands

TABLE OF CONTENTS

PREFACE

The Twenty Fourth Jerusalem Symposium reflected the high standards of these distinguished scientific meetings, which convene once a year at the Israel Academy of Sciences and Humanities in Jerusalem to discuss a specific topic in the broad area of quantum chemistry and biochemistry. The topic at this year's Jerusalem Symposium was mode selective chemistry, which constitutes a truly interdisciplinary subject of central interest in the areas of chemical physics, photochemistry and photobiology.

The main theme of the Symposium was built around the exploration of the possibility and conditions for non-statistical reaction dynamics in molecules, van der Waals molecules, clusters and condensed phases. The main issues addressed photoselective and coherent excitation modes, bottlenecks for intramolecular vibrational energy redistribution, the consequences of the internal structure of many-atom systems and of rotational-vibrational level structure for intramolecular dynamics, bond selective photodissociation, ultrafast chemical clocks for energy disposal, coherent control of photochemical reactions and nonstatistical unimolecular reaction dynamics. The interdisciplinary nature of this research area was deliberated by intensive and extensive interactions between theory and experiment. This volume provides a record of the invited lectures at the Symposium.

Held under the auspices of the Israel Academy of Sciences and Humanities and the Hebrew University of Jerusalem, the Twenty Fourth Jerusalem Symposium was sponsored by the Institut de Biologie Physico-Chimique (Fondation Edmond de Rothschild) of Paris. We wish to express our deep thanks to Baron Edmond de Rothschild for his continuous and generous support, which makes him a true partner in this important endeavour. We would also like to express our gratitude to the Administrative Staff of the Israel Academy and, in particular, to Mrs. Avigail Hyam for the efficiency and excellency of the local arrangements.

Joshua Jortner
Raphael D. Levine
Bernard Pullman

HOW WE AND MOLECULES EXPLORE MOLECULAR LANDSCAPES

R. STEPHEN BERRY
Department of Chemistry and the James Franck Institute
The University of Chicago
5735 South Ellis Avenue
Chicago, Illinois 60637
U. S. A.

ABSTRACT. The multidimensional potential surfaces of clusters and polyatomic molecules can now be mapped by efficient techniques for finding minima and saddle points and, from them, "reaction paths". How the geometry and topology of the surface arise from interactions between individual particles can be learned by studying the changes in the maps as the elementary forces are varied systematically. The influence of the form of the multidimensional potential on the dynamics can be inferred from dynamic simulations on those systematically varied surfaces. For example the conditions required for a classical system to exhibit bottlenecks or coexisting phases can be inferred in this way. The extent of chaotic and ergodic character of the atomic motion within an isolated molecule or cluster can be estimated, respectively, from the Liapunov spectrum and K-entropy and from the fractal dimension of the trajectory of the system in its phase space.

1. Introduction

Mode Selective Chemistry is one of the oldest desires and newest subjects in Chemistry. It became a real topic when the theoretical and experimental tools seemed at last to be almost at hand, and recent experience has begun to show that the long-standing hopes are being fulfilled (Scherer et al., 1990, 1991; Park et al., 1991). One of the necessary--but far from sufficient--tools to carry out mode-selective processing is an understanding of the potential surface or surfaces on which the molecular system is to be guided. Mapping the multidimensional potential surfaces of systems of more than three or four atoms is still a major challenge, particularly if we are called upon to construct a surface from first principles or some imitation-approximation thereof. (We shall return to this point.) However if we have some knowledge of the primitive forces between the particles of the molecule or cluster if interest, we can at least compute the effective potential energy of the system in any desired configuration. Until rather recently, this gave us the power to find minima on the potential surface, sometimes all and sometimes only an undetermined fraction of the total, but little more.

The combination of a growing arsenal of algorithms and ever-more-powerful computational tools has changed the situation so that it is now feasible to find not only minima but saddles and reaction paths, to classify the saddles by rank, and thereby to construct moderately accurate regional and global maps of multidimensional potential surfaces. It is now practical to find the dozen or two of most important minima and saddles of a cluster of 50 or 60 atoms. In fact we have the power to generate far more information about potential surfaces than we know how to use, particularly for molecules or clusters of more than about 15 or 20 atoms.

The next section of this paper reviews the state of our knowledge of how to map potential

J. Jortner et al. (eds.), Mode Selective Chemistry, 1–15.

surfaces that we know how to construct, and mentions how such maps can be used to study molecules and their surfaces. The section closes with a brief discussion of some approaches to constructing unknown surfaces that require nontrivial calculations.

Multidimensional surfaces have a variety of complexities that depend on the elementary interactions of pairs (and sometimes triples) of particles. By varying the basic pair interactions systematically in a model and studying the consequent variations of the multidimensional contours, we can learn what factors govern such characteristics as bottlenecks, the onsets of instabilities of vibrational modes and even the appearance or disappearance of minima. We can look forward to a time when such information can be used in mode-selective chemistry to vary species systematically to achieve desired characteristics for purposes of controlling a reaction. The third section discusses how such explorations have been carried out and applied.

With a surface in hand, we can carry out simulations of the behavior of molecules and clusters. A variety of methods are available for this task: we may use classical or quantum mechanics, Monte Carlo (MC) or molecular dynamics (MD), constant temperature, constant energy, constant pressure or other constraining conditions. We may combine MC or MD simulations with other probes such as quenching and descent, ascent or some other traversal of the potential, to explore, for example, where the system's dynamics have carried it. From the simulations on a single surface, we may well learn about interesting observables. However by studying simulations on related, systematically varied surfaces, we can learn how the form of the surface governs observable characteristics of an ensemble such as the appearance of distinct isomeric or phase-like forms, or fluxionality, or heat capacities. The fourth section of the paper deals with this general subject.

In the fifth section, we turn even more to examine how molecules or clusters explore their own potential landscapes, in terms of the extent of the ergodic and chaotic character of their trajectories. The extent of ergodicity can be expressed in terms of a fractal dimension of the trajectory of the system in phase space. The extent of chaotic behavior is quantifiable for a classical system by the spectrum of positive Liapunov exponents and their sum, the Kolmogorov (or K-) entropy. Clusters are proving to be particularly useful in enabling us to explore the way the extent of chaotic and ergodic character evolve with time.

Finally we summarize some of the most challenging open questions concerning the exploration of multidimensional potential surfaces.

2. Mapping a Potential Surface: Minima, Saddles and Reaction Paths

Finding potential surfaces for polyatomic systems can be mechanical and almost trivial, as it is whenever the surface can be represented adequately by a sum of simple pair potentials, or it can be exceedingly complex, for example in the case of heterogeneous metal carbonyl clusters or even homogeneous clusters of transition metal atoms. Pair potentials, such as the ubiquitous Lennard-Jones potential (Hoare and Pal, 1971; McGinty, 1973), the Morse potential (c.f. Hoare, 1979) and the Born–Mayer potential for alkali halides (Tosi and Fumi, 1964; Martin, 1983), and extensions of these, such as the Gupta potential (Gupta, 1981; Sugano and Sawada, 1989a,b) and the Embedded Atom potential for homogeneous systems (Nørskov, 1982) and some refinements of the Born–Mayer potential (Welch et al., 1978) retain the simplicity of central forces. Angular contributions to the interparticle forces can be introduced through polarization terms, as in the Axilrod–Teller potential (Wales, 1990) and the extended alkali halide potential used by Welch et al. (1978). All of these are convenient for high–speed numerical computation and easily lend themselves to molecular dynamics and Monte Carlo methods.

Mapping a potential surface is primarily a matter of finding its most significant points, specifically its minima and saddles, then finding how these are connected, then how the global surface divides into catchment basins around the minima, how regions of connected catchment basins are built from individual basins, and finally how the global topology arises from the way permutationally different, geometrically equivalent regions link together to determine the global topology of the surface. It has become practical to carry out the first of these steps, locating minima and saddles, on any known surface forclusters or molecules containing as many as 8 or 10 atoms, and in many cases more than that. The reliability of the minima and saddles is, far and away, determined by the reliability of the surface, not of the method used to fine the minima and saddles. The development of this subject is discussed by Mezey (1987) and, for small molecules, by Schatz (1989). A review of available potential surfaces was given by Truhlar et al. (1987).

Finding minima of multidimensional functions by numerical computation is now straightforward. The most obvious method is probably the method of steepest descents, the numerical solution of the coupled equations for the coordinates q_i of particle i with mass m_i, moving in the potential $V(\mathbf{q})$ in time steps τ:
$dq_i/d\tau = - m_i^{-1}\nabla_i V(\mathbf{q})$. This method is efficient enough for relatively small systems but converges slowly, becoming slower and slower as a stationary point is approached. It is particularly efficient when carried out with low-order algorithms (Garrett et al., 1988). More efficient methods, notably conjugate gradient methods such as the Fletcher-Reeves and Polak-Ribiere (Press et al., 1986) are much faster; for the 7-atom argon cluster, the conjugate gradient methods find minima two orders of magnitude faster than steepest descents. Programs are given by Press et al. for these methods.

Finding saddles has been more problematic. A general method, in principle, is a kind of hill-climbing in which a course is found to move upward from a minimum, starting in one of the normal modes of that minimum. As the point climbs upward on the energy surface, the directions of the "locally normal" modes, i.e. the principal directions of the Hessian of the surface, are fequently recomputed and the direction of the chosen mode is readjusted with each recalculation of the eigenvalues of the Hessian so that the same mode is followed continuously. If no fork in the path is encountered, the walk up the surface takes the point to a saddle and then down again. If there is a fork, it is probably most useful to keep track of its location and follow *both* of the branches above their joining point. This method has been proposed in several forms (Cerjan and Miller, 1981; Simmons et al., 1983; O"Neal et al., 1984; Banerjee et al., 1985; Baker, 1986, 1987); the form we have adapted is that of Cerjan and Miller (1981) which has proven quite efficient. The hill-climbing method nevertheless has certain limitations: it does not necessarily find all the saddles up to a chosen energy, and it does not necessarily find the most important saddles. It does, partly because of its efficiency, find most of the saddles at least for systems of as many as 7 particles.

Another way to find saddles that complements hill-climbing nicely is a sort of skiing-down procedure called "the method of slowest slides" (Berry et al., 1988). This method is derived from a combination of molecular dynamics and quenching. One carries out classical molecular dynamics, keeping track of the potential energy along the way. When a potential maximum along the trajectory is found, the kinetic energy is removed and the representative point is moved by steepest descents. Note that here conjugate gradient methods are not successful; steepest descents must be used. The rate of descent decreases as the saddle is approached and then increases as the point moves off the saddle region and downward toward a minimum. The point of minimum rate of descent, or of "slowest slide", is in the vicinity of the saddle. It may be necessary to iterate once or even twice to locate the saddle as accurately as needed. This method is typically three to eight times slower than the climbing method but it does find saddles of known importance and can be used to find saddles between specific wells, which may or may not be possible by the climbing method depending on the amount of prior information one has about the geometry of the transition state, i.e. the molecule at its cluster configuration.

In situations requiring extensive computation to generate reliable potential surfaces, as with metal, strongly-bound insulator and semiconductor clusters, it may be helpful to use a combination of methods, a crude, approximate method based on effective potentials to locate stationary points approximately by the mapping techniques just described and on more accurate computations in the vicinity of stationary points and perhaps other important regions. The more reliable results can be used to refine the parameters of the approximate effective potentials in the important regions, so that dynamics calculations can be done with the improved, approximate potential surface. This approach has been used to treat clusters of copper atoms (Berry et al., 1990a) with the embedded atom (EA) method (Nørskov, 1982; Daw and Baskes, 1984; Foiles et al., 1986) was chosen as the approximate method and the discrete variable method (DVM) was the chosen version of the Xα local density approach used for the more accurate calculations (Baerends et al., 1973; Rosén and Ellis, 1975; Rosén et al., 1976; Delley and Ellis, 1982; Delley et al., 1983). By refining the parameters of the EA potential for different regions of the potential surface of the six-atom cluster, an EA potential for Cu_8 and Cu_9 was generated which yielded the same minimum-energy structures as the DVM calculation.

3. Exploring the Map

The climbing and skiing methods were compared (Davis et al.1990) in explorations of the Lennard-Jones surface for Ar_7 and the Handy-Carter surface for formaldehyde (Handy and Carter, 1981a,b). The argon cluster was studied to understand more about the mechanism of isomerization in this cluster, on the basis that the potential is probably fairly reliable. The formaldehyde surface was studied to study its range of validity, particularly because it was developed primarily for the region around the equilibrium geometry of formaldehyde but is tempting to use for problems such as reactive scattering that involve other parts of the surface.

The exploration of the argon cluster's surface revealed the four previously-known structures corresponding to local minima (Hoare and Pal, 1972, 1975) and eight of the saddles. These are shown in Fig. 1; Fig. 2 is a schematic representation of a cross section through the energy surface, showing the minima (solid circles) and saddles connecting them (open circles). The horizontal axis in this figure has no significance, but the vertical relations are quantitatively accurate. The shape of the surface, with relatively deep wells separated from other, higher wells by rather high barriers, is consistent with this cluster being the smallest that exhibits distinct solid-like and liquid-like forms. This topic will be discussed later in the context of dynamics.

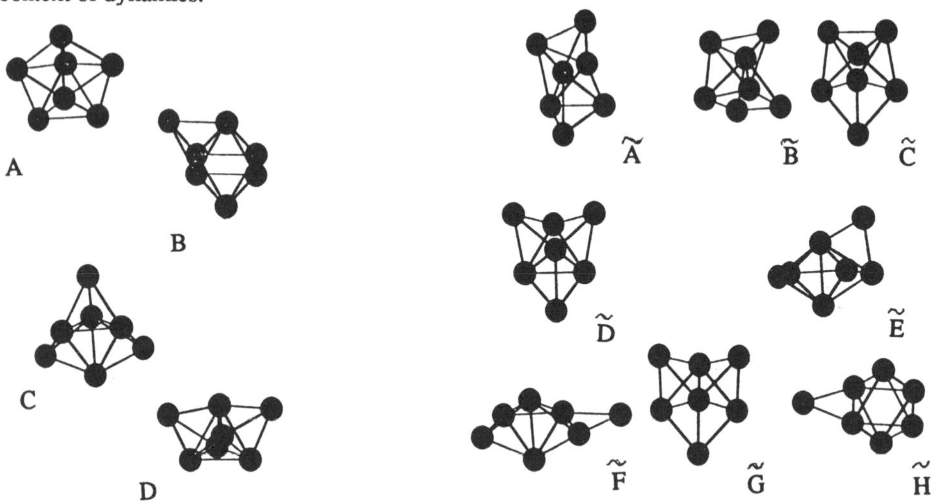

Figure 1. The four stable structures (a) and eight of the structures at saddles on the potential surface (b) of Ar_7, modeled by a Lennard-Jones potential. The same structures occur for the Morse potential with $\rho_0 \cong 6$.

The analysis of Lennard-Jones potentials was extended to those with anisotropic interactions through the introduction of induced moments and their interaction, specifically with Axilrod-Teller interactions (Wales, 1990); the resulting structures at the minima are not very different from those on the Lennard-Jones surface when the polarization terms are of the small magnitude characteristic of the rare gases but quite different structures become stable if the anisotropic terms are large enough.

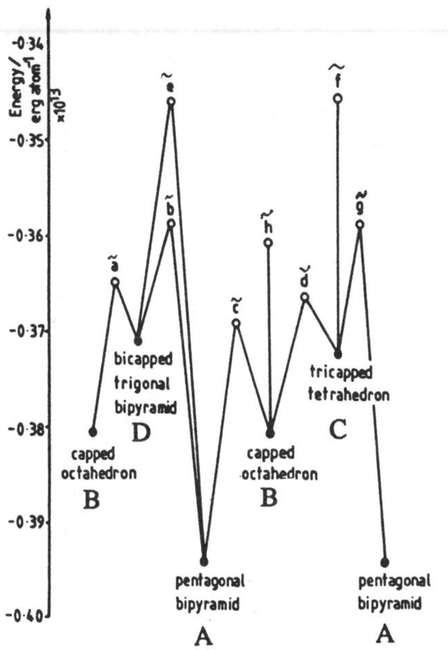

Figure 2. A schematic cross-section through the potential surface of Ar_7, showing the energies of the four minima and the saddles whose structures are in Fig. 1. The horizontal axis has no quantitative meaning.

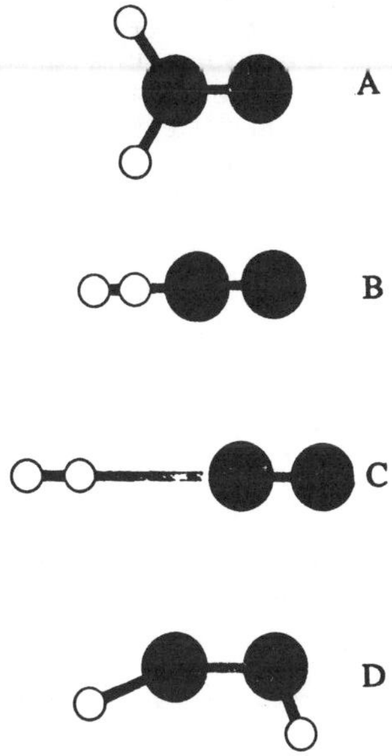

Figure 3. Geometries at the minima of the Carter-Handy surface of $H_2C{=}O$, formaldehyde; the first two are physically correct, the third is plausible but in fact should be *trans*, not *cis*, and the fourth is impossible.

Mapping of the formaldehyde surface revealed interesting limitations of the Handy-Carter surface. Although it represents the ground state of formaldehyde rather well, and has a local minimum at a structure completely plausible for a van der Waals molecule of CO with H_2, this potential has a minimum for a *cis*- HCOH structure rather than a trans, and, more dramatic, has a minimum at a linear H-H-C=O structure with an H-H distance of approximately 0.5 bohr, an impossibly short distance. The structures of the minima are shown in Fig.3. The analysis of the surface reveals, in this case, a shortcoming of that particular surface and therefore a limitation on the range of problems for which it is valid.

Another instance in which the surfaces of relatively complex species have been mapped is the study of the tetramer and pentamer of KCl (Rose and Berry, 1991). The stable geometries of these and other alkali halide clusters had been determined by Martin (1983). Using Born-Mayer pair potentials to represent the pair interactions in these ionic compounds, Rose and Berry mapped the minima and saddles of the surfaces of these 8-atom and 10-atom clusters and studied the dynamics on the surfaces. Both species showed two important regions of configuration space, a cuboidal region and a planar region, in which the species can exist for extended intervals of time over easily obtainable ranges of energy. At higher energies, the tetramer simply evaporated; the pentamer shows liquid-like behavior. Both clusters have some high-energy minima as well as the cuboidal and planar forms. Fig. 4 shows the geometries (a) and schematic cross section of the potential surface (b) for $(KCl)_4$. As Fig. 4b shows, the barriers among the planar, string-of-beads structures are relatively low, so that the clusters can pass readily among them, but in so doing, the order of the ions around the ring is not changed. Permutations of equivalent ions among sites occur only if the cluster passes into a cuboidal geometry. The study of these clusters led to a definition of a liquid cluster that requires not only that, to be liquid-like, the cluster must not only pass among minima at a rate not very much slower than it vibrates in one well; it must also pass among permutational isomers of its higher-symmetry stable structures at a rate not very much slower than it passes among wells generally.

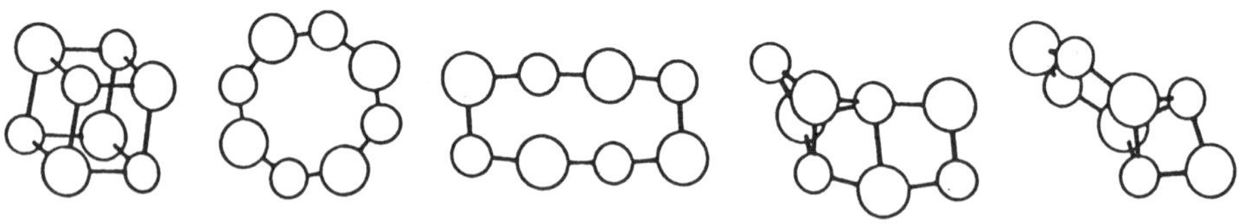

Figure 4. (a) Five stable structures of $(KCl)_4$.

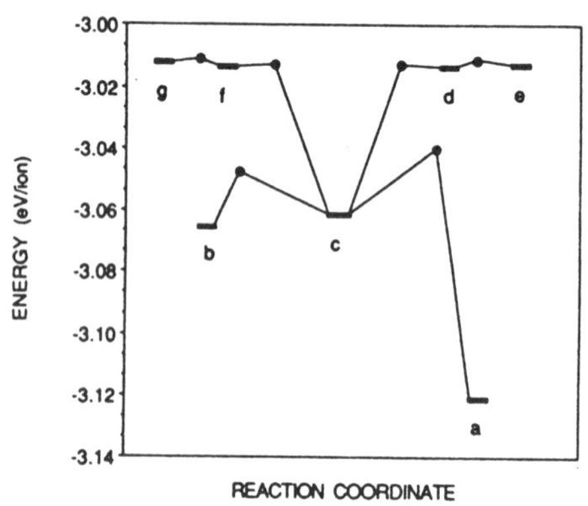

Figure 4. (b) the energies of seven of the stable structures of $(KCl)_4$ and of the saddles that link them. The letter indices in b) correspond to the structures in (a).

Another kind of study of potential surfaces is the exploration of how the geometry of the surface is determined by the range of the pair potential (Braier et al., 1990). This study was done in a spirit similar to the study of the role of anisotropic forces (Wales, 1990) insofar as both introduced systematic variations in the parameters of the pair potential and then probed the minima and saddles of the global potential as the pair potential changed. Braier et al. used the Morse potential, and in a related approach Stillinger and Stillinger (1990) used an inverse power potential; both are potentials of the form $V(r) = \varepsilon ([g(r) - 1]^2 - 1)$ with three variable parameters or, if the energy is scaled to units of ε and the distance is in units of the equilibrium distance r_e, one variable. The Morse potential takes $g(r) = \exp[-\beta(r- r_e)] = \exp[-(\rho-\rho_0)]$; the Stillinger-Stillinger or extended Lennard-Jones potential takes $g(r) = \varepsilon (r_e/r)^n$. These can both be considered one-parameter potentials whose range can be varied by varying

ρ_0 in the Morse case and the exponent n in the extended Lennard-Jones. The Lennard-Jones potential itself, with n=6, gives potential surfaces nearly indistinguishable from the Morse with ρ_0=6; these two have the same curvature at the minimum of the pair potential. The Morse potentials that give best fits to the known diatomic molecules all have range parameters ρ_0 between 2 and 7. For small values of ρ_0 or n, or long ranges of the potential, the multidimensional potential is simple, typically with only one kind of minimum. For large values of ρ_0 or n, corresponding to short ranges of the potential, the multidimensional potential has various other kinds of potential minima, at energies higher than that of the global minimum. Each kind of minimum has its own corresponding critical value of the range parameter at which the potential surface has a stationary point of zero curvature; for longer ranges, that minimum does not exist; for shorter ranges, it is a true minimum.

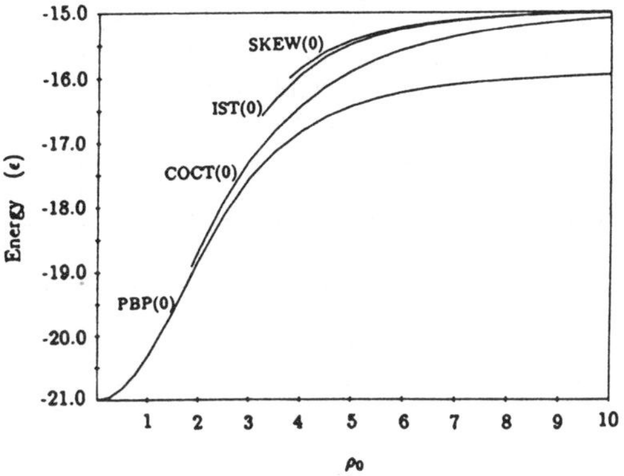

Figure 5. Energies of the four stable structures of the M_7 or 7-particle, pairwise-Morse cluster as functions of the range parameter ρ_0. The structures are not stable to the left of (i.e. for potentials of longer range than) the termini of the curves.

An example is the 7-particle Morse system which exhibits four kinds of minima if the potential is sufficiently short-ranged. The geometries are the same as those of the minima in Fig. 1; in fact the saddles are also identical to those of the Lennard-Jones 7-particle cluster. The energies of the minima, as functions of the range parameter, are shown in Fig. 5. The bounds on the energy (in negative reduced units) are the number of pair interactions in the two limiting cases. In the long-range limit, it is N(N-1)/2 for an N-particle cluster; in the short-range limit, it is the number of nearest-neighbor pairs in the lowest-energy structure. For the 7-particle cluster these are 21 and 15, respectively.

The saddles can also be shown in a useful way as functions of the range of the potential. The six-particle cluster, which has only two kinds of minima and four kinds of saddles, is somewhat richer than its 7-particle counterpart here. These are shown in Fig.6a and 6b, respectively. Fig. 6a shows the energies of the saddles (dashed curves) and the energy of the distorted octahedron or "incomplete pentagonal bipyramid" (IPB) *relative to the energy of the regular octahedron* which is the structure of minimum energy for the six-particle cluster. At values of $\rho_0 < 4.1$, only the octahedral structure is stable. However there is another kind of critical value of the range parameter at ρ_0=1.95 at which the mechanism of isomerization of one octahedron to another changes. For longer ranges, the saddle is a right trigonal prism, reached by twisting one triangular face relative to its opposite partner, transforming their projection from a six-pointed star into an equilateral triangle. For $\rho_0<1.95$, the trigonal prism is a simple saddle of rank 1, meaning that it has only one negative force constant, and that is in the direction of twisting the opposite faces clockwise or counterclockwise into an octahedron. For shorter ranges of the potential, the trigonal prism has another set of unstable modes: it can slide into a slewed trigonal prism in any of three ways, to make a figure that can be called either a slewed trigonal prism or a capped, square-based pyramid (CSBP). The right trigonal prism becomes a rank-3 saddle and therefore not the transition-state saddle. That role is taken over by the capped, square-based pyramid which is a simple saddle at an energy lower than that of the right trigonal prism (TP). This can be seen in Fig. 6a by the change of the TP saddle from rank 1 to rank 3 (3 negative eigenvalues of the Hessian matrix of the potential at that point, corresponding to the original twisting mode of symmetry A_2' and the doubly-degenerate slewing mode of symmetry E") and the appearance of the rank 1 CSBP at ρ_0=1.95. Another way to see this is to look at the force constants that become negative at critical values of the range parameter. Fig. 6b illustrates this for just a few of the relevant modes. The dashed curves are those of the TP and show how, for small range parameter values (long range potentials), this form has only one negative force constant, but at 1.95 the force constant of a second mode, in fact the slewing mode, also becomes negative. Furthermore at this value of the range, the CSBP first becomes a saddle; for longer-range potentials that structure is an ordinary, nonstationary point on the surface. The solid curves of Fig. 6b show the negative force constant of the unstable A" mode and the positive force constants of the stable A' (steeper) and A" modes of the CSBP which derive from the three unstable modes of the TP.

One generalization that can be made from the mapping of the surfaces of the 6-particle and 7-particle clusters concerns their mechanisms of rearrangement. Most of the important rearrangements from one geometry to another occur by mechanisms that can be classified under the broad rubric recognized by Lipscomb (1966) as "diamond-square-diamond" or DSD processes: four atoms forming two triangles with a shared edge, e.g. between atoms 1 and 3, go through a transition state in which all four atoms are equivalent to a new pair of triangles with a shared edge linking atoms 2 and 3.The atoms may or may not be coplanar. This is the mechanism that connects the 6-cluster's octahedron with its excited IPB structure. In general, DSD processes involving the breaking and making of only one bond tend to be the processes with the lowest barriers. Seven of the saddles found for the 7-cluster are transition states of DSD processes. The other common, generic rearrangement process is edge-bridging (EB) whose generality was pointed out relatively recently by Johnson (1986). This process accounts for four of the saddles of the 7-cluster and is the mechanism for interconversion among the lowest excited geometries of the 13-particle Lennard-Jones and short-range Morse cluster, all formed from the "magic number" icosahedron by promoting one atom from the shell of twelve onto one of the triangular faces.

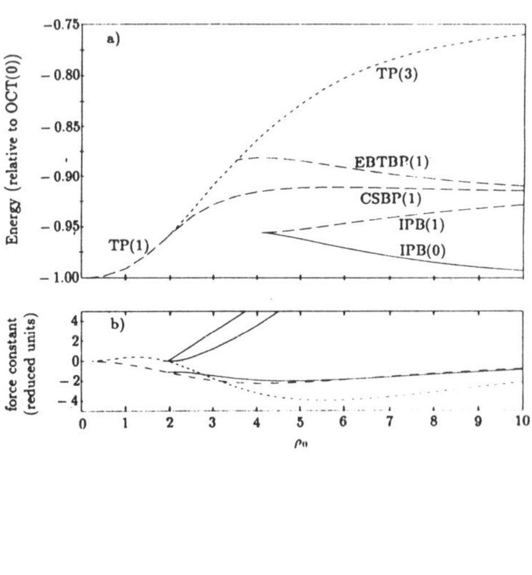

Figure 6. (a) Energies of stationary points on the potential surface of the M_6 or 6-particle, pairwise-Morse cluster as functions of the range parameter ρ_0 relative to the energy of the regular octahedron; the horizontal axis is thus the energy of the octahedron, the solid curve terminating at ρ_0=4.1 is the relative energy of the IPB structure, the dashed curves are the energies of the saddle points and the parenthesized numbers are the ranks of the stationary points, i.e. the number of negative eigenvalues of the Hessian at that point; (b) some of the most relevant force constants of the saddle structures TP (trigonal prism) and capped, square-based pyramid (CSBP) or slewed trigonal prism. Note how certain force constants change from positive to negative.

Exploring surfaces in this way will inevitably play an important part in clarifying how lattice-based geometries evolve to be the most stable structures for large clusters, even though smaller clusters are most stable in polyhedral geometries. In fact the fcc cubooctahedral structure of the 13-particle Morse (or Lennard-Jones) cluster is a first-order saddle for the entire physical range of ρ_0. So is the cubooctahedral structure of the 55-atom Lennard-Jones cluster, the next "magic" number, whose lowest-energy structure is an icosahedron. But for the next "magic" number, 147, the cubooctahedron and decahedron, both fcc, are local minima, although well above the icosahedral structure in energy (Uppenbrink and Wales, 1991). The decahedron is a local minimum for the 55-cluster.

At least one further use can be made of the map of a potential surface. From the map, one can determine the connectivity of all the catchment basins of the surface, i.e. the topology of the surface. For example the 6-particle Morse, Lennard-Jones or Gupta surface has 30 geometrically equivalent, permutationally distinct minima at octahedral configurations. Around each of these are 12 geometrically equivalent, permutationally distinct IPB local minima. Four of the twelve distorted octahedra around one regular octahedron connect by first-order saddles to four other distorted octahedra around a (permutationally) different regular octahedron, so that each "clock", consisting of an octahedron and its set of twelve distorted neighbors, is linked to three other "clocks". To make a smallest circuit from clock #1 to clock #2 and eventually back to clock #1, it is necessary to visit only one more clock; the smallest closed circuit among clocks is three (Berry et al., 1990b). This construction allows us to build the topology of the entire potential surface. From the connectivity, we find the symmetry appropriate to describe the fine splittings of vibrational levels of this complicated set of communicating wells. Whether this process is invertible, so that one could infer the pattern of connectivity from the pattern of level splittings, is still an open question for real, rearranging systems.

4. Dynamics and Phase Changes

Equilibrium properties of molecules and clusters can be determined computationally by solving the equations of motion, step by step, doing what is now known as "molecular dynamics," (MD), or by sampling randomly from what is equivalent to an ensemble; this procedure was given the name "Monte Carlo" or MC. Traditionally, MD was used for Hamiltonian systems, conservative systems. Consequently the time history of a single mechanical system was the principal result, but if the system is at least quasi-ergodic, the results of the time trajectory of the single system gives average properties equivalent to those of a microcanonical distribution. Monte Carlo calculations were originally developed for simulating properties of systems in thermal equilibrium, and hence give properties characteristic of canonical ensembles. In recent years, it has become feasible to carry out MD calculations that yield average values corresponding to other ensembles than the microcanonical; these are notably constant-pressure ensembles (Anderson, 1980) and constant-temperature ensembles (Nosé, 1984a,b; Jellinek, 1988; Jellinek and Berry, 1988, 1989).

It is well established now that classical simulations of polyatomic systems, notably clusters, show solid-like behavior when the energy or temperature of the system, whichever is constrained, is low enough and that at high enough energies or temperatures, provided the atoms do not evaporate before equilibrium properties can be established, polyatomic systems exhibit characteristics we associate with liquid-like behavior in bulk matter (Berry et al., 1988, and Berry, 1990a,b, review this subject). In short, cold molecules and clusters have the small-amplitude vibrations and fixed atomic positions characteristic of solids, and hot or energetic clusters may have the compliance and freedom of atomic motion characteristic of liquids--or they may break bonds and evaporate. At energies or temperatures between these bounding regions, the systems may show gradual softening and slush-like behavior (Beck et al., 1987), they may show coexistence of solid-like and liquid-like forms or they may show coexistence of two or more forms, none of which can be justifiably be called liquid-like (Sawada and Sugano, 1989; Rose and Berry, 1991). If the systems are treated quantum-mechanically, the results are essentially the same as the classical for systems whose atoms are as massive as Ar atoms, differ in quantitative detail for atoms with masses comparable to that of neon, and are qualitatively different, as one expects, if the individual particle mass is as light as helium (Beck et al., 1989).

The diagnosis of whether a cluster is solid-like, liquid-like or something else is much more directly done by examining dynamics than by studying an equilibrium distribution, particularly because of difficulties in distinguishing liquids from amorphous or glassy solids by static sampling. Hence MD investigations are likely to be more frequently cogent or unambiguous in probing "phases" of clusters.

The judgment itself has been made according to several criteria used traditionally for bulk matter: the mean square displacement as a function of time, whose slope is 6D, where D is the diffusion coefficient; the mean relative fluctuation of nearest-neighbor distances, which, according to Lindemann's criterion, implies liquid-like behavior if it is greater than about 0.1; the velocity autocorrelation function and its Fourier transform, the density of vibrational states, whose value at zero frequency is proportional to the diffusion coefficient. A more novel criterion has been the frequency of passage among different catchment basins on the potential surface, as can be obtained by quenching the cluster during an MD run (Stillinger and Weber, 1982, 1983) from time to time, and then, by any of the minimum-finding algorithms, finding the catchment basin within which the system is moving. Fig. 7 shows how the $(KCl)_5$ cluster visits different catchment basins by showing the quenched energies along a classical MD trajectory. A refinement of this method determines the distribution of dwell times in each basin (Amar and Berry, 1986). The last of these two criteria have become particularly significant for characterizing liquid-like and other kinds of nonrigid behavior in molecules and clusters because they probe the question of whether the system passes among catchment basins at a rate not much slower than the rate of vibration within a well. It seems that a mean rate of interwell passage of about 1/100 the vibration frequency is quite compatible with liquid-like behavior.

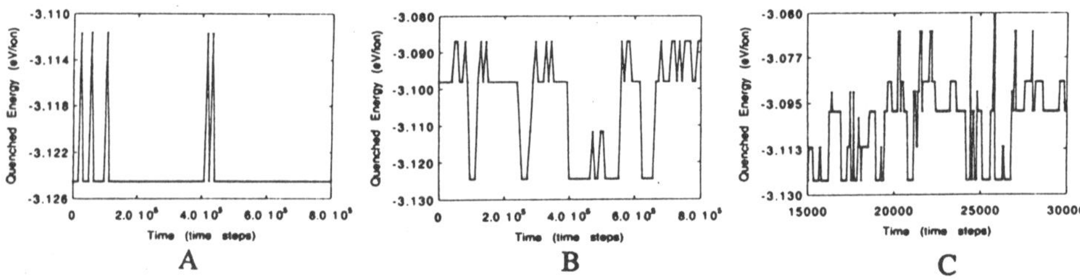

Figure 7. Energies of the $(KCl)_5$ cluster after quenching, i.e. energies of the local potential minima, along time trajectories at (a) –3.04 eV/ion, (b) –2.9575 eV/ion, (c) –2.845 eV/ion. The first two were done with quenches every 8000 time steps of $3x10^{-15}$ s; the third was done with quenches every 50 steps.

It turns out, interestingly, that none of these criteria are quite adequate to provide sufficient conditions for a cluster to be called liquid, although they all unquestionably provide necessary criteria. The reason for their inadequacy becomes apparent when one looks at the 6-particle clusters such as Cu_6 (Sawada and Sugano, 1989) and at some of the alkali halide clusters such as $(KCl)_4$ and $(KCl)_5$ (Rose and Berry, 1991). In all these systems, the criteria just listed imply, for a range of energies, that the clusters are liquid-like. However close examination of how the individual atoms move shows that "liquid-like" is too strong a description. There is a range of energy or temperature for each of these examples within which the cluster seems soft, compliant, diffusive, but the atoms merely move in large-amplitude motions around a single "home-base" structure. The Cu_6 cluster, like the 6-particle Lennard-Jones and Morse clusters, goes from a regular octahedron into any of the 12 distorted octahedra or IPB's and back, or on to others among the 12, within one energy range. At higher energies this cluster passes among all the permutationally different octahedra. The small alkali halide clusters, in their planar forms, show quantitative diagnostics that imply liquid-like behavior, but in the energy range where passage between cuboid and planar forms is very slow, the planar forms would not be appropriately called "liquid-like"; their motion is more like that of a string of beads on a table: the ring of ions can open and close and form closed loops of different shapes, but the sequence of ions around the ring cannot change by any accessible mechanism except returning to the cuboidal structures and reopen into a differently ordered ring.

This problem has recently led to the introduction of a new criterion for liquid-like behavior of clusters, which, together with the condition of relatively fast passage among wells, will, we believe, be sufficient to specify when a cluster is liquid-like (Rose and Berry, 1991). The new criterion is based on an index called the "permutation index", which measures the mean distance, after a selected number of steps such as 100,000, between particles which began the interval as nearest neighbors of a chosen kind. For example in examining KCl clusters, one can compare K-K distances, Cl-Cl distances and K-Cl distances. The selected distance is then compared with the mean distance between all particle-pairs of that kind, and with the mean distance of nearest neighbors, at that same energy. Fig. 8 shows the behavior of the permutation index for the anion-cation distances of $(KCl)_5$ as a function of energy.

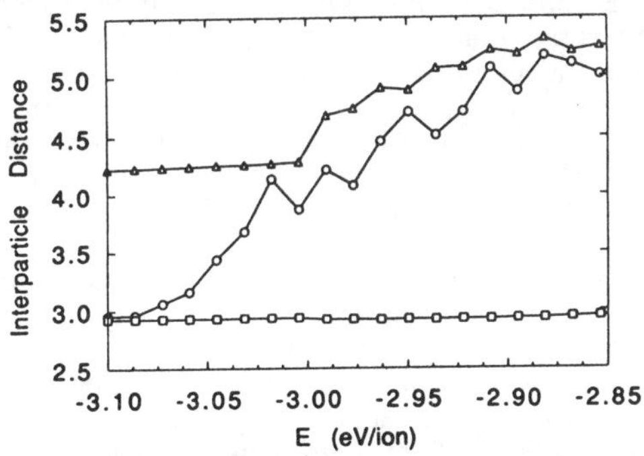

Figure 8. Permutation index (middle curve) for $(KCl)_5$ as a function of energy; the lower curve is the average nearest-neighbor, anion-cation distance; the upper curve is the average anion-cation distance in the cluster.

One other kind of criterion is not particularly relevant to classification or analysis according to "phase" but has a great deal to do with the chemistry associated with motion of a cluster or a molecule on its potential surface. This comes from considering the passage among basins, when it is infrequent, as a unimolecular, isomerization reaction. This viewpoint immediately suggests that the results of simulation be compared with predictions from a statistical theory, and the first that inevitably comes to mind is the Rice-Ramsberger-Kassel-Marcus (RRKM) theory. Such comparisons have recently been made for $(KCl)_4$ and $(KCl)_5$, particularly for their rates of passage between cuboidal and planar forms (Rose and Berry,1991). The results, shown in Figs. 7a and b, tell us that the RRKM model, even with the harmonic approximation for the densities of vibrational states, gives a very good qualitative and semiquantitative representation of the results of classical simulation for these relatively heavy alkali halide clusters. However the two examples shown here also have very systematic quantitative deviations between RRKM and simulated results. The data in one case imply that the harmonic levels are too close together compared with the "real" levels, and those of the other example imply the opposite.

What is the relevance of the analysis of dynamics of the sort we have been describing to photocontrolled reactions? Photocontrol of hydrogen and other light atoms cannot be assisted much by knowledge of their classical motion on potential surfaces; only a quantum description is reliable for them. However, as Beck et al. showed, the dynamics of particles heavier than neon are indeed reliably described by classical simulations. Analyses of this sort can be expected to give useful indications of the time scales for passage between different catchment basins and different, locally-connected sets of catchment basins, and of the dwell times in particular basins and regions. Moreover these time scales come as functions of energy or temperature. Hence we can use such analyses to tell what energy or temperature ranges should be chosen for molecules that will be targets for photocontrol, and what time scale will be relevant to the photocontrol processes.

5. Classical Chaotic and Ergodic Explorations

Exhibition of identifiable phases and determination of dwell times and rearrangement rates are not the only probes of how a molecule or cluster explores its phase and configuration spaces. These focus on the structure of the potential energy surface, leading to questions about identifiable regions on that surface, about how to characterize regions that can be given identities of their own. Other probes that have been used in recent years measure global properties of the trajectories on the surface; most notable among this second category are the measures of chaotic and ergodic behavior. We use the Liapunov spectrum and Kolmogorov entropy (the sum of the positive Liapunov exponents) as measures of the extent of chaos and fractal dimensions of the trajectory as measures of the extent of ergodicity, i.e. of how thoroughly the trajectory fills its supporting phase space. One of the tempting directions for the future is the possibility of developing local and regional information about how chaos and ergodicity develop, information that would tie together the approach through chaos and ergodicity and the approach through characterization of regions of the space and of transitions among those regions.

The smallest free, isolated molecules and clusters have three internal degrees of freedom. Model systems with fewer than three degrees of freedom may give insight into questions of chaos and ergodicity but must be considered strictly as models. Liapunov spectra, K-entropies and fractal dimensions of trajectories of systems of three degrees of freedom have been determined for a few cases (Contopoulos et al., 1978; Hamilton and Brumer, 1981; Stine and Noid, 1983a,b; Martens and Ezra, 1984; Pettini and Vulpiani, 1984;Kuz'min et al., 1986) The Ar_3, Lennard-Jones cluster was used to study clusters specifically (Beck et al., 1988). This cluster cannot be said to explore a complicated potential surface but it does have one characteristic that distinguishes it from the other studied systems with 3 internal degrees of freedom. Its geometry of minimum energy is an equilateral triangle and it has a saddle at each of its three permutationally distinct linear configurations. This gives Ar_3 some of the character of a cluster that can melt; at energies below an equivalent temperature of about 30 K, the cluster cannot cross the saddle, but at 30 K, it just can. At this energy and above, even Ar_3 satisfies some of the criteria of being liquid-like. How, then, do the ergodicity and chaotic behavior change as the cluster's energy reaches and passes the equivalent of 30 K?

The degree of ergodicity increases slowly, as measured by the fractal dimension D_2, from just above its minimum possible value of 3 (3 degrees of vibrational freedom, each of which would contribute exactly one dimension in phase space if the three modes were completely separable) at about 2 K, increases much more rapidly in the 18-20 K range and in the neighborhood of 29 K reaches essentially its maximum value of 8 (really about 7.6±0.6; the maximum dimensionality is 6x3 total degrees of freedom – 10 constraints: energy, momentum, angular momentum and the location of the center of mass, hence 18–10=8). Then, if the energy is increased to correspond to 30 K, the fractal dimension drops; it increases again if the energy is raised to 31-32 K.

The positive Liapunov spectrum and K-entropy show the same behavior: the K-entropy increases slowly at low energies, then faster, up to 29 K, and then drops as the system reaches approximately 30 K. At still higher energies, it increases again.

This behavior is entirely traceable to what happens when the system has enough energy to penetrate a saddle region and enough probability to find the saddle. When these conditions are met, the system enters the saddle, but with very low kinetic energy. Hence it moves slowly and spends long intervals of time crossing (or not succeeding in crossing) the saddle. Moreover the low-kinetic-energy trajectories in the saddle region are smooth, nearly parallel, like flow lines, not strongly divergent like highly chaotic trajectories. The algorithms for finding Liapunov spectra, K-entropies and fractal dimensions take, in effect, time averages over the trajectory of local contributions and follow these to very long times. Just below 30 K, the trajectories are essentially kept out of the saddle regions and correspond to rapid, chaotic motion; just at 30 K, the trajectories penetrate the saddle regions and rather abruptly, as the energy is increased, get large contributions to their global average behavior from regions of high regularity. This was first shown by Beck et al. (1988) and confirmed later by Leitner (1989) and by Wales and Berry (1991).

The Ar_3 cluster is too small to be the basis for generalization, and it seems that the K-entropy and fractal dimension of the Ar_7 cluster do not show anything more than a minimum or perhaps zero slope in the energy range at which this cluster just penetrates its lowest saddle region. However Ar_7 is a particularly important test case because it is the smallest cluster that shows distinct, coexisting solid-like and liquid-like phases (Briant and Burton, 1975). We conjecture at this time that the critical degree of freedom, the mode that becomes the soft mode and allows the system to cross the saddle, and any modes coupled closely with that mode, will show the same behavior as the trimer. We are now investigating that question.

It is tempting to speculate about the possibility of developing a kind of local measure of chaotic or ergodic behavior. This has been explored by several groups (Brumer and Duff, 1976; Duff and Brumer, 1977; Kosloff and Rice, 1981), most notably and successfully by Sepúlveda et al. (1988), who showed how local Liapunov exponents can be defined and used in a system of two degrees of freedom. A diagnostic that is a sort of local K-entropy, a sum of local or regionally averaged positive Liapunov exponents can be defined (Wales and Berry, 1991) whose average mirrors in its energy dependence the converged, global K-entropy, but its local behavior exhibits how regions of the potential surface can be associated with high or low degrees of chaotic behavior. Such a diagnostic would become an index of the extent to which a region of the potential surface scatters a prepared wave packet chaotically or simply and regularly. Hence we may look forward, if such an index is successfully developed, to using it to help find the regions of potential surfaces best suited for applying particular kinds of photocontrol.

6. Acknowledgments

The author is indebted to his students and coworkers whose works are cited here for their essential and central efforts that made our contribution to this work happen. He also wishes to acknowledge the support of the National Science Foundation and the Department of Energy, for Grants under which the work was carried out.

References

Amar, F. and Berry, R. S. (1986) J. Chem. Phys. **85**, 5943.
Andersen, H. C. (1980) J. Chem. Phys. **72**, 2384.
Baerends, E. J., Ellis, D. E. and Ros, P. (1973) Chem. Phys. **2**, 41.

Baker, J. (1986) J. Comput. Chem. **7**, 385.
Baker, J. (1987) J. Comput. Chem. **8**, 563.
Banerjee, A., Adams, N. and Simons, J. (1985) J. Phys. Chem. **89**, 52.
Beck, T. L., Jellinek, J. and Berry, R. S. (1987) J. Chem. Phys. **87**, 545.
Beck, T,L., Leitner, D. M. and Berry, R.S. (1988) J. Chem. Phys. **89**, 1681.
Beck, T. L., Doll, J. D. and Freeman, D. L. (1989) J. Chem. Phys. **90**, 3181.
Berry, R. S. (1990) in *The Chemical Physics of Atomic and Molecular Clusters,* G. Scoles, ed.; Proc.Int'l. Sch. Phys."Enrico Fermi", Course 57, North-Holland, Amsterdam, p.3, p.23.
Berry, R. S., Beck, T. L., Davis, H. L. and Jellinek, J. (1988) in *Evolution of Size Effects in Chemical Dynamics, Part II,* I. Prigogine and S. A . Rice, eds., Adv. Chem. Phys. **60**, Part 2, 75.
Berry, R. S., Davis, H. L. and Beck, T. L. (1988), Chem. Phys. Lett. **147**, 13.
Berry, R. S., Cheng, H.-P. and Rose, J. (1990a) High Temp. Sci. **27**, 61.
Berry, R. S., Braier, P. A., Hinde, R. J. and Cheng, H.-P. (1990b) Israel J. Chem. **30**, 39.
Braier, P. A., Berry, R. S. and Wales, D.J. (1990) J. Chem. Phys. **93**, 8745.
Brumer, P. and Duff, J. W. (1976) J. Chem. Phys. **65**, 3566.
Cerjan, C. J. and Miller, W. H. (1981) J. Chem. Phys. **75**, 2800.
Contopoulos, G., Galgani, L. and Georgilli, A. (1978) Phys. Rev. A **18**, 1183.
Davis, H. L., Wales, D. J. and Berry, R. S. (1990) J. Chem. Phys. **92**, 4308.
Daw, M. S. and Baskes, M. I. (1984) Phys. Rev. B **29**, 6443.
Delley, B. and Ellis, D. E. (1982) J. Chem. Phys. **76**, 1949.
Delley, B., Ellis, D. E., Freeman, A. J., Baerends, E. J. and Post, D. (1983), Phys. Rev. B **27**, 2132.
Duff, J. W. and Brumer, P. (1977) J. Chem. Phys. **67**, 4898.
Foiles, S. M., Baskes, M. I. and Daw, M. S. (1986) Phys. Rev. B **33**, 7983.
Garrett, B. C., Regmon, M. J., Steckler, R., Truhlar, D. G., Baldridge, K., Bartol, D., Schmidt, M.W. and Gordon, M. S. (1988) J. Chem. Phys. **92**, 1476.
Gupta, R. P. (1981) Phys. Rev. B **23**, 6265
Hamilton, I. and Brumer, P. (1981) A **23**, 1941.
Handy, N. C. and Carter, S. (1981) Chem. Phys. Lett. **79**, 118; ibid. **83**, 216.
Hoare, M. R. (1979) Adv. Chem. Phys. **40**, 49.
Hoare, M. R. and Pal, P. (1971) Adv. Phys. **20**, 161.
Hoare, M. R. and Pal, P. (1972) J. Cryst. Growth **17**, 77.
Hoare, M. R. and Pal, P. (1975) Adv. Phys. **24**, 645.
Jellinek, J. (1988) J. Phys. Chem. **91**, 3163.
Jellinek, J. and Berry, R. S. (1988), Phys. Rev. A **38**, 3069.
Jellinek, J. and Berry, R. S. (1988), Phys. Rev. A **40**, 2816.
Johnson, B. F. G. (**1986**) J. Chem. Soc. Chem. Commun., 27.
Kosloff, R. and Rice, S. A. (1981) J. Chem. Phys. **74**, 1947.
Kuz'min, M. V., Nemov, I. V., Stuchebrukhov, A. A., Bagratashvili, V. N. and Letokhov, V. S. (1986) Chem. Phys. Lett. **124**, 522.
Leitner, D. M. (1989) private communication; unpublished.
Lipscomb, W. N. (1966) Science **153**, 373.
Martens, C. C. and Ezra, G. S. (1984) Chem. Phys. Lett. **108**, 573.
Martin, T. P. (1983) Phys. Rep. **95**, 167.
McGinty, D. J. (1973) J. Chem. Phys. **58**, 4733.
Mezey, P. G. (1987) *Potential Energy Hypersurfaces,* Elsevier, Amsterdam.
Nørskov, J. K. (1982) Phys. Rev. B **26**, 2875.
Nosé, S. (1984a) Mol. Phys. **52**, 244.
Nosé, S. (1984b), J. Chem. Phys. **81**, 511.
O'Neal, D., Taylor, H. and Simmons, J. (1984) J. Phys. Chem. **88**, 1510.
Park, S. M., Lu, S.-P. and Gordon, R. J. (1991) J. Chem. Phys., in press.

Pettini, M. and Vulpiani, A. (1984) Phys. Lett. A **106**, 207.
Press, W.H., Flanery, B.P., Tueklolsky, S. and Vetterling, W. T. (1986) *Numerical Recipes* (Cambridge University Press, Cambridge, U. K., 1986).
Rose, J. and Berry, R. S. (1991) (submitted).
Rosén, A. and Ellis, D. E., (1975) J. Chem. Phys. **62**, 3039.
Rosén, A., Ellis, D. E., Adachi, H. and Averill, F. W. (1976) J. Chem. Phys. **65**, 3629.
Sawada, S. and Sugano, S. (1989a) Z. Phys. D **12**, 189.
Sawada, S. and Sugano, S. (1989b) Z. Phys. D **14**, 247.
Schatz, G. C. (1989) Revs. Mod. Phys. **61**, 669.
Scherer, N. F., Ruggiero, A. J., Du, M. and Fleming, G. R. (1990) J. Chem. Phys. **93**, 856.
Scherer, N. F., Carlson, R. J., Matro, A., Du, M., Ruggiero, A. J., Romero-Rochin, V., Cina, J.,Fleming, G. R. and Rice, S. A. (1991) J. Chem. Phys. (in press).
Simmons, P., Jorgensen, P., Taylor, H. and Ozment, J. (1983) J. Phys. Chem. **87**, 2745.
Sepúlveda, M. A., Badii, R. and Pollak, E. (1989) Phys. Rev. Lett. **63**, 1226.
Stillinger, F. H. and Weber, T. A. (1982) Phys. Rev. A **25**, 978.
Stillinger, F. H. and Weber, T. A. (1983) Phys. Rev. A **28** 2408.
Stillinger, F.H. and Stillinger, D. K. (1990) J. Chem. Phys.**93**, 6106.
Stine, J. R. and Noid, D. W. (1983a) Chem. Phys. Let. **100**, 282.
Stine, J. R. and Noid, D. W. (1983b) J. Phys. Chem. **87**, 3038.
Tosi, M. P., and Fumi, F. (1964) J. Phys. Chem. Solids **25**, 45.
Truhlar, D. G., Steckler, R. and Gordon, M. S. (1987) Chem. Revs. **87**, 217.
Uppenbrink, J. and Wales, D. W. (1991) J. Chem. Soc.Farad. Trans. **87**, 215.
Wales, D. J. (1990) J. Chem. Soc. Farad. Trans. **86**, 3505.
Wales, D. J. and Berry, R. S. (1991) (submitted).
Welch, D. O., Lazareth, O. W., Dienes, G. J. and Hatcher, R. D. (1978) J. Chem. Phys. **68**, 2159

INTRAMOLECULAR VIBRATIONAL REDISTRIBUTION ON THE NANOSECOND TIME SCALE: ν_1 AND $2\nu_1$ SPECTRA OF SUBSTITUTED ACETYLENE COMPOUNDS

K.K. LEHMANN, B.H. PATE AND G. SCOLES
Department of Chemistry
Princeton University
Princeton, NJ 08544-1009 USA

ABSTRACT. Intramolecular vibrational energy redistribution (IVR) continues to be a subject of wide interest to chemical physics. Most previous work has found that IVR is rapid (on the time scale of 10 psec or less) whenever the density of ro-vibrational states is on the order of 100 states per wavenumber or greater. We have been studying IVR in a series of substituted acetylene compounds using optothermal detection of the acetylenic C-H stretch or its first overtone absorption spectrum in a cold (few degrees K), collimated molecular beam. This gives a resolution of 10 MHz, which allows us to study IVR on times scales up to 30 nsec. Our results show that these molecules undergo statistical IVR (energy spreads over all vibrational modes) but with lifetimes of 200 psec or greater. This is clearly long enough to allow bimolecular collisions before reactions, and thus possible mode specific chemistry even in the gas phase at modest pressure. Chemical modification is allowing us to make interesting correlations between structure and relaxation lifetime.

1. Introduction

Intramolecular vibrational energy redistribution (IVR) is one of the most elementary and common transformations of a molecular system. The highly successful statistical theories of molecular reactions assume that IVR is both fast and complete (producing a microcanonical distribution in vibrational phase space) on a time scale short compared to intra and intermolecular chemical reactions[1]. What would be desirable, but so far lacking, is a general understanding of what aspects of a molecule's structure and its potential energy surface determine the time scale and the extent of IVR following excitation in a localized coordinate of the molecule. Earlier work, by Sibert et al[2], and Quack and coworkers[3], has demonstrated that Fermi resonances control the fastest IVR that has been observed. The time scale for interchange of vibrational excitation between a localized hydrogen stretch and its own bending vibration, ~50 fsec, are remarkably similar in a number of related CX_3H compounds. This similarity of rates reflects the similar strength of the equivalent bend-stretch anharmonic coupling in these molecules.

Specific low order resonances are known to determine the rate of IVR in a number of other molecules undergoing subpicosecond IVR[4]. Such IVR is now well understood, at least at modest energy levels compared to dissociation. Much less well understood is how IVR proceeds even in the absence of low order resonances. In particular, the IR emission experiments of McDonald and coworkers[5] has shown that essentially all molecules will undergo IVR when their hydrogen stretching fundamentals are excited as long as the density of ro-vibrational levels exceeds 10-100

J. Jortner et al. (eds.), Mode Selective Chemistry, 17–23.

per cm^{-1}. A similar "universal threshold" for the onset of IVR has been observed on the S_1 surface of many aromatic molecules[6]. Such work argues that there are sufficient anharmonic couplings between essentially all ro-vibrational levels to produce IVR lifetimes on the order of 100 psec or less whenever there are isoenergetic states to relax into. Isoenergetic is defined by the homogeneous width for the IVR lifetime.

A theoretical understanding of such behavior is seriously lacking. The most fundamental question, whether or not the same physical model used to describe fast IVR is appropriate here, has not been established in our opinion. Given the highly averaged nature of most of the previous measurements, the phenomenon has not been well characterized experimentally either. Of the many methods that have been used, we believe that when it is practical, eigenstate resolved, high resolution spectroscopy provides the greatest insight into this problem as the pioneering work of Perry and coworkers has demonstrated[7].

2. Experimental

In the last two years, we have been studying the ν_1 and $2\nu_1$ absorption spectra of a number of substituted acetylene compounds[8]. The ν_1 mode is the acetylenic CH stretching fundamental near 3330 cm^{-1}. The spectra are obtained in a collimated molecular beam using the optothermal detection method[9] This has given us a substantially better signal-to-noise ratio and resolution than direct absorption, even in a slit jet[10]. For the fundamental spectrum of butyne (CH_3CH_2CCH), for example, we have recently obtained spectra with a S/N ratio of 250 to 1 and a resolution of 6 MHz. Resolution is important in these studies since one often has dense, overlapping spectra from many rotational transitions and because the resolution determines the length of time we can follow the dynamics. Sensitivity is important because it determines how fractionated a spectrum can become and still be observed. Many transitions also have small intensities and may be missed without sufficient signal-to-noise. We have chosen to study substituted acetylene compounds for several reasons, some practical, some theoretical. One interesting reason is that the frequency of the ν_1 and $2\nu_1$ absorptions hardly changes as we change molecules. This supports our view that the initial excitation is essentially the same as we change molecule, only the "bath" of dark states and their couplings to the "bright" state change. Table 1 shows the calculated density of states at ν_1 and $2\nu_1$ for a number of molecules which we have studied or are studying. We see that we can, in principle, compare similar molecules with the same density of states but different levels of excitation which should allow us to isolate the effects of vibrational amplitude, density of states, and molecular complexity. Another advantage of these molecules is that the observed spectroscopic structure due to IVR is less than 2B, the spacing of the symmetric top rotational lines. This dramatically simplifies the interpretation of the spectrum since, at least for symmetric tops, we can rather cleanly separate inhomogeneous structure, due to rotations, from the interesting homogeneous structure, which relates to the dynamics. Even for asymmetric tops, the close clustering of lines makes spectral assignments by combination-differences easier and more secure.

3. Discussion

Our discussion here will focus on the IVR lifetimes observed for molecules in the intermediate or large molecule limits where a lifetime can be defined. We refer the reader to reference 8c for a more complete discussion of the theoretical issues involved in what we mean by a lifetime in the present context. The operational definition is that each bright ro-vibrational state must be split into at least 5-10 transitions so that the power spectrum of the observed spectrum (which gives the survival probability) has a decay at early times instead of just ringing. Table 2 presents the lifetimes we have thus far determined. Many other molecules that have been studied are in the sparse to intermediate limit; for them we obtain a wealth of information about the couplings, but not an unambiguous lifetime. In a few cases, such as the overtone of butyne,we have failed to observe spectra that we had expected to see with high sensitivity, likely reflecting a much more rapid IVR that obliterates the high resolution structure in the spectrum.

Several general comments need to be made about our results. First, the lifetimes given in Table 2 are equivalent to T_1 times, i.e. they describe the decay of the excited state population. In the statistical limit, Lorentzian lines are observed the width of which directly gives T_2 (which we incorrectly gave as "the lifetime" in our preliminary publication![8b]), but in the present case just as in spontaneous emission, $T_2=2T_1$.

Secondly, our lifetime results are consistent with the relaxation rate having no dependence on rotational state (with a possible exception of CF_3CCH $2\nu_1$) and clearly rules out a strong rotational dependence. The lack of a systematic decrease in the IVR lifetime with rotational excitation demonstrates that it is determined by anharmonic interactions, not Coriolis or centrifugal couplings, for at least the range of rotational energy (0-20 cm^{-1} typically) that we have explored in our beam spectra. However, it may well be that at 300 cm^{-1} of rotational energy, typical of a room temperature molecule, the rotationally induced couplings may become more important and even dominate the relaxation rate. Saturation spectroscopy (in the frequency or time domain) could easily resolve this issue. A further point is that while Coriolis interactions do not contribute to the linewidth, we do have evidence of Coriolis interactions in some of the smaller molecules we have studied. Thus, our results are consistent with K scrambling of the eigenstates (the so called IVRET)[11], as long as the time scale for tumbling is longer than the IVR lifetimes we have observed. Because the IVR lifetime is long compared to (A-B) in these molecules it is clear that each eigenstate can have significant contriutions from at most one bright ro-vibrational basis state, thus K scrambling can only be directly observed by measurements that are sensitive to the nature of the dark states that make up each eigenstate. Resonance fluorescence would be expected to be as polarized as if there was no IVRET.

What is perhaps the most surprising result is that for the molecules with a large density of states, the trimethyl substituted molecules[8b,d], we do not even observe fluctuations in the linewidth as a function of J or K, while our work on molecules with a small to intermediate density of states shows large changes in the spectrum as a function of rotational state[8c]. As a function of J, different dark vibrational states will tune through resonance with the bright state because of different effective B values. For the symmetric tops, the tuning as a function of K is due largely to parallel Coriolis interactions, which are on the order of the A rotational constant itself causing new states to be in resonance for each single value of K. In the same molecule, therefore, the observation that the IVR rate appears independent of J and K appears to imply that the relaxation rate is independent of the states that the energy relaxes into!

If we compare the series of molecules: butyne, cis and trans pentyne (work of McIlroy and Nesbitt[10b]) and t-butyl acetylene, which all have the C-C-C≡CH linkage, we find a remarkable consistency of linewidth as well. Any specific resonance must detune by orders of magnitude more than the observed linewidth (0.02 cm^{-1}) as we move in this series, yet the relaxation rate stays the same. We see systematic variation in linewidths in the series $(CH_3)_3XC{\equiv}CH$, X=C, Si, Sn suggesting a heavy atom blocking effect[12] (though the blocking may reflect longer bond length and decrease in bond stiffness as much as increased mass). It is easy to rationalize that the strength of anharmonic coupling should decrease in this series, but in a resonance model the relaxation rate depends upon the specific resonance that is involved and its detuning. Given the complexity of these molecules one would expect that the near resonances would be completely different from one molecule to another. Once again, one would expect erratic linewidths.

Both the J,K and the molecule independence of the IVR rate in these molecules appear to be difficult to rationalize with the idea of a chain of specific resonance as controlling the IVR process. One is tempted to try to rescue the resonance picture by invoking a specific "doorway state" as discussed by McIlroy and Nesbitt for some of these molecules[10b]. The difficulty is that if the doorway state is localized on the acetylene end of the molecule (which is required for it not to be completely detuned as the substituted part of the molecule is changed), then we have only changed the problem to why the coupling of the doorway state to the bath of states does not depend upon the vibrational character of the specific states of the bath. In a formal sense, a doorway state can always be defined[13] but unless this is dominated by some specific normal mode state, nothing appears to be gained by its introduction into the theory.

Perhaps the bath states are chaotically mixed over an interval that is much larger than the width of our ultra-narrow resonance. There are both experimental and theoretical reasons for supposing this, although the matter has not been conclusively demonstrated. Our own work on CF_3CCH argues that the bath states are not strongly mixed at ν_1, but are at $2\nu_1$[8c]. This model implies that couplings of the bright state to the bath should fluctuate, but with a constant mean square matrix element over the width of the resonance. This model is consistent with our observations of molecules with intermediate to dense spectra. In order to relate this to the resonance model, however, we must assume that coupling between the bright state and the bath states comes from coupling to one or a few specific states that are strongly mixed with all the others. But the spectral density of these doorway, or "first tier", states will change over a frequency interval of their own relaxation rates. This could rationalize the observed stability to changes in rotational quantum numbers if the lifetimes of the doorway states are on the 1 psec or less time scale. However, this model does not predict the stability of our observed lifetime to changes in the molecule. We would likely need to invoke unreasonably short lifetimes inorder that the resonances not completely detune (relative to the linewidth of the doorway states) when we change from one molecule to another. Instead, we are presently of the opinion that the long lifetimes of the acetylene chromophore is due to the absence of any low order resonances. The bath states isoenergetic with ν_1 are found (based upon harmonic frequencies) to be made up of a total of at least 7-10 quanta distributed in other modes. We expect weak, high order couplings to these states which should be completely different when we change molecules.

Resonances between states in quantum mechanics are really of two distinct kinds. Some of these resonances are the quantum mechanical equivalent of classical resonances between modes. The bend-stretch Fermi resonance case has been shown to be of this form. If we ran classical trajectories we would expect to find IVR on a similar time scale. But specific state resonances in quantum mechanics can also reflect quantum tunneling through a potential or a dynamical barrier. The local mode doubling observed in many symmetric hydrides[14] and asymmetry doubling of the rotational levels of an asymmetric top[15] are familiar examples of the latter. While in the language of quantum mechanics it is difficult to distinguish between the two, they are qualitatively different when we attempt to understand what is happening in the dynamics. The fastest of our observed lifetimes is still four orders of magnitude longer than the vibrational period and thus it is unlikely that this can be a classically allowed relaxation. However, extremely small rates could naturally be expected if the rate limiting step is quantum tunneling through a dynamical barrier.

Consider the following model, the outline of which was proposed to us by Howard Taylor. Each of these molecules has a stable periodic orbit that is essentially a pure ≡C-H stretch with a rapidly decreasing motion of atoms as one goes down the chain. Quantization along this periodic orbit is the state referred to as ν_1. Excitation in the directions perpendicular to this periodic orbit produces combination states with the other, lower frequency modes of the molecule. But this periodic orbit is the center of a small island of stability, moving away we will come to a separatrix and a chaotic sea beyond. Taylor has stressed the fact that "extreme motion states", where all the action is localized along a single mode, are often the last regions of periodic motion to be destroyed classically[16]. Getting to that separatrix requires excitation in one or more perpendicular modes without taking energy out of ν_1, thus producing a dynamical barrier. Beyond the separatrix, motion in ν_1 is unstable and energy will flow from ν_1 into other modes of the molecule. If the total volume of phase space in the molecule (the density of states) is too small, amplitude will reflect back to the well, and destructive interference will prevent most of the amplitude of ν_1 from leaving the well. In this case, one eigenstate of the molecule will be mostly ν_1. As the density of states increases, we will first get an intermediate and then a large molecule limit spectrum as the ν_1 state converts into a resonance. To get relaxation we need a high density of states, but the rate of relaxation is determined by tunneling through the dynamical barrier. If this barrier does not change, then the relaxation rate will be the same.

Ultimately, the separatrix and thus the dynamical barrier are related to classical resonances between the modes. But since the periodic orbit is presumed stable, the important resonance could be a low order (and therefore broad) resonance of a highly anharmonically shifted mode. Arguments based upon the need for precise matching of energy within a narrow resonance no longer apply. Presently, what is needed is a way to turn this "cartoon" into a real calculation. This picture is attractive because if we can learn how to locate and calculate the dynamical barrier we may be able to make a semiclassical estimate of the rate without the need to run trajectories (or worse, quantum wavepackets!) for 10^5 periods.

4. Conclusion

Our present results raise fundamental questions about the nature of intramolecular dynamics. Our experimental work is continuing and we hope to have a more complete empirical picture in the near future. One key feature we want to continue to explore is the relationship of chemical structure to relaxation lifetimes. We would also like to perform experiments that use the nsec lifetimes to do laser selective chemistry of the type Crim and coworkers[17] have been doing on HDO, where IVR is not an issue. In many of molecules, we can truely heat one bond of a large molecule for a chemically significant time, something everything but hope told us was impossible until recently. Our theoretical understanding of IVR on the nsec time scale is poorly developed and calls out for more help. In this talk we have presented our own speculations on this matter, which are admittedly very crude. I hope some members of the theoretical community will take up the fight!

Acknowledgements

We would like to acknowledge the assistance of Erik Kerstel and Thomas Mentel who have been collaborators in our IVR measurements at Princeton, Howard Taylor who suggested tunneling as the key to understanding our lifetimes, and the National Science Foundation for financial support.

References

1. Robinson, P.J. and Holbrook, K.A. (1972) Unimolecular Reactions, Wiley-Interscience, New York, Chapter 4.
2. Sibert III, E.L., Reinhardt, W.P. and Hynes, J.T. (1984) J. Chem. Phys., 81, 1115.
3. Quack, M. (1982) Adv. Chem. Phys. 50, 395 .
4. Lehmann, K.K. and Coy, S.L. (1988) J. Chem. Soc. Faraday Trans. 2, 84, 1389.
5. Kim, H.L., Kulp, T.J. and McDonald, J.D. (1987) J. Chem. Phys., 87, 4376 .
6. Parmenter, C.S. (1983) Faraday Disscus. Chem. Soc., 75, 7.
7. deSouzz, A.M., Kaur, D. and Perry, D.S. (1988) J. Chem. Phys., 88, 4569 .
8. a) Lehmann, K.K., Pate, B.H. and Scoles, G. (1990) J. Chem. Soc. Faraday Trans, 86, 2071 ; b) Lehmann, K.K., Pate, B.H. and Scoles, G. (1990) J. Chem. Phys., 93, 2152; c) Pate, B.H., Lehmann, K.K. and Scoles, G, J. Chem. Phys., to be published; d) Kerstel, E.R.Th., Mentel, T.F., Lehmann, K.K., Pate, B.H. and Scoles, G. J. Phys. Chem., to be published.
9. Gough, T.E., Miller , R.E. and Scoles, G. (1977) Appl. Phys. Lett., 30, 338 .
10. a) McIlroy, A. and Nesbitt, D.J. (1989) J. Chem. Phys., 91, 104 ; b) McIlroy, A. and Nesbitt, D.J. (1990) J. Chem. Phys., 92, 2229 .
11. Fredrick, J.H., McClelland, G.M. and Brumer, P. (1985) J. Chem. Phys., 83, 190.
12. a) Lopez, V., Fairen, V., Lederman, S.M. and Marcus, R.A. (1986) J. Chem. Phys., 84, 5494 ; b) Uzer, T. and Hynes, J.T. (1986) J. Phys. Chem., 90, 3524; c) Swamy, K.N. and Hase,W.L. (1985) J. Chem. Phys., 82, 123 .
13. Cable, R. and Rhodes, W. (1980) J. Chem. Phys., 73, 4736.
14. Child, M.S. and Halonen, L. (1980) Adv. Chem. Phys., 42, 1.
15. Harter, W.G., Patterson, C.W. and da Paixao, F.W. (1978) Rev. Mod. Phys., 56, 37 [Symmetric Tops]
16. Hose, G. and Taylor, H.S. (1984) Chem. Phys., 84, 375 .
17. Sinha, A., Hsiao, M.C. and Crim, F.F. (1990) J. Chem. Phys., 92, 6333.

Table I.

Density of States[a] for Several Substituted Acetylenes at the Fundamental and First Overtone

Molecule	$\rho(v=1)$	$\rho(v=2)$
propyne	0.3	10
trifluoropropyne	10	10^3
tertbutylacetylene	1×10^4	2×10^7
trimethylsilylacetylene	4×10^5	1×10^9
trimethylstannylacetylene	1×10^7	6×10^{10}
propynal	1.5	49
propiolyl fluoride	4	193
vinyl acetylene	4	320
propargylamine	16	1.8×10^3
Propargyl alcohol	20	1.5×10^3
1-butyne	75	$3. \times 10^3$

[a] The reported density of states are the number of vibrational states per cm^{-1} expected in transitions from the ground state, calculated in the harmonic approximation through a direct count algorithm. By ground vibrational state, we include all tunneling components and assume that the same are resolved in the excited state. Selection rules have been used to reduce the density of states where some symmetry species should not mix with the bright states and states degenerate by symmetry are only counted once.

Table II.

IVR Lifetimes for Some Substituted Acetylenes Determined by High Resolution Infrared Spectroscopy

Molecule	Vibrational Level	Lifetime (nsec)
1-butyne	v = 1	0.2
tertbutylacetylene	v = 1	0.2
trimethylsilylacetylene	v = 1	2.0
trimethylstannylacetylene	v = 1	6.0
propynal	v = 2	1.2
trifluoropropyne	v = 2	2.0
tertbutylacetylene	v = 2	0.1
trimethylsilylacetylene	v = 2	>4.
trimethylstannylacetylene	v = 2	>5.

Multiple Laser Probes of Intramolecular Dynamics at Chemically Significant Energies

Patrick R. Fleming, Xin Luo, and Thomas R. Rizzo
Department of Chemistry
University of Rochester
River Station
Rochester, NY 14627

ABSTRACT. Optical-infrared and infrared-optical double resonance schemes incorporating vibrational overtone excitation provide a new view of intramolecular vibrational energy redistribution (IVR) in dissociating molecules. In the first approach, an infrared predissociation spectrum of a molecule is recorded subsequent to light atom stretch overtone excitation. Such spectra directly reveal the extent of vibrational state mixing and the nature of the dark states involved in IVR at the excited level. The second approach employs infrared-optical double resonance to generate fully rotationally resolved vibrational overtone spectra of small molecules at energies above their unimolecular dissociation threshold. The spectral splittings and linewidths of single rotational transitions to dissociative states provide information about the rotational motion of a vibrationally excited molecule as it moves toward dissociation. These studies also reveal a marked vibrational mode dependence to the coupling matrix elements between the zeroth-order bright state and nearly isoenergetic dark states. This paper describes applications of these double resonance techniques to HOOH, $HONO_2$ and $(CH_3)_3COOH$.

Introduction

Any successful scheme for selective photochemical bond breaking must take appropriate measures to avoid, inhibit, or counteract the inherent loss of mode selectivity caused by intramolecular vibrational energy redistribution (IVR). This requires understanding the detailed couplings between the zeroth-order state that provides oscillator strength for photoexcitation (*i.e.,* the "bright" state) and the nearly isoenergetic zeroth-order "dark" states. The prospect of using highly excited light atom stretching vibrations in selective bond breaking schemes has motivated intense effort toward understanding the mechanism of energy redistribution from these levels [1-22]. Much of the experimental work has focussed on determining the homogenous contribution to vibrational overtone linewidths, since this reflects the coupling matrix elements, and hence the vibrational energy redistribution rates, of the zeroth-order light atom stretch mode [1-16]. Theoretical studies have sought to determine the specific mechanisms that couple light atom stretch oscillators to other vibrational modes of a polyatomic molecule [17-22].

J. Jortner et al. (eds.), Mode Selective Chemistry, 25–45.

Two fundamental problems have hampered experimental studies of IVR in overtone excited molecules. As in any experiment that employs spectroscopic techniques, inhomogeneous spectral congestion can prevent clean excitation of molecules to single rovibrational states, and this obscures the spectral splittings that result from the coupling between zeroth-order vibrational levels. Spectral congestion must be eliminated before unambiguous conclusions can be drawn from spectral linewidth measurements. A second problem arises from the inherent difficulty of studying the zeroth-order dark states involved in IVR at high energies on the ground potential surface. Because they carry no oscillator strength from the ground state, direct absorption or excitation spectroscopy provides no information on their vibrational character. Understanding the nature of these dark states is at the heart of understanding the mechanism of vibrational energy flow.

Although supersonic molecular beams can eliminate spectral congestion in vibrational overtone transitions and reveal detailed spectroscopic and dynamical information about the excited molecule [10, 11, 15, 16], this approach has a serious drawback. Because molecules in the cold environment of a supersonic expansion begin in states with low rotational quantum numbers, selection rules for photoexcitation restrict access to low rotational states of the excited molecule. An increasing amount of evidence has implicated the importance of rotational motion in vibrational energy redistribution [23-28], and the limitations of supersonic expansion techniques preclude their use for systematic investigation of this phenomenon.

An alternative method for simplifying thermally congested spectra is to require the molecules to simultaneously satisfy two resonance conditions [15, 23, 29]. Double resonance methods can select molecules from any rotational state that has appreciable population at room temperature, and when implemented in a spectroscopic scheme that accesses highly excited vibrational states, this approach permits the detailed investigation of rotational effects on intramolecular processes. Coy and Lehmann were the first to use this general approach to simplify vibrational overtone transitions in their microwave-optical double resonance study of the congested N–H stretch bands of NH_3 [29]. We have recently developed an infrared-optical double resonance scheme that serves both as a spectroscopic tool for simplifying high vibrational overtone transitions and as a reactant preparation technique for unimolecular decomposition studies [30-32]. This report summarizes a few of our results that are pertinent to the topic of mode selective chemistry.

Double resonance techniques possess the added advantage that they permit spectroscopic access to vibrational states that are not directly accessible from the ground state. Although dark states that mix with the zeroth-order bright state carry no oscillator strength from the ground state, after preparation of a mixed eigenstate *via* vibrational overtone excitation, all zeroth-order components of the eigenstate can give rise to infrared transitions to higher excited states. Infrared absorption spectra subsequent to vibrational excitation can directly provide information on the identity of the coupled zeroth-order dark states. We have implemented an optical-infrared double resonance scheme in which we detect infrared absorption after vibrational overtone excitation by monitoring the predissociation products from the upper level involved in the infrared transition [33-35].

Double resonance methods can thus overcome both of the aforementioned problems associated with IVR studies of high vibrational overtone levels. By eliminating spectral congestion *via* infrared-optical double resonance, we can examine transitions to single rotational states of high vibrational overtone levels and infer dynamical information from the observed spectral splittings and linewidths. Alternatively, by using optical-infrared double resonance to first populate a vibrational overtone level and then measure an infrared spectrum, we can probe the vibrational character of the zeroth-order dark states that mix with the bright state.

Experimental Approach

The details of the experimental approach for both our infrared-optical and optical-infrared double resonance experiments have been described previously [33, 36]. Figure 1 shows a schematic of these two excitation schemes for an HOX molecule where X is OH, NO_2, NH_2, or $(CH_3)_3CO$.

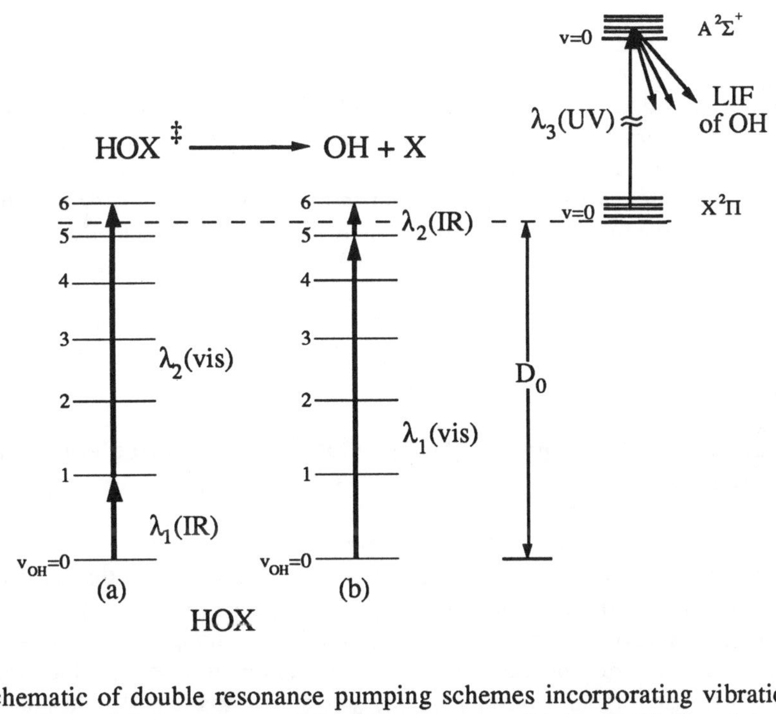

Figure 1. Schematic of double resonance pumping schemes incorporating vibrational overtone excitation. (a) Infrared-optical double resonance; (b) Optical-infrared double resonance. In each case, the absorption of the second photon (λ_2) is detected by monitoring the OH dissociation product *via* laser induced fluorescence.

The infrared-optical double resonance excitation scheme, labeled (a), uses an infrared pulse from an optical parametric oscillator (OPO) to first promote molecules from the ground state to the v_{OH}=1 level (or any combination level containing one quantum of OH stretch). This excitation can prepare sufficiently small molecules in a single or small subset of rotational states. A second pulse from a visible dye laser promotes only the molecules in v_{OH}=1 to a highly excited OH stretch level above the O–X dissociation threshold. Even in cases where a few rotational transitions overlap in the first step, the requirement that molecules satisfy two resonance conditions can allow single rotational states to be prepared *via* the vibrational overtone excitation step. Once excited above its dissociation threshold, an HOX molecule decomposes to produce an OH fragment which is subsequently detected *via* laser induced fluorescence (LIF) in the A-X band. Fixing the infrared excitation wavelength (λ_1) and scanning the overtone excitation laser wavelength (λ_2) while probing a particular OH rovibrational level produces an overtone excitation spectrum which for sufficiently small molecules is completely free from rotational congestion. Spectra generated in this way provide information on both IVR and chemical bond breaking from highly excited OH stretch levels.

In the optical-infrared excitation scheme, labeled (b) in Fig. 1, an intense dye laser pulse first promotes molecules to a high OH stretch level *via* a vibrational overtone transition. An infrared pulse from an OPO subsequently excites an O-H stretch transition of the excited molecule that terminates on a state above the dissociation threshold. The infrared absorption is monitored by LIF detection of the OH product. In this scheme, the overtone excitation laser wavelength (λ_1) is held fixed and the infrared wavelength (λ_2) is scanned to generate an infrared predissociation spectrum of the highly excited molecule. As is demonstrated below, the infrared spectrum of a vibrationally excited species can reveal the nature of the dark states that mix with the zeroth-order bright state. It is important to note that the rotational selectivity in this latter scheme is not nearly as good as in the infrared-optical double resonance scheme. This is a consequence of the greater spectral congestion in the first excitation step when going to a high vibrational level. The conclusions about the intramolecular dynamics drawn from the infrared predissociation spectra therefore pertain to a weighted average over a distribution of rotational states.

Infrared Spectroscopy of Dark States *via* Optical-Infrared Double Resonance

The infrared absorption or emission spectrum of a molecule in a mixed eigenstate reveals the vibrational character of the zeroth-order components of that eigenstate. Because the zeroth-order dark states that mix with the bright state for an $\mathbf{n}v_{OH}$ vibrational overtone transition can have O-H stretch quantum numbers ranging from 0 to $\mathbf{n}$-1, the infrared absorption spectrum subsequent to overtone excitation may consist of a series of bands in the O-H stretch region. Consider for example a polyatomic molecule excited *via* a 4←0 O-H stretch transition. If the zeroth-order bright state for this transition is unmixed, the infrared spectrum in the O-H stretch region will contain a single band at the 5 ←4 O-H stretch frequency. If the zeroth-order $4v_{OH}$ stretch level couples to nearly isoenergetic dark states, the infrared predissociation spectrum may also contain vibrational bands at the 4←3, 3←2, 2 ←1, and 1←0 O-H stretch frequencies insofar as the dark states may have OH stretch quantum numbers ranging from 0-3. The members of this vibrational progression will be separated by approximately twice the O-H stretch anharmonicity, or ~180 cm^{-1}. An infrared spectrum subsequent to excitation of a vibrational overtone transition can therefore classify the coupled zeroth-order states by their O–H stretch quantum number and determine, in an average way, how much they contribute to the mixed eigenstate.

We have measured infrared predissociation spectra of HOOH, $HONO_2$, and $(CH_3)_3COOH$ subsequent to $4\nu_{OH}$ stretch overtone excitation [33-35]. While $\Delta v_{OH}=1$ transitions from the $4\nu_{OH}$ level are sufficient to dissociate $HONO_2$ and $(CH_3)_3COOH$, $\Delta v_{OH}=2$ transitions are required to ensure that HOOH molecules dissociate upon infrared absorption. Comparisons between the vibrational state mixing characteristics at the respective $4\nu_{OH}$ levels of these three molecules provides a measure of the effect of increasing density of states and molecular complexity on the vibrational redistribution process.

HYDROGEN PEROXIDE

Figure 2 shows an infrared predissociation spectrum of HOOH measured 15 ns after excitation to the $4\nu_{OH}$ level [34].

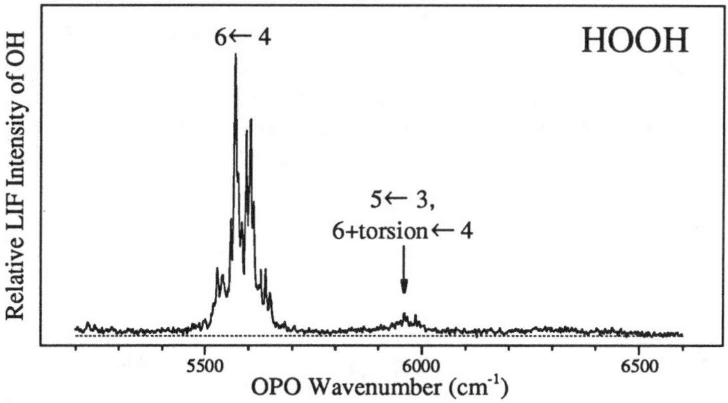

Figure 2. Infrared predissociation spectrum of HOOH in the $\Delta v_{OH}=2$ region after excitation to the $4\nu_{OH}$ vibrational overtone level.

The strong band centered at 5585 cm^{-1} occurs precisely at the 6←4 OH stretch frequency predicted by the Birge-Sponer parameters for that mode [37]. The structure in this band is rotational in origin and results from incomplete state selection in the 4←0 overtone excitation step [34]. A second, smaller vibrational band occurs at ~5970 cm^{-1}, 385 cm^{-1} to the blue of the 6←4 OH stretch band. While this small band is close to the expected frequency for a 5←3 OH stretch transition, we believe it is a combination band involving low frequency torsional motion built upon the 6←4 transition. A weak torsional combination band has been observed ~390 cm^{-1} to the blue of the 6←0 OH vibrational overtone transition in HOOH, and this shift reflects the torsional level spacing in the excited OH stretch state [38]. We therefore expect to observe a transition with the same shift in the vibrational predissociation spectrum of Fig. 2. Because of the overlap of this torsional combination band and the predicted frequency of a 5←3 OH stretch transition, we would not be able to detect small contributions from the latter if present. If we assume all of the intensity in this region corresponds to 5←3 OH stretch transitions we can determine a lower limit to the amount of bright state character in the eigenstates prepared by vibrational overtone excitation. After correcting for the relative oscillator strengths of the 6←4 and 5←3 bands, we determine that >85% of the integrated spectral intensity in the infrared predissociation spectrum originates from the zeroth-order state with 4 quanta of OH stretch (*i.e.*, the zeroth-order bright state). Recall that this represents a weighted average over all the rotational states within the 0.2 cm^{-1} overtone excitation bandwidth. Since we expect the torsional combination band to make a major contribution to the spectral intensity in the region of the 5←3 OH stretch transition, it is likely that the eigenstates prepared by $4v_{OH}$ vibrational overtone excitation are almost entirely composed of the zeroth-order bright state.

The spectrum of Fig. 2 indicates that properly symmetrized vibrational eigenfunction prepared by $4v_{OH}$ overtone excitation is therefore

$$|j\rangle \cong \frac{1}{\sqrt{2}}|4\rangle|0\rangle|0\rangle|0\rangle|0\rangle|0\rangle \pm \frac{1}{\sqrt{2}}|0\rangle|0\rangle|0\rangle|0\rangle|0\rangle|4\rangle$$

where the first and last kets in each product are the OH stretch functions for each end of the molecule. If this is indeed the eigenstate, the infrared predissociation spectrum should show two OH stretch transitions: one from the OH oscillator that has 4 quanta of excitation, which we designate as OH*, and one from the unexcited OH oscillator, designated OH^o. While the 2 ←0 OH^o transition is beyond the wavelength range of our OPO, the 1←0 OH^o stretch transition falls in an easily accessible spectral region and supplies the molecules with just enough energy to exceed the dissociation threshold. Figure 3 shows an infrared predissociation spectrum of $4v_{OH}$ excited hydrogen peroxide in the Δv_{OH}=1 region.

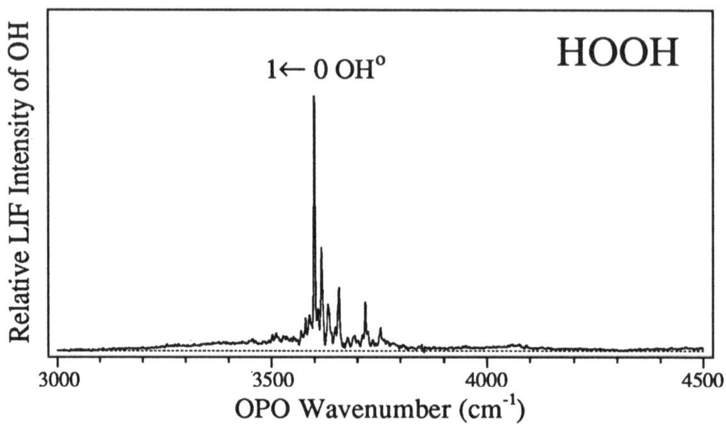

Figure 3. Infrared predissociation spectrum of HOOH in the Δv_{OH}=1 region after excitation to the $4v_{OH}$ vibrational overtone level. The notation OH^o denotes the initially unexcited OH stretch mode.

The primary band in this spectrum occurs at ~3610 cm^{-1}, almost precisely the 1 ←0 OH stretch frequency, and we assign it to a 1←0 OH^o stretch transition. The rotational structure in this band is simpler than that of the 6←4 OH* stretch transition because only the higher J,K transitions supply the molecule with enough energy to dissociate. An infrared predissociation spectrum of HOOD excited to the $4\nu_{OH}$ level shows only a 6←4 OH* stretch transition, since deuteration of the unexcited oscillator shifts its 1←0 stretch frequency out of the range of our OPO. This demonstrates that the 3610 cm^{-1} absorption arises from the initially unexcited OH oscillator. To our knowledge this is the first example of sequential local mode-local mode excitation.

The near equivalence between the zeroth-order $4\nu_{OH}$ stretch level in HOOH and the vibrational eigenstate makes this molecule a good candidate for selective chemical bond breaking schemes that requires the energy to remain in the OH stretch [39].

NITRIC ACID

Figure 4 shows an infrared predissociation spectrum of $HONO_2$ subsequent to $4\nu_{OH}$ vibrational overtone excitation [33].

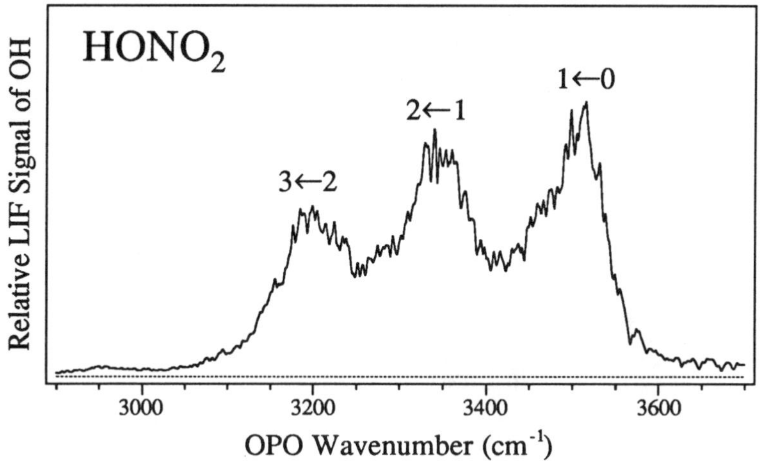

Figure 4. Infrared predissociation spectrum of $HONO_2$ after excitation to the $4\nu_{OH}$ vibrational overtone level.

This spectrum contains three strong bands spaced by twice the OH stretch anharmonicity, each of which occur slightly to the red of the the expected positions of the 3←2, 2←1, and 1←0 OH stretch bands. These correspond to infrared transitions that derive oscillator strength from zeroth order states with 2, 1, and 0 OH stretch quanta respectively. Zeroth-order dark states nearly isoenergetic with the $4\nu_{OH}$ bright state with less than 4 quanta of OH stretch must have at least several quanta in other modes, and since most off-diagonal anharmonicities are negative, this should produce a net red-shift in the OH stretch frequency as observed.

The lack of spectral features beyond the $3 \leftarrow 2$ OH stretch band results in part from the closeness of the molecules to the 16,700 cm^{-1} dissociation threshold after absorption of the infrared photon. The requirement that the infrared excitation put molecules above the dissociation threshold imposes a cutoff frequency below which infrared absorption will not contribute to the predissociation spectrum. As we pump different parts of the vibrational overtone band in the first step and provide the molecules with more rotational energy, they can dissociate after absorbing a less energetic infrared photon. This extends the spectrum out to lower frequency, although at a decreased overall signal level.

To derive meaningful information from the relative intensities in the infrared predissociation spectrum of Fig. 4 we must make corrections for the different infrared transition strengths of the OH stretch bands and the infrared frequency dependence of our detection sensitivity [33]. We estimate the relative strengths of the $\Delta v_{OH}=1$ transitions using a harmonic oscillator model and linear dipole moment function. Our detection sensitivity varies across the infrared spectrum because we detect the infrared absorption by monitoring single quantum states of the dissociation products. The product quantum state distribution changes with increasing energy above threshold, and this has the effect of enhancing the intensity at the low frequency end of the spectrum where the signal is concentrated into only a few states. To correct for this we must integrate over all the product states formed.

After making these corrections, a clear pattern emerges in the spectral intensities. Shown in column 2 of Table 1 are the corrected spectral intensities classified by the OH stretch quantum number of the zeroth-order state that supplied the oscillator strength.

TABLE 1. Corrected spectral intensities and density of states for $HONO_2$

Number of OH stretch quanta	Corrected spectral intensities	states per cm^{-1}	relative states per cm^{-1}
4	**<0.003**	-	-
3	**<0.003**	0.3	**0.003**
2	**0.08**	3	**0.03**
1	**0.25**	27	**0.23**
0	**1.0**	120	**1.0**
		~150 per cm^{-1}	

In the context of the framework we have discussed, one can see that vibrational state mixing at the $4v_{OH}$ level of $HONO_2$ is very different from that in HOOH. The lack of intensity at the $5 \leftarrow 4$ frequency indicates that the eigenstates at the $4v_{OH}$ energy are so extensively mixed that the bright state is now only a minor component. Inasmuch as the OH stretch carries the oscillator strength for overtone absorption, the eigenstates prepared contain little $4v_{OH}$ character. States with *zero* quanta of OH stretch and many quanta in lower frequency modes make the biggest collective contribution to the eigenstates prepared by overtone excitation. Thus, although on the basis of the overtone absorption frequencies one might judge the OH stretch to be a good local mode, the eigenstates prepared by overtone excitation are not at all localized on the OH bond.

Comparison of these spectral intensities with the relative density of zeroth-order states at the $4\nu_{OH}$ energy is quite revealing. Columns 3 and 4 of Table 1 show the density of states at the $4\nu_{OH}$ energy classified by their OH stretch quantum number. There is a remarkably good correspondence between the relative spectral intensities and the relative densities of zeroth-order states. The infrared predissociation spectrum of Fig. 4 is precisely what one expects if the eigenstates prepared by vibrational overtone excitation were statistical mixtures of all zeroth-order states at the $4\nu_{OH}$ energy. This does not imply that the eigenstates are completely (or statistically) mixed in that we do not know if every nearly isoenergetic zeroth-order state contributes equally, but rather it indicates that the vibrational mixing doesn't discriminate on the basis of the OH stretch quantum number. Since we only observe spectra in the OH stretch region of the infrared, our results provide no information on the distribution of other vibrational quantum numbers in the mixed $4\nu_{OH}$ eigenstates. However, if the vibrational mixing does not discriminate on the basis of the OH stretch quantum number, it is likely not to discriminate with respect to the number of quanta in lower frequency vibrational modes.

The difference in state mixing characteristics between HOOH and $HONO_2$ most likely reflects the difference in density of states. Hydrogen peroxide has a state density of only ~4 per cm^{-1} at the $4\nu_{OH}$ energy, which is below the IVR threshold of 5-30 per cm^{-1}. suggested by McDonald and coworkers. The 150 states per cm^{-1} at the $4\nu_{OH}$ energy of nitric acid appears to be sufficient for extensive IVR to occur, although the apparent insensitivity of the state mixing to the OH stretch quantum number is somewhat surprising.

TERT-BUTYL HYDROPEROXIDE

A molecule the size of t-butyl hydroperoxide should clearly represent an example of statistical case IVR. Table 2 shows the density of states of t-butyl hydroperoxide at the $4\nu_{OH}$ energy classified by their OH stretch quantum number.

TABLE 2. Density of zeroth-order states for t-butyl hydroperoxide

Number of OH stretch quanta	states per cm^{-1}	rclativc states per cm^{-1}
4	-	-
3	4.4×10^2	3×10^{-8}
2	4.3×10^5	3×10^{-5}
1	1.2×10^8	0.007
0	1.6×10^{10}	0.993
	~1.6×10^{10} per cm^{-1}	

If the relative band intensities in the infrared spectrum of vibrationally excited t-butyl hydroperoxide reflect a statistical mixture of all the nearly isoenergetic zeroth-order states, then the spectrum should resemble an ordinary infrared spectrum of the ground state molecule, since greater than 99% of the zeroth-order states at the $4\nu_{OH}$ energy have an OH stretch quantum number of zero.

Figure 5(a) shows an infrared spectrum of t-butyl hydroperoxide subsequent to $4\nu_{OH}$ excitation [35]. Figure 5(b) shows a photoacoustic spectrum of ground state t-butyl hydroperoxide taken over the same spectral region at the same resolution.

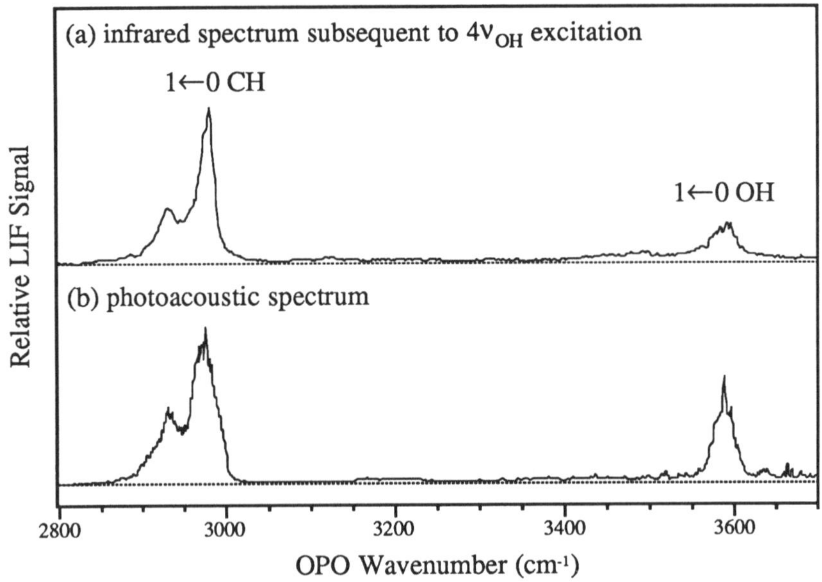

Figure 5. (a) Infrared predissociation spectrum of t-butyl hydroperoxide after excitation to the $4\nu_{OH}$ vibrational overtone level; (b) Photoacoustic spectrum of ground state t-butyl hydroperoxide.

The infrared predissociation spectrum exhibits 1 ←0 OH and 1 ←0 CH stretch transitions at precisely the same frequencies as the ground state molecule. The similarity of the infrared predissociation spectrum and the photoacoustic spectrum confirms the assignment of the former and indicates that the structure in the CH stretch region is characteristic of the infrared spectrum of the t-butyl moiety and is not a consequence of the vibrational excitation. We observe no progression of OH stretch bands originating from zeroth-order states with OH stretch quantum numbers between 1 and 4. The spectrum of Fig. 5 is consistent with $4\nu_{OH}$ eigenstates that are statistical mixtures of all the nearly isoenergetic zeroth-order states, since most of those states have zero OH and CH stretch quanta.

It is important to note that these experiments reflect the end result of IVR inasmuch as they do not use coherent excitation and they probe the molecule in the long time limit (30 ns). Experiments which coherently excite a vibrational overtone level and probe the prepared state with a short infrared pulse should be able to unravel the hierarchy of the vibrational couplings.

Infrared-Optical Double Resonance of High Vibrational Overtone Levels

The optical-infrared double resonance studies described above reveal the extent of vibrational mixing and the nature of the zeroth-order dark states but provide no information on the timescale of vibrational energy redistribution. Moreover, since the experiment is performed at room temperature, the vibrational overtone excitation step prepares a distribution of rotational states, and the information we derive from the predissociation spectra are rotationally averaged.

Information about the timescale of IVR at high vibrational overtone levels is available either directly from time domain experiments or from measurements of the vibrational splitting of the zeroth-order bright state. For the second approach to be unambiguous, inhomogeneous spectral congestion must be completely eliminated. We have used infrared-optical double resonance to prepare small molecules in single rovibrational levels at energies above their dissociation threshold. This scheme allows us to examine the linewidths and spectral splittings resulting from vibrational state mixing and their dependence on the rotational state of the molecule. After presenting the basic framework in which we interpret the overtone linewidths of dissociating molecules, we describe the results of infrared-optical double resonance studies of HOOH. We emphasize the implications of the spectra for determining the nature of the rotational motion of a molecule as it proceeds towards dissociation and the dependence of IVR on the vibrational character of the bright state.

COUPLING MODEL

Rotationally resolved vibrational overtone spectra of molecules excited above their dissociation threshold potentially contain information about both IVR and the unimolecular decomposition rate. The schematic of Fig. 6 illustrates the potential spectral consequences of these two processes.

In the absence of a dissociation channel, the zeroth-order bright state $|s\rangle$ couples to some set of dark states $\{|l\rangle\}$ through matrix elements V_{sl}, spreading the bright state character among several molecular eigenstates $\{|j\rangle\}$. The coupling matrix elements may result from purely vibrational (anharmonic) interactions or from vibration-rotation interactions. At energies sufficient for dissociation, these states $\{|j\rangle\}$ are not eigenstates of the bound molecule in that they are coupled to the dissociative continuum. This coupling imposes a width upon these levels, designated as Γ_c, which is related to the dissociation rate of the excited molecule. The observed structure of a single zeroth-order rotational line is determined by the number of coupled states and the relative magnitudes of the coupling matrix elements V_{sl} and the coupling strength to the continuum, Γ_c.

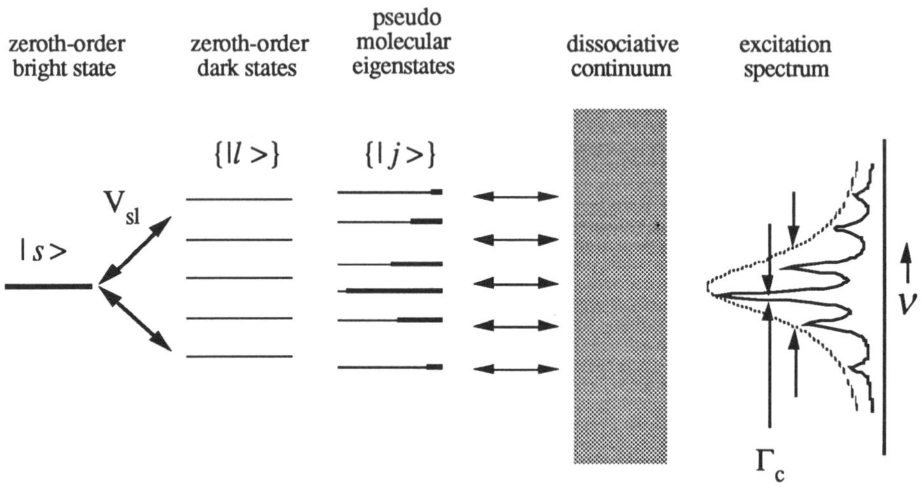

Figure 6. Schematic of coupling model illustrating the effect of vibrational coupling and unimolecular dissociation on high vibrational overtone spectra. V_{sl} is the average coupling matrix element between the zeroth-order bright state and zeroth-order dark states. Γ_c represents the width imposed on a spectral feature by coupling to the dissociative continuum. In the case shown here $V_{sl} > \Gamma_c$.

If coupling to the continuum is much weaker than the coupling among zeroth-order vibrational states (*i.e.*, $V_{sl} >> \Gamma_c$), each zeroth-order bright state splits into a multitude of sharp components, the number of which is determined by the number of dark states $|l>$ with significant V_{sl}. Physically, this corresponds to IVR occurring much faster than unimolecular dissociation. In this case the overall clump width is related to the rate of IVR, and the width of an individual component of that clump is related to the dissociation rate.

If coupling to the continuum is comparable to or greater than the coupling among zeroth-order vibrational states (*i.e.*, $\Gamma_c \geq V_{sl}$), a zeroth-order J,K transition will appear as a single broadened line, the width of which contains information on both processes. In the limit that Γ_c is much greater than V_{sl}, the linewidth reflects the unimolecular dissociation rate.

The assignability of the spectra to zeroth-order J and K quantum numbers depends upon the magnitudes of these two coupling terms relative to the rotational constants of the molecule. If either Γ_c or V_{sl} is greater than several times the B rotational constant, the spectra will be inherently difficult to assign. As is described more fully below, the magnitudes of the coupling terms relative to the rotational constants provides information on the nature of rotational motion of the dissociating molecule.

ROTATIONALLY RESOLVED VIBRATIONAL OVERTONE SPECTRA

Figure 7 shows the degree of spectral simplification achieved by infrared-optical double resonance excitation of a high vibrational overtone transition by comparing a $1 \leftarrow 0$, $6 \leftarrow 1$

double-resonance spectrum of HOOH (Fig. 7(b)) with a direct 6←0 overtone excitation spectrum (Fig. 7(a)) [36].

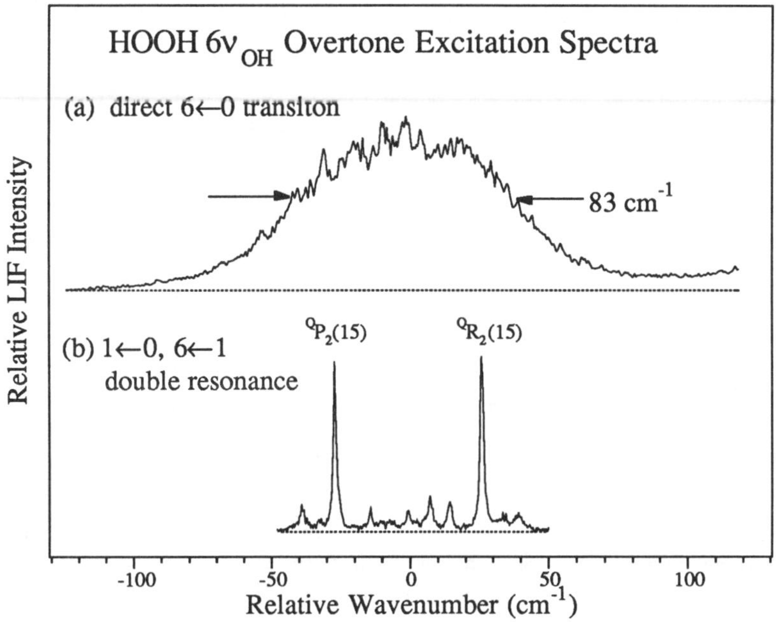

Figure 7. Vibrational overtone excitation spectra of the $6\nu_{OH}$ level of HOOH. (a) direct 6←0 excitation; (b) 1 ←0, 6←1 double resonance.

The vibrational overtone excitation spectrum of Fig. 7(b) results from tuning the optical parametric oscillator to coincide with the ${}^{P}P_3(16)$ transition of the ν_5 band of HOOH, and scanning the visible dye laser. The two strongest features of this spectrum correspond to the ${}^{Q}P_2(15)$ and ${}^{Q}R_2(15)$ transitions of a parallel polarized vibrational overtone band of a near prolate symmetric top. The smaller features arise from transitions with other J and K that coincidentally overlap the ${}^{P}P_3(16)$ transition of the ν_5 band. We can prepare excited molecules in a wide range of rotational states by changing OPO frequency to excite a different rotational transitions in the ν_5 band and rescanning the visible dye laser. In each case, the rotationally resolved overtone spectrum appears exactly as predicted for a parallel vibrational band of near symmetric top and is completely assignable to zeroth-order J and K quantum numbers.

Figure 8 shows a single zeroth-order rotational transition in the $6\nu_{OH}$ band that terminates on a level with J=12, K=2. The consistency between measured widths for pairs of transitions that originate from a different lower state but terminate on a common upper state (e.g. ${}^{Q}R_2(11)$ and ${}^{Q}P_2(13)$) verify that we are observing single J, K states. The 1.05 cm^{-1} width of the ${}^{Q}R_2(11)$ transition is an order of magnitude greater than the laser linewidth and a factor of 30 greater than the doppler width and calculated asymmetry splitting. In general, the linewidths of the $6\nu_{OH}$ transitions range from 1-3 cm^{-1} but exhibit no systematic dependence upon J and K. The fact that the spectral features appear as single lines and exhibit widths that are are smaller than or comparable to several times the B rotational constant facilitates their assignment to zeroth-order J, K levels.

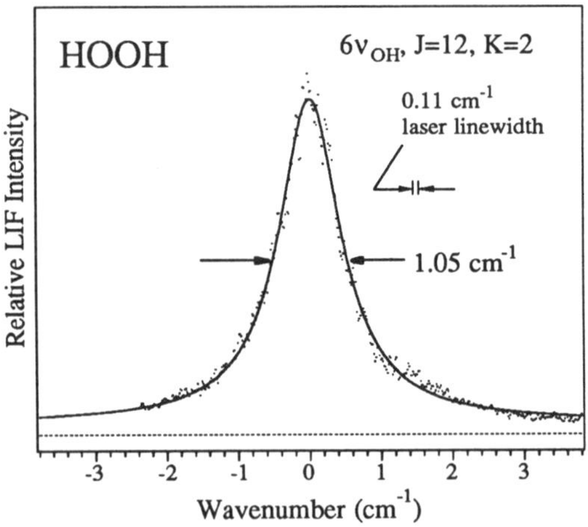

Figure 8. $^{Q}R_2(11)$ transition of the 6←1 vibrational overtone band of HOOH. The J and K indicated in the figure are the rotational quantum numbers in the $6\nu_{OH}$ level. The points are the experimental data and the solid line is a Lorentzian fit to the data.

The observation of vibrational overtone lines that follow the expected pattern for a symmetric top (*i.e.*, are assignable to zeroth-order J, K, quantum numbers) indicates that K remains a good quantum number for at least a time $t = (2\pi c\Delta\tilde{\nu})^{-1}$ where $\Delta\tilde{\nu}$ is the width of a single J, K feature. Physically, conserving K means that the molecules rotate with a fixed projection of J on the top axis, and our spectra indicate that at the $6\nu_{OH}$ level, this projection is maintained for at least 5 ps. The quantum number K may be conserved considerably longer than this depending upon the magnitude of the interactions that mix states of different K (*i.e.*, what fraction of the overall linewidth results from Coriolis or centrifugal coupling).

What is most important from a chemical standpoint is the relative timescales of K conservation and unimolecular dissociation. Statistical theories of unimolecular dissociation dynamics assume that K rotational energy freely exchanges with vibrational energy on the timescale of chemical bond breaking. Violation of this assumption can make an order of magnitude difference in predictions of the dissociation rate at energies near threshold and can strongly influence the predicted product energy partitioning. Information about the relative timescales of these processes is contained in the ratio of V_{sl} to Γ_c. Although we cannot directly determine the contribution to the linewidth due to the finite lifetime of the molecule, the smoothness of the vibrational overtone transitions and their Lorentzian lineshapes suggest that the individual components overlap to within their natural (lifetime broadened) linewidth.

Measuring vibrational overtone linewidths at energies much closer to the dissociation threshold should greatly increase the lifetime of the excited molecules and hence decrease the lifetime broadening component to the observed linewidths. Explicitly observing structure resulting from vibrational state mixing would allow us to assess the relative timescales of IVR and unimolecular dissociation. We have employed two different excitation schemes to access vibrational bands closer to the dissociation threshold. By first exciting hydrogen peroxide molecules to a combination band involving one quantum of OH stretch and one quantum of OO stretch at ~4474 cm^{-1} in the infrared and then adding 4 additional quanta to the same OH oscillator *via* vibrational overtone excitation, we can prepare reactant molecules containing 5 quanta of OH stretch and one quantum of OO stretch [32]. The $5\nu_{OH}+\nu_{OO}$ combination level is 61 cm^{-1} above the dissociation threshold, and at this excess energy the lifetime of the excited molecules should be significantly slower than at 2000 cm^{-1} above threshold (*i.e.*, the excess energy at the $6\nu_{OH}$ level).

The other excitation scheme for preparing molecules closer to the dissociation threshold involves distributing the excitation energy among the two identical OH oscillators. By first exciting a 1←0 OH stretch transition of one oscillator and then inducing a 4←0 transition in the initially unexcited OH stretch, we access the 4,1 local-local combination level which we have determined to be 88 cm^{-1} *below* the 17,052 cm^{-1} dissociation threshold. Transitions that prepare the molecule with greater than 88 cm^{-1} of rotational energy can be detected by monitoring the OH dissociation products.

Figure 9 shows ${}^QP_0(2)$ line of a vibrational overtone transition to the $5\nu_{OH}+\nu_{OO}$ combination level [32].

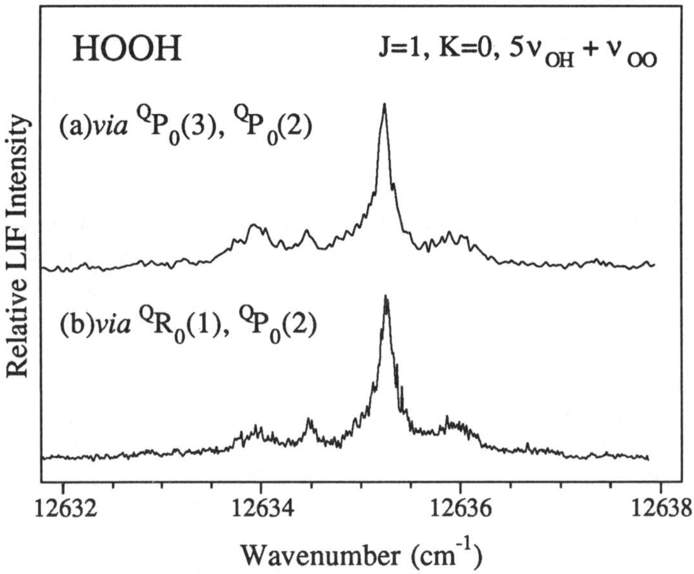

Figure 9. ${}^QP_0(2)$ vibrational overtone transition to the J=1, K=0 rotational state of the $5\nu_{OH}+\nu_{OO}$ level of HOOH *via* different transitions in the infrared excitation step. (a) *via* ${}^QP_0(3)$; (b) *via* ${}^QR_0(1)$.

The OPO excites hydrogen peroxide molecules to the $\nu_{OH}+\nu_{OO}$ combination level, and the spectra are generated by scanning the visible dye laser over the $5\nu_{OH}+\nu_{OO} \leftarrow \nu_{OH}+\nu_{OO}$ band. The spectra of Figs. 9(a) and 9(b) differ in the rotational transition used to reach the $\nu_{OH}+\nu_{OO}$ intermediate state ($^{Q}P_0(3)$ and $^{Q}R_0(1)$ respectively). By observing the same spectral splittings in transitions to a common final state *via* two different routes we ensure that the structure arises from the distribution of the single zeroth-order bright state among several molecular eigenstates and not from residual rotational congestion.

The breakup of a single J, K feature into several components of finite width indicates that the matrix elements V_{sl} are greater than or equal to the coupling strength to the dissociative continuum, Γ_c (see Fig. 6). The overall clump width results from coupling of the bright state to the dark states and hence defines a timescale for IVR. The width of a single feature results from coupling to the dissociative continuum and puts a lower limit on the lifetime of the dissociating molecule of 30 ps. Because the entire clump arises from a single zeroth-order J, K state and the ratio of the clump width to the width of an individual line is about 5-10, these spectra indicate that the molecule rotates with fixed projection for *at least* a fifth to a tenth of the lifetime of the dissociating molecule. If the primary source of spectral splitting at low J is anharmonic in nature, K may remain a good quantum number for considerably longer than this. This indicates that statistical and dynamical models of the unimolecular dissociation rate must include at least partial conservation of K rotation.

VIBRATIONAL STATE DEPENDENCE OF THE OVERTONE LINEWIDTHS

The ability to prepare excited HOOH molecules in different vibrational states above the unimolecular dissociation threshold allows us to examine the vibrational state dependence of the coupling between the bright state and zeroth-order dark states. Figure 10 shows transitions to single rotational levels in three different vibrational bands: $6\nu_{OH}$, $5\nu_{OH}+\nu_{OO}$, and $4\nu_{OH}+\nu_{OH'}$. Each of these transitions arises from a single zeroth-order bright state with well defined J and K.

Although the $5\nu_{OH}+\nu_{OO}$ level is ~2000 cm^{-1} below the $6\nu_{OH}$ level, the overall clump widths of transitions to the former is on the same order as the $6\nu_{OH}$ transitions. If we assume that the lifetime broadening component at $6\nu_{OH}$ is at least as much as at the $5\nu_{OH}+\nu_{OO}$ level (and it is almost certainly greater), then the component of the $6\nu_{OH}$ transitions resulting from vibrational state mixing is clearly smaller than that at the combination level, even though the energy is 2000 cm^{-1} higher!

The comparison between transitions to the $5\nu_{OH}+\nu_{OO}$ level and those to $4\nu_{OH}+\nu_{OH'}$ are particularly enlightening in that the excess energies are very nearly the same. The $4\nu_{OH}+\nu_{OH'}$ transitions show what appear to be single lines that have widths comparable to the single components of the $5\nu_{OH}+\nu_{OO}$ transitions. The widths of the $4\nu_{OH}+\nu_{OH'}$ transitions are at least a factor of 5 smaller than the overall clump width of the $5\nu_{OH}+\nu_{OO}$ transitions at equivalent excess energy above the dissociation threshold.

The spectra of Fig. 10 suggest that the vibrational coupling of the $5\nu_{OH}+\nu_{OO}$ level to the zeroth-order dark states is significantly stronger than that of the $6\nu_{OH}$ and $4\nu_{OH}+\nu_{OH'}$ levels. The $5\nu_{OH}+\nu_{OO}$ level contains one quantum in the low frequency O-O stretch mode whereas the others have all the energy in high frequency OH stretch modes. These results seem to confirm the general expectation that high frequency light atom stretch vibrations are more poorly coupled to the rest of the molecule than modes with lower frequency vibrational components. In this case the difference may be as much as a factor of 10.

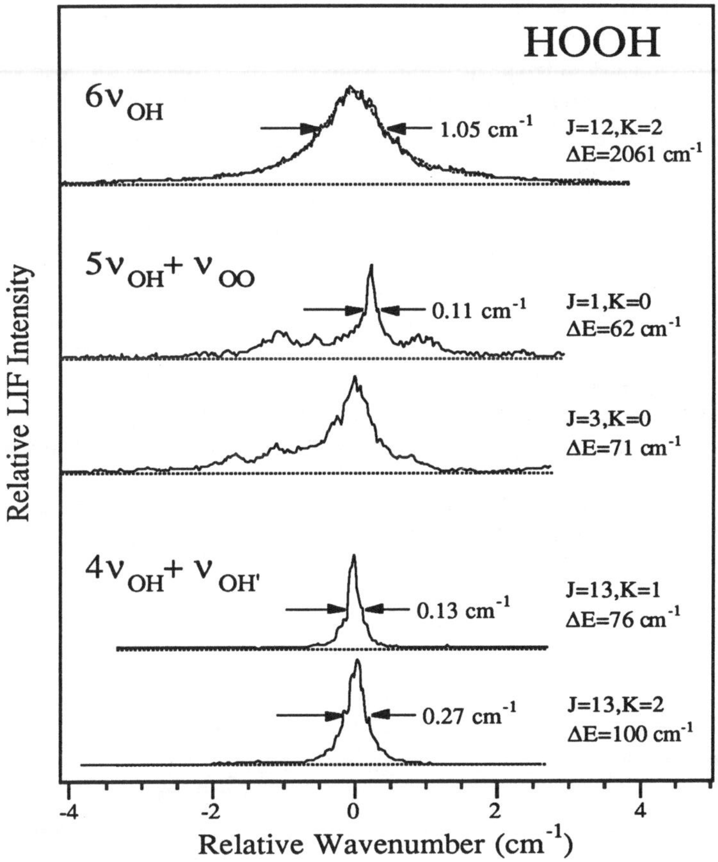

Figure 10. Single rotational transitions to the $6\nu_{OH}$ (top), $5\nu_{OH}+\nu_{OO}$ (middle two), and $4\nu_{OH}+\nu_{OH'}$ (lower two) vibrational levels. The quoted linewidths are values after deconvolution of the 0.11 cm^{-1} laser linewidth.

Summary

Optical-infrared and infrared-optical double spectroscopy of high vibrational overtone levels provide a new view of intramolecular processes in dissociating molecules. Infrared predissociation spectra recorded subsequent to overtone excitation of a light atom stretching vibration directly reveals the extent of vibrational state mixing and classifies the zeroth-order dark states by their OH stretch quantum number. The $4\nu_{OH}$ level of HOOH is virtually unmixed, and this permits sequential excitation of the two local OH stretch oscillators. In contrast the $4\nu_{OH}$ levels of $HONO_2$ and $(CH_3)_3COOH$ are extensively mixed. The infrared predissociation spectrum of these two molecules subsequent to $4\nu_{OH}$ excitation indicates that vibrational eigenstates are composed of zeroth-order vibrational states with OH stretch quantum numbers ranging from 0 to 4 in amounts proportional to their relative density at the $4\nu_{OH}$ energy.

Infrared-optical double spectroscopy generates fully rotationally resolved vibrational overtone spectra of HOOH molecules excited to energies above their unimolecular dissociation threshold. The assignability of the rotationally resolved overtone spectra to zeroth-order J and K quantum numbers indicates that although the molecules are highly excited, they rotate with a fixed projection of J on the top axis for a significant fraction of the time until dissociation. This suggests that statistical and dynamical models of the unimolecular dissociation dynamics of this molecule must incorporate at least partial conservation of K to accurately predict the dissociation rate.

The infrared-optical double resonance spectra also reveal a marked vibrational dependence to the coupling matrix elements between the zeroth-order bright state and nearly isoenergetic dark states. A zeroth-order $5\nu_{OH}+\nu_{OO}$ level, which has one quantum of low frequency vibration in addition to the OH stretch excitation, is coupled to the zeroth-order dark states significantly stronger than states that have all the excitation energy in high frequency OH stretches.

These results help define the conditions that must be met in order to use high vibrational overtone excitation in mode selective chemical bond breaking schemes.

Acknowledgements

We gratefully acknowledge the support of this work by the Office of Basic Energy Sciences of the Department of Energy and the donors of the Petroleum Research Fund, administered by the ACS.

References

1. Heller, D. F. (1979) "Molecular overtones as local modes", Chem. Phys. Lett. **61**, 583.

2. Burberry, M. S. and Albrecht, A. C. (1979) "Local mode combination bands and local mode mixing", J. Chem. Phys. **71**, 4631.

3. Henry, B. R., Hung, I.-F., MacPhail, R. A. and Strauss, H. L. (1980) "A Local Mode Description of the CH-Stretching Overtone Spectra of the Cycloalkanes and Cycloalkenes", J. Am. Chem. Soc. **102**, 515.

4. Reddy, K. V., Heller, D. F. and Berry, M. J. (1982) "Highly vibrationally excited benzene: Overtone spectroscopy and intramolecular dynamics of C_6H_6, C_6D_6 and partially deuterated or substituted benzenes", J. Chem. Phys. **76**, 2814.

5. Lehmann, K. K., Scherer, G. J. and Klemperer, W. (1982) "Classical chaos and quantum simplicity: Highly excited vibrational states of HCN", J. Chem. Phys. **77**, 1982.

6. Chuang, M. C., Baggott, J. E., Chandler, D. W., Farneth, W. E. and Zare, R. N. (1983) "Unimolecular decomposition of t-butylhydroperoxide by direct excitation of the 6-0 OH stretching overtone", Faraday Discuss. Chem. Soc. **75**, 301.

7. Crim, F. F. (1984) "Selective excitation studies of unimolecular reaction dynamics", Ann. Rev. Phys. Chem. **34**, 657; and references cited therein.

8. Dübal, H.-R. and Quack, M. (1984) "Tridiagonal Fermi resonance structure in the infrared spectrum of the excited CH chromophore in CF_3H", J. Chem. Phys. **81**, 3779.

9. Chuang, M. C. and Zare, R. N. (1985) "Evidence for inhomogeneous broadening in vibrational overtone transitons: Formatin of 1, 3-cyclohexadiene form *cis*-1, 3, 5-hexatriene", J. Chem. Phys. **82**, 4791.

10. Butler, L. J., Ticich, T. M., Likar, M. D. and Crim, F. F. (1986) "Vibrational overtone spectroscopy of bound and predissociative states of hydrogen peroxide in a supersonic expansion", J. Chem. Phys. **85**, 2331.

11. McGinley, E. S. and Crim, F. F. (1986) "Homogeneous and inhomogeneous structure in the vibrational overtone spectrum of tetramethyldioxetane", J. Chem. Phys. **85**, 5741.

12. Scherer, N. F., Doany, F. E., Zewail, A. H. and Perry, J. W. (1986) "Direct picosecond time resolution of unimolecular reactions initiated by local mode excitation", J. Chem. Phys. **84**, 1932.

13. Green, W. H., Jr., Lawrance, W. D. and Moore, C. B. (1987) "Kinetic anharmonic coupling in the trihalomethanes: A mechanism for rapid intramolecular redistribution of CH stretch vibrational energy.", J. Chem. Phys. **86**, 6000.

14. Crofton, M. W., Stevens, C. G., Klenerman, D., Gutow, J. H. and Zare, R. N. (1988) "Overtone spectra of C-H oscillators in cold molecules", J. Chem. Phys. **89**, 7100.

15. Page, R. H., Shen, Y. R. and Lee, Y. T. (1988) "Local modes of benzene and benzene dimer, studied by infrared-ultraviolet double resonance in a supersonic beam", J. Chem. Phys. **88**, 4621.

16. Foy, B. R., Casassa, M. P., Stephenson, J. C. and King, D. S. (1989) "Dissociation lifetimes and level mixing in overtone-excited HN_3 (X^1A')", J. Chem. Phys. **90**, 7037.

17. Sibert, E. L., Reinhardt, W. P. and Hynes, J. T. (1984) "Intramolecular vibrational relaxation and spectra of CH and CD overtones in benzene and perdeuterobenzene", J. Chem. Phys. **81**, 1115.

18. Uzer, T., Hynes, J. T. and Reinhardt, W. P. (1985) "Overtone-induced dissociation of hydrogen peroxide: A classical trajectory study", Chem. Phys. Lett. **117**, 600.

19. Sumpter, B. G. and Thompson, D. L. (1985) "Intramolecular vibrational energy flow in model four-atom systems", J. Chem. Phys. **82**, 4557.

20. Holme, T. A. and Hutchinson, J. S. (1986) "Dynamics of overtone excitation processes for a polyatomic molecule", J. Chem. Phys. **84**, 5455.

21. Lu, D.-H., Hase, W. L. and Wolf, R. J. (1986) "The sensitivity of IVR in benzene to bend-stretch potential energy coupling", J. Chem. Phys. **85**, 4422.

22. Getino, C., Sumpter, B. G., Santamaria, J. and Ezra, G. S. (1989) "Unimolecular decay lifetimes and intramolecular energy redistribution in HOOH: Sensitivity to potential energy surface", J. Phys. Chem. **93**, 3877.

23. Dai, H. L., Corpa, C. L., Kinsey, J. L. and Field, R. W. (1985) "Rotation-induced vibrational mixing in X^1A_1 formaldehyde: Non-negligible dyanamical consequences of rotation", J. Chem. Phys. **82**, 1688.

24. Minton, T. K., Kim, H. L. and McDonald, J. D. (1988) "Rovibrational state mixing in the C-H stretch fundamental region of norboradiene", J. Chem. Phys. **88**, 1539.

25. Dolson, D. A., Holtzclaw, K. W., Moss, D. B. and Parmenter, C. S. (1986) "Chemical Timing 4. The Rovibronic Level Structure Associated With Intramolecular Vibrational Redistribution In S_1 P-Difluorobenzene", J. Chem. Phys. **84**, 1119.

26. Go, J., Bethardy, G. A. and Perry, D. S. (1990) "Rotationally-Mediated IVR in Jet-Cooled trans-Ethanol", J. Phys. Chem. **94**, 6153.

27. Nathanson, G. M. and McClelland, G. M. (1986) "Fluorescence Polarization From $^1B_1$ Pyrimidine: Evidence For Intramolecular Vibration-Rotation Energy Transfer In A Highly Excited Molecule", J. Chem. Phys. **84**, 3170.

28. Schubert, U., Riedle, E., Neusser, H. J. and Schlag, E. W. (1986) "Lifetimes Of Single Rotational States In The 'Channel Three' Region Of C_6H_6", J. Chem. Phys. **84**, 6182.

29. Coy, S. L. and Lehmann, K. K. (1986) "Rotational structure of ammonia N-H stretch overtones: Five and six quanta bands.", J. Chem. Phys. **84**, 5239.

30. Luo, X., Rieger, P. T., Perry, D. S. and Rizzo, T. R. (1988) "A new technique for state-to-state studies of unimolecular reactions", J. Chem. Phys. **89**, 4448.

31. Luo, X., Fleming, P. R., Seckel, T. A. and Rizzo, T. R. (1990) "Broad Vibrational Overtone Linewidths in the $7\nu_{OH}$ Band of Rotationally Selected NH_2OH.", J. Chem. Phys. **93**, 9194.

32. Luo, X. and Rizzo, T. R. (1991) "Unimolecular Dissociation of Hydrogen Peroxide from Single Rovibrational States Near Threshold", Journal of Chemical Physics **94**, 889.

33. Fleming, P. R., Li, M. and Rizzo, T. R. (1991) "Infrared Spectroscopy of Vibrationally Excited $HONO_2$: Shedding Light on the Dark States of IVR", J. Chem. Phys. **94**, 2425.

34. Fleming, P. R., Li, M. and Rizzo, T. R. (1991) "Local modes of HOOH probed by optical-infrared double resonance", Journal of Chemical Physics **95**, in press.

35. Fleming, P. R. and Rizzo, T. R. (1991) "Infrared spectrum of t-butyl hydroperoxide excited to the $4\nu_{OH}$ vibrational overtone level", Journal of Chemical Physics **95**, (in press).

36. Luo, X. and Rizzo, T. R. (1990) "Rotationally resolved vibrational overtone spectroscopy of hydrogen peroxide at chemically significant energies", Journal of Chemical Physics **93**, 8620.

37. Rizzo, T. R., Hayden, C. C. and Crim, F. F. (1983) "Product energy partitioning in the decomposition of state-selectively excited HOOH and HOOD", Faraday Discuss. Chem. Soc. **75**, 223.

38. Dübal, H.-R. and Crim, F. F. (1985) "Vibrational overtone predissociation spectroscopy of hydrogen peroxide", J. Chem. Phys. **83**, 3863.

39. Sinha, A., Hsiao, M. C. and Crim, F. F. (1991) "Controlling bimolecular reactions: Mode and bond selected reaction of water with hydrogen atoms", Journal of Chemical Physics **94**, 4928.

MODE SELECTIVE VIBRATIONAL REDISTRIBUTION AND UNIMOLECULAR REACTIONS DURING AND AFTER IR–LASER EXCITATION

MARTIN QUACK
Laboratorium für Physikalische Chemie
ETH Zürich (Zentrum)
CH–8092 Zürich
Switzerland

ABSTRACT. We summarize recent evidence from work of our group for mode selective intramolecular vibrational redistribution in polyatomic molecules. This is an essential phenomenon for unimolecular reactions during and after infrared laser multiphoton excitation and in general. It is found that on a simple level mode selective coupling can be related easily to structural features, such as the dominant Fermi resonance between bending and stretching motions in the alkyl CH chromophore and its quenching in acetylenic CH. The experimental evidence for the dynamics of the CH group tied to a heavy frame with sp^3, sp^2 and sp character is discussed. The molecular hamiltonian for anharmonic vibrational motion can be derived from a combination of high resolution spectroscopy with ab initio potential surfaces. Quantum wave packet motion for realistic hamiltonians shows distinct differences from quasiclassical wave packet motion if the anharmonicity of the coupled oscillator system is sufficiently large. Time irreversible equations such as the Pauli equation, implying an increase of entropy with time by de facto symmetry breaking in coupled quantum oscillators may be related to strongly anharmonic motion in real molecules. General aspects of symmetry and symmetry reduction in relation to intramolecular vibrational redistribution are discussed with the example of alkyl CH compounds.

1. Introduction: Why Understanding Intramolecular Vibrational Redistribution?

Intramolecular vibrational redistribution (IVR) lies at the heart of both thermally activated [1] and infrared multiphoton induced unimolecular reactions [2–4] as well as overtone induced chemistry [5–7] and many other primary processes in chemistry. To name but a few specific examples we may consider the reaction of hydrogen atoms with alkyl radicals, important for pyrolysis and combustion:

$$H + C_2H_5 \longrightarrow H\overset{*}{-}CH_2{-}CH_3 \longrightarrow (CH_3{-}CH_3)^{*} \longrightarrow (CH_3\overset{*}{-}CH_3) \longrightarrow 2\,CH_3 \qquad (I)$$

J. Jortner et al. (eds.), Mode Selective Chemistry, 47–65.

It is obvious that the initially excited C–H bond must ultimately transfer its energy to the C–C bond, which breaks to give two methyl radicals and we shall consider the primary process for this migration of energy below. Similarly, in multiphoton excitation of CF_3I one might think of initially exciting the CF_3 stretching vibration near 1070 cm^{-1} with a CO_2 laser, if the excitation takes place on a very short time scale. But ultimately it is the CI bond (fundamental wavenumber near 286 cm^{-1}), which breaks [8]

$$CF_3I \xrightarrow{\geq\, 17h\nu} (CF_3I)^* \longrightarrow CF_3 + I \qquad \text{(II)}$$

One photon overtone excitation of the OH stretching vibration in hydrogen peroxide ultimately leads to breaking of the O–O bond [5–7]

$$\mathrm{H{\diagdown}O{-}O{\diagdown}H} \xrightarrow{h\nu} 2\,OH \qquad \text{(III)}$$

In the much more weakly bound hydrogen bond molecule $(HF)_2$, one infrared photon near the HF stretching fundamentals (about 3900 cm^{-1}) suffices to break the hydrogen bond [9,10], but overtone excitation leads, of course, also to reaction ($D_0 \simeq 1065$ cm^{-1}) [11,12]

$$\mathrm{H{\diagdown}F\ldots H{-}F} \xrightarrow{h\nu} 2\,HF$$

A more subtle example is provided by IR–multiphoton excitation of the bichromophoric, isotope labeled difluorobutane [13]

$$F{-}CH_2{-}CH_2{-}CH_2{-}CHDF \xrightarrow{nh\nu} (C_4H_7DF_2)^* \begin{cases} \nearrow CH_2{=}CH{-}CH_2{-}CHDF & \text{(IVa)} \\ \searrow CH_2F{-}CH_2{-}CH{=}CHD & \text{(IVb)} \end{cases}$$

Laser excitation at the wavenumber of the –CHDF chromophore near 930 cm^{-1} leads to HF elimination with equal probability from both ends of the difluorobutane on time scales of 10 ps, from which it has been concluded that energy migration occurs at least within this period (the $R{-}CH_2F$ chromophore absorbs around 1050 cm^{-1} only [13]). This example for such a simple chain molecule already gives an indication of the possible importance of energy migration processes in the macromolecules of biochemistry.

The role of energy migration in polyatomic molecules should thus be evident and it is the aim of the present article to summarize some of the concepts and results from the work of our group, with stress on recent observations on time dependent molecular motion and mode selective redistribution as derived from the analysis of high resolution spectra (obtained under stationary conditions). We shall start by a somewhat historical summary of the basic concepts that provide a first level of understanding and are very widely used in the literature following the pioneering work on radiationless electronic transitions [14,15]. We shall mention some simple mode coupling ideas, which provide physical insight and then penetrate more deeply into the analysis of the real, time dependent phenomena in polyatomic molecules under intense infrared radiation.

2. A First Level of Understanding: Level Schemes for IVR as a Radiationless Transition

Many of the basic concepts for intramolecular vibrational redistribution can be taken over from the theory of electronic radiationless transitions and relaxation (internal conversion and intersystem crossing [14–16]). The analogy can be made even more compelling, if use is made of the vibrational adiabatic approximation, similar to the electronic Born–Oppenheimer approximation [11,12,17]. Coupling is commonly represented by an electric dipole allowed transition to a "bright" zero order state, which may evolve by anharmonic or Coriolis interaction with a background of "dark" states, to which direct dipole transitions are forbidden. Depending on the relation of coupling strengths and density of background states different dynamical limiting situations have been distinguished (for example "small molecule limit" and "dense statistical" limit as well as "intermediate level structure" [14–16]). These concepts have been complemented for IVR by certain ideas from the statistical theory of spectra and random coupling [18,19] and sequential coupling schemes [20,21], subsequently called "tier models" [22]. We refer to the references cited for detailed discussions of these concepts. An obvious limitation of such general ideas, is that molecular properties, structure and dynamics, enter only in an extremely crude way into the proposed mechanisms for IVR. A further level of understanding thus clearly must relate to specific molecular properties.

3. Simple Specific Molecular Coupling Concepts for Mode Selective IVR Involving the CH–Chromophore

It can be expected that transferable and universal structural features in large polyatomic molecules can be related to dynamical properties of vibrational motion. We have emphasized such concepts in systematic investigations of IVR in molecules involving different types of CH chromophores over the last decade [23–31]. Table 1 summarizes an overview of the most important results in a semi–quantitative manner. The general findings are:

(i) In all sp^3 C–H hydrocarbon molecules there is a strong, selective coupling between the CH stretching (s) and bending (b) modes with effective Fermi resonance coupling constants ranging from about 30 to 100 cm^{-1}. This corresponds to IVR times of less than 100 femtoseconds for these two types of modes. The CX_3 (or CFX_2) frame vibrations show effective couplings of less than 10 cm^{-1} and thus one has a distinct separation of time scales for the two types of IVR (with a great variety of X substituents), the slower processes occuring on time scales of more than a ps, potentially much longer in some cases. A similarly long time scale applies to coupling of the CH stretch bend system to the vibrational–rotational motion characterized by the vibrational angular momentum quantum number ℓ in the symmetric top CHX_3 molecules. ℓ is therefore an approximate, good quantum number on short time scales for C_{3v} molecules. In the asymmetric top molecules ℓ

TABLE 1. Summary of results for mode selective vibrational redistribution in C–H containing compounds

	Structural feature		Coupling constants k/cm^{-1}	Characteristic times for IVR τ/ps	Ref.
sp^3	X_3 C–H	s–bb	30 to 100	< 0.1	[23–26]
		s–b–ℓ	< 5	> 1	
		frame	< 10	> 1	
	X_2 F C–H	s–b'b'/–b"b"	50 to 100	< 0.1	[28,29]
		b'b'–b"b"	≃ 10		
		frame	< 10	> 1	
sp^2	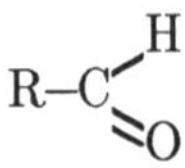	i.p. s–b	~ 100	< 0.1	[27]
	($R=CD_3$)	i.p/o.p.b.	< 10		
		frame	< 10	> 1	
sp	R–C≡C–H		adiabatically decoupled		
	($R=CH_3,CF_3,(CF_3)_3,(SiH_3)_3$)		($x_{sb} \simeq 20cm^{-1}$)		[17,24
		frame	< 1 to 10 cm^{-1}	>> 1	30,31]

is no good quantum number, even on short time scales and there is very similar selective coupling of the C–H stretching motion to both CH bending vibrations (characterised by sb'b' and sb"b" couplings of the order of 50 to 100 cm^{-1}). There is also a Darling Dennison resonance between the two bending modes (about 10 cm^{-1}). Again, coupling to the CFX_2 frame vibrations leads to distinctly longer time scales, of the order of a ps and more.

(ii) In sp^2 CH molecules it seems that the strong Fermi resonance coupling is restricted to the CH stretching and the in plane bending mode. The out of plane bending mode has a substantially lower frequency (about half the in plane frequency), which quenches the CH stretch bend Fermi resonance. The resonance between the in plane and out of plane bending vibrations has a small coupling constant of less that about 10 cm^{-1} and even weaker couplings seem to apply for the frame vibrations.

(iii) Finally in sp ≡C–H molecules both CH bending vibrations (potentially degenerate) have low frequency (600 to 700 cm^{-1}), which totally quenches the strong coupling. Residual coupling to the frame can arise by an off resonant mechanism with the CH bending overtones. The coupling matrix elements for this are large (about 20 to 100 cm^{-1}), but the resonance defect is of the order of 2000 cm^{-1} (compared to 0 to 200 cm^{-1} in sp^3 and sp^2 CH). The coupling of the CH bending motion to the frame is also reduced because of the geometrical isolation. Whereas in X_3C–H the CH bending mode at large amplitude can couple to the CX_3 frame by a repulsive, "collisional" mechanism, particularly for the bulky $X=CF_3$ groups, this interaction is largely absent in R_3C–C≡C–H, as pointed out in [24]. Thus alternative mechanisms for coupling of the CH stretching motion to the frame may become dominant, such as the coupling through the linear C–C≡C–H chain, which is expected to be rather ineffective, indicating long isolation times for the acetylenic ≡C–H chromophore, of more than 10 ps even for very large molecules such as $(CF_3)_3$ C–C≡CH [24] and $(SiH_3)_3$ C–C≡C–H [31] as has been recently confirmed by very high resolution measurements on related molecules [32].

These qualitative and semiquantitative findings and concepts can be very useful in planning further investigations. However, they still do not provide a true picture of the time evolution, molecular motion in intramolecular vibrational redistribution. In order to obtain this, we have to consider both the time dependent processes of state preparation and evolution and the derivation of the molecular hamiltonian, which governs this evolution.

4. Time Dependent Quantum Dynamics of Simple Systems: Fundamental Concepts

We consider a molecular system without or with excitation by a strong quasi-classical, quasimonochromatic laser radiation field of circular frequency $\omega = 2\pi\nu$. The time dependent Schrödinger equation describes its motion:

$$i \left[\frac{h}{2\pi}\right] \frac{\partial \psi (r,t)}{\partial t} = \hat{H}(t) \psi (r,t) \tag{1}$$

$$\hat{H}(t) = \hat{H}_{mol} + \hat{V} \cos (\omega t + \eta) \tag{2}$$

$\hat{H}_{mol}$ is the molecular hamiltonian in the absence of a radiation field. $\hat{V} \cos (\omega t + \eta)$ describes the interaction with the classical radiation of frequency ω and phase η. We may in general expand $\psi(t)$ explicitly in terms of the stationary states $\phi_k(r)$, which depend only upon the coordinates r. In the case of a one–dimensional oscillator r is just the oscillator coordinate such as the bond extension in a diatomic molecule, but in general we denote by "r" the *collection* of all spatial (x, y, z) coordinates and spin coordinates of all particles:

$$\psi (r,t) = \sum_k b_k (t) \phi_k (r) \tag{3}$$

The complex coefficients $b_k(t)$ depend only upon time. The general solution of Eq. (1) is conveniently formulated in terms of the time evolution operator U or its matrix representation **U** in the basis ϕ_k, for instance (we shall write $\mathbf{U}^b$ to stress this). **U** satisfies a differential equation similar to Eq. (1):

$$i \frac{h}{2\pi} \frac{d\,\mathbf{U}^b}{dt} = \{\mathbf{W} + \mathbf{V} \cos(\omega t + \eta)\}\, \mathbf{U}^b \tag{4}$$

The diagonal matrix **W** contains the eigenfrequencies of the molecular hamiltonian

$$\hat{H}_{mol}\, \phi_k = \frac{h}{2\pi} W_{kk}\, \phi_k \tag{5}$$

The matrix **V** contains the coupling matrix elements, which in the electric dipole approximation take the following form, for example with z–polarized radiation

$$V_{kj} = -\langle \phi_k \mid \mu\, e_z \mid \phi_j \rangle \left[\frac{2\pi}{h}\right] |E_0| \tag{6}$$

where μ is the dipole moment vector and e_z the unit vector in the z direction, E_0 being the electric field amplitude.

U provides a solution to all the relevant quantum equations of motion in the following form:

$$\mathbf{b}(t) = \mathbf{U}^b\, \mathbf{b}(0) \tag{7}$$

$$\mathbf{Q}(t) = \mathbf{U}^\dagger(t)\, \mathbf{Q}(0)\, \mathbf{U}(t) \tag{8}$$

$$\mathbf{P}(t) = \mathbf{U}(t)\, \mathbf{P}(0)\, \mathbf{U}^\dagger(t) \tag{9}$$

Eq. (7) provides the explicit solution to the Schrödinger equation (1) for the wavefunction ψ, by means of the expansion in Eq. (3), Eq. (8) provides the solution for the Heisenberg equations of motion for some observable Q with matrix representation **Q** (for instance the generalized coordinates q_i and momenta p_i of the particles), and Eq. (9) gives the solution for the statistical mechanical Liouville – von Neumann equation for the density operator ρ with matrix representation **P**. We have omitted in Eqs. (8) and (9) the exponent b on all the matrices, because the equations remain valid after transformation to another basis. For instance, it is often convenient to transform to the *quasiresonant basis* by means of the following equations

$$X_{kk} = W_{kk} - n_k\, \omega \tag{10}$$

$$-\omega/2 < X_{kk} < +\omega/2 \tag{11}$$

The matrix elements X_{kk} define a resonance defect with respect to the nearest n–photon excitation (integers n_k) for a molecular level k. The transformation to a new basis vector with coefficient matrix **a** instead of **b** is given by the diagonal transformation matrix

$$S_{kk} = \exp(i\, n_k \omega t) \tag{12}$$

$$U^a = S\, U^b\, S^{-1} \tag{13}$$

The transformation to the quasiresonant basis is useful for the approximate numerical solution of the equations of motion for laser excitation. In the *quasiresonant approximation* [19] the time evolution matrix $\mathbf{U}^a$ takes the *approximate* form

$$U^a = \exp[-i\,(\mathbf{X} + \tfrac{1}{2}\,\mathbf{V})\, t] \tag{14}$$

with a time independent effective hamiltonian matrix

$$H_{eff} = \mathbf{X} + \tfrac{1}{2}\,\mathbf{V} \tag{15}$$

For the discussion of relative merits of the quasiresonant and the Floquet–Liapounoff approximation see [1,2,19,36].

For the isolated molecule without laser field the quasiresonant transformation (and approximation) is obviously irrelevant, as we have a time independent molecular, field free hamiltonian:

$$\text{isolated molecule:} \qquad U = \exp(-i\,\mathbf{W}\, t) \tag{15}$$

Closed analytical solutions for coherent optical excitation exist only for the harmonic oscillator, so far (to our knowledge [33,34]). In general numerical solutions can be obtained, for instance by means of the published program package URIMIR [35–37]. Graphical representations for the evolution of one dimensional oscillators show regular and irregular motion, which, in contrast to the all too often represented stationary states, are not readily available in the textbook literature but give considerable introductory insight into dynamical phenomena in molecules. Molecular reality requires, however, consideration of coupled oscillator systems.

5. The Dynamics of Coupled Anharmonic Oscillators and Molecular Fermi Resonance

Whereas the hamiltonian and the resulting dynamics are known by definition for simple one–dimensional oscillator systems, a major first problem in treating the quantum dynamics of polyatomic molecules is the discovery of the relevant "true" molecular hamiltonian. From an *ab initio* point of view one might introduce the hamiltonian by representing molecules as collections of nucleï and electrons, perhaps using some simplification from the adiabatic separation of electronic and nuclear degrees of freedom. However, in practice the best current ab initio calculations for polyatomic molecules are still far from providing sufficient accuracy for the analysis of the true molecular quantum dynamics. *Spectroscopy* (in the infrared for vibrational motion) is the tool to provide such accuracy, but the analysis of molecular spectra in terms of a hamiltonian is often ambiguous, leading to multiple solutions. The best current approach seems to be the analysis of high resolution molecular spectra in terms of hamiltonians approximately consistent in general shape with ab initio theory, but more accurate. The general scheme for the approach to the analysis of molecular motion from high resolution spectroscopy can thus be summarized as follows

High Resolution Molecular Spectroscopy — MOLECULAR SPECTRA

Fourier Transform Spectroscopy

Laser Spectroscopy

↓ ↑

Effective Hamiltonians

↓ ↑ Rovibrational Schrödinger Equation — Electronic Schrödinger Equation ↙

Molecular Hamiltonian ← **Ab initio potential hypersurfaces**

↓

Time Evolution Operator (Matrix)

↓ — MOLECULAR MOTION

Molecular Rate Processes and Statistical Mechanics

We have described the most important experimental and theoretical techniques mentioned in this scheme and have reviewed our spectroscopic approach to molecular dynamics recently [38]. As described in chapter 3, a considerable data base exists now for the universal molecular Fermi resonance and femtosecond dynamics in the alkyl R_3CH chromophore in organic compounds and we can summarize the conclusions for time dependent intramolecular dynamics from this research. (The first, stationary state spectroscopic interpretation of this type of anharmonic resonance was given by *Fermi* [39]). Wave packet calculations quantitatively confirm the qualitative conclusion for mode selective redistribution mentioned in chapter 3 [40,41]. A particularly intriguing question for current theoretical modeling of molecular dynamics concerns the validity of classical, Newtonian dynamics for the heavy particle nuclear or atomic motion in polyatomic molecules, "classical trajectory calculations", which are the standard tool in current molecular dynamics investigations [42,43]. We have shown that molecular quantum dynamics introduces certain nontrivial true quantum effects [41,44] which must be considered in future investigations of molecular dynamics. A most intriguing finding concerns the role of quantum dynamics for molecular chaos. This question has found repeated interest, usually stressing the classical dynamical point of view [45]. We have established that a modest increase of the anharmonic Fermi resonance coupling leads to a distinct transition from quasiclassical wave packet dynamics to quantum spreading [41].

For strong resonance we have for the subset of coupled modes the following, most interesting, approximate equalities

$$|\psi(r,t)|^2 \simeq \langle |\psi(r,t)|^2 \rangle_t \equiv \langle P(r,t) \rangle_t \tag{16a}$$

$$\langle P(r,t) \rangle_t \simeq P_{microcanonical}(r) \tag{16b}$$

The last equation (16b) between the time average of the spatial probability density $\langle P(r,t)\rangle_t$ and the microcanonical average $P_{microcanonical}(r)$ would also apply for an ergodic classical mechanical system. However, the first equality in Eq. (16a) is a quantum relation, impossible in strict classical mechanics. In the quantum mechanical case the approximate equality is possible with rare exceptions (and recurrences), which are irrelevant for almost all times. One might simulate such a phenomenon by taking a classical mechanical ensemble corresponding to the quantum uncertainty in the initial conditions; then, for a classical chaotic system, the first equality would, indeed, hold for the ensemble, but still not for an individual trajectory, whichever one might choose. These findings suggest an intimate relation between molecular chaos (in the physical sense) and highly anharmonic quantum vibrational motion.

We note that the considerations are all based on experimental findings for a very limited class of vibrational modes in a limited class of molecules. On the one hand they do not depend on simplistic theoretical assumptions. However, on the other hand they cannot, without further evidence, be generalized to other molecules or other vibrational modes.

6. Molecular Chaos and the Statistical Mechanics of Coherent Motion in Isolated Molecules and in IR–Multiphoton Excitation

The quantum dynamical equations of motion discussed in section 4 show *time reversal symmetry.* However, the analysis of complex spectra and the resulting dynamics of polyatomic molecules both in isolation and when coupled to coherent laser radiation in IR–multiphoton excitation shows *de facto irreversibility and time asymmetry.* Under certain conditions intramolecular kinetics can be described by a master equation for coarse grained level populations p_L with a time independent rate coefficient matrix **K** [2,46]:

$$\frac{d\mathbf{p}}{dt} = \mathbf{K}\,\mathbf{p} \tag{17}$$

$$\mathbf{p}(t) = \exp(\mathbf{K}\,t)\,\mathbf{p}(0) \tag{18}$$

$$p_L = \sum_{k=1}^{N_L} P_{kk(L)} \tag{19}$$

These equations for intramolecular kinetics provide also the starting point for the theory of unimolecular reactions, the long time changes of molecular structure on a larger scale, and, perhaps, phase transitions in crystals. A major goal of spectroscopy is to derive rate coefficient matrices **K**, using approximate first principle equations, from spectral properties of the molecular hamiltonian. A number of applications have been discussed [2–4,11,12,46,47], including the possible role of the parity violating weak nuclear interaction on molecular motion and reaction dynamics as well as biochemical evolution.

An interesting consequence from the use of time irreversible statistical mechanical equations, such as the Pauli equation [48], is the increase of the parameter entropy S with time, tending towards a maximum [49]

$$\frac{dS}{dt} \geq 0 \qquad + \text{fluctuations} \tag{20}$$

$$S_{t \longrightarrow \infty} \simeq S_{max} \qquad + \text{fluctuations} \tag{21}$$

These equations have been derived in a straightforward way from Eq. (17) by Pauli [48] in his first discussion of the second law in the framework of quantum mechanics. Pauli used the following definition for microcanonical entropy ("fine grained")

$$S_f = -k \sum_j p_j \, \ell n \, p_j \tag{22}$$

where strictly speaking he made no distinction between fine grained (p_j) and coarse grained (p_L) populations. We have investigated the possibility, whether the approach to local equilibrium (of a limited number of strongly coupled oscillators) in a molecular system can be described by a suitably defined entropy rather than by use of Eq. (16), which would become very awkward for high dimensional systems with many oscillators. For the system of three strongly coupled CH oscillators in $CHFX_2$ molecules we have calculated, from integration of the time dependent Schrödinger equation, the time dependent entropy in Eq. (22) as well as further, coarse grained quantum entropies

$$S_Q = -k \sum_j \langle p_j \rangle \, \ell n \, \langle p_j \rangle \tag{23}$$

As we have argued before [46], it is really more consistent to introduce coarse grained observables (and level populations) into the statistical mechanical equations (17)–(19), and this applies also to entropy. One definition of the average population $\langle p_j \rangle$ will be (from Eq. (19)):

$$\langle p_j \rangle = P_J/N_J = \sum_{k=1}^{N_J} P_{kk(J)}/N_J \tag{24}$$

Other averaging procedures can be used as well. It turns out, however, that different choices lead to very similar numerical results for the time dependent entropy [38]: For the example of highly excited $CHC\ell F_2$ the entropy S approaches about 90% of its maximum value within less than 100 fs and remains at this value over extended periods of time with "fluctuations" as expected from the very limited subset of oscillators. We should stress that this result was obtained from a completely dynamical calculation using molecular data derived from spectroscopic experiment, we believe for the first time. It is not the result of a statistical model. The "fluctuations" are the consequences of the deterministic time evolution from the Schrödinger equation for the (isolated) molecule. Their size can be modelled by a statistical model, following the Ehrenfests [52], but not their individual behaviour. We note that the deterministic (irreversible) master equations (17) and (18) remove the fluctuations in the limit of very large statistical systems. Fig. 1 illustrates the time evolution of entropy for $CHC\ell F_2$.

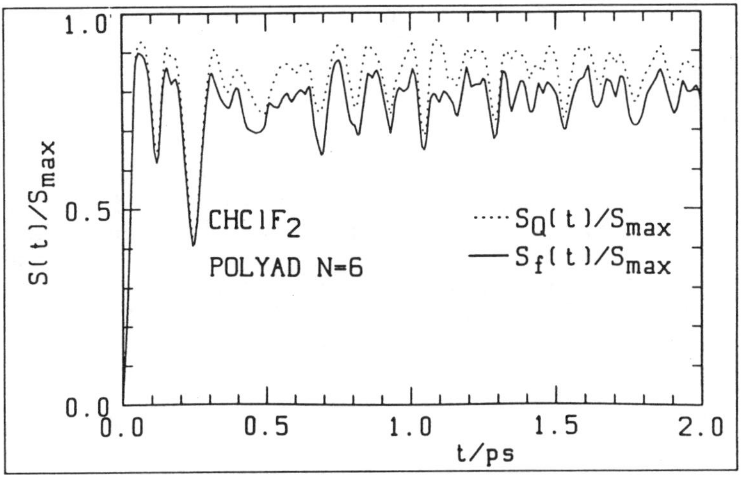

FIGURE 1. *Time dependent "entropy" S for a subset of three coupled oscillators in $CHC\ell F_2$ from quantum mechanical time evolution as derived from spectroscopic experiment (see ref. 38 and literature cited therein).*

There have been numerous discussions of the possibility to use "Pauli like" entropy definitions as in Eqs. (22), (23) for time dependent phenomena, rather than the *time independent* Gibbs–von Neumann equilibrium entropy, the latter being often preferred [53].

$$S_N = -k\,\mathrm{Tr}\,(\varrho\,\ell n\,\varrho) \tag{25}$$

A very common objection to the Pauli definition is its basis dependence (it would become a trivially time independent entropy, if energy eigenstate populations were used instead of some other basis, for instance). While it is not the aim of the present paper to resolve these long standing controversies [50,51], we mention that clearly Eq. (25) *cannot* be used for the time dependent formulation of the second law for irreversible processes, which is certainly one of the historically legitimate applications of entropy [49]. Our results indicate that a resolution of the dilemma lies in the possibility that "most" definitions (in the set theoretical sense) of entropy according to Eqs. (22), (23) will lead to similar numerical results for sufficiently large physical systems. Thus, at least as a heuristic measure, such entropy will be useful to describe relaxation of physical observables and a thorough mathematical foundation for this use may well be in reach.

7. The Role of Symmetry and Symmetry Reduction in Vibrational Redistribution

We have noted in the previous sections that the conservation of the approximately good vibrational angular momentum quantum number ℓ (on short time scales!) in C_{3v} type CHX_3 molecules can be removed by reducing the symmetry group to C_s by approximate substitution. This is just an example for the general role of symmetry and symmetry reduction in vibrational redistribution, which we shall address now. Table 2 provides an appropriate survey for the CH chromophore. Starting from R–C≡C–H, we have the infinite group of three uncoupled oscillators, two of them degenerate in C_{3v}, with quantum numbers v_s, v_b, ℓ_b and, including inversion splittings parity + or – [47,54] (see below). There appears just one "bright" state in the spectrum. The first symmetry reduction leads to C_{3v} type X_3CH molecules. Now one of the good quantum numbers v_s and v_b must go and we replace the couple by the polyad quantum number N [38]. In each polyad there appear N+1 bright states. The vibrational quantum number ℓ and parity (as discussed) are retained. This situation seems to conform to the short time dynamics observed for D_3CH, F_3CH and others that we have investigated. We still have an infinite group, with irreducible representations given by the ℓ. The quantum number N is in essence only an energy ordering (mentioned in parenthesis hereafter) as we allow, in general, interpolyad couplings. In principle C_{3v} symmetry in X_3CH would correspond to the molecular symmetry group S_3^* of order 12, mentioned in the third step. We note in passing that S_3^* is of a larger order than C_{3v}, not isomorphous to it, as we retain parity as a good quantum number, which certainly for some purposes is appropriate (see [54,55] for a detailed discussion of concepts and nomenclature). It turns out that none of the observed C_{3v} molecules shows the reduced symmetry in the spectra, i.e., the couplings are too weak to be easily observed and are not relevant for the short time dynamics, to our present knowledge. Our theoretical analysis, however, suggests further experimental search for the symmetry breaking, which is underway.

More easily observed is the consequence of the symmetry reduction in step 4, where we obtain a reduction to a group S_2^* of order 4. As discussed in the previous sections we have numerous experimental examples which clearly show the consequences of this symmetry reduction. The spectrum of bright states now becomes much richer, the number being roughly of the order $\frac{1}{2}N^2$ or in continuum language, the density of states increases proportional to E^2 as for three coupled oscillators, compared to a proportional increase with E (or N) in step 2 (effectively two coupled oscillators). The next step 5 roughly doubles the number of bright levels by removing the A', A" symmetry of C_s. Again this is a strong symmetry breaking which is relevant for the short time dynamics, the remaining group for the CH chromophore being of order 2, S_1^*, the reflection group.

In the order of strength of interaction we would now have to consider the coupling to other vibrational modes and to nuclear spin, which will be relevant

TABLE 2. Symmetry Reduction in Vibrational Redistribution in the CH chromophore (Coupling to other vibrational modes, rotation, and nuclear spin is not explicitly indicated, see text).

Case No.	Example	Approximately good quantum numbers (symmetry labels)	Group (Order)	Levels visible in absorption spectrum (schematic)
1.	R–C≡C–H	v_s, v_b, ℓ (+,–)	(∞)	
2.	X_3CH (X=D,F...)	N, ℓ, (+,–)	(∞)	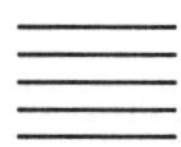
3.	X_3CH (—)	$A_1^{\pm}$, $A_2^{\pm}$, $E^{\pm}$ (N)	S_3^* (12)	
4.	FX_2CH (X=D, Cℓ) F_2XCH	$A'^{(\pm)}$, $A''^{(\pm)}$ (N)	S_2^* (4)	
5.	FXYCH (X=Cℓ, Y=Br) (X=^{35}Cℓ, Y=^{37}Cℓ)	A^+, A^- (N)	S_1^* (2)	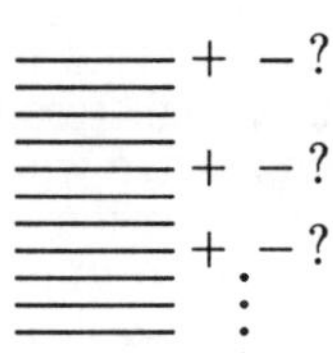
6.	FXYCH (as 5.)	A (N)	S_1 (1)	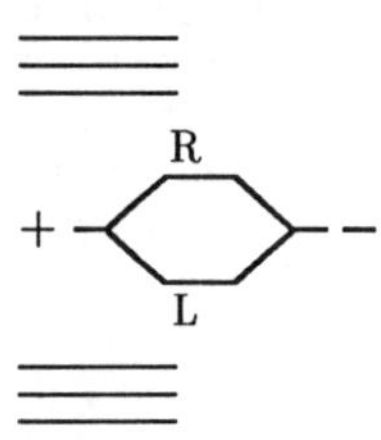

from the ps to ns time scale and ns to ms time scale, respectively. For the purpose of the present, simplified discussion of the CH chromophore dynamics only we pass over these steps (indicated by a line in the table, which also indicates that we are passing now to unobserved properties). Already in step 5 we have marked the + or – parity levels with a question mark. Certainly on short time scales (ps to ns) parity remains a good quantum number for molecules. However, on long time scales (hours to days), the parity violating weak nuclear interaction will break this symmetry, which again doubles the number of coupled levels. Instead of a closely degenerate tunneling doublet (+ –) we obtain a split pair of left (L) and right (R) handed states, as illustrated by the greatly magnified example in the level scheme of table 2. These effects have to date never been observed in molecules, but we have predicted ways to observe them and we refer to [47] for a detailed discussion.

8. Summary and Conclusions

We have provided a relatively broad survey of intramolecular vibrational energy transfer as relevant to many primary processes in unimolecular reactions. Two decades ago, the general picture would have been that in highly excited molecules IVR is essentially a phenomenon restricted to time scales of ps, after which it would be "completed", resulting in a statistical distribution. This picture resulted mainly from RRKM theory and very limited experimental evidence from reaction kinetics. At the same time, many spectroscopists would have even denied the very existence of the phenomenon of IVR as a time dependent process (this attitude can still be encountered today).

What has emerged largely from investigations of the last decade is an extremely rich set of phenomena in intramolecular energy redistribution with time scales ranging from femtoseconds (as for the alkyl CH chromophore) to nanoseconds (as we have recently shown by direct time resolved experiments on CF_3I after multiphoton excitation [56]) and further on to milliseconds and days if we include nuclear spin symmetry and parity violation.

We believe that our approach to intramolecular time evolution as obtained from stationary high resolution molecular spectroscopy (sections 4 and 5) is more than just competitive with direct femtosecond experiments [57]. Often, the route of the diagram in section 5 will be the fastest and best (but this will not always be so). The most important conclusions, so far, are

(i) *Highly anharmonic quantum molecular dynamics is strongly non–classical.* Quantum spreading may lead to a new understanding of intrinsic exponential decay in unimolecular reactions of polyatomic molecules. It may also result in a new understanding of time dependent molecular statistical mechanics and entropy.

(ii) *Molecular symmetry is a fundamental principle governing not just molecular structure but also molecular dynamics including intramolecular vibrational redistribution.* While many of the symmetries involved are only approximate, they are active for short time motion.

(iii) *Detailed features of molecular structure lead to well defined, transferable and thus predictable consequences for intramolecular dynamics.* For the chemist this finding opens up a very large number of new avenues of investigation, contradicting the long standing false dogma of "universally fast vibrational redistribution on the ps time scale". The new concepts have qualitative, predictive power in a very simple chemical sense, which we shall illustrate with what we believe was the first clearcut and systematically investigated case from our work: The alkyl– and acetylenic CH chromophore [58]

Consider the bichromophoric molecule

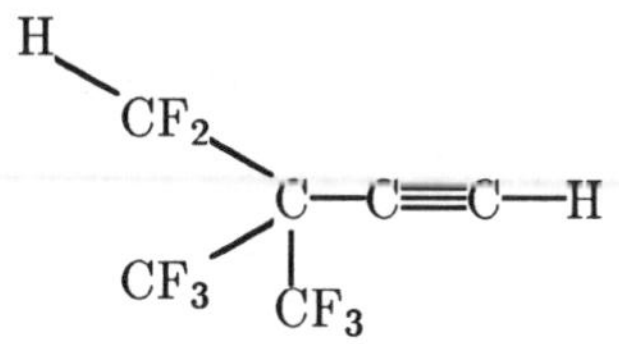

This molecule has not yet been studied. Predictions on the overtone spectra are: The acetylenic CH overtones will be narrow, isolated structures. The alkyl CH overtones will be very broad, with some superstructure extending over hundreds of cm^{-1}. The time evolution will show fast subpicosecond decay after excitation of the alkyl CH chromophore. The acetylenic CH overtones will be long lived, certainly many picoseconds. This has obvious consequences for unimolecular and bimolecular reactivity in the picosecond to nanosecond time range. We will consider a somewhat special, also hypothetical example for the sake of the argument. Let us assume that we could form (in a supersonic jet) a hydrogen bond with some HX or X_2 at the CH (obviously CF will compete, in the opposite sense for HX)

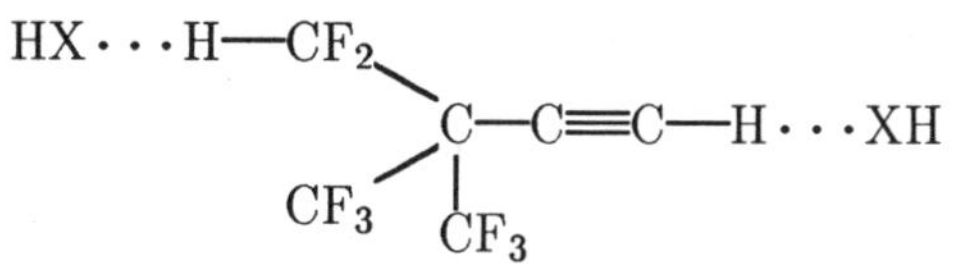

If we excite an alkyl CH overtone, the energy will be lost very quickly from the local environment into the frame of the molecule. Then the weak hydrogen bond may survive a long time. If we excite the acetylenic CH, the energy will remain localized and might be efficiently used for dissociation of the hydrogen bond on the ps to ns time scale. Similarly, if we try to stimulate photoreaction with an X fragment from UV photodissociation of HX by previous IR–excitation of high CH overtones, this will be efficient on long times for the acetylene CH but only on short time scales for alkyl CH.

Many more examples can be devised and we believe that some of them may become practical in the near future. Understanding IVR clearly should be worth the effort.

Acknowledgement: The work summarized here has profited from the contributions of many coworkers as evident from the list of references. Thanks to them also for many discussions. Our work is supported financially by the Schweizerischer Nationalfonds and the Schweizerischer Schulrat. The hospitality of the Israel Academy of Sciences and the Fondation Edmond de Rothschild are gratefully acknowledged. This paper is dedicated to Professor Dr. K. H. Büchel on the occasion of his 60th birthday.

References

[1] Quack, M. and Troe, J. (1981) 'Current aspects of unimolecular reactions', Int. Rev. Phys. Chem. **1**, 97–147.

[2] Quack, M. (1982) 'Reaction dynamics and statistical mechanics of the preparation of highly excited states by intense infrared radiation', Advances in Chemical Physics **50**, 395–473 (K. Lawley, I. Prigogine and S.A. Rice, eds.)

[3] Lupo, D.W. and Quack, M. (1987) 'IR laser photochemistry', Chem. Rev. **87**, 181–216.

[4] Quack, M. (1989) "Infrared laser chemistry and the dynamics of molecular multiphoton excitation', Infrared Physcis **29**, 441–466.

[5] Crim, F.F. (1984) 'Selective excitation studies of unimolecular reaction dynamics', Annu. Rev. Phys. Chem. **35**, 657–691.

[6] Rizzo, T.R., Hayden, C.C., and Crim, F.F. (1984) J. Chem. Phys. **81**, 4501.

[7] Ticich, T.M., Rizzo, T.R., Dübal, H.R., and Crim, F.F. (1986) 'Unimolecular reactions near threshold, the overtone vibration initiated decomposition of HOOH ($5\nu_{OH}$)', J. Chem. Phys. **84**, 1508.

[8] Quack, M. and Seyfang, G. (1982) 'Absolute rate parameters for infrared photochemistry: $CF_3I \longrightarrow CF_3 + I$', J. Chem. Phys. **76**, 955–965.

[9] Pine, A.S., Lafferty, W.J., and Howard, B.J. (1984) 'Vibrational predissociation, tunneling, and rotational saturation in the HF and DF dimers', J. Chem. Phys. **81**, 2939–2950.

[10] Miller, R.E. (1990) 'Vibrationally induced dynamics in hydrogen bonded complexes', Acc. Chem. Res. **23**, 10–16.

[11] Puttkamer, K. v. and Quack, M. (1989) 'Vibrational spectra of $(HF)_2$, $(HF)_n$ and their D–isotopomers: Mode selective rearrangements and nonstatistical unimolecular decay', Chem. Phys. **139**, 31–53.

[12] Quack, M. and Suhm, M. (1991) 'Potential energy surfaces, quasiadiabatic channels, rovibrational spectra and intramolecular dynamics of $(HF)_2$ and its isotopomers from quantum Monte Carlo calculations', J. Chem. Phys. (in press).

[13] Quack, M. and Thöne, H.J. (1987) 'Absolute and relative rate coefficients in the IR–laser chemistry of bichromophoric fluorobutanes: Tests for inter– and intramolecular selectivity'. Chem. Phys. Lett. **135**, 487–494.

[14] Bixon, M. and Jortner, J. (1968) 'Intramolecular radiationless transitions', J. Chem. Phys. **48**, 715.

[15] Jortner, J., Rice, S.A., and Hochstrasser, R.M. (1969) 'Radiationless transitions in photochemistry', Adv. Photochem. **7**, 149.

[16] Jortner, J. and Levine, R.D. (1981) 'Photoselective chemistry', Adv. Chem. Phys. **47**, 1–114.

[17] Dübal, H.R. and Quack, M. (1982) 'Sequence structure in the high resolution infrared spectrum of trifluoropropyne', Chem. Phys. Lett. **90**, 370–374.

[18] Mehta, M.L. (1967) Random Matrices and the Statistical Theory of Energy Levels, Academic Press, New York and London.

[19] Quack, M. (1978) 'Theory of unimolecular reactions induced by monochromatic infrared radiation', J. Chem. Phys. **69**, 1282–1307.

[20] Quack, M. (1981) 'On local and global vibrational states', Faraday Disc. Chem. Soc. **71**, 359–384.

[21] Quack, M. (1982) 'The role of intramolecular coupling and relaxation in IR–photochemistry', Jerusalem Symposia **15**, 371–390 (J. Jortner and B. Pullman (eds) Reidel, Dordrecht).

[22] Sibert, E.L., Reinhardt, W.P., and Hynes, J.T. (1984) J. Chem. Phys. **81**, 1115.

[23] Dübal, H.R. and Quack, M. (1981) 'High resolution IR–spectrum of fluoroform: A close resonance', Chem. Phys. Lett. **80**, 439–444;
Dübal, H.R. and Quack, M. (1984) 'Tridiagonal Fermi resonance structure in the IR–spectrum of the excited CH–chromophore in CF_3H', J. Chem. Phys. **81**, 3779–3792.

[24] Puttkamer, K. v., Dübal, H.R., and Quack, M. (1983) 'Time–dependent processes in polyatomic molecules during and after intense infrared irradiation', Faraday Disc. Chem. Soc. **75**, 197–210.

[25] Segall, J., Zare, R.N., Dübal, H.R., Lewerenz, M., and Quack, M. (1987) 'Tridiagonal Fermi resonance spectrum of the CH–chromophore in CF_3H, II. Visible spectra', J. Chem. Phys. **86**, 634–646.

[26] Lewerenz, M. and Quack, M. (1988) 'Vibrational spectrum and potential energy surface for the CH chromophore in CHD_3', J. Chem. Phys. **88**, 5408–5432.

[27] Amrein, A., Hollenstein, H., Quack, M., Zenobi, R., Segall, J., and Zare, R.N. (1989) 'Fermi resonance structure in the CH vibrational overtones of CD_3CHO' J. Chem. Phys. **90**, 3944–3951.

[28] Amrein, A., Dübal, H.R., and Quack, M. (1985) 'Multiple anharmonic resonances in the vibrational overtone spectra of $CHC\ell F_2$', Mol. Phys. **56**, 727–735.

[29] Luckhaus, D. and Quack, M. (1991) 'The rovibrational analysis of the high resolution IR–spectrum of CD_2HF', Chem. Phys. Lett. (in press) and further work to be published.

[30] Amrein, A., Quack, M., and Schmitt, U. (1987) 'High resolution interferometric Fourier transform infrared spectroscopy in supersonic free jet expansions: A new technique for ultracold gaseous samples', Z. Physik. Chem. **154**, 59–72.

[31] Puttkamer, K. v. and Quack, M. (1983) 'Homogeneous and inhomogeneous structure in the vibrational spectrum of the CH–chromophore in $(SiH_3)_3$ C–C≡C–H' (unpublished);
Puttkamer, K. v. Diplomarbeit (1983).

[32] Lehmann, K.K., Pate, B.H., and Scoles, G. (1991) 'Intramolecular vibrational redistribution on the nanosecond time scale, ν_1 and $2\nu_1$ spectra of substituted acetylene compounds' (draft for Jerusalem Symposium).

[33] Marquardt, R. and Quack, M. (1989) 'IR–multiphoton excitation and wavepacket motion of the harmonic and anharmonic oscillators: Exact solutions and quasiresonant approximation', J. Chem. Phys. **90**, 6320–6327.

[34] Marquardt, R. and Quack, M. (1989) 'Molecular motion under the influence of a coherent IR–laser field', Infrared Physics **29**, 485–501.

[35] Quack, M. and Sutcliffe, E. (1985) 'Primary photophysical processes in infrared multiphoton excitation: Wavepacket motion and state selectivity', Infrared Physics **25**, 163–173 (1985).

[36] Quack, M. and Sutcliffe, E. (1985) 'On the validity of the quasiresonant approximation for molecular infrared multiphoton excitation', J. Chem. Phys. **83**, 3805–3812.
[37] Quack, M. and Sutcliffe, E. (1986) 'URIMIR: Programs for the calculation of the quantum dynamics of IR–multiphoton excitation and dissociation (unimolecular reactions induced by monochromatic infrared radiation)', Quantum Chemistry Program Exchange, Program 515, QCPE Bulletin **6**, 98.
[38] Quack, M. (1990) 'Spectra and dynamics of coupled vibrations in polyatomic molecules. Annual. Rev. Phys. Chem. **41**, 839–874.
[39] Fermi, E. (1931) 'Über den Ramaneffekt des Kohlendioxids', Zeitschrift für Physik **71**, 250.
[40] Marquardt, R. and Quack, M. (1987) 'The dynamics of polyatomic molecules during and after coherent excitation', Jerusalem Symposia **20**, 31–44 (J. Jortner, A. Pullman and B. Pullman (eds) Reidel, Dordrecht).
[41] Marquardt, R. and Quack, M. (1991) 'The wave packet motion and intramolecular vibrational redistribution in CHX_3 molecules under IR—multiphoton excitation', J. Chem. Phys. (in press).
[42] Bunker, D.L. (1971) 'Classical trajectory methods', Meth. Comp. Phys. **10**, 287–325.
[43] van Gunsteren, W.F. and Berendsen, H.J.C. (1990) 'Computer simulation of molecular dynamics', Angewandte Chemie Int. Ed. Engl. **29**, 992.
[44] Quack, M. (1990) 'The role of quantum intramolecular dynamics in unimolecular reactions', Phil. Trans. Roy. Soc. London A **332**, 203–220.
[45] Rice, S.A. (1981) 'An overview of the dynamics of intramolecular transfer of vibrational energy', Adv. Chem. Phys. **47**, 117–200.
[46] M. Quack (1981) 'Statistical mechanics and dynamics of molecular fragmentation', Il Nuovo Cimento **63 B**, 358–377.
[47] Quack, M. (1989) 'Structure and dynamics of chiral molecules', Angewandte Chemie **101**, 588–604; Angewandte Chemie Int. Ed. Engl. **28**, 571–586
[48] Pauli, W. (1928) 'Über das H–Theorem vom Anwachsen der Entropie vom Standpunkt der neuen Quantenmechanik' in "Probleme der modernen Physik", Festschrift zum 60. Geburtstage von Arnold Sommerfeld, S. 30–45, Hirzel Verlag, Leipzig.
[49] Clausius, R. (1865) 'Über verschiedene für die Anwendung bequeme Formen der Hauptgleichungen der mechanischen Wärmetheorie', Poggendorf Ann. Physik **125**, 353–400.
[50] Tolman, R.C. (1967) The Principles of Statistical Mechanics, Oxford University Press.
[51] Jancel, R. (1963) Les Fondements de la Mécanique Statistique Classique et Quantique, Gauthier–Villars, Paris.
[52] Ehrenfest, P. and Ehrenfest, T. (1911) 'Über zwei bekannte Einwände gegen das Boltzmannsche H–Theorem', Pys. Z. **8**, 311.
[53] Alhassid, Y. and Levine, R.D. (1977) 'Entropy and chemical change III', J. Chem. Phys. **67**, 4321–4339;
but see also Remacle, F. and Levine, R.D. 'The maximum entropy of an optical spectrum and the redistributon in phase space', preprint 1991.

[54] Quack, M. (1977) 'Detailed symmetry selection rules for reactive collisions', Mol. Phys. **34**, 477–504.
[55] Quack, M. (1983) 'Detailed symmetry selection rules for chemical reactions', Stud. Phys. Theor. Chem. **23**, 355–378.
[56] Quack, M., Schwarz, R., and Seyfang, G. To be published.
[57] Gruebele, M., Roberts, G., and Zewail, A.H. (1990) 'Femtochemistry of the reaction of IHgI: theory versus experiment', Phil. Trans. Roy. Soc. London **A 332**, 35–56.
[58] Quack, M. (1985) 'Trieste winter school lectures' (unpublished)

VIBRATIONAL LEVEL STRUCTURES AND RESONANCES IN TRIATOMIC MOLECULES: H_2O

John C. Light and Seung E. Choi
James Franck Institute
University of Chicago
Chicago, IL 60637

ABSTRACT. As the vibration and rotation energy of triatomic molecules is increased, the interaction of different degrees of freedom leads classically to chaotic motions. Quantum mechanically these interactions lead to energy level structures which, due to the coupling of large amplitude motions, cannot be decomposed easily into zero order vibrations and rotations. In this article we present a method of calculating, accurately and efficiently, a large number of highly excited vibration-rotation quantum states of triatomic molecules. We apply these methods to H_2O, and examine several questions concerning the energy level structures, in particular the normal mode/local mode "transition", the energy level spacing distribution statistics, and effects of angular momentum on these. Coriolis coupling and asymmetric top coupling terms are included exactly, and modifications of the potential energy surface of Carter and Handy are made to obtain physically correct behaviour in the high energy regions.

1. Introduction

The vibration-rotation dynamics of highly excited molecules such as H_2O have been studied with great precision both in the frequency[1,2,3,4] and the time domains[5]. Such experimental studies provide information relevant to a number of important theoretical and (perhaps) practical questions, such as, the possibility of control of dissociation via coherent excitation, selective photodissociation via initial state preparation, and, more abstractly, to provide physical (i.e. quantum) realizations of classical non-linear dynamical systems which have transitions from regular to chaotic behaviour.

Classical spectroscopic analysis[6,7,8,9] of the vibration-rotation motion of molecular systems starts, of course, with separable zero order approximations to these motions, and then proceeds to resolve the complexities introduced by the break down of these approximations, giving explanations, for example, in terms of Fermi resonances between modes, Coriolis couplings, normal mode to local mode transitions, etc. However, as the energy of the molecule increases, the couplings become sufficiently strong that the zero order pictures lose relevance, and the spectra and dynamics become progressively more difficult to understand and assign. [10]

One purpose of accurate quantum calculations of high vibration-rotation states is to help in the analysis of the spectra and the assignment of states for a given assumed potential energy surface. After approximate levels are calculated and assigned, we can use

J. Jortner et al. (eds.), Mode Selective Chemistry, 67–86.

the comparison with experiment to improve the potential energy surface [11,12,13,14], thus permiting improved fits of the new calculated states to the experimental data and assignment of higher energy states. This "bootstrap" procedure is particularly important for "floppy" molecules in which the vibration-rotation levels are strongly coupled. A more important purpose, perhaps, of such accurate quantum calculations of vibration-rotation energy levels is to provide realistic theoretical molecular models of highly coupled, multidimensional quantum systems. These will correspond, classically, to highly non-linear dynamical systems.[15,16,17] Indeed, one of the interesting questions, to date not particularly well understood, is what to expect as the quantum manifestations of classical chaos[18,19,20,21,22,23], the role of quantization of different degrees of freedom (e.g. angular momentum), and the role (if any) of resonances (e.g. Fermi resonances) in these quantum systems as the classically chaotic regime is approached.

Recently sufficient progress has been made in the methods of solution of the Schrödinger equation for vibration-rotation motions [24,25,26,14,27,28,29], and sufficiently improved (and inexpensive) computers have become available, that we can examine some of the questions above for real triatomic systems using realistic potential energy surfaces. In particular we can now solve for the vibration-rotation states of systems such as H_2O to energies well above the barrier to the linear H-O-H configuration and for total angular momentum up to $J = 25$. These calculations can yield several hundred accurate eigenvalues for each J, and the agreement with experiment for measured vibration-rotation states is quite good. Therefore, we can assume that the dynamical properties of this model, with the accurate quantum solution for vibrational motions on accurate potential energy surfaces, are qualitatively realistic and very largely quantitatively accurate.

Thus solutions for this system (full three vibrational degrees of freedom plus rotation) will be sufficient to yield quantitative answers to some of the above questions of chaos,etc. Although quantitative, the results of analysis of the energy level spacing distributions may be ambiguous if the real system is intermediate, i.e. does not conform to either of the limiting cases of a Poisson or Wigner distribution [20,18,30,31,32,33,34,35]. This would indicate that the limiting distributions, in this case the Poisson for the energy level spacing distribution resulting from several independent modes; and the Wigner resulting from strong interactions among all modes, are both poor approximations for the real molecule, H_2O.

The Discrete Variable Representation (DVR) has greatly enhanced the efficiency of the numerical treatment of diverse quantum mechanical problems [36], including the evaluation of vibration-rotation states of molecules. Accurate computations of highly excited rovibrational states of triatomic systems such as HCN, LiCN [37], and H_3^+ [29] have been done using the DVR. Because of the ease of evaluation of $\mathbf{H}_{\mathrm{RV}}$-matrix elements in the DVR, an optimum basis for each internal coordinate can be chosen from a wide class of available mathematical functions. The sparse nature of the multi-dimensional $\mathbf{H}_{\mathrm{RV}}$-matrix in the DVR also makes this approach extremely efficient numerically. Since in a DVR the coordinate matrix is diagonal, the potential energy surface is not restricted to be expanded in a particular basis; any analytical form of potential energy surface can be used effectively. The DVR methods developed for this study of H_2O and reported below appear to be roughly two orders of magnitude more efficient computationally than the methods used in a related recently reported study[38].

A general triatomic rovibrational Hamiltonian is derived for Radau internal coordinates in the total angular momentum representation. The characteristic rotational and vibrational coupling terms in the Hamiltonian are easily identified in the standard body-fixed reference frame. The Hamiltonian is further simplified for X_2Y triatomic molecules (C_{2V}

symmetry) when the body-fixed axes are defined in the standard fashion to be the instantaneous inertial tensor principal axes. This compromise minimizes the combined Coriolis coupling and asymmetric-top coupling effects. We use mass-weighted linear combinations of R_1 and R_2, the Radau stretching coordinates to form R and r, which are symmetric and antisymmetric with respect to reflection through the symmetry plane of the molecule. Since a C_{2V} system consists of four symmetry groups (e.g. A_1, A_2, B_1, and B_2), we can divide the molecular states into even and odd parity states, each with symmetric and asymmetric states. The DVR allows the accurate calculation of energy levels of highly excited states for a given angular momentum and symmetry for a given potential energy surface.

For obvious reasons water is one of the most widely studied molecules, both theoretically and experimentally. The rich nature of various spectroscopic interactions has been a particularly interesting feature of H_2O. High resolution spectra of water vapor have been measured for pure vibrational states (J=0) with energy levels upto ~18000cm^{-1} , which is well above the energy barrier (~12500cm^{-1}) for the H-O-H linear configuration ($\theta = \pi$). For higher total angular momentum states, most of first 20 vibrational energy levels have been measured [39,40,1]. The observed states, however, do not include highly excited bending vibrational states. In this paper we use the general approach above to study the ro-vibrational states of H_2O to very high vibrational levels (to ~25,000cm^{-1} above the zero point energy) and to reasonably high total angular momentum.

In the following section, we present a brief summary of the coordinate system, the Hamiltonian operator, the basis functions, the DVR transformations, and the method of solution for the rovibrational states of H_2O. A more detailed description of the method is given elsewhere [41]. We also discuss briefly the different potential energy surfaces available for the H_2O system. In section 3 computational results are presented. We give brief comparisons with experiment and other calculations, then focus primarily on the analysis of the highly excited states of the system. In particular, we discuss briefly the local mode picture and the effects of increasing energy and angular momentum, J, and its body fixed projection, K, on it. We also look at measures of basis set mixing as E, J, and K are increased, i.e. the quantum manifestations of classical chaos ("quantum chaos") via the usual statistical measures of energy level spacing distributions.

2. Numerical Methods (Summary)

The full rovibrational Hamiltonian operator, $\hat{H}_{RV}$, of a triatomic molecule is a function of three internal coordinates and three rotational coordinates (ξ, χ, ϕ). The three Euler angles (ξ, χ, ϕ) specify the instantaneous orientation of the body-fixed reference frame, $(\hat{x}, \hat{y}, \hat{z})$, with respect to the space-fixed reference frame [42]. The total angular momentum, J, is written in terms of its body-fixed $(\hat{x}, \hat{y}, \hat{z})$ components. Following Johnson and Reinhardt [43], and Bacic, Watt, and Light [27], we choose Radau coordinates, (R_1, R_2, θ), as the internal coordinates, where R_1, R_2, and θ are similar to the internal valence coordinates, but orthogonal.

The Radau coordinates, (R_1, R_2, θ), are related to the two bond distances and angle, (r_1, r_2, ρ), as follows:

$$\vec{R}_1 = (\mu + 1 - g) \cdot \vec{r}_1 + (g - 1) \cdot \vec{r}_2 \tag{1}$$
$$\vec{R}_2 = (\mu - g) \cdot \vec{r}_1 + g \cdot \vec{r}_2 \tag{2}$$
$$\cos\theta = \frac{\vec{R}_1 \cdot \vec{R}_2}{\mid \vec{R}_1 \cdot \vec{R}_2 \mid} \tag{3}$$

where $\mu = \sqrt{\frac{\mathrm{m}_3}{\mathrm{m}_1 + \mathrm{m}_2 + \mathrm{m}_3}}$; $g = \frac{\mathrm{m}_1 + \mathrm{m}_2\mu}{\mathrm{m}_1 + \mathrm{m}_2}$

and m_1, m_2, and m_3 are the masses of three atoms, respectively.

Linear combinations of the mass-weighted Radau stretching coordinates (R_1, R_2) define a set of symmetrized coordinates, (R, r), which are symmetric and anti-symmetric under the $\mathrm{C_{2V}}$ symmetry operation (i.e., reflection) through the body-fixed $\hat{x}$-$\hat{y}$plane. These are:

$$R = \sqrt{\tfrac{1}{2}}\,(\,g_1 R_1 \,+\, g_2 R_2\,) \tag{4}$$
$$r = \sqrt{\tfrac{1}{2}}\,(\,g_1 R_1 \,-\, g_2 R_2\,) \tag{5}$$

where

$$g_1 = \sqrt{\frac{\mathrm{m}_1}{\mathrm{m}_1 + \mathrm{m}_2}}\ ; \qquad g_2 = \sqrt{\frac{\mathrm{m}_2}{\mathrm{m}_1 + \mathrm{m}_2}}$$

in a $\mathrm{C_{2V}}$ molecule where $\mathrm{m}_1 = \mathrm{m}_2$, $g_1 = g_2 = \sqrt{1/2}$.

The Hamiltonian operator, $\hat{\mathrm{H}}_{\mathrm{RV}}$, is now expressed in the symmetrized Radau coordinates, (R, r, θ), and the total angular momentum components with respect to the body-fixed axes, $(\hat{\mathrm{J}}_x, \hat{\mathrm{J}}_y, \hat{\mathrm{J}}_z)$. We define the body-fixed $\hat{z}$-axis as the instantaneous principal axis with the least moment of inertia and the $\hat{x}$-axis to be perpendicular to it in the triatomic plane. Then the Hamiltonian can be written:

$$\begin{aligned}\hat{\mathrm{H}}_{\mathrm{RV}}(R, r, \theta, \hat{\mathrm{J}}_x, \hat{\mathrm{J}}_y, \hat{\mathrm{J}}_z) \;=\; & -\frac{1}{\mathrm{m}_1}\left(\frac{\partial^2}{\partial R^2} + \frac{\partial^2}{\partial r^2}\right) + \hat{\mathrm{V}}(\mathrm{R}, \mathrm{r}, \theta) \\ & - \frac{1}{2I}\left(\frac{1}{\sin\theta}\frac{\partial}{\partial\theta}\sin\theta\frac{\partial}{\partial\theta}\right) \\ & + G^{xx}\cdot\hat{\mathrm{J}}_x^2 \,+\, G^{yy}\cdot\hat{\mathrm{J}}_y^2 \,+\, G^{zz}\cdot\hat{\mathrm{J}}_z^2 \\ & + G^{xz}\cdot(\hat{\mathrm{J}}_x\hat{\mathrm{J}}_z + \hat{\mathrm{J}}_z\hat{\mathrm{J}}_x) \,+\, G^{\mathrm{cor}}\cdot(-i\hbar\hat{\mathrm{J}}_y)\end{aligned} \tag{6}$$

The elements of the contravarient metric **G**-tensor can be evaluated simply as

$$
\begin{aligned}
G^{xx} &= \frac{1}{8I \sin^2(\theta/2)} \\
G^{yy} &= \frac{1}{8I} \\
G^{zz} &= \frac{1}{8I \cos^2(\theta/2)} \\
G^{xz} &= \frac{1}{4 \sin\theta}(\frac{1}{I_1} - \frac{1}{I_2}) \\
G^{\rm cor} &= -\frac{1}{4}(\frac{1}{I_1} - \frac{1}{I_2})\left(\cot\theta + 2\frac{\partial}{\partial\theta}\right)
\end{aligned}
$$

in which $I_1 = \mathrm{m}_1 R_1{}^2$, $I_2 = \mathrm{m}_2 R_2{}^2$ and $1/I = (1/I_1) + (1/I_2)$.

It is clear from equation (6) and the **G**-tensor that the kinetic energy operators exhibit singular behaviour at the collinear configurations, $\theta = 0, \pi$; and as r_1 or r_2 approach zero. Although this is a natural consequence of the choice of this (and most) internal coordinate system, it has caused substantial problems and comment[10,38,44,45] in earlier treatments. In particular discussions of approximate Hamiltonian formulations such as normal coordinate Hamiltonians, "rigid bender" Hamiltonians [46], etc. have been concerned about these terms. With proper attention to the physics near these singularities, basis functions can be chosen which eliminate or minimize the problems associated with these singularities.

The optimal choice of basis functions depends on the coordinate system used, the potential energy surface, the number of accurate eigenvectors desired, and the method of solution of the Schroedinger equation. Because we want many rovibrational states and we want to use the DVR, we choose direct product bases in the internal coordinates (one for $K = 0$ and another for $K = 1$) [41]. Small values of the symmetrized coordinate, R, correspond to the three atoms close to each other, while large R and r correspond to dissociation. These are regions of very high potential energy. Therefore a uniform basis (Chebychev polynomials [36]) over some range of positive R is appropriate which yields an evenly spaced set of DVR points. For r, however, the potential is symmetric about $r = 0$ and so a symmetrized harmonic oscillator basis in r is appropriate. The angle, θ, ranges from 0 at the O-H-H linear configuration to π at the linear H-O-H configuration. Since the potential differs substantially at the two linear configurations, they are not equivalent, and our basis in angle should behave differently in these limits. The appropriate basis functions are Jacobi polynomials, used earlier in our studies of H_3^+ [29]. The parameters of the Jacobi polynomials can be chosen to eliminate separately the effective potential singularities occurring at each linear configuration. Finally, parity-adapted symmetric-top rotor functions are used as a basis for the overall rotation.

These basis functions then have the desired characteristics of treating the effective potential singularities well and concentrating the nodes of the basis functions in regions where the true vibrotor states have the most rapidly varying amplitudes. We note that this basis can be partitioned easily into separate bases for different parity and symmetry (for H_2O). This full symmetry decomposition is utilized in our method although we calculate states of all symmetries and parity.

The DVR's corresponding to the bases for each degree of freedom, (R, r, θ), and the transformations from the function basis representations, (FBR), to the DVR's are obtained by diagonalizing R, r, and $\cos\theta$ in the Chebychev basis for R, the (symmetrized) harmonic oscillator basis for r, and the Jacobi polynomial basis for $\cos\theta$, respectively. Since different parameters of the Jacobi polynomials are used for even and odd K, they generate separate direct product bases for $J > 0$. These are coupled only by Coriolis terms (which are included exactly), for our choice of the body fixed $\hat{z}$-axis and symmetric top rotor functions. We specify the basis in detail elsewhere [41].

Although a number of potential energy surfaces, accurate near the equilibrium configuration of H_2O, have been presented [14,12,11], none was very reasonable for short H-H distances in the very high energy region (above 30,000cm^{-1} above the potential minimum). The Sorbie-Murrell type surfaces [13], on the other hand, behave well in the dissociative limits but were not of spectroscopic accuracy near the equilibrium geometry. We chose to modify the Carter-Handy PES [14] to yield physically reasonable behaviour in this high energy region. As will be shown in the next section the comparison of calculated energy levels using this potential with experiment is sufficiently good that we expect the dynamics to be well represented over a large energy range.

Finally, we represent the full vibration-rotation Hamiltonian in the DVR/parity-adapted symmetric top rotational basis. The Hamiltonian can be represented in the rotation (J, K) basis as

$$\begin{aligned} \mathbf{H}^{J}_{K,K'}(R,r,\theta) &= \mathbf{H}^{J,INT}_{K,K'}\,\delta_{K,K'} \\ &+ \mathbf{H}^{J,ASYM}_{K,K'}\,\delta_{K,K\pm2} \\ &+ \mathbf{H}^{J,COR}_{K,K'}\,\delta_{K,K\pm1} \end{aligned} \tag{7}$$

The diagonal $\mathbf{H}^{J,INT}_{K,K}$ blocks contain the full symmetric top type Hamiltonians for vibrations in (R, r, θ) including effective potentials for rotation depending on J, K. Within these blocks the Hamiltonian is very sparse since the potentials and effective potentials are diagonal in the DVR, and the kinetic energy operators for (R, r, θ) couple separately the DVR bases in (R, r, θ), respectively [36]. The off-diagonal blocks, $\mathbf{H}^{J,ASYM}_{K,K\pm2}$ and $\mathbf{H}^{J,COR}_{K,K\pm1}$ contain the asymmetric top and Coriolis coupling terms, respectively.

For a given J the rovibrational solutions are obtained by sequential diagonalization/-truncation [36]. Specifically, for given J and $K = 0$ and $K = 1$ separately, we find the 3-dimensional vibrational states of each symmetry by sequential solution. Starting with a typical basis size in (R, r, θ) of $N_R = 30$, $N_r = 24$, and $N_\theta = 36$, $(N_{TOT} = N_R N_r N_\theta)$ we can determine about $150 - 200$ $(J, K = 0, 1)$ vibrational states to $\sim$ 1cm^{-1} accuracy for each symmetry (given the PES). This includes energies more than 25,000cm^{-1} above the zero point energy. A large subset of these states is then used as a basis to construct all the required K, K' blocks of the full Hamiltonian matrix, namely the diagonal $\mathbf{H}^{J}_{K,K}$ $(K > 1)$ blocks containing the effective rotational potentials not included in the original diagonalization (for $K = 0, 1$), and the $\mathbf{H}^{J,ASYM}_{K,K\pm2}$ and $\mathbf{H}^{J,COR}_{K,K\pm1}$ blocks. The accurate solutions to the full vibration-rotation Hamiltonian are then obtained by diagonalization of this matrix. Because the residual coupling included here is relatively weak, up to 70% of the eigenvalues resulting from this diagonalization are accurate to 1cm^{-1} . This may yield more than 1200 accurate rovibrational states for higher $J (J > 10)$.

For analysis purposes, we also diagonalize the $\mathbf{H}^{J,INT}_{K,K}$ (diagonal) blocks separately to provide a set of approximate reference eigenvalues and eigenfunctions, separable in K. Comparison of these to the exact solutions provides K labels for the exact states as well as information on the effects of the asymmetric top coupling and Coriolis coupling.

Finally we note the computational efficiency of this approach. A typical calculation for $J = K = 0$ with $N_R = 30$, $N_r = 24$, and $N_\theta = 36$ yielded about 160 symmetric and 150 anti-symmetric states accurate to ~1cm^{-1} to energies ~25,000cm^{-1} above the zero point energy. Such a calculation required about 45 minutes on a Stardent GS-2000 single processor. A comparable calculation for $J = 10$ to yield some 1000 states of each parity/symmetry to comparable accuracy required about 3 hours.

3. Results and Discussion

Although the aim of the discussion here is not to present results of spectroscopic accuracy for use by spectroscopists, but rather to examine the role of various couplings in the high energy vibrational level structure, it is useful to make brief comparison with experiment and prior work in order to confirm the reliability of our model PES and calculational method. Since our potential differs from that of Carter and Handy (CH) only at high potential energy, our results can be compared to theirs and to experiment for the energy levels in the literature. In Table 1 we present some of our calculated energy levels, the differences from experiment, and the CH results [14] (where available), for selected rovibrational states for $J = 0$, $J = 2$ and $J = 10$ to energies more than 17000cm^{-1} above the ground vibrational level. Our assignments, $(\nu_1\, \nu_2\, \nu_3)$, as symmetric stretch, bend, and asymmetric stretch are also given. Agreement with the CH results is uniformly excellent (as it should be) and agreement with experiment is also excellent. (The CH results, where available, are taken from their Table IV which disagrees with their Table II [14]).

For $J > 0$ CH show excellent agreement (~2cm^{-1}) with experiment for the lowest 12 vibrational states to $J = 9$ and for the lowest 23 vibrational states to $J = 2$. Rather than presenting a large table of calculated and experimental energy levels, we can state that our agreement with the CH results indicates that our model for H_2O is equally accurate, certainly sufficiently accurate to give quantitative answers to the questions of separability, etc. which were posed earlier. We therefore turn to this analysis.

In extending the range of accurate ($J = 0$) vibrational states from the ~12,000cm^{-1} total energy limit of CH to about 30,000cm^{-1} , we increase the number of accurately known eigenvalues and eigenfunctions from 14 per symmetry/parity to about 150 with, of course, corresponding increases for $J > 0$. Thus we can examine the "non-spectroscopic" questions such as the level structure for bending states above the collinear energy barrier, the adequacy of local modes and the effects of rotation on the local mode picture; the effects of rotation on separability in an excellent zero order basis (the vibrational eigenfunctions for $J = 0$), and finally, the effect of rotation on the energy level spacing distributions. Only the latter questions will be examined here.

The simplest way to examine the effect of rotation on separability of motions is to find a basis, such as local modes [47,48,49,11,50,51], in which the dynamics are nearly separable (for some states) for $J = 0$, and examine the effect of rotation on these. This is particularly easy for local modes which are linear combinations of degenerate symmetric and anti-symmetric vibrational states. Since we compute the symmetry classes separately, appropriate degeneracies imply the local mode dynamics is a good approximation for excitation of a bond. Experiments such as those of Crim [52] probe the dynamics of such wave packets quite directly.

Table 1: a. (J=0) Symmetric Vibrational Energy Levels, given in cm^{-1}.

(ν_1, ν_2, ν_3)	Calc[†]	CH[‡]	Calc − Expt[§]
(0 1 0)	1595.401	1595	0.654
(0 2 0)	3151.960	3152	0.321
(1 0 0)	3656.321	3656	−0.732
(0 3 0)	4665.488	4666	−1.305
(1 1 0)	5234.457	5235	−0.528
(0 4 0)	6129.171	6129	−4.859
(1 2 0)	6771.970	6772	−3.125
(2 0 0)	7204.355	7204	2.815
(0 0 2)	7442.629	7443	−2.441
(0 5 0)	7531.422	7532	
(1 3 0)	8266.474	8267	−7.503
(2 1 0)	8763.977	8764	2.398
(0 6 0)	8849.958	8851	
(0 1 2)	8997.490	8997	−2.650
(2 2 0)	10281.317		−3.083
(0 2 2)	10517.475		−6.825
(3 0 0)	10608.341		8.681
(1 0 2)	10856.922		11.938
(3 1 0)	12146.999		7.799
(1 1 2)	12392.794		14.826
(0 4 2)	13442.984		−5.016
(3 2 0)	13641.525		−0.475
(2 0 2)	13840.435		12.135
(1 2 2)	13889.189		21.611
(4 0 0)	14210.378		10.765
(0 0 4)	14531.714		−5.156
(3 3 0)	15097.331		−9.669
(2 1 2)	15341.736		−2.763
(4 1 0)	15726.458		16.329
(2 2 2)	16819.962		−5.268
(3 0 2)	16915.215		16.795
(4 2 0)	17196.546		31.154
(5 0 0)	17451.843		33.360
(1 0 4)	17717.690		30.383

† Energy levels (above the zero-point energy) of J=K=0 symmetric vibrational states calculated using the method described in the text.

‡ Calculated energy levels given by Carter and Handy. (See Ref. [8].)

§ Experimentally observed energy levels. (See Ref. [3] references therein.)

Table 1: b. (J=0) Antisymmetric Vibrational Energy Levels (cm^{-1}).

(ν_1, ν_2, ν_3)	Calc†	CH‡	Calc − Expt§
(0 0 1)	3756.243	3756	0.313
(0 1 1)	5332.541	5333	1.272
(0 2 1)	6872.581	6873	1.060
(1 0 1)	7248.806	7249	−1.016
(0 3 1)	8373.327	8375	−0.521
(1 1 1)	8806.118	8806	−0.884
(0 4 1)	9829.415	9841	−4.165
(1 2 1)	10324.462	10327	−4.258
(2 0 1)	10616.752	10617	3.342
(0 0 3)	11029.553	11030	−2.847
(0 5 1)	11232.096	11281	
(1 3 1)	11803.630	11816	−9.560
(2 1 1)	12153.924	12156	2.664
(0 6 1)	12559.977		
(0 1 3)	12565.793	12561	0.793
(1 4 1)	13239.695		16.305
(2 2 1)	13648.162		−4.488
(3 0 1)	13841.050		10.128
(0 2 3)	14057.829		−8.364
(1 0 3)	14301.283		17.519
(1 5 1)	14624.736		15.264
(2 3 1)	15104.894		14.132
(3 1 1)	15356.367		8.418
(1 1 3)	15810.266		22.491
(3 2 1)	16817.958		−3.668
(2 0 3)	16914.908		16.080
(1 2 3)	17280.652		31.888
(4 0 1)	17474.829		20.688

† Refer to footnotes in Table 1.a.

‡ Refer to footnotes in Table 1.a.

§ Refer to footnotes in Table 1.a.

Table 1: c. Comparison of J=2 rotational energy levels (cm^{-1}) of excited bending rovibrational states

ν_2	K_a	K_c	$(0\ \nu_2\ 0)$			$(1\ \nu_2\ 0)$			$(0\ \nu_2\ 1)$		
			Calc	C−E†	CH‡	Calc	C−E	CH	Calc	C−E	CH
0	0	2	70.200	0.11	70	68.992	0.10	69	70.596	0.17	70
	1	2	79.520	0.02	80	77.749	−0.09	78	92.251	0.01	92
	1	1	95.272	0.10	95	93.450	0.04	94	110.546	0.13	110
	2	1	134.766	−0.14	134	131.118	−0.52	131	188.189	−0.33	188
	2	0	136.046	−0.12	136	132.430	−0.49	133	189.128	−0.31	189
1	0	2	70.371	0.15	71	69.230	0.21	69	69.751	0.28	69
	1	2	82.329	0.01	83	80.360	−0.16	79	80.155	0.01	80
	1	1	99.027	0.12	99	97.102	0.07	97	97.131	0.23	97
	2	1	147.308	−0.25	148	142.821	−0.95	142	140.549	−0.53	140
	2	2	148.513	−0.23	149	144.069	−0.89	144	141.906	−0.48	141
2	0	2	70.507	0.19	71	69.389		69	69.967	0.32	69
	1	2	86.296	0.02	86	83.977		84	83.536	−0.04	83
	1	1	103.843	0.14	104	101.650		102	101.426	0.20	101
	2	1	164.191	−0.05	164	158.373	−1.54	158	154.987	−0.85	155
	2	2	165.280	−0.29	165	159.497	−1.51	160	156.241	−0.79	156
3	0	2	70.596	0.17	70	69.482	0.27	69	70.153	0.35	71
	1	2	92.251	0.01	92	89.472	−0.61	89	88.512	−0.10	89
	1	1	110.546	0.13	110	107.990	−0.27	108	107.181	0.19	107
	2	1	188.189	−0.33	188	180.435	−2.54	180	174.951	−1.17	176
	2	2	189.128	−0.31	189	181.324	−2.46	181	176.064	−1.11	177
4	0	2	70.611		70	69.456		69	70.297		70
	1	2	102.111		102	99.517		99	94.637		95
	1	1	121.030		121	118.787		119	115.736		115
	2	1	224.853		225	214.765		215	204.324		204
	2	2	225.607		226	217.039		216	205.260		205
5	0	2	70.505		70						
	1	2	121.530		121						
	1	1	140.932		141						
	2	1	286.577		286						
	2	2	287.116		287						
6	0	2	70.108		70						
	1	2	169.465		170						
	1	1	189.291		190						
	2	1	402.109		281						
	2	2	402.402		283						

† E(Calculated) − E(Experimental)

‡ See Ref. [8].

Table 1: d. Comparison of calculated and experimental rotational energies (cm^{-1}) for J=10 rovibrational states.

K_a, K_c		(0 0 0) Calc†	C−E‡	(1 0 0) Calc	C−E	(0 0 1) Calc	C−E
0	10	1115.518	0.98	1093.383	0.07	1098.910	2.09
1	10	1115.536	0.99	1093.399	0.06	1098.918	2.09
1	9	1294.121	1.10	1268.909	0.17	1273.971	2.64
2	9	1294.722	1.09	1269.391	0.09	1274.284	2.65
2	8	1439.498	1.53	1412.848	0.81	1418.546	3.42
3	8	1447.486	1.36	1419.486	0.27	1423.196	3.17
3	7	1540.249	2.10	1513.866	1.88	1522.057	4.35
4	7	1582.870	1.53	1551.303	0.55	1551.976	3.18
4	6	1618.583	2.13	1590.833	1.08	1601.978	2.64
5	6	1720.157	1.44	1679.358	1.42	1681.396	2.84
5	5	1726.322	1.61	1694.589	0.23	1689.298	3.13
6	5	1876.092	1.12	1838.395	−1.5	1827.066	2.17
6	4	1876.601	1.14	1838.580	−1.58	1827.817	2.22
7	4	2055.028	0.68	2010.075	−3.49	1994.077	1.34
7	3	2055.052	0.68	2010.089	−3.49	1994.116	1.35
8	3	2254.433	0.15	2202.934	−5.62	2179.722	−0.18
8	2	2254.434	0.15	2202.934	−5.62	2179.723	−0.18
9	2	2470.756	−0.50	2414.744	−8.18	2381.738	−1.66
9	1	2470.756	−0.50	2414.744	−8.18	2381.738	−1.66
10	1	2700.603	−1.29	2602.368	−5.31	2596.266	−3.54
10	0	2700.603	−1.29	2602.368	−5.31	2596.266	−3.54

† Calculated energy levels (cm^{-1})

‡ Calculated – Experimental

In Table 2 we give the calculated energy differences between the pairs of exact states which would be combined in a local mode representation. Only pairs for which the splitting was less than 100 cm^{-1} were identified and only a small subset (low bend states) is presented. (The bending state has little effect on the degeneracy). The major implications of this table are:

a) For $J = 0$ a normal mode-local mode "transition" occurs at about the (3,0,0);(2,0,1) vibrational level, 10600cm^{-1} above the zero point; however,

b) the local mode description, even for $J = 0$ is better for certain types of local mode pairs (such as (n,0),(0,n) in our notation), than for others;

c) the local mode separability is sometimes severely damaged by even small rotations ($J = K = 1$);

d) the above is relatively independent of the bend mode excitation; and

e) perhaps surprisingly, the local mode picture appears to *improve* for higher vibrational states when the full rovibrational coupling is included.

Although it is possible to trace the degeneracies required for a good local mode description to some high energy states, most of the states which would be combined in a zero order local mode description have splittings of 100cm^{-1} or more which is comparable to the

Table 2: Splitting of near degenerate symmetric and antisymmetric vibrational states corresponding to local mode states (cm^{-1}) for various (J,K).

$E^{\dagger}_{vib}$	$(\nu_1, \nu_2, \nu_3)^{\ddagger}$	$(n, m)^{\S}$	(J,K) (0,0)	(5,0) Diag-K*	(5,0) VR(E)**	(5,5) Diag-K	(5,5) VR(E)
3656.3	(1 0 0)	(1,0)	>100	>100	>100	>100	>100
7204.4	(2 0 0)	(2,0)	44.45	46.342	45.6	36.815	36.23
10608.4	(3 0 0)	(3,0)	8.41	9.226	8.9	>100	6.29
13840.4	(4 0 0)	(4,0)	0.61	0.807	1.2	>100	-0.20
16915.2	(5 0 0)	(5,0)	-0.31	-0.318	-0.1	>100	-0.21
19786.3	(6 0 0)	(6,0)	-0.10	-0.107	-0.2	-78.758	-0.05
10859.9	(1 0 2)	(2,1)	>100	>100	>100	>100	>100
14210.4	(2 0 2)	(3,1)	90.90	94.513	93.2	>100	75.43
17451.8	(3 0 2)	(4,1)	22.98	25.369	24.2	59.189	14.74

† Energy of J=K=0 symmetric vibrational states above the zero-point energy.

‡ Normal mode assignments.

§ Local mode assignment (n>m).

* Diagonal K-blocks only (no asymmetric top or Coriolis coupling).

** Fully coupled Hamiltonian.

overall energy spacings above about $15000cm^{-1}$. Thus above these energies the separability of local modes appears to lose validity.

Finally we turn to the question of the classically chaotic behaviour of H_2O in these energy ranges, and whether this is manifested in the quantum energy level spacing distributions. Although no complete 3-dimensional classical studies of the dynamics of H_2O have been reported, there have been a number of reduced dimensionality studies. In the model studies of Frederick and McClelland [19] and of Ezra [22], the dynamics of a "rigid bender" model of H_2O was studied, with rotation-vibration interactions. With "reasonable", but simplified potentials the dynamics of these reduced dimensionality models became largely chaotic by $J = 20$ and a total energy of about $30{,}000cm^{-1}$. [22]

We are now in a position to examine, for the first time, the effect of rotation on the energy level spacing distributions. Two possibilities suggest outcomes themselves. First, if the Coriolis coupling and asymmetric top coupling are very weak, then the different K manifolds for a given J should correspond to independent but shifted vibrational distributions and together would tend to yield a Poisson distribution, even if the distributions for each K block separately were like the Wigner distribution. If, on the other hand, the Coriolis and asymmetric top couplings were very strong, this would tend to couple vibrations which otherwise might not be coupled, thus moving an "independent vibration" model (which would yield a Poisson distribution) toward a coupled (Wigner) distribution.

In Figure 1a-b we plot the histogram of the normalized energy level spacing distribution for $J = 0$, for the symmetric and anti-symmetric vibrations. For $J = 0$ first 175 and 121 states are included for each symmetry, respectively. In Figures 2a-d, we plot the second moments of the energy level distributions versus median energy for the four symmetry/parity combinations for J =10, 20, and 25. For given angular momentum and symmetry/parity group, we generated energy level spacing distributions (normed to an average spacing of unity) for rovibrational energy states within the statistical averaging window. To obtain the second moments, σ, as a function of energy, we used a "window" centered at the desired energy and included $N_w/2$ states above and below this energy in the distribution. N_w was 120 for Figures 2a-d. More than 1000 states are included for each parity/symmetry. The results can be compared with the two model distributions corresponding to independent degrees of freedom (Poisson distribution, σ = 2.00) and to fully coupled degrees of freedom (Wigner distribution, σ = 1.27).

As can be seen from the figures, the results, although quantitative, do not appear definitive, at least in the sense of giving a clear trend from one distribution to the other as a function of energy and angular momentum, i.e. from σ = 2.00 to σ = 1.27. On closer inspection, however, it is possible to discern "noisy" trends; as the angular momentum is increased, the trend toward the Wigner distribution is increased as a function of energy and angular momentum. This would indicate that as the density of rovibrational states is increased and the rotational energy is increased, the rovibrational coupling is also increased. Eventually the mixing in the quantum representations is very large.

Overall, however, no striking "approach to chaotic distributions" is evident, in strong contrast to the classical trajectory findings. Thus although as J, K are increased there is substantial mixing of the exact $J = K = 0$ vibrational basis, the overall quantum dynamics retains a significant degree of separability. Indeed, it is often feasible to assign quantum numbers to the vibration-rotation states, even in the J = 10 and 25,000cm^{-1} range. (However, even for J = 10, there are some states even below the 20,000cm^{-1} range which cannot be assigned quantum numbers due to specific resonances.)

Our tentative conclusions, therefore, are that the quantum dynamics are significantly more separable than the general classical dynamical picture would suggest. However, the recent focus of quasi-classical studies on determining isolated periodic orbits [53] or of determining the low resolution spectra from relatively short time quasi-periodicities, have also shown that these regular motions embedded in a sea of chaotic trajectories do have direct connections to the quantum states of systems such as H_3^+ and Na_3^+ [54]. Thus it seems at least, that the initial appearance of classical chaos is not a reliable guide to the coupling of rovibrational quantum states. Perhaps the disappearance of all the "stability islands" in classical trajectory studies may correlate better with the quantum energy level spacing statistics. This study however, brings into question the calculation of intramolecular rovibrational energy transfer rates from classical trajectory studies, since the quantum states appear not to be as fully coupled.

We hope to extend these studies by comparison with accurate 3-dimensional classical studies of H_2O on this surface and to examine other measures of the quantum manifestations of classical chaos.

Figure 1: a-b. Histogram of the normalized locally averaged energy level spacing distribution for $J = 0$, for the symmetric and anti-symmetric vibrational states, overlaped with Wigner and Poisson distribution functions.

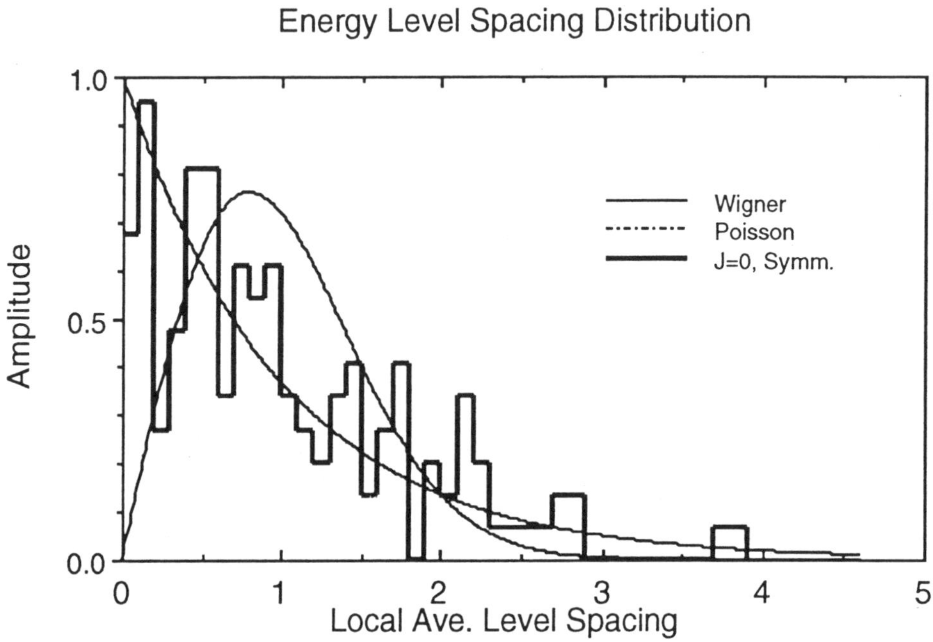

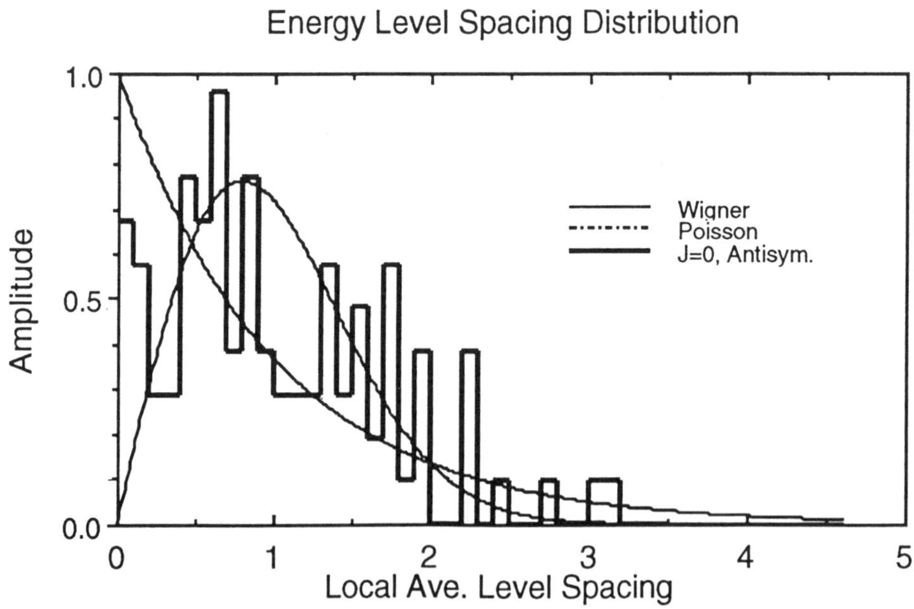

Figure 2: a-b. For J=10, 20, and 25, second moments of the locally averaged spacing distributions of rovibrational energy levels which lie within specific energy ranges are plotted against the median energy of the specified range: a=SYMM/EVEN and b=ASYM/EVEN, where SYMM and ASYM denote symmetric and antisymmetric states, respectively, for EVEN parity states.

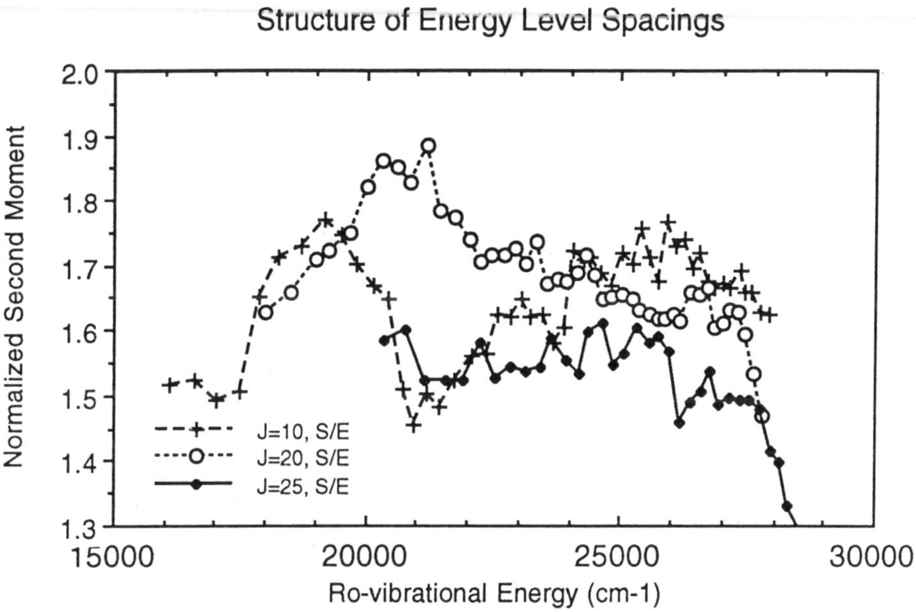

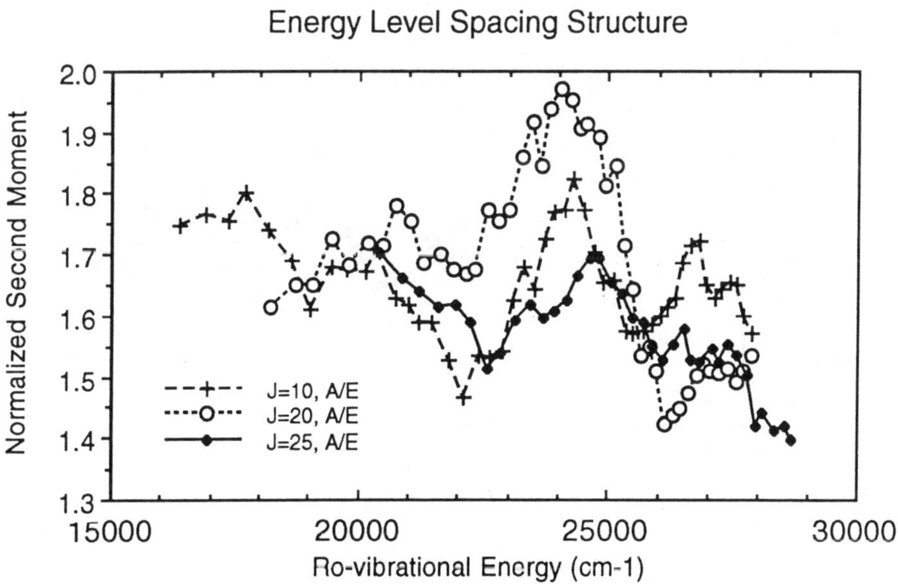

Figure 2: c-d. (Same and Figure 2a-b), except c=SYMM/ODD and d=ASYM/ODD, which contain ODD parity states.

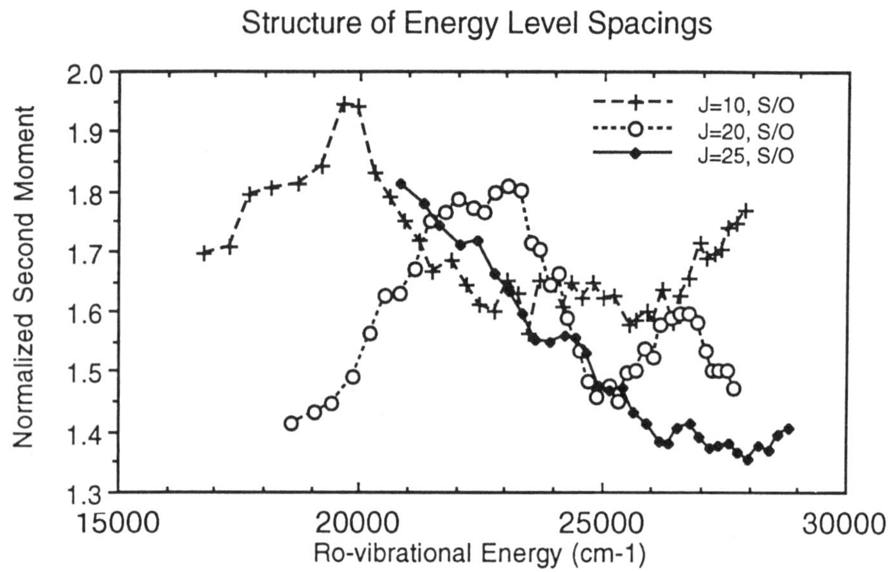

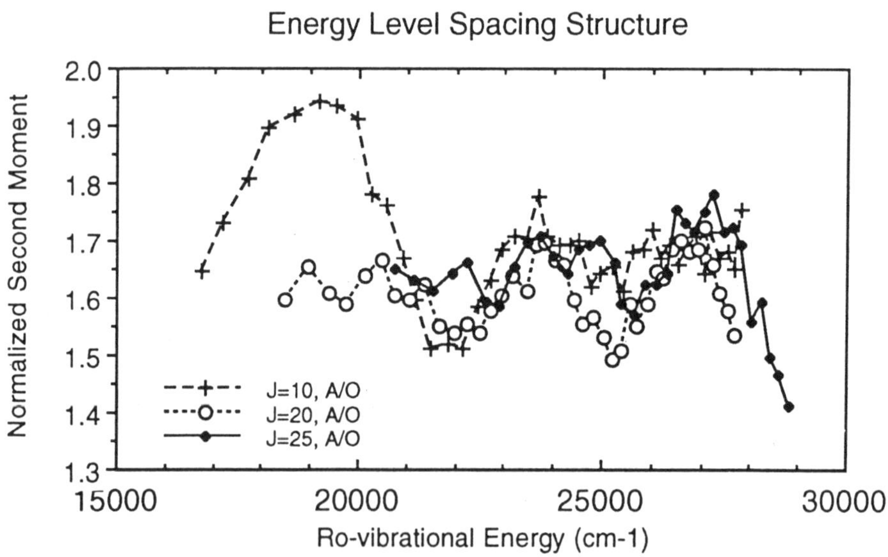

Acknowledgements: This research was supported by a grant from the National Science Foundation, NSF CHE-8806514.

References

[1] J.M. Flaud, C. Camy-Peyret, and R.A. Toth. Water vapor line parameters from microwave to medium infrared. In *International Tables of Selected Constants*, chapter 19. Pergamon, 1981.

[2] X. Luo, P.T. Rieger, D.S. Perry, and T.R. Rizzo. "A new technique for state-to-state studies of unimolecuar reactions"; *J. Chem. Phys.*, **89**(7); 4448, (1988). H2O2 photodissociation.

[3] Randall L. Vander Wal and F. Fleming Crim. "Controlling the pathways in molecular decomposition: The vibrationally mediated photodissociation of water"; *J. Phys. Chem.*, **93**; 5331, (1989). photodissociation H2O.

[4] R.L. Vander Wal, J.L. Scott, and F.F. Crim. "States resolved photodissociation of vibrationally excited water: Rotations, stretching vibrations, and relative cross sections"; *J. Chem. Phys.*, **94**(3); 1859, (1991). photodissociation of H2O.

[5] N.F. Scherer and A.H. Zewail. "Picosecond photofragment spectroscopy. II. The overtone initiated unimolecular reaction H_2O_2 ($\nu_{OH} = 5$)$\rightarrow$2OH"; *J. Chem. Phys.*, **87**; 97, (1987). H2O2 overtone photodissociation.

[6] A.D. Bykov, Yu.S. Makushkin, and O.N. Ulenikov. "The Vibrational Analysis of H_2O"; *J. Mol. Spect.*, **99**; 221, (1983). H2O rovib spec.

[7] J.M. Flaud and C. Camy-Peyret. "The Interacting States (020), (100), and (001) of H_2O"; *J. Mol. Spect.*, **51**; 142, (1974).

[8] C. Camy-Peyret and J.M. Flaud. "The Interacting States (030), (110), and (011) of H_2O"; *J. Mol. Spect.*, **59**; 327, (1976). H2O rovib spec.

[9] J.M. Flaud, C. Camy-Peyret, K.N. Rao, D.W. Chen, Y.S. Hoh, and J.P. Maillard. "Spectrum of Water Vapor between 8050 and 9370 cm^{-1} "; *J. Mol. Spect.*, **75**; 339, (1979). H2O rovib spec.

[10] B.T. Sutcliffe and J. Tennyson. Understanding floppy molecules. In J. Avery *et al.*, editor, *Understanding Molecular Properties*, page449. Reidel Publishing Company, 1987. floppy triatomic molecules.

[11] J.R. Reimers and R.O. Watts. "A local mode potential function for the water molecule"; *Mole. Phys.*, **52**; 357, (1984).

[12] A.R. Hoy, I.M. Mills, and G. Strey. "Anharmonic force constant calculations"; *Mole. Phys.*, **24**; 1265, (1972).

[13] K.S. Sorbie and J.N. Murrell. "Analytical potentials for triatomic molecules from spectroscopic data"; *Mole. Phys.*, **29**; 1387, (1975).

[14] S. Carter and N.C. Handy. "A theoretical determination of the rovibrational energy levels of the water molecule"; *J. Chem. Phys.*, **87**; 4294, (1987).

[15] John H. Frederick, Gary M. McClelland, and Paul Brumer. "Classical trajectory study of vibration-rotation interaction in highly excited triatomic molecules"; *J. Chem. Phys.*, **83**; 190, (1985). Classical rovib of triatom.

[16] Bobby G. Sumpter and Donald L. Thompson. "Influence of Rotation on the Intramolecular Dynamics of Hydrogen Peroxide"; *Chem. Phys. Lett.*, **153**; 243, (1988). classical H2O2.

[17] Stephen K. Gray and Michael J. Davis. "Coriolis induced vibration and rotation mixing in Formaldehyde"; *J. Chem. Phys.*, **90**; 5420, (1989). Classical dynamics study of CH2O.

[18] David Farrelly. "Does an irregular sequence of level spacings provide a new representation of quantum chaos?"; *Phys. Lett.*, **104A**; 63, (1984). chaos.

[19] John H. Frederick and Gary M. McClelland. "Nonlinear dynamics of vibration-rotation interactions: Rigid bender H_2O"; *J. Chem. Phys.*, **84**(8); 4347, (1986). classical rigid bender H2O.

[20] John S. Hutchinson and Robert E. Wyatt. "Quantum Ergodicity and the Wigner Distribution"; *Chem. Phys. Lett.*, **72**; 378, (1980). chaos.

[21] R.Benito, F.Borondo, J.Kim, B.Sumpter, and G.S.Ezra. "Comparison of Classical and Quantum Phase Space Structure of Nonrigid Molecules, LiCN"; *Chem. Phys. Lett.*, **161**(1); 60, (1989). Classical rovib of LiCN.

[22] Gregory S. Ezra. "Interaction between Bending Vibrations and Molecular rotation: A Model Study"; *Chem. Phys. Lett.*, **127**(5); 492, (1986). classical study of rovib.

[23] Klaus Weide, Klaus Kühl, and Reinhard Schinke. "Unstable periodic orbits, recurrences, and diffuse vibrational structures in the photodissociation of water near 128nm"; *J. Chem. Phys.*, **91**; 3999, (1989). class and expt of H2O photodissociation.

[24] S. Carter and N.C. Handy. "A variational method for the calculation of vibrational levels of any triatomic molecule"; *Mole. Phys.*, **47**; 1445, (1982).

[25] C.L. Chen, Bärbel Maessen, and Max Wolfsberg. "Variational calculations of rotationa-vibrational energy levels of water"; *J. Chem. Phys.*, **83**; 1795, (1985).

[26] J. Tennyson and B.T. Sutcliffe *Mole. Phys.*, **58**(5); 1067, (1986). LAM.

[27] Z. Bačić, D. Watt, and J.C. Light. "A variational localized representation calculation of the vibrational levels of the water molecules up to 27000 cm^{-1}"; *J. Chem. Phys.*, **89**; 947, (1988).

[28] J. Tennyson and J.R. Henderson *J. Chem. Phys.*, **91**; 3815, (1989).

[29] Robert M. Whitnell and J.C. Light. "Efficient pointwise representations for vibrational wave functions: Eigenfunctions of $H_3{}^+$"; *J. Chem. Phys.*, **90**; 1774, (1989). H3+ vibrational states.

[30] E.B. Stechel and E.J. Heller. "Quantum ergodicity and Spectral Chaos"; *Ann. Rev. Phys. Chem.*, **35**; 563, (1984). Chaos.

[31] Michael J. Davis. "Phase Space Structure in Classically Chaotic Regions and the Nature of Quantum Eigenstates"; *J. Phys. Chem.*, **92**; 3124, (1988). classical chaos.

[32] J.A. Bentley, J-P. Brunet, R.E. Wyatt, R.A. Freisner, and C. Leforestier. "Quantum Mechanical Study of the $\tilde{A}\,^1A'' \rightarrow \tilde{X}\,^1\Sigma^+$ SEP Spectrum of HCN"; *Chem. Phys. Lett.*, **161**; 393, (1989). SEP spec of HCN.

[33] Tetsuyuki Yukawa and Tadashi Ishikawa. "Statistical Theory of Level Fluctuations"; *Prog. Theo. Phys. Supp.*, **98**; 157, (1989). chaotic level spacing.

[34] Craig C. Martens. "Vibration-rotation interaction in the rigid bender: A quantum mechanical phase space view"; *J. Chem. Phys.*, **90**; 7064, (1989). quantum phase space study of a rigid-bender.

[35] David E. Logan and Peter G. Wolynes. "Quantum localization and energy flow in many-dimensional Fermi resonant systems"; *J. Chem. Phys.*, **93**; 4994, (1990).

[36] J.C. Light, R.M. Whitnell, T.J. Park, and S.E. Choi. Quantum dynamics of small systems using direct variable representations. In A. Lagana, editor, *NATO ASI Series*, page187. Reidel Publishing Company, 1989.

[37] Z. Bačić and J.C. Light. "Theoretical Methods for Rovibrational states of Floppy Molecules"; *Ann. Rev. Phys. Chem.*, **40**; 469, (1989).

[38] D.M. Leitner, J.E. Hunter III, G.A. Natanson, R.S. Berry, P. Villarreal, and G. Delgado-Barrio. "Variational calculation of the rovibrational spectrum of H_2O using the coupled-rotor basis"; *J. Chem. Phys.*, **94**(9); 5917, (1991).

[39] D.F. Smith and J. Overend *Mole. Phys.*, **28**; 471, (1972).

[40] L.S. Rothman. "Update of the AFGL atmospheric absorption line parameters compilation"; *Applied Optics*, **17**; 3517, (1978). H2O absorp rovib spec.

[41] S.E. Choi and J.C. Light. "Highly excited rovibrational states of H_2O"; *J. Chem. Phys.*, (to be published).

[42] Herbert M. Pickett. "Vibration-Rotation Interactions and the Choice of Rotating Axes for Polyatomic Molecules"; *J. Chem. Phys.*, **56**; 1715, (1972).

[43] Bruce R. Johnson and William P. Reinhardt. "Adiabatic separations of stretching and bending vibrations: Application to H_2O"; *J. Chem. Phys.*, **85**(8); 4538, (1986). Radau coord H2O vib states.

[44] Edwin L. Sibert III. "Rotationally induced vibrational mixing in formaldehyde"; *J. Chem. Phys.*, **90**; 2672, (1989). perturbation calculation of rovib CH2O.

[45] Gregory A. Natanson. "The Eckart-frame hamiltonian of a triatomic molecule in terms of the bond lengths and the intrabond angle"; *Mole. Phys.*, **66**; 129, (1989).

[46] Gregory A. Natanson. "Wilson-Howard rotation-vibrational hamiltonian in the limiting case of a linear triatomic molecule"; *Mole. Phys.*, **67**; 173, (1989).

[47] Randall B. Shirts. "Decoupling of the local mode stretching vibrations of water through rotational excitation. I. Quantum mechanics"; *J. Chem. Phys.*, **85**; 4949, (1986).

[48] Lauri Halonen and Tucker Carrington, Jr. "Fermi resonances and local modes in water, hydrogen sulfide, and hydrogen selenide"; *J. Chem. Phys.*, **88**; 4171, (1988). H2O H2S H2Se vib states local-modes Fermi.

[49] John S. Hutchinson and Kenneth T. Marshall. "Local mode Overtones"; *jcsft2*, **84**; 449, (1988).

[50] I.M. Mills and A.G. Robiette. "On the relationship of normal modes to local modes in molecular vibrations"; *Mole. Phys.*, **56**; 743, (1985).

[51] Jan Makarewicz. "Rovibrational Hamiltonian of a triatomic molecule in local and collective internal coordinates"; *J. Phys.*, **B21**; 1803, (1988). rovib Hamiltonian of triatoms.

[52] R.Schincke, R.L.Van del Wal, J.L. Scott, and F.F. Crim. "The effect of bending vibrations on product rotations in the fully state-resolved photodissociation of A state of H_2O"; *J. Chem. Phys.*, **94**; 283, (1991).

[53] D. Wintgen and H. Friedrich *Phys. Rev.*, **A36**; 131, (1987).

[54] J.M. Gomez-Llorente and H.S. Taylor. "Spectra in the chaotic region: A classical analysis for Na_3"; *J. Chem. Phys.*, **91**; 953, (1989).

MODE SELECTIVE VAN DER WAALS CHEMISTRY

JOSEPH C. ALFANO, SELSO J. MARTINEZ III, and DONALD H. LEVY
The James Franck Institute and the Department of Chemistry
The University of Chicago, Chicago, Illinois 60637, USA

ABSTRACT. Three vibronic levels of similar energy ($16b^2$, $16a^2$, and $6a^1$) have been excited in the van der Waals molecules aminotetrazine-Ar and aminotetrazine-Ar_2. The dissociation rates observed for aminotetrazine-Ar were found to vary by over two orders of magnitude depending on the initially excited level. The product branching ratio observed when aminotetrazine-Ar_2. dissociates can be changed from pure aminotetrazine-Ar to almost pure aminotetrazine by changing the initially excited level.

1. Introduction

The study of mode selective chemistry, the subject of this meeting, has been a long, hard struggle. Ideally chemists would like to be able to direct the course of a broad class of chemical reactions simply by adjusting the various buttons and knobs of their lasers, and we are still a long way from this goal. The ability to adjust the rate or better yet the path of a photochemical reaction by changing the properties of the light incident on the system has been restricted to a relatively small number of favorable systems, and in even in many of these systems the effects are often small and difficult to observe.

Those of us who have been content to study van der Waals photochemistry have in some sense sacrificed the great goal to the expedient of being able to get results. [1] Although it is unlikely that mode selective van der Waals chemistry will directly open up vast new areas of synthetic chemistry, one might expect that if large scale mode selective effects were to be observed anywhere, they would be observed in van der Waals molecules. Because of the relative weakness of the van der Waals bond as compared to the chemical bond, only small amounts of energy are necessary to achieve van der Waals chemistry, the breaking of van der Waals bonds. This in turn means that at the necessary levels of excitation there need not be extensive mixing of all modes, and individual couplings may still play a dominant role without being overwhelmed by the statistical effects produced by extensive mixing. The molecule might then be able to remember the nature of the initially excited state for times approaching its dissociation lifetime.

Even here, however, nature has been resistant. It is true that mode selective effects have been observed in a fairly large number of van der Waals molecules in the sense that dissociation rates can be altered in a given molecule by exciting different initial states. On the whole these effects have been larger than those observed in chemically bound molecules, but they have still been smaller than might be expected in these most favorable systems. Differences in rate of factors of two or three have been fairly common, but differences of orders of magnitude have been hard to come by. [2,3,4,5] There are in fact a number of studies which have interpreted all apparent mode selective effects in van der

J. Jortner et al. (eds.), Mode Selective Chemistry, 87–95.

Waals molecules as being entirely due to manifestations of statistical behavior. [6] The models used in these studies attribute the observed photochemisty to statistical IVR from the initially excited chemical mode to the van der Waals modes followed by RRKM dissociation from the van der Waals modes. In these models any difference in rate or product state distribution observed when different initial modes are excited is attributed only to the different energies of the initial modes.

In this paper we report large mode selective effects in complexes of aminotetrazine (AT) with argon. [7] Although the mechanism of this mode selectivity is not understood, it would seem hard to attribute the observations to purely statistical behavior.

2. Experimental

The experimental methodology and data analysis used in this paper is essentially identical to that used previously in a number of studies of the photochemistry of van der Waals complexes of tetrazine. [2] The complexes were made and cooled in a continuous supersonic free jet, and photochemistry was initiated with an argon ion pumped tunable dye laser operating in the region 18,000-20,000 cm^{-1}. The results of the photochemical reaction were monitored by observing the emission spectrum produced by both the initially excited complex and the electronically excited species generated in the reaction. The amount of excitation energy and the initially excited vibrational state could be adjusted by tuning the laser, and the relative yield of various product states could be measured from the intensity of various features in the emission spectrum. If emission from the initially excited state could be observed, relative yields could be converted to absolute rates using the measured fluorescence lifetime of the complex. In effect we clock the unknown dissociation rates against the known fluorescence lifetime.

3. Results and Discussion

The fluorescence excitation spectrum of uncomplexed aminotetrazine is shown in fig. 1. The origin is at 18,411 cm^{-1}, 285 cm^{-1} to the blue of the origin transition of tetrazine. In the first 1400 cm^{-1} to the blue of the origin, we observe progressions in the in-plane totally symmetric modes 6a and 1 and the out-of-plane modes 16b and 16a. In addition we observe a sequence band of the inversion mode, I, a combination band involving modes 1 and 6a, and several weak, unassigned features. [8]

When a few percent argon is added to the helium carrier gas, new features grow in to the red of each vibronic band which are assigned to complexes of an aminotetrazine molecule with one or more argon atoms. The first feature to the red, shifted 25 cm^{-1}, is assigned to the complex AT-Ar while the second feature, shifted by 50 cm^{-1}, is assigned to the complex AT-Ar_2. Larger complexes do not produce constant spectral shifts per atom, and complexes as large as AT-Ar_6 have been observed.

The photophysics and photochemistry of the complexes have been studied by observing the emission spectrum produced by exciting a single vibronic level of a particular complex. Dissociation of a complex produces emission from either smaller complexes or from free aminotetrazine, while those molecules that do not dissociate produce emission from the initially excited complex, either from the initially excited vibrational state or from other vibrational states if IVR has occurred. The relative intensities of the various bands observed in the emission spectrum provides a quantitative measure of the dissociation and IVR lifetimes.

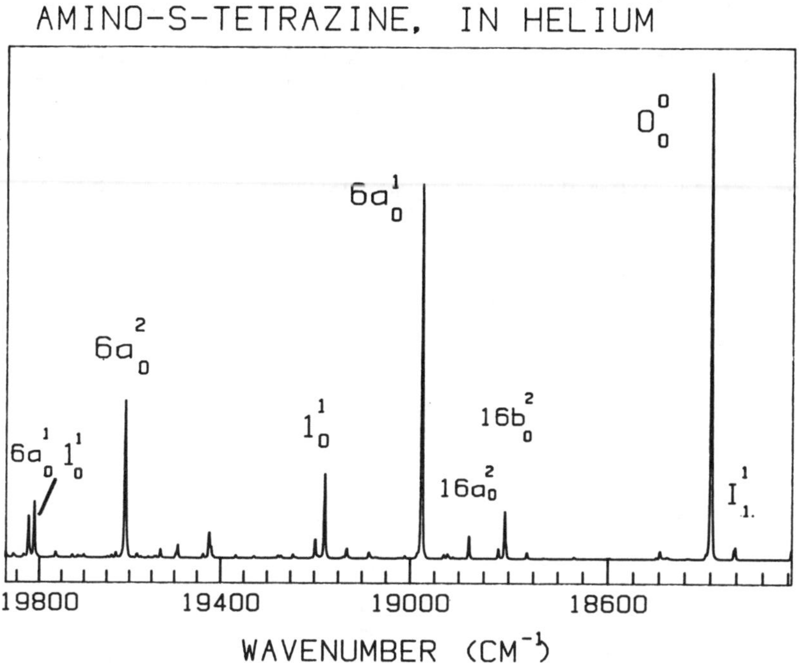

Fig. 1. The fluorescence excitation spectrum of aminotertrazine in a helium supersonic free jet.

The measurement actually produces the ratio of the dissociation or IVR lifetimes relative to the fluorescence lifetime; that is, the fluorescence lifetime is the clock against which the other dynamical processes are timed. The fluorescence lifetimes of many levels of aminotetrazine have been measured [9] and have been found to vary from 154 ns for the zero-point level to 54 ns for the $6a^1 1^1$ level, the lifetimes decreasing monotonically with increasing vibrational energy. It is important to note that these fluorescence lifetimes are much longer than the corresponding lifetimes in tetrazine. For example, the fluorescence lifetime of the zero-point level of tetrazine is around 0.5 ns, more than two orders of magnitude smaller than the corresponding lifetime in aminotetrazine. Thus, the timing clock in aminotetrazine is running very slowly, and complexes that survive for the full fluorescence lifetime have lived for quite a long time.

In this paper we would like to concentrate on the results obtained from exciting three vibronic levels: $16b^2$ at 446 cm^{-1}, $16a^2$ at 505 cm^{-1}, and $6a^1$ at 606 cm^{-1} above the zero point level. Because the energies of these levels are not too different, it is likely that any major differences in the dynamics out of the various levels will be due to mode-selective effects and not to energetic effects.

3.1. DISSOCIATION OF AT-Ar

The emission spectrum produced when the $16b^2$ level of AT-Ar is excited is shown in fig. 2. All of the features in this spectrum can be assigned to emission from the zero-point vibrational level of the AT fragment that is produced when the molecule dissociates. Since we see no emission from the AT-Ar complex, we can infer that the dissociation rate is much faster than the fluorescence rate, and we can estimate a lower limit for the dissociation rate of 4×10^8 sec^{-1}. The dissociation rate of tetrazine-Ar excited to various vibrational levels was of the order of $1\text{-}2 \times 10^8$ sec^{-1}, and therefore the limit of the dissociation rate of AT-Ar from $16b^2$ is similar to that of tetrazine-Ar.

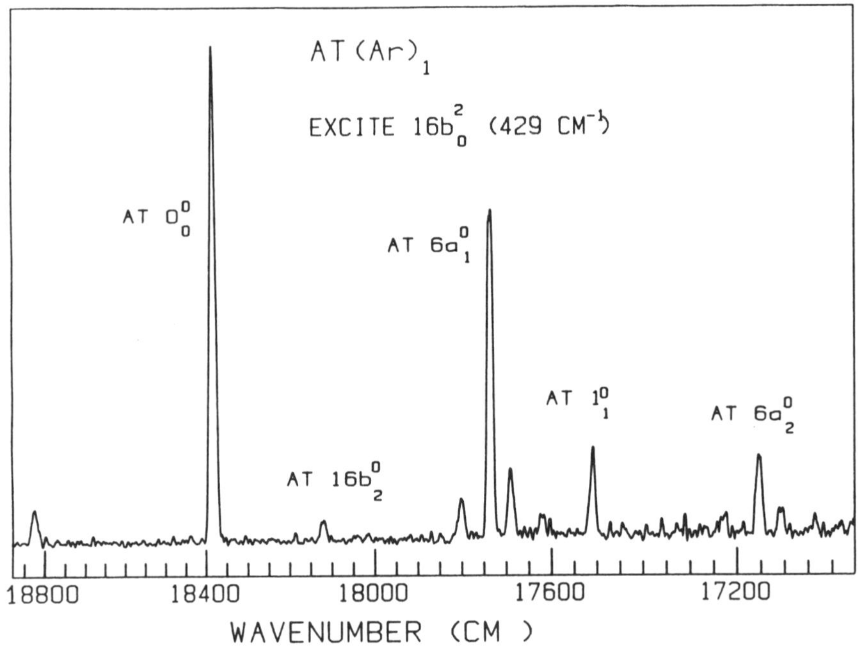

Fig. 2. The emission spectrum produced by exciting the $16b^2$ band of aminotetrazine-Ar at 18815 cm^{-1}.

The emission spectrum resulting from $16a^2$ excitation of AT-Ar is shown in fig. 3, and it contains features from both the zero-point level of the AT fragment as well as features which can be assigned to emission from the originally excited level of the complex. Because we now see emission from the complex, we can infer that dissociation is slower from the $16a^2$ level than it was from the $16b^2$ level. From the relative intensities of the fragment and complex features and from the previously measured fluorescence lifetime of the $16a^2$ level, we can extract a dissociation rate from $16a^2$ of 6×10^7 sec^{-1}. Note that $16a^2$ is 59 cm^{-1} higher in energy than $16b^2$, and a purely energetic effect would be expected to be small and in the opposite direction.

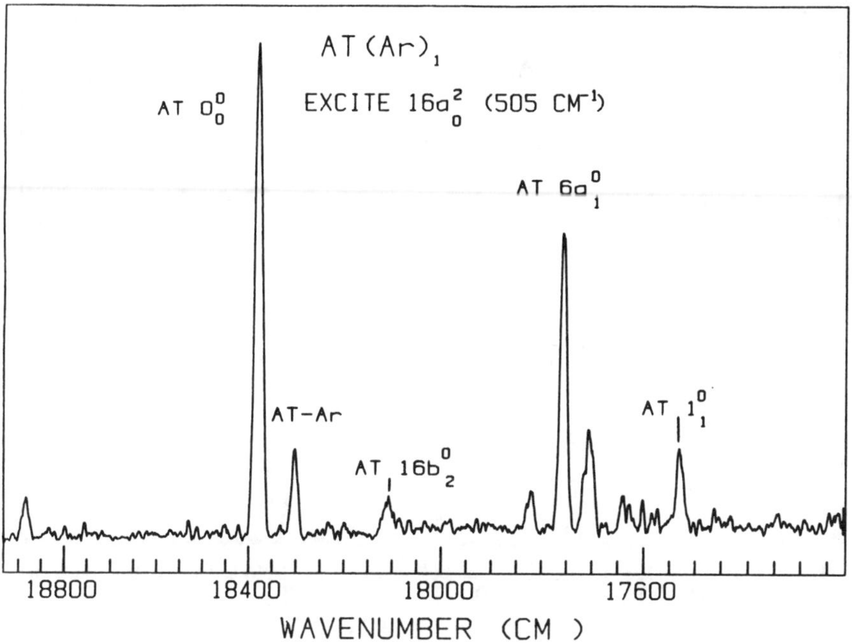

Fig. 3. The emission spectrum produced by exciting the $16a^2$ band of aminotetrozine-Ar at 18891 cm^{-1}.

The third level which we excited, $6a^1$, produces the emission spectrum shown in fig. 4. Here we do see some emission from the zero-point level of the AT fragment, but most of the intensity in the emission spectrum is due to resonance fluorescence from the $6a^1$ level of the AT-Ar complex. For this level dissociation is slow compared to fluorescence, and since the fluorescence lifetime of this level is 90 nsec., dissociation is very slow indeed. We measure a rate of 3×10^6 for dissociation from the $6a^1$ level. Thus, the dissociation lifetime of AT-Ar can be changed by at least two orders of magnitude by exciting different vibrational modes. Since the levels are of similar energy, it is unlikely that the changes in dissociation rate are due to energy effects. The fact that the higher energy states have the smaller dissociation rates further indicates that this is a mode-selective effect.

3.2. DISSOCIATION OF AT-Ar_2

In the case of AT-Ar, we clearly observed mode-selectivity. However, if mode selective chemistry is taken to mean controlling the branching between two possible reaction channels, dissociation of AT-Ar cannot produce mode selective chemistry since there is only one possible set of products, AT and Ar. In the case of AT-Ar_2, there is the possibility of doing mode-selective van der Waals chemistry since there are now two possible sets of products: AT + 2Ar and AT-Ar + Ar. We have observed that the

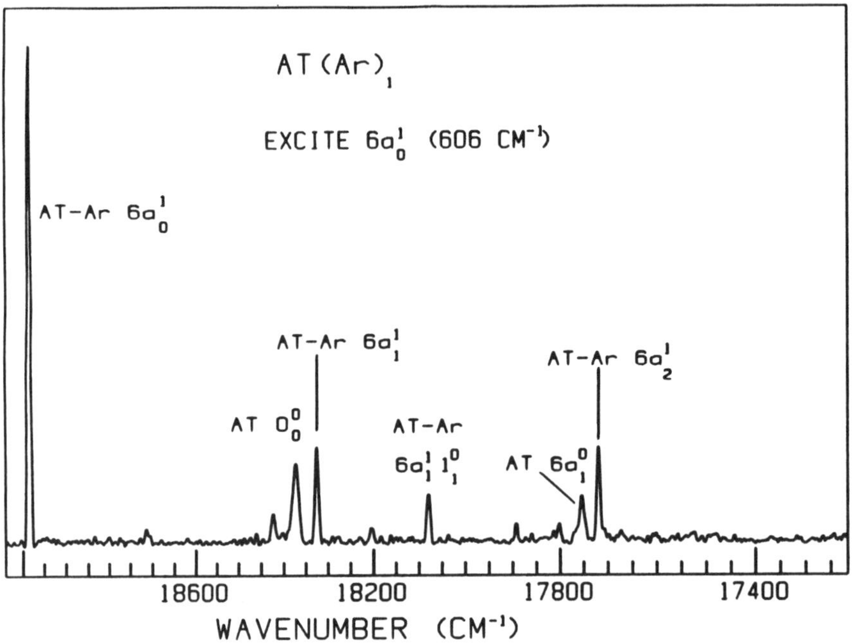

Fig. 4. The emission spectrum produced by exciting the $6a^1$ band of aminotetrozine-Ar at 18992 cm^{-1}.

branching ratio between these two product channels can be significantly changed by selecting the mode that is originally excited, and we will now describe these observations. Once again, we will concentrate on the same three closely spaced levels to separate mode-selective from energetic effects.

In fig. 5 we show the emission spectrum produced when the level $16b^2$ of AT-Ar_2 is excited. Most of the emission in this figure comes from the completely dissociated product, AT, although some emission from the product AT-Ar can be seen. Just as in the case of $16b^2$ excitation of AT-Ar, no emission from the initially excited complex can be seen. It should be noted that the energy contained in the level $16b^2$, 446 cm^{-1}, is enough to break both AT-Ar bonds. Therefore, the upper limit for the average van der Waals bond strength in the excited electronic state is 223 cm^{-1}, less than the bond strength in tetrazine-Ar which was in the range 277-306 cm^{-1}. [2]

The emission spectrum produced by excitation of $16a^2$ of AT-Ar_2 is shown in fig. 6. In this case, all of the observed emission comes from the intermediate product AT-Ar and no emission from either AT or AT-Ar_2 is observed. Thus, by shifting between two nearly isoenergetic levels one can shift the dissociation products from mostly AT to entirely AT-Ar. To the extent that there are energetic effects, they are in the wrong direction with the more completely dissociated product coming form the lower energy level.

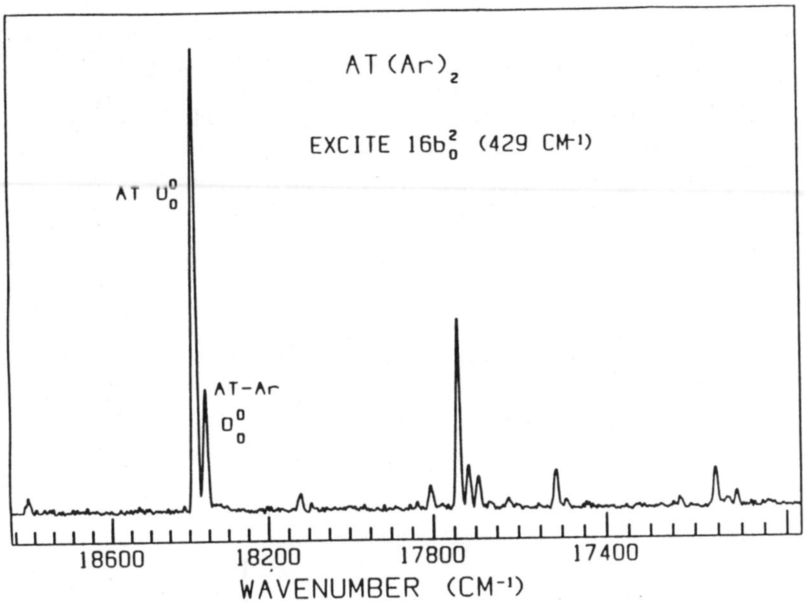

Fig. 5. The emission spectrum produced by exciting the $16b^2$ band of aminotetrazine-Ar_2 at 18790 cm^{-1}.

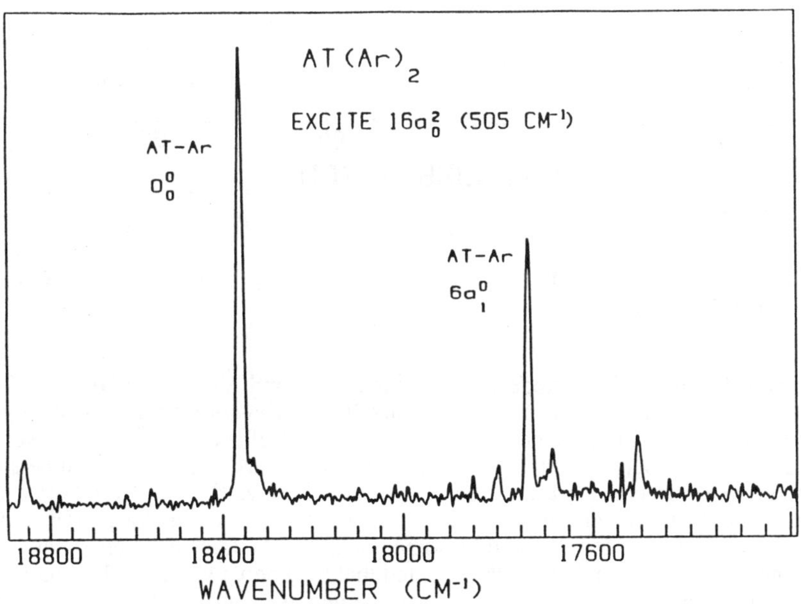

Fig. 6. The emission spectrum produced by exciting the $16a^2$ band if aminotetrazine-Ar_2 at 18866 cm^{-1}.

Excitation to the $6a^1$ level of AT-Ar_2 produces the emission spectrum shown in fig. 7. Most of the intensity comes from the zero-point level of AT-Ar, while there are some weaker features due to resonance fluorescence from the intially excited level of AT-Ar_2. There is also a small, broad feature which is probably due to internal vibrational relaxation in AT-Ar_2. Thus excitation of AT-Ar_2 to $6a^1$ yields only the single dissociation product AT-Ar, but the observation of resonance fluorescence indicates that the dissociation rate into this one product channel is less than the dissociation rate from the $16a^2$ level.

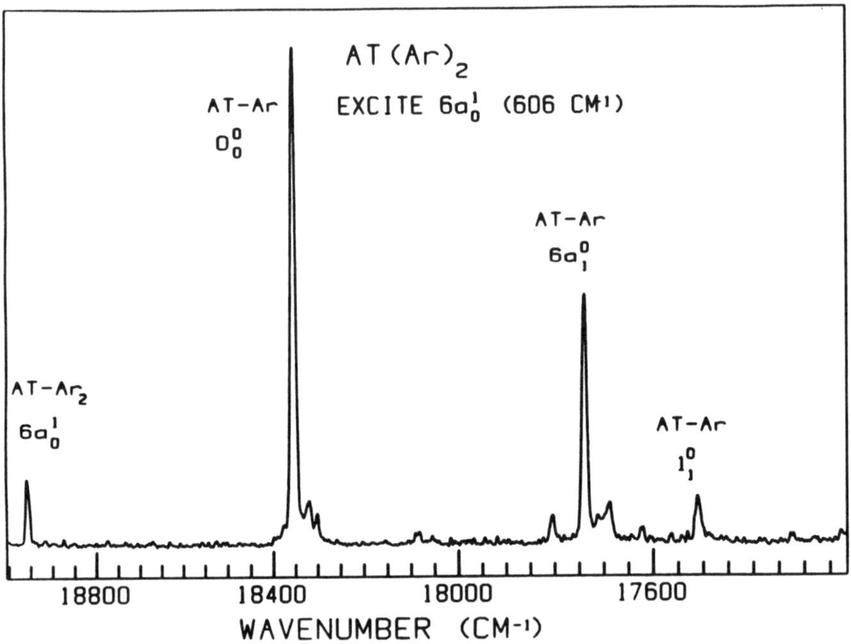

Fig. 7. The emission spectrum produced by exciting the $6a^1$ band of aminotetrazine-Ar_2 at 18965 cm^{-1}.

The reasons for the extreme mode selectivity in AT-Ar and AT-Ar_2 are not at all understood. This selectivity is particularly puzzling in view of the fact that the argon van der Waals molecules of tetrazine show very little mode selectivity. At first we supposed that the differences between tetrazine and AT van der Waals molecules might be attibuted to differences in structure. More recent measurements have shown that in both cases the structures are the same with the argon atoms centered over the ring above and below the plane of the molecule. [10] It should also be noted that the transition frequency of AT is only slightly shifted from that of tetrazine and that the known vibrational frequencies are similar in both molecules with the differences being attributable to the additional mass of the amino group in AT. Therefore there seems to be little difference in electronic or vibrational structure in the two molecules.

Acknowledgements

This work was supported by the National Science Foundation under grant CHE-8818321. J. C. A. gratefully acknowledges support under the National Sciences Foundation Graduate Fellowship Program. S. J. M. acknowledges support under the Illinois Minority Graduate Fellowship Program. We are indebted to Dr. D. D. Yang for synthesis of the sample of aminotetrazine.

References

[1] Of course, in addition to expediency, studies of van der Waals molecules are driven by the hope that the information and understanding produced by these studies might be carried over to studies of chemical bonds.

[2] Brumbaugh, D.V., Kenny, J.E. and Levy, D.H. (1983) J. Chem. Phys. 78, 3415.

[3] Ramaekers, J.J.F., van Dijk, H.K., Langelaar, J. and Rettschnick, R.P.H., (1983) Faraday Discuss. Chem. Soc. 75, 183 (1983); Ramaekers, J.J.F., Krignen, L.B., Lips, H.J., Langelaar, J., and Rettschnick, R.P.H. (1983) Laser Chem. 2, 125; Heppener, M., Kunst, A.G.M., Bebelaar, D., and Rettschnick, R.P.H. (1985) J. Chem. Phys. 83, 5341; Heppener, M., and Rettschnick, R.P.H., in Weber, A. (ed.), Structure and Dynamics of Weakly Bound Molecular Complexes, Reidel, Dordrecht, 1987.

[4] Stephenson, T.A., and Rice, S.A. (1984) J. Chem. Phys. 81,1083.

[5] Butz, K.W., Catlett, Jr., D.L., Ewing, G.E., Krajnovich, D., and Parmenter, C.S. (1986) Phys. Chem. 90, 3533; Jacobson, B.A., Humphrey, S.,and Rice, S.A. (1988) J. Chem. Phys. 89, 5624; O, H.-K., Parmenter, C.S., and Su, M.C. (1988) Ber, Bunsenges. Phys. Chem. 92, 253.

[6] Kelley, D.F., and Bernstein, E.R. (1986) J. Phys, Chem. 90, 5164; Nimlos, M.R., Young, M.A., Bernstein, E.R. (1989) J. Phys. Chem. 91, 5268; Knee, J.L., Khundkar, L.R., and Zewail, A.H. (1987) J. Chem. Phys. 87, 115; Semmes, D.H., Baskin, J.S., and Zewail, A.H. (1990) J. Chem. Phys. 92, 3359.

[7] Alfano, J.C., Martinez III, S.J., and Levy, D.H., (1990) J. Chem. Soc. Faraday Trans. 86, 2503; Alfano, J.C., Martinez III, S.J., and Levy, D.H. (1989) J. Chem. Phys. 91, 7302.

[8] Alfano, J.C., Martinez III, S.J., Yang, D.D., and Levy, D.H. (1990) J. Molec. Spectr. 143, 366.

[9] Alfano, J.C., Martinez III, S.J., and Levy, D.H. (1991) J. Chem. Phys. 94, 2475.

[10] Alfano, J.C., Martinez III, S.J., and Levy, D.H. (1991) J. Chem. Phys. 94, 1673.

THEORETICAL STUDIES ON PHOTOFRAGMENTATION OF RARE GAS–INTERHALOGEN VAN DER WAALS COMPLEXES [1]

G. Delgado–Barrio and P. Villarreal
Instituto de Física Fundamental, Centro Mixto C.S.I.C.–U.C.M.
Serrano 123, 28006 Madrid, SPAIN

Abstract

Several quantum mechanical treatments to study the Vibrational Predissociation (VP) of rare gas $-ICl$ van der Waals complexes are presented and discussed. Time–independent converged three–dimensional calculations provide a detailed description of the fragmentation dynamics in these systems. In addition, they constitute a useful benchmark data to test other methods and approximations and compare with the available experimental results.

I Introduction

Recently a great deal of detailed experimental information on intermolecular forces and also on intramolecular energy transfer and redistribution has stimulated the theoretical research on intermolecular potential energy surfaces (PES) as well as the dissociation dynamics of van der Waals complexes[1]. In particular, several experimental studies on photofragmentation in electronically excited states of triatomic systems involving a rare gas atom and a halogen or interhalogen diatomic molecule have been conducted[2]. They provided valuable data on VP of these molecules: lifetimes and linewidths, spectral shifts, as well as final state vibrational and rotational distributions of the diatomic fragment.

Recently[3], Lester an co. have studied the $Ne, He - ICl$ complexes in the region of the $ICl(B^3\Pi_{0^+}, v' = 2)$ excited state. A high dominance of the $\Delta v' = -1$ channel together with highly inverted ICl fragment rotational distributions centered at $j \sim 7 - 8(He)$ and $j \sim 17(Ne)$ have been found.

[1]Work supported by CICYT under Grant $N^oPB87 - 0272$ and EEC Grant No. SC1.145.C

J. Jortner et al. (eds.), Mode Selective Chemistry, 97–112.

Several time–independent treatments have been used to study the VP of these molecules:

(a) Converged three–dimensional calculations, involving the ground and excited electronic states, to obtain the photofragmentation cross sections depending on the energy of the photon absorbed and, hence, to determine energies and half–width of vibrational predissociated levels[4].

(b) Vibrational diabatic approximations (VDA) based on the decoupling between the (usual) fast stretching diatomic motion and the other ones, once the predissociating level has been prepared. In the Fermi's Golden Rule frame, the associated half–widths are obtained through the matrix element of the potential interaction between the initial (discrete) and final (continuum) states[4,5]. A further simplification can be introduced by describing the final states within the rotational infinite order sudden approximation (VD+RIOSA)[4f].

One of the bottle–necks to perform dynamical calculations is the lack of knowledge of the relevant PES. In order to get these intermolecular potentials from *first principles*, it should be necessary to account for the correlation energy and relativistic effects in the case of heavy atoms. Due to the weakness of the vdW bond, a small error from these two sources leads to a PES not accurate enough to study the dissociation dynamics. An alternative way to get the potential would be to assume a functional form, estimating its corresponding parameters through dynamical calculations by a multiproperty fitting of the available experimental data. Of course, this procedure needs to use a cheap dynamical calculation, i.e. an approximate one, but reliable to describe the process.

In addition, the approximate treatments of VP allow to implement time–dependent versions, for instance through the wave packet approach, using the same techniques as those used to treat optical line shapes and Raman spectra. In particular, the Golden Rule treatment in the time domain will lead to a much shorter propagation than a full three–dimensional wave packet, that has to propagate in time for an interval of the order of the lifetime, i.e., thousands of the monomer vibrational periods. Further inclusion of an IOS approximation greatly reduces the computational time and, as it will be shown later, can help to incorporate the VD approximation in trajectory calculations to study VP in larger systems like those composed by a fast vibrating diatomic molecule and several rare gas atoms.

The paper is organized as follows: Section II is devoted to briefly describe the *exact* and approximate vibrational diabatic based treatments for VP of $X \cdots ICl$ systems in the energy domain. In Section III, the time–dependent version of the Golden Rule plus RIOS approximations is shown together with a very quick way to obtain angular dependent half–widths. Also, the possible application in the classical frame of the VD approximation is included in this Section, Finally, some preliminary results are presented and discussed in Section IV.

II Time independent treatments

II.1 Three–Dimensional quantal model

The Hamiltonian, after separation of the center of mass, for triatomic systems of the type $X \cdots ICl$, may be written in Jacobi coordinates as

$$H = \frac{\hbar^2}{2\mu}\left(-\frac{\partial^2}{\partial R^2} + \frac{\mathbf{l}^2}{R^2}\right) + \frac{\hbar^2}{2m}\left(-\frac{\partial^2}{\partial r^2} + \frac{\mathbf{j}^2}{r^2}\right) + U_{ICl}(r) + V(R, r, \theta) \qquad (1)$$

where R is the distance between X and the center of mass of ICl, r is the intermolecular distance for ICl, and θ is the angle between the two vectors $\mathbf{r}$ and $\mathbf{R}$, while the μ and m factors are the corresponding reduces masses: $\mu = m_x m_{ICl}/(m_x + m_{ICl}), m = m_I m_{Cl}/(m_I + m_{Cl})$. In Eq. (1) $\mathbf{l}$ and $\mathbf{j}$ are the angular momentum operators (in units of $\hbar$) associated with $\mathbf{R}$ and $\mathbf{r}$, respectively. The $U_{ICl}(r)$ and $V(r, R, \theta)$ functions are the intramolecular (ICl) and van der Waals intermolecular potentials, respectively.

In the framework of the first order perturbation theory for electric dipole transitions, the cross section for the excitation from an initial bound state $\mid \Psi_i >$ with energy E_i, to a final dissociative state $\mid \Psi_{fE} >$ of energy E , is given by:

$$\sigma_{fE \leftarrow i} \;\; \propto \;\; |< \Psi_{fE} \mid \mathbf{M} \cdot \mathbf{e} \mid \Psi_i >|^2 \;\; \text{with } E = E_i + \hbar\omega \qquad (2)$$

where $\mathbf{e}$ is the polarization vector of the incident photon with energy $\hbar\omega$, $\mathbf{M}$ is the transition dipole moment and f is a set of quantum numbers which indicates the final quantum state of the fragments after the dissociation. $\mid \Psi_i >$and $\mid \Psi_{fE} >$belong to different electronic states and are eigenfunctions of the corresponding Hamiltonians for nuclear motion like Eq. (1).

For the initial bound state $\mid \Psi_i >$the wave function is expanded as:

$$\Psi_i(\mathbf{r}, \mathbf{R}) = \sum_{v'} \sum_{n'_s, n'_b} a^{(i)}_{v', n'_s, n'_b} \, \chi^{(i)}_{v'}(r) \, \varphi^{(i)}_{n'_s}(R) \, F^{(i)}_{n'_b}(\hat{\mathbf{r}}, \hat{\mathbf{R}}) \qquad (3)$$

where $\chi^{(i)}_{v'}(r)$ belong to an intramolecular basis set for the isolated ICl and are obtained by numerical integration. φ and F are radial and bending basis set functions, respectively. The coefficients $a^{(i)}_{v', n'_s, n'_b}$ and the energy E_i of $\mid \Psi_i >$are obtained by diagonalizing the corresponding Hamiltonian matrix.

For the dissociative state $\mid \Psi_{fE} >$the wave function is expanded as:

$$\Psi^{vj\Omega}_{fE}(\mathbf{r}, \mathbf{R}) = \sum_{v''} \sum_{j''\Omega''} \chi^{(f)}_{v''}(r) \, \phi^{(f)vj\Omega}_{Ev''j''\Omega''}(R) \, F^{(f)}_{j''\Omega''}(\hat{\mathbf{r}}, \hat{\mathbf{R}}) \qquad (4)$$

Introduction of this expression in the time independent Schrödinger equation with the Hamiltonian given in Eq. (1) leads to a set of close coupled equations to obtain the $\phi^{(f)vj\Omega}_{Ev''j''\Omega''}(R)$, that are being numerically solved after the usual outgoing conditions are imposed. In order to calculate the photodissociation cross section we need to take into account that in our case the transition dipole operator lies along the ICl nuclear axis (parallel transition) and we assume that this component M_o only depends on the internuclear distance r of the diatomic molecule and is not affected by complexation. By keeping all this in mind the partial photofragmentation cross section at energy E to produce diatomic fragments in the vibrational state v and the rotational state j will be given by[4f]:

$$\sigma_{fvjE\leftarrow i} \propto \sum_{\Omega} \left| \sum_{v'j'\Omega'} \sum_{v''j''\Omega''} C^{j'\Omega'J'}_{j''\Omega''J''} < \phi^{(f)vj\Omega}_{Ev'j'\Omega'} \mid \varphi^{(i)}_{v''j''\Omega''} >< \chi^{(f)}_{v'} \mid M_o(r) \mid \chi^{(i)}_{v''} > \right|^2 \tag{5}$$

where the $C's$ include geometrical factors expressed in terms of $3j$ symbols[4f].

The total cross section provides, as a function of E, the photofragmentation line--shape and will be given by:

$$\sigma_{fE\leftarrow i} = \sum_{vj} \sigma_{fvjE\leftarrow i} \tag{6}$$

Finally, the distribution of the fragments is obtained through the ratio between the expresions given by Eq. (5) and (6).

II.2 Vibrational Diabatic Approximation (VDA)

The resolution of the close coupled equations that appear in the three–dimensional quantal model involves a lot of channels and in some cases the calculation becomes prohibitive. In most cases, the franck–Condon factors strongly favour the optical transition from the initial state $\mid \Psi_i >$ to a discrete, quasibound, state of the excited electronic surface $\mid \Psi_{fv} >$, meanwhile the corresponding continuum states are very weakly reached by direct optical excitation. This is the situation of vibrational predissociation with pure Lorentzian line shape, defined by a half–width at half–maximum of the resonance and its location (usually close to the energy position of the quasibound state). Moreover, in weakly bound complexes like $X \cdots ICl$, there is a natural decoupling between the *fast* vibrational motion of ICl and the intermolecular *slow* vibrations[1d]. In this case the discrete and the continuum states can be approximated by

$$\Psi_{fv}(\mathbf{r},\mathbf{R}) = \chi^{(f)}_{v}(r) \sum_{n'_s n'_b} a^{(f,v)}_{n'_s,n'_b} \varphi^{(f,v)}_{n'_s}(R)\, F^{(f,v)}_{n'_b}(\hat{\mathbf{r}},\hat{\mathbf{R}})$$

$$\Psi^{j,\Omega}_{f,v',E}(\mathbf{r},\mathbf{R}) = \chi^{(f)}_{v'}(r) \sum_{j'\Omega'} \phi^{(f,v')j,\Omega}_{E,j',\Omega'}(R)\, F^{(f,v')}_{j',\Omega'}(\hat{\mathbf{r}},\hat{\mathbf{R}}) \tag{7}$$

with $v' < v$ and E being the on–shell energy. In this approximation the coupling between the stretching and bending motions is treated explicitly. Therefore this is an exact $2D$ calculation.

The half-width corresponding to the quasibound level is given in the Golden Rule scheme by:

$$\Gamma_v = \pi \sum_{\substack{v'<v \\ j,\Omega}} \left| < \Psi^{j\Omega}_{f,v'E} \mid V(R,r,\theta) \mid \Psi_{f,v} > \right|^2 \tag{8}$$

and the final state distribution will then be given by:

$$P^v_{v',j} = \sum_{\Omega} \frac{\pi \,|< \Psi^{j\Omega}_{fv'E} \mid V(R,r,\theta) \mid \Psi_{f,v} >|^2}{\Gamma_v} \tag{9}$$

II.3 Vibrational Diabatic and Rotational Infinite Order Sudden Approximation (VD + RIOSA)

Since in the final dissociative channels the motion along the stretching coordinate (R) is presumably faster than the rotation of ICl, the rotational infinite order sudden approximation[6] can be applied to produce a more decoupled scheme. The discrete function is given by:

$$\Psi_{f,v,n_s,n_b}(\mathbf{r},\mathbf{R}) = \chi^{(f)}_v(r)\phi^{(f)}_{v,n_s}(R;\theta)F^{(f)}_{v,n_s,n_b}(\hat{\mathbf{r}},\hat{\mathbf{R}}) \tag{10}$$

The functions $\phi^{(f)}_{v,n_s}$ are discrete solutions, at each θ value of the 1D Schrödinger equation:

$$\left[-\frac{\hbar}{2\mu}\frac{\partial^2}{\partial R^2} + V_{vv}(R;\theta) - E_{v,n_s}(\theta)\right]\varphi^{(f)}_{v,n_s}(R;\theta) = 0 \tag{11}$$

where the eigenvalue $E_{v,n_s}(\theta)$ will act as an effective potential for the bending motion.

Then, neglecting the effect of the angular momentum operator on the $\varphi^{(f)}_{v,n_s}(R;\theta)$ function, it results that the $F^{(f)}_{v,n_s,n_b}(\hat{\mathbf{r}},\hat{\mathbf{R}})$ are solutions of the equation:

$$\left[\frac{\hbar^2}{2\mu\bar{R}^2}\mathbf{l}^2 + B_v\mathbf{j}^2 + E_{v,n_s}(\theta) - E_{v,n_s,n_b}\right]F^{(f)}_{v,n_s,n_b}(\hat{\mathbf{r}},\hat{\mathbf{R}}) = 0 \tag{12}$$

On the other hand the continuum solutions are written, in the RIOSA frame, as:

$$\Psi_{f,v,j,\Omega,E}(\mathbf{r},\mathbf{R}) = \chi^{(f)}_v(r)\phi^{(f)}_{v,E}(R;\theta)F^{(f)}_{j\Omega}(\hat{\mathbf{r}},\hat{\mathbf{R}}) \tag{13}$$

where $\phi^{(f)}_{v,E}(R;\theta)$ are continuum solutions satisfying similar outgoing asymptotic conditions than in the 3D case, except the absence of angular functions.

In order to get the linewidths we use now the "Golden Rule" formula

$$\Gamma_{v,n_s,n_b} = \pi \sum_{v',j,\Omega} |< \Psi_{f,v',j,\Omega,E} \mid V(r,R,\theta) \mid \Psi_{f,v,n_s,n_b} >|^2 \tag{14}$$

being straighforward to get the final state distributions.

III The Golden Rule approximation in the time domain

For a total angular momentum $J = 0$ and once the VD approximation has been applied, the rate of decay of an isolated bound level $\varphi_i^{(v)}$ to a single final continuum $\varphi_{fE}^{(v')}$ is expressed through the Fermi's Golden Rule as

$$\Gamma_{vv'}(E) = \pi \mid < \varphi_{fE}^{(v')}(R,\theta) \mid V \mid \varphi_i^{(v)}(R,\theta) > \mid^2 \tag{15}$$

where the potential coupling V represents the quadrature

$$V = < \chi_{v'}(r) \mid V(R,r,\theta) \mid \chi_v(r) > \tag{16}$$

Let us consider the following initial wave packet

$$\mid \Phi(R,\theta,t=0) > = \mid V\varphi_i^{(v)} > \tag{17}$$

We may expand this wave packet in terms of the bound and continuum states of the final Hamiltonian according to

$$\mid \Phi(R,\theta;t=0) > = \sum_k < \varphi_{fk}^{(v')} \mid V \mid \varphi_i^{(v)} > \mid \varphi_{fk}^{(v')} > + \int dE < \varphi_{fE}^{(v')} \mid V \mid \varphi_i^{(v)} > \mid \varphi_{fE}^{(v')} > \tag{18}$$

the second term representing its projection onto the continuum part of the spectrum,

$$\mid \Phi_c(R,\theta;t=0) > = \int dE < \varphi_{fE}^{(v')} \mid V \mid \varphi_i^{(v)} > \mid \varphi_{fE}^{(v')} > =$$

$$= \mid V\varphi_i^{(v)} > - \sum_k < \varphi_{fk}^{(v')} \mid V \mid \varphi_i^{(v)} > \mid \varphi_{fk}^{(v')} > \tag{19}$$

This *continuum* wave packet evolves with time according to

$$\mid \Phi_c(t) > = e^{-iH_f t/\hbar} \mid \Phi_c(0) > = \int dE < \varphi_{fE}^{(v')} \mid V \mid \varphi_i^{(v)} > e^{-iEt/\hbar} \mid \varphi_{fE}^{(v')} > \tag{20}$$

where the dependence on R, θ is implicitely assumed.

Taking into account the energy normalization of continuum wavefunctions, as well as the integral representation of the δ–distribution

$$\delta(E - E') = \frac{1}{2\pi\hbar} \int dt e^{-i(E-E')t/\hbar} \tag{21}$$

we obtain the following expression for the autocorrelation function $< \Phi_c(0) \mid \Phi_c(t) >$:

$$< \Phi_c(0) \mid \Phi_c(t) >= \frac{1}{\pi} \int dE e^{-iEt/\hbar} \Gamma(E) \tag{22}$$

i.e., the rate of decay can be obtained straighforwardly from the Fourier transform of that autocorrelation function,

$$\Gamma_{vv'}(E) = (1/2\hbar)\mathcal{F}[< \Phi_c(0) \mid \Phi_c(t) >] \tag{23}$$

(Note that this expression is going to be evaluated on the energy-shell, as the energy conservation requires).

III.1 Bidimensional approach

We may expand the wave packet in terms of an angular basis set[8]

$$\Phi_c(R, \theta, t) = \sum_j C_j(R, t) F_j(\cos\theta) \tag{24}$$

where,

$$F_j(\cos\theta) = \sqrt{\frac{2j+1}{2}}\, P_j(\cos\theta)$$

with P_j being Legendre polynomials. In this way, the Schrödinger equation

$$i\hbar \frac{\partial \mid \Phi_c(R, \theta, t) >}{\partial t} = H_f \mid \Phi_c(R, \theta, t) > \tag{25}$$

leads to the following set of first order coupled equations,

$$\begin{aligned} i\hbar\frac{\partial C_j(R,t)}{\partial t} &= \left\{ -\frac{\hbar^2}{2\mu}\frac{\partial^2}{\partial R^2} + \hbar^2 j(j+1)\left[\frac{1}{2m} < \chi_{v'} \mid r^{-2} \mid \chi_{v'} > + \frac{1}{2\mu R^2}\right]\right\} C_j(R,t) + \\ &+ \sum_{j'} < F_j \mid V \mid F_{j'} > C_{j'}(R,t) \end{aligned} \tag{26}$$

The autocorrelation function becomes

$$< \Phi_c(0) \mid \Phi_c(t) >= \sum_j < C_j(R,t) \mid C_j(R,0) > \tag{27}$$

and, due to the linearity of the Fourier transform, partial half-widths can be obtain as

$$\Gamma_{vv'}^{(j)} = (1/2\hbar)\mathcal{F}[< C_j(R,t) \mid C_j(R,0) >] \tag{28}$$

from which rotational distributions of the diatomic fragment can be compared with the experiment,

$$d_j = \Gamma_{vv'}^{(j)}/\Gamma_{vv'} = \Gamma_{vv'}^{(j)}/\sum_j \Gamma_{vv'}^{(j)} \tag{29}$$

III.2 One dimensional approach

Since a RIOS approximation has monstrated its reliability in time–independent calculations, we may consider, for each θ–value fixed, the following initial wave packet,

$$\mid \Phi_c(R,t=0;\theta) >= \int dE < \varphi_{fE}^{(v')}(R;\theta) \mid V \mid \varphi_i^{(v)}(R;\theta) >\mid \varphi_{fE}^{(v')}(R;\theta) > \tag{30}$$

and solve the Schrödinger equation

$$i\hbar\frac{\partial \mid \Phi_c(R,t;\theta) >}{\partial t} = H_f^{(\theta)} \mid \Phi_c(R,t;\theta) > \tag{31}$$

obtaining θ–dependent rates,

$$\Gamma_{vv'}(E,\theta) = (1/2\hbar)\mathcal{F}[< \Phi_c(R,t=0;\theta) \mid \Phi_c(R,t;\theta) >] \tag{32}$$

that are going to be averaged on the initial angular distribution,

$$\bar{\Gamma}_{vv'}^{RIOS}(E) = \int_0^{\pi} d\theta D^{in}(\theta)\Gamma_{vv'}(E,\theta) \tag{33}$$

III.2.1 Exponential approximation

The one dimensional approach has been numerically tested[7] on the $Ne\cdots ICl(B^3\pi_{0^+}, v=2) \rightarrow Ne + ICl(B^3\pi_{0^+}, v'=1)$ predissociation for the equilibrium angular configuration, $\theta_{eq} \simeq 140^o$. The autocorrelation function, obtained through wave packet propagation, could be fitted by a two parameter function —with surprisingly good accurancy— as follows

$$< \Phi_c(0) \mid \Phi_c(t) >\simeq \frac{A}{\pi(\beta^2 + t^2/\hbar^2)}(\beta - i\,t/\hbar) \tag{34}$$

leading to a simple exponential behavior of the rate,

$$\Gamma_{vv'}(E,\theta_{eq}) \simeq A\,e^{-\beta E} \tag{35}$$

The square norm N^2, as well as the mean energy $\bar{E}$, of the initial wave packet can then be expressed in terms of the A and β parameters as follows,

$$N^2(\theta_{eq}) =< \Phi_c(0) \mid \Phi_c(0) >= \frac{1}{\pi}\int dE\, \Gamma_{vv'}(E,\theta_{eq}) = \frac{A}{\pi\beta}$$

$$\bar{E}(\theta_{eq}) = \frac{1}{N^2(\theta_{eq})} < \Phi_c(0) \mid H_f^{(\theta_{eq})} \mid \Phi_c(0) >= \frac{1}{\beta} \tag{36}$$

or, equivalently, the rate may be rewritten as

$$\Gamma_{vv'}(E,\theta_{eq}) \simeq \pi \frac{N^2(\theta_{eq})}{\bar{E}(\theta_{eq})} e^{-E/\bar{E}(\theta_{eq})} \tag{37}$$

allowing its calculation by simplely determining both quantities, N^2 and $\bar{E}$, for the initial *continuum* wave packet.

III.2.2 Classical VD approximation

The presence of fast and slow modes in the complex allows, as it was mentioned, to carry out natural separations of motions leading to a great simplification of quantal treatments, for instance, the VD approximation. The situation is just the contrary to apply classical trajectory (CT) based treatments, since the corresponding Hamilton equations need to be integrated using a time step small enough as to accurately describe the fastest motion within the complex, and during an interval of the order of its lifetime. Hence, CT methods, already used to study not only three–atomic[9] but also polyatomic systems[10] –where quantal treatments become almost forbidden– present a great difficulty if the diatomic partner has a high vibrational frequency. Therefore, it would be both interesting and valuable to incorporate the adventageous features of the VD approximation into the classical frame:

a) A larger time step is going to achieve convergency, owing to the disappearing of the highest frequency mode,

b) The total time of integration reduces because only the exiting, final surface is accounted for, and

c) The set of Hamilton equations is also reduced.

In the following, we suggest a possible way to do this starting with the one–dimensional time–dependent VD + RIOS Golden Rule approximation.

Let us define the autocorrelation time, τ, as the elapsed time to get one half of the initial wave packet square norm. Within the exponential approximation, we get

$$|< \Phi_c(0) \mid \Phi_c(\tau) >| = \frac{N^2}{2} = \frac{A}{\pi(\beta^2+\tau^2/\hbar^2)^{1/2}} \tag{38}$$

i.e.,

$$\tau = \frac{\hbar\sqrt{3}}{\bar{E}} \tag{39}$$

and the rate can be rewritten again as

$$\Gamma_{vv'}(E) \simeq \pi\, N^2 \frac{\tau}{\hbar\sqrt{3}}\, e^{-E\tau/\hbar\sqrt{3}} \tag{40}$$

where the dependence on the angle θ is implicitely assumed.

The task now consists in determining τ from CT calculations on the exiting surface. In principle, it seems reasonably to estimate τ as an averaged time expended by the system to reach a distance between the fragments of the order of the outer classical turning point corresponding to the initial state. In fact, to largely exceed that point would imply the disappearing of such state, i.e., the breaking up of the weak bond. However, we have to be careful concerning with the choosing of this limiting distance, because of the dependence on τ of the rate, Eq. (40).

IV Results

IV.1 Time–Independent

In order to apply time–independent as well as time–dependent treatments to the $Ne - ICl$ complex, we have described the PES as an addition of pairwise Morse potentials. The relevant parameters for the ICl interaction were taken from the literature[11] meanwhile the $Ne - I$ and $Ne - Cl$ parameters were determined by a multiproperty fitting procedure[4f]. All of them are listed in Table I of Reference 4f.

In figure 1 we display the final rotational distribution of the ICl fragment produced in the predissociation of the ground van der Waals level

$$Ne \cdots ICl(B, v' = 2, n_s' = 0, n_b' = 0) \longrightarrow Ne + ICl(B, v = 1, j)$$

obtained through the different treatments, from the *exact* 3D line shape calculations, through the diabatic vibrational Golden Rule approximation, up to the DVGR plus RIOSA approximation. They are being compared with the experimental data of Lester and co.[3a,b]. As can be seen, a good agreement with the experiment is achieved by means of all the calculations, showing a pronounced peak at $j \sim 17, 18$. Also, 3D and DVGR results become almost the same, while RIOSA produces a general shift to higher rotational excitations due to the independence of the continuum wave functions on the fragments rotational energy and, in addition, all final rotational channels are open. However, the accord between the RIOSA distribution and the others, and also with the experiment, advises to use this sort of very cheap calculation to get potential parameters, as it was done. In this way, the *dynamical* fitting predicts an equilibrium configuration in which the Ne atom is situated near the Cl end of ICl, at $\sim 140^o$

relative to the $Cl-I$ bond. This fact, added to final state interactions as the fragments separate, can be thought as responsible for the highly inverted rotational distribution found.

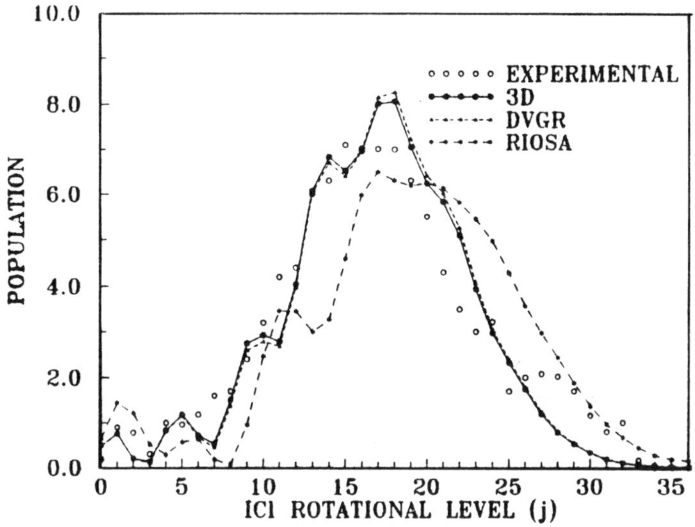

Fig. 1 Comparison of the final ICl rotational distribution obtained with different calculations and the experiment.

IV.2 Time–Dependent

Recently[7], we have performed a time–dependent Golden–Rule calculation on this system for the angular equilibrium configuration , $\theta_{eq} \simeq 140^{o}$, by propagating a *continuum* wave packet on the final surface. The relevant autocorrelation function, from which half–widths can be obtained through Fourier transform, exhibited a simple monotonic decay, all recurrences being washed out since discrete components were taken off the initial wave packet. In Figure 2, we plot the real and imaginary parts of the autocorrelation function obtained through numerical propagation of the *continuum* wave packet. As can be realized, they correspond to a Lorentzian function and a Fano–type one, respectively, showing the feasibility of applying a further exponential approximation, were only the square norm and the mean energy of the initial wave packet are needed.

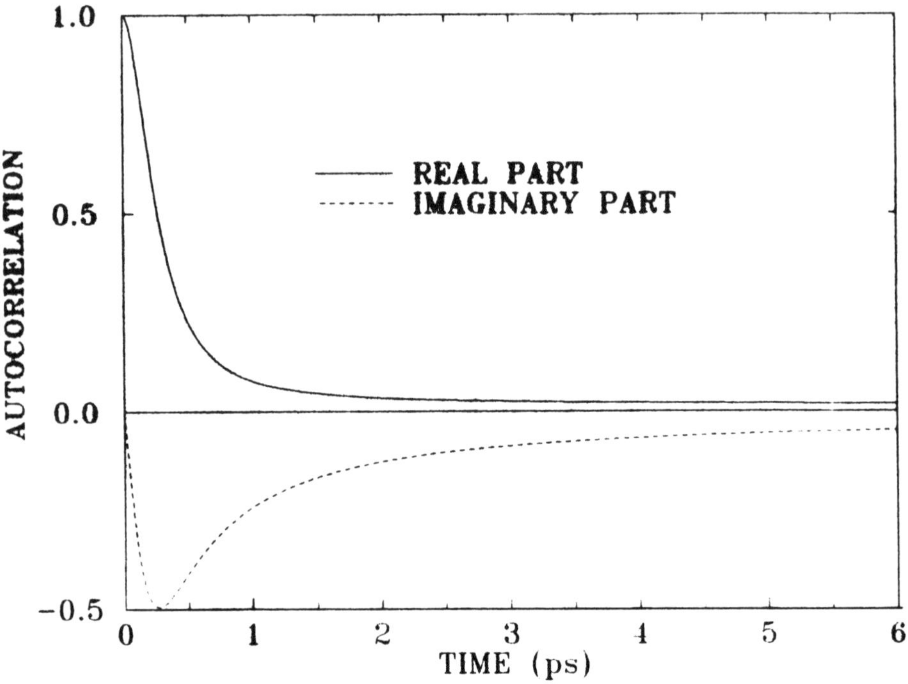

Fig. 2 Time dependence of the autocorrelation function $< \Phi_c(0) \mid \Phi_c(t) >$ for an initially normalized wave packet.

Relaxing the angle, figure 3 displays the half–widths obtained by propagation and those coming after applying the exponential approximation. All of them are calculated for the angular dependent energy that corresponds to the difference between the $v = 2$ and 1 levels of $ICl(B)$ minus the vdW ground energy at each configuration. As can be seen, the exponential approximation works fairly well in all the range except at $\theta > 150^o$, i.e., for configurations where the Ne atom is in the neighborhood of the Cl atom. However, and in order to estimate averaged half–widths, this region has not a great importance since the initial state is peaked at 140°, as it is also plotted in this figure.

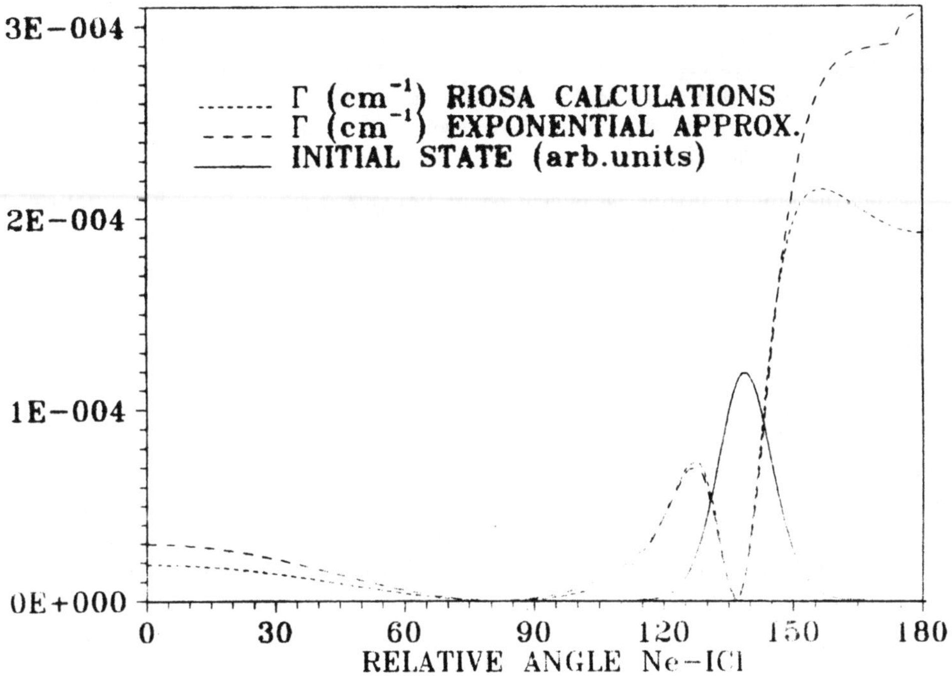

Fig. 3 Angular dependent half-widths. The angular distribution of the initial state is also included.

Finally, and in order to suggest the possible application of a VD approximation in the classical frame, we have carried out simple one-dimensional calculations at the equilibrium angle $\theta_{eq} = 140^o$.

By fitting to the final curve a Morse potential

$$< \chi_{v=1} \mid V(R, r, \theta_{eq}) \mid \chi_{v=1} > \simeq D\left\{e^{-2\alpha(R-\bar{R})} - 2e^{-\alpha(R-\bar{R})}\right\}$$

with the parameters

$$\begin{aligned} D &= 60.7cm^{-1} \\ \alpha &= 1.824\mathring{A}^{-1} \\ \bar{R} &= 4.95\mathring{A} \end{aligned}$$

classical trajectories at a given energy become analytical[12]. In principle, we should have to calculate several trajectories, corresponding to different initial conditions, and consider an average of them. As regard to the energy, we have considered a normal distribution centered at the mean energy of the initial continuum wave packet, i.e., $\bar{E} = 20.43cm^{-1}$, and with a standard deviation that, according to the exponential approximation, corresponds to the same value, $\sigma = \bar{E}$. For a given energy, we can realize that starting at the inner turnig point becomes the same than averaging trajectories for a different initial $Ne - ICl$ distance accounting for its two possible characters, incoming or outgoing. In fact, the time expended by the system from the inner turning point till a fixed limiting distance equals to the averaged time of two trajectories, one incoming and the other outgoing, to reach the same final distance. In this way, we plot in figure 4 the mean trajectory, energy averaged, that is, the mean distance between the fragments versus time.

From the numerical time–dependent quantal calculations, a value of $\tau = 0.45ps$ is obtained for the autocorrelation time. As it was already mentioned, it is reasonable to assume a limiting distance in the neighborhood of the classical outer turning point corresponding to the initial state, that is, RFIN $\simeq 5.40\mathring{A}$. For this value, the classical autocorrelation time is of $\tau_{CLAS} \simeq 0.33ps$, and this disagreement of ~ 25 % leads to a factor of 5 in the corresponding half-width, Eq. (40), between the quantal and classical results, if an energy $E \simeq 140cm^{-1}$ is used. However, we have to consider not only the initial state, neither its product by the coupling, but the projection of such a product onto the continuum states of the final Hamiltonian, as it was done regarding the energies. We estimate the mean position of the initial *continuum* wave packet as $R_m \simeq 5.04\mathring{A}$, its standard deviation getting a value $\sigma_R \simeq 0.72\mathring{A}$. For a limiting distance $RFIN = R_m + \sigma_R = 5.76\mathring{A}$, the value obtained for the classical autocorrelation time is $\tau_{CLAS} = 0.47ps$, very close to the actual one of $0.45ps$, being depicted in figure 4. Of course this new estimate yields the correct half-width value.

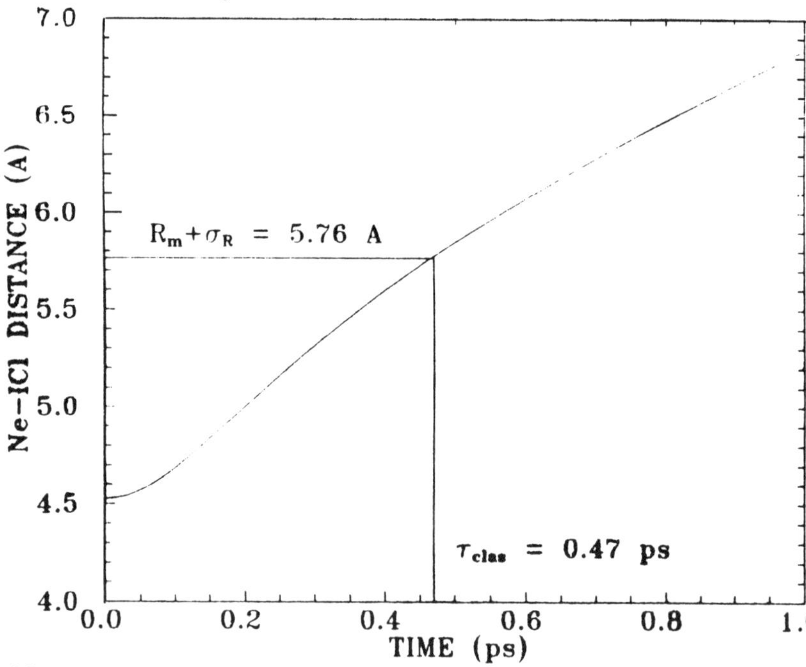

Fig. 4 Mean trajectory on the final potential curve, averaged through a normal distribution of energies around the mean energy of the initial wave packet. That potential was approximated by a Morse oscillator to manage analytical trajectories[12].

V Concluding Remarks

We have presented *exact* 3D line shape calculations on the photofragmentation of the $Ne \cdots ICl(B, v = 2)$ complex, showing a very good agreement with the experimental results. They constitute quantum mechanical benchmark data very useful to test several approximations. In particular, the vibrational diabatic plus Golden–Rule approximation, and the further inclusion of the RIOSA treatment has revealed as a powerful tool to estimate reasonable potential energy surfaces. In addition, a time–dependent version of the Golden Rule scheme opens a different kind of approximations. Starting with this quantal treatments, we suggest a possible way to include the VD approximation in classical trajectory calculations, within a one–dimensional model. Of course, it is not straight to extend the method to higher dimensionality, but this is a challenge needing more investigation. It should allow to treat diatomic molecule–rare gas atoms complexes containing high frequency modes.

References

[1] a) A. van der Avoid, P.E.S. Wormer, F. Mulder and R.M. Berns, Top. Curr. Chem., **93**, (1980) 1.

b)J.H. van Lenthe, J.G.C.M. van Duijneveldt-van de Rijdt and F.B. van Duijneveldt, Adv. Chem. Phys., **69**, (1987) 521.

c) A.D. Buckingham, P.W. Fowler and J.M. Hutson, Chem. Rev., **88**, (1988) 963.

d) J.A. Beswick and J, Jortner, Adv. Chem. Phys., **47**, (1981) 363.

e) D.H. Levy, Adv. Chem. Phys., **47**, (1981) 323.

f) K.C. Janda, Adv. Chem. Phys., **60**, (1985) 201.

g) Faraday Discuss. Chem. Soc., **73**, (1982).

h) "Structure and Dynamics of Weakly Bound Molecular Complexes", NATO ASI Series C: Mathematical and Physical Sciences, Vol.**212**, Reidel, Dordrecht, 1987.

i) "Dynamics of Polyatomic van der Waals Complexes", NATO ASI Series B: Physics, Vol. **227**, Edited by N. Halberstadt and K.C. Janda, Plenum Press, New York, 1990.

[2] For a review see, K.C. Janda and C.R. Bieler in "Atomic and Molecular Cluster" Edited by E.R. Bernstein (Elsevier, Amterdam, 1989).

[3] a) J.M. Skene, J.C. Drobits and M.I. Lester, J. Chem. Phys. **85**, (1986) 2329.

b) J.M. Skene, Ph. D. Thesis, University of Pennsylvania, 1988

c) R.L. Waterland, J.M. Skene and M.J. Lester, J. Chem. Phys. **89** (1988) 7272.

[4] a) N. Halbesrtadt, J.A. Beswick and K.C. Janda, J. Chem Phys. **87** (1987) 3966.

b) J.I. Cline, B.P. Reid, D.D. Evard, N. Sivakumar, N. Halberstadt and K.C. Janda, J. Chem. Phys. **89** (1988) 3535.

c) J.I. Cline, N. Sivakumar, D.D. Evard, C.R. Bieler, B.P. Reid, N. Halberstadt, S.R. Hair and K.C. Janda, J. Chem. Phys. **90** (1989) 2605.

d) N. Halberstadt, O. Roncero and J.A. Beswick, Chem. Phys. **129** (1989) 23.

e) O. Roncero, J. A. Beswick, N. Halberstadt, P. Villarreal and G. Delgado–Barrio, Bull. Soc. Roy. des Sciences de Liège, **3–4** (1989) 227.

f) O. Roncero, J. A. Beswick, N. Halberstadt, P. Villarreal and G. Delgado–Barrio, J. Chem. Phys. **92** (1990) 3348.

[5] R.L. Waterland, M. Lester, and N. Halberstadt, J. Chem. Phys. **92** (1990) 4261.

[6] a) D. Secrest, J. Chem. Phys. **62** (1975) 710.

b) L.W. Hunter, J. Chem. Phys. **62** (1975) 2855.

c) L. Eno and G.G. Balint–Kurti, J. Chem. Phys. **71** (1979) 1447.

d) J.A. Beswick and G. Delgado–Barrio, J. Chem. Phys. **73** (1980) 3653.

[7] P. Villarreal, S. Miret–Artés, O. Roncero, G. Delgado–Barrio, J.A. Beswick, N. Halberstadt, and R.D. Coalson, J. Chem. Phys. **94** (1991) 4230.

[8] a) S.K. Gray and C.E.Wozny, J. Chem. Phys. **91** (1989) 7671.

b) ibid. **94** (1991) 2817.

[9] a)S.B. Woodruff and D.L. Thompson, J. Chem. Phys., **71**, (1979) 376. b) G. Delgado-Barrio, P. Villarreal, P. Mareca, and G. Albelda, J. Chem. Phys. **78** (1983) 280.

[10] a) A. García-Vela, P.Villarreal, and G. Delgado-Barrio, J. Chem. Phys. **92** (1990) 6504.

b) A. García-Vela, P.Villarreal, and G. Delgado-Barrio, J. Chem. Phys. (in press).

[11] K.P. Huber and G. Herzberg in "Molecular Spectra and Molecular Structure IV. Constants of Diatomic Molecules", (Van Nostrand Reinhold, New York, 1979).

[12] M.M. Maricq, J. Chem. Phys. **93** (1990) 2460.

MODE SPECIFIC VIBRATIONAL PREDISSOCIATION DYNAMICS IN FRAGILE MOLECULES

David J. Nesbitt*
Joint Institute for Laboratory Astrophysics, University of Colorado and National Institute of Standards and Technology, and Department of Chemistry and Biochemistry, Boulder, CO 80309-0440 USA

Abstract -- State resolved vibrational dynamics in weakly bound complexes is investigated via high resolution IR absorption spectroscopy in a slit supersonic jet expansion. The tunable frequency near IR source can readily excite the high frequency intramolecular vibrations of the subunits, as well as the low frequency, intermolecular vibrations via combination band excitations built on the high frequency modes. These upper state energies lie far above the van der Waals binding energy limit, and yet predissociate typically on the nsec time scale or longer (i.e. many millions of vibrational periods). This exceptional metastability in such small systems is highly non-statistical, and proves to be a rich arena of vibrational mode mixing and mode-specific predissociation behavior. We select recent examples from our group effort to illustrate these concepts.

I. INTRODUCTION

There has been a long standing and vigorous interest in the chemical physics community in the feasibility of "mode specific chemistry", i.e. the opportunity to influence the products or rate of evolution to products via selective laser excitation of specific vibrational modes. The efficiency of such a process hinges sensitively on the time scale on which this often quite expensive, selective excitation propagates into other modes of the molecular framework. These intramolecular rates can be quite fast, particularly at the levels of vibrational excitation corresponding to severing a chemical bond. Hence it has proven an ongoing challenge to find systems where the vibrational selectivity is sufficiently high to achieve true mode selective behavior. Nonetheless, there have been some spectacular successes in the recent past, for example the work of Crim and coworkers on vibrationally directed photodissociation and chemistry of HOD molecules.[1,2]

Near IR studies of van der Waals or hydrogen bonded complexes offer some particular advantages for investigation of such mode specific vibrational dynamics for

*Staff Member, Quantum Physics Division, National Institute of Standards and Technology.

J. Jortner et al. (eds.), Mode Selective Chemistry, 113–126.

several reasons. First of all, by virtue of the weak coupling, the initial vibrational excitation can be well localized in one or the other of the subunits. Secondly, vibrational excitation of the subunits in the near infrared prepares the complex with energy well in excess of the dissociation limit, even at the single quantum level. Finally, a wide range of metastable intermolecular and intermolecular vibrations can be accessed, permitting a detailed study of predissociation dynamics as a function of initial state wavefunction and eigenenergy.

In this report we present results on mode specific vibrational predissociation in a series of van der Waals and hydrogen bonded complexes. The spectroscopic method used here is direct absorption of a tunable IR laser in a slit supersonic beam. The technique offers high sensitivity, due to the long absorption path lengths (50 cm), and high spectral resolution (0.001 cm^{-1}), due to frequency stabilization of the lasers as well as the natural velocity collimation, and hence reduction of Doppler broadening, that occurs naturally in the slit jet expansion geometry. Consequently, state dependent predissociation rates can be measured in these species via line broadening of individual rovibrational transitions.

II. EXPERIMENT

The experimental apparatus has been described in previous reports,[3,4] to which the interested reader is directed. A pulsed (200-500 μsec), planar supersonic jet (4 cm by 100 μm) is used to prepare weakly bound complexes in an intense, 5 K expansion. The long absorption path lengths accessible in the slit jet geometry are further augmented by White cell optics which multipass the tunable IR laser light up to 50 cm through the dense region of the expansion. Narrow band IR light between 2.2 and 4.2 μm is generated via difference frequency generation of frequency stabilized Ar^+ and ring dye lasers in a 5 cm $LiNbO_3$ crystal, heated to achieve 90° phase matching conditions.

Recently, we have augmented this technique to cover the 1.2 to 2.2 μm region by coherent subtraction of frequency stabilized dye or Ti:sapphire laser from a single mode, diode pumped 300 mW Nd:YAG laser.[5] The combination of these two laser schemes permits us to tune readily over both the "fundamental" ($v=1\leftarrow0$) and "overtone" ($v=2\leftarrow0$) region of HX stretches in these complexes, as well as a manifold of combination band transitions of intermolecular vibrations built on top of these high frequency modes. Differential transient absorption due to molecules in the slit jet is detected on either a pair of liquid nitrogen cooled InSb or room temperature intrinsic Ge photodiodes, carefully balanced to eliminate the common mode amplitude noise on the IR laser. The absorption sensitivity is $10^{-6}/\sqrt{Hz}$, which is near the shot noise level of photon arrival for the 1 mW difference frequency light source. Due to a combination of high absorption sensitivity, long path lengths and sub-Doppler linewidths in the slit jet, the detection sensitivity in the slit jet is $10^8/cm^3$/quantum state for complexes containing absorbers such as HF, HCl, and H_2O.

III. RESULTS

The slit jet spectrometer achieves a sufficient sensitivity to observe a wide variety of weakly bound complexes demonstrating vibrationally mode specific predissociation behavior. In this section we illustrate this with several recent examples from our laboratory.

A. Mode Specific Internal Rotor Predissociation in Rare Gas-HX Complexes

The rare gas van der Waals complexes[6] have recently been intensively studied in the near and far-IR, with the rare gas hydrogen halide species ArHCl/DCl and ArHF/DF being particular foci of attention. In each of these rare gas-HX systems, the potential surface exhibits a double minimum at the two linear configurations, with the H-bonded form (M-H-X) being somewhat more stable than the anti-H-bonded form (M-X-H). Excitation in the near IR on $\nu_1=1\leftarrow0$ prepares the complex at an energy 30-40 times above the dissociation limit. Were this energy rapidly distributed throughout the complex, the system would be anticipated to predissociate on a time scale of a few HX vibrational periods, which would eliminate any hope of rotationally resolved structure.

Dramatically to the contrary, all rare gas hydrogen halide complexes (HeHF/DF,[7] HeHCl,[7] NeHF/DF,[8,9] NeHCl/DCl,[10,11] ArHF/DF,[12,13] and ArHCl/DCl[11,14]) observed in the slit jet exhibit beautifully structured, rovibrational spectra on the $\nu=1\leftarrow0$ bands. Indeed, the widths are apparatus limited ($\lesssim 30$ MHz), and from a Voigt analysis of the sub Doppler slit jet line shapes, predissociative broadening contributes at less than the 1 MHz level (see Fig. 1). This is true even for the extremely weakly bound He complexes,[7] with binding energies of only 7-10 cm^{-1}. In fact, it is likely that the weakness of the binding only enhances the decoupling of ν_1 from the ν_3 van der Waals stretch corresponding to dissociation. An upper limit of 1 MHz corresponds to these excited states surviving for more than 0.2 μsec, which translates into more than 10^8 vibrational periods of the HX. Clearly, the standard RRK assumption of rapid redistribution among <u>all</u> vibrational modes is qualitatively invalid in such small, weakly bound systems. The essential reason for this complete breakdown of statistical behavior lies in the large amount of vibrational energy that must be deposited unimolecularly into rotation and translation of the HX products, or equivalently, the large energy mismatch between the excited ν_1 mode and both the ν_2 and ν_3 intermolecular modes.

In van der Waals systems with sufficiently weak binding energies, predissociation via intermolecular excitation of the ν_2 bend becomes energetically open. For example, in the series of Ar,[12] Ne,[8] and He[7] complexes with HF, the binding energies decrease from 110 cm^{-1} to 35 cm^{-1} to 7 cm^{-1}. As the intermolecular potential surface becomes progressively more shallow it also becomes more isotropic with respect to rotation of the HX subunit. Hence the ν_2 bending vibration can be more accurately described as an

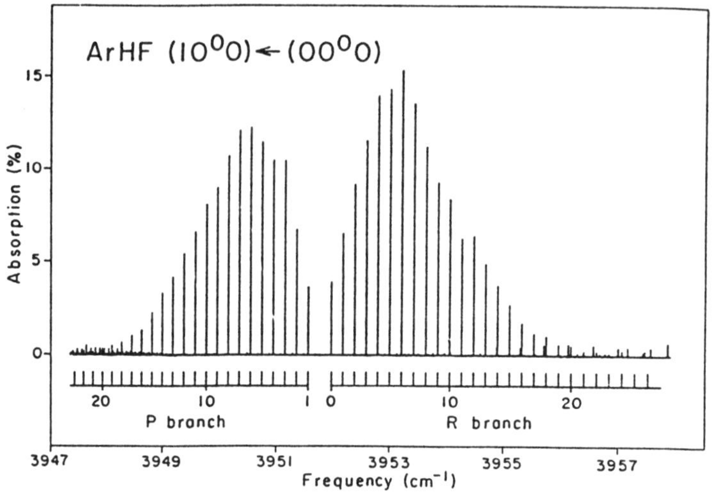

Fig. 1 Raw data absorption spectra in the slit jet expansion for ArHf $v_1=1 \leftarrow 0$. Note the extremely narrow linewidths and extensive rotational structure, despite the fact that the HF excitation energy is $\gtrsim$ 35 times above the dissociation limit.

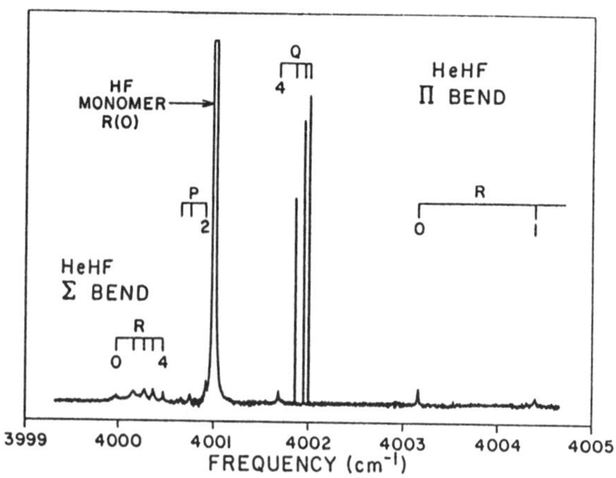

Fig. 2 Direct absorption spectra of HeHF complexes excited to the $j_{HF}=1$ internal rotor level, built-in combination with the high frequency HF stretch. Note the broad linewidths for the Σ component, and instrumentally limited linewidths for the Π^f component (seen in the Q branch). The Π^e internal rotor component (seen in P/R branch) exhibits a strong J dependence to the linewidths due to Coriolis mixing with the rapidly predissociated Σ bend state. The abrupt termination of all transitions at J"=4 is due to centrifugal predissociation of the complex in the ground state. The clear line broadening in the Π^f component of J'=4 (seen in Q(4)) is due to tunneling through a centrifugal barrier in the upper state, i.e., a shape resonance.

internal rotor state with a well defined internal angular momentum, j, and projection along the body fixed axis, m_j. For j=1, the m_j=0 component and the e/f linear combinations of $m_j=\pm1$ components form the Σ and doubly degenerate Π^e/Π^f states, respectively. From this perspective, one can easily explain the trend in ν_2 bend frequencies from 70 cm^{-1} to 42 cm^{-1} to 40 cm^{-1} for Ar to Ne to He complexes, respectively, simply as a smooth progression from a hindered "librator" to a j=1 internal rotor state in HF, for which $E_{rot}=Bj(j+1)=40$ cm^{-1}. Classically, the Σ states correspond to "pinwheel" (i.e. in plane) excitation of the rotor, whereas the two Π states correspond to "helicoptor" (i.e. out of plane) excitation, in a clockwise and counterclockwise sense. With the slit jet levels of sensitivity, all three of these Σ and Π states can be observed via combination band excitation, built on the $\nu_1=1\leftarrow0$ HX stretch fundamental.

More relevantly to the issue of mode specific dynamics, both the Σ and Π internal rotor j=1 states are in excess of the binding energies in NeHF (D_0=35 cm^{-1}) and HeHF (D_0=7.1 cm^{-1}) complexes. Hence, predissociation to form HF(j=0) + M is an energetically open channel, and by selection of the initial state can be investigated as a function of rotor alignment. An example of this is shown from HeHF (Fig. 2), showing the trend of rapid predissociation from Σ states, to the slower, J dependent predissociation from the Π^e states (P/R branch transitions). This trend of J dependent linewidths in the Π^e states can be seen even more clearly in NeHF (Fig. 3).

The J dependent predissociation behavior in the corresponding Π^e manifold is due to rotationally induced mixing (Coriolis) of the rapidly dissociating Σ states into the more predissociatively stable Π states. Clear evidence of this strong Σ-Π Coriolis mixing can be seen in the anomalous line intensities for the different branches. For example, in the Σ band in HeHF (Fig. 2), only the R branch is observed, with the P branch down by $\gtrsim$ 20 fold in intensity!

Even more remarkable, perhaps, are the extremely narrow Π^f transitions observed in the Q branch for J=1,2,and 3 ($\lesssim$ 1 MHz homogeneous broadening) for Q(1)-Q(3) corresponding to >0.2 μsec lifetimes. Quantum mechanically, this is due to conservation of parity in the predissociation event, since the lowest channel with f parity is HF(j=1) + M, which is energetically closed in the absence of V-R,T relaxation of the high frequency HF stretch.

The careful observer will note that Q(4) in HeHF[7] (Fig. 2) is at much lower peak intensity and significantly broadened. This occurs because only J=0,1,2, and 3 states of f parity are bound, with the J=4 state lying slightly above the dissociation limit due to the opening of a pure centrifugal predissociation channel to form HF(j=1) + M, i.e. without any relaxation of the internal rotor. Expressed in the language of a scattering event, Q(4) reflects a "shape" resonance and provides detailed information about the long range part of the potential surface. The absence of the corresponding Q(5) transition is

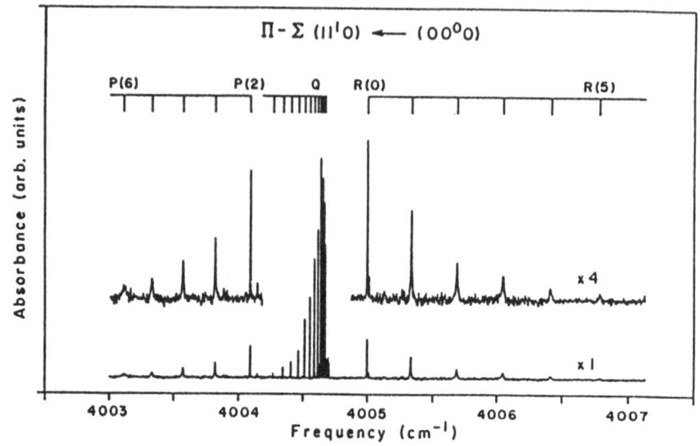

Fig. 3 J dependent predissociation in the Π_{bend} internal rotor state of NeHF. The Π^e component, observed in the P/R branches, predissociates more rapidly with increasing J due to Coriolis mixing with the Σ_{bend} internal rotor state. The Π^f component, on the other hand, is of the wrong symmetry to mix with the Σ_{bend} and exhibits completely instrument limited linewidths. Extra structure in the spectra is due to ^{22}Ne complexes in natural abundance.

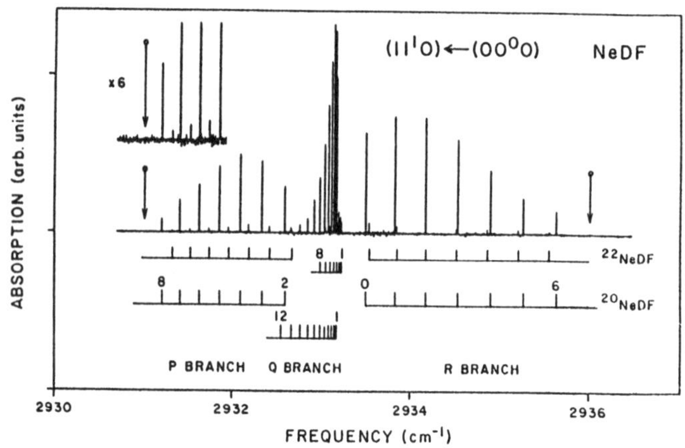

Fig. 4 Π_{bend} internal rotor excitation of NeDF complexes. Since the j=1 DF internal rotor states are below the dissociation limit, the complexes slow to predissociation broadening for low J. Above J'=7, however, the upper states can fall apart via internal rotor and centrifugal predissociation, leading to abrupt cut-off in the spectra.

the result of centrifugal dissociation of the J=5 lower state, also evident in the abrupt cutoff in the R branch of the Σ band at R(4).

B. The Role of Open Vibrational Channels in Photoproducts via Isotopic Substitution Tests

If, as in NeDF,[9] we "slow down" the internal rotor motion via isotopic substitution, the j=1 levels are now energetically bound with respect to predissociation, at least for low values of total J. Consequently, the corresponding Σ, Π^e and Π^f bands now exhibit instrument limited linewidths. However, closer investigation indicates that there is an abrupt cutoff in the spectra at an overall J value due to a combination of end-over-end tumbling and internal rotor energy exceeding the binding energy (Fig. 4). If we examine each of the bound vibrations of e parity (van der Waals stretch, Σ and Π bends) in NeDF, we find the spectra terminate at different values of J but at a common value of energy corresponding to the dissociation limit (see Fig. 5). The Π^f states, once again, remain bound above this energy due to the symmetry constraints on the HF(j=1) + M product state.[15] Since predissociation requires transfer of energy into the van der Waals stretch coordinate, this observation strongly indicates that each of the intermolecular modes is highly mixed on the time scale of the rupturing of the van der Waals bond. From the S/N of lines for the highest J observed, this predissociation lifetime can be estimated to be 100 psec or shorter, which implies intermolecular stretch-bend coupling in excess of 10^{10}/sec at energies near dissociation.

From the studies of NeHF and NeDF one can see the power in isotopic substitution as a probe of predissociation dynamics on the same Born-Oppenheimer potential surface. This trick can be even more powerfully applied in double rotor, 4 atom complexes, where the isotopic substitution can be made away from the site of the initial excitation. Hence, one can study the effects of intermolecular resonances and open product channels explicitly on the predissociation event.

An excellent example system is that of H_2HX and D_2HX, which we have been studying[16,17] as a prototype for oriented H + H_2 reactions initiated in a complex. In H_2HF, the line widths obtained from exciting the HF stretch fundamental have been shown to be extremely narrow[18] (6 MHz). This may be readily attributed to the large amount of energy necessarily released into rotation and translation of photoproducts. There is a mounting body of experimental and theoretical evidence that excess energy in a predissociation event is channeled least efficiently into translation. Hence, the slow predissociation in H_2HF might require up to 7 or 13 quanta of rotation in the H_2 or HF products, respectively. This large "quantum gap" requires correspondingly higher order terms in the potential anisotropy, which for such a weakly bound complex are bound to be small.

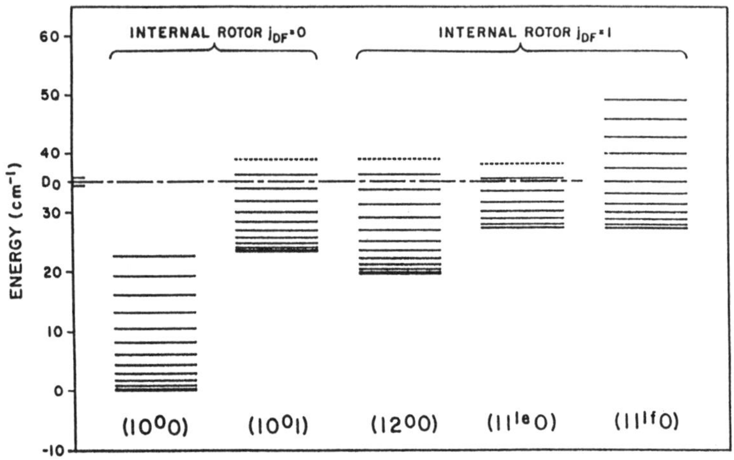

Fig. 5 J dependent predissociation for a series of intermolecular excited states in NeDF. Note that for Π_{bend}, Σ_{bend} and van der Waals stretch levels, the cut-off in the observed J levels occurs essentially as soon as the dissociation channel becomes energetically open. This is evidence for strong mixing of intermolecular vibrations near the dissociation threshold.

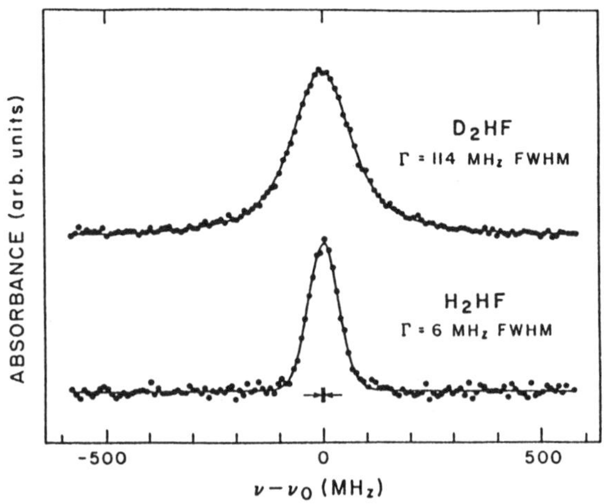

Fig. 6 Predissociative line broadening in H_2HF and D_2HF complexes excited in the HF stretch, obtained from Voigt analysis of the observed line shape. The nearly 20 fold increase in predissociation rate for D_2HF versus H_2HF strongly suggests the importance of an energetically open $D_2(v=1)$ + HF channel.

In the isotopically substituted D_2HF complexes,[19] a similar line of argument would suggest even slower predissociation due to the smaller rotational constant of D_2 versus H_2. On the other hand, there is a near resonant V-V channel to form $D_2(v=1)$ + HF(v=0) that becomes energetically open in HF stretch excited D_2HF complexes due to the lower vibrational frequency of D_2 (2991 cm^{-1}) versus H_2 (4155 cm^{-1}). This would in turn require only 3 or 6 quanta, respectively, in D_2 or HF rotation to accommodate the excess energy. Indeed, experimentally one observes a 20 fold <u>increase</u> in the predissociation rates (see Fig. 6), strongly suggesting that near resonant V-V transfer in the complex is an important accelerating influence on the predissociation dynamics. This is further supported by studies of H_2DF, in which instrument limited linewidths for the DF stretch band are once again observed.

The importance of open vibrational channels in the photoproducts in promoting predissociation is also demonstrated in the series of HCl dimer complexes and isotopic permutations, and is summarized in Table I. If we excite the free HCl stretch in HCl-HCl in which there is no open vibrational channel, the predissociation lifetimes are in excess of 0.1 μsec (i.e., instrument limited). On the other hand, in free HCl stretch excited HCl-DCl complexes, where the DCl(v=1) is now energetically open, the predissociation lifetimes are reduced to 9.6 nsec. A comparable situation is found for excitation of the H-bonded HCl stretch in these two complexes. In HCl-HCl the lifetime is 31 nsec, whereas in DCl-HCl the lifetime has been reduced by an order of magnitude to 3.6 nsec. These results strongly suggest that final state resolved experiments would reveal a high level of vibrational excitation in the DCl photoproducts, and which could be probed via sensitive 2+1 MPI detection schemes.[22] In any event, it is clear that highly mode specific predissociation can occur in these systems, and that the rates can be quite sensitive to the presence of <u>open</u> or <u>closed</u> vibrational channels in the photoproducts.

Table I. HCl Dimer Predissociation Rates

	HCl-HCl[20]	HCl-DCl[21]	DCl-HCl[21]
Free HCl Stretch	τ_{ν_1} > 100 nsec	τ = 9.6 nsec	----
H-bonded HCl Stretch	τ_{ν_2} = 31 nsec	---	τ = 3.6 nsec

C. Predissociation Dependence on Intermolecular Vibrational State Dependence

In addition to isotopic substitution, one can also probe predissociation dynamics on the same Born Oppenheimer potential surface by observing a series of intermolecular vibrational levels. Of course, this is essentially what accounts for the effects described above for the triatomic rare gas-hydrogen halide complexes. These effects can also be quite dramatic polyatomic complexes as well, as demonstrated in the hydrogen bonding OCO-HF system.[23,24] In this system, the H bond proves to be remarkably floppy, supporting a manifold of vibrational levels, approximately 10 cm^{-1} spacing and extremely anharmonic due to large amplitude bending motion in a shallow, "box-like" angular potential. In such a floppy molecule, this bending excitation can lead to significant increases in the rotational constant due to foreshortening onto the intermolecular axis. Studies of the linewidths in this series of states in OCO-HF indicates a systematic increase in predissociation with bending excitation.[23]

This potential link between predissociation and vibrationally averaged geometry of the complex is highlighted in Fig. 7, where a plot of predissociation rate exhibits a remarkably linear dependence on the observed rotational constant. This would suggest that there is some strong correlation between <u>bending</u> the H-bond and <u>predissociating</u> the complex. From the above discussion, this link could be related to the ability of the photofragments to access high rotational levels only from <u>bent</u> transition state configurations. Once again, the capability of final state probing of fragments could provide enormous insight into the predissociation dynamics of these simple but challenging systems.

D. Unimolecular Resonances and the Role of "Doorway" States in Predissociation

It is worth making a distinction between rovibrational channels in the photoproducts that can help limit the amount of energy deposited into other degrees of freedom, and a near resonant vibrational state in the complex that can facilitate the intermolecular transfer of energy into the dissociation coordinate. Both clearly can depend quite sensitively on the vibrational level structure of the complex. However, the role of the so-called near resonant "doorway" state in the pre-dissociation process has been difficult to assess. In this final section we briefly describe two cases where the impact of such a resonant intermediate can be clearly seen in the data.

Our first example is the nitrous oxide-hydrogen fluoride complex,[25] bound in the ONN-HF versus NNO-HF configuration. As shown in Fig. 8, the predissociation rates of the HF stretch excited complexes exhibit a dramatic, non-monotonic on J dependence. Clear "peaks" in the predissociation rate are observed at $J=2$ and $J=16$, that rise up to nearly a factor of 2 over the "baseline" rate seen at many other J values. These "peaks" are unambiguously verified by sampling both P and R branch linewidths that end on a

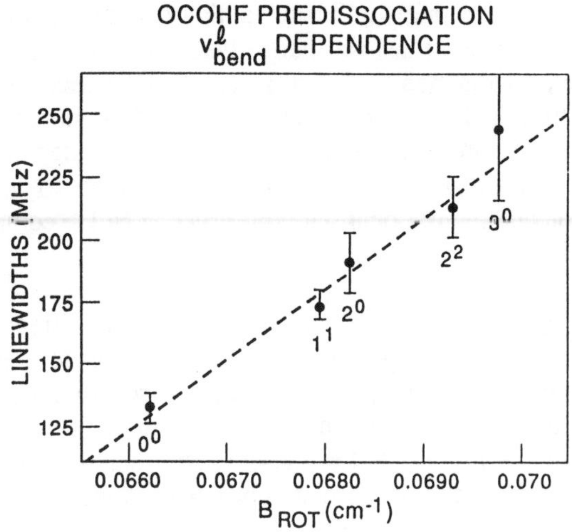

Fig. 7 Geometry dependent predissociation rates in OCO-HF. The plot of Δv versus B_{rot} for the HF stretch excited state indicates a clear sensitivity to the vibrationally averaged geometry of the species.

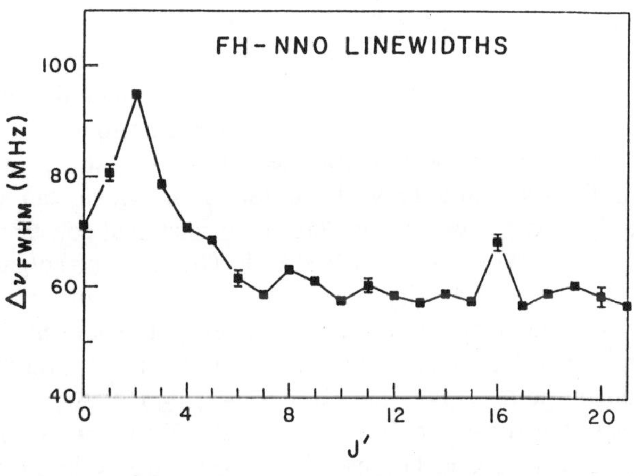

Fig. 8 Nonmonotonic predissociation linewidths for ONN-HF, indicating the resonance features in predissociation continuum reflecting to possible "doorway" states for intramolecular relaxation fo the HF. The J dependence arises from a "tuning" of rotational energy levels in and out of resonance.

common upper J state. The data strongly indicate that there are near resonant states in the complex with slightly different B values and hence that "tune" in and out of resonance with the HF stretch state as a function of J. For a typical difference between rotational constants of different vibrational levels of $<1\%$, the simplest modeling of the low J resonance (J=2) would predict these bright state-dark state couplings to be on the order of 0.01 cm^{-1}. These values are quite comparable, for example, to the magnitude of coupling between intermolecular vibrational states observed in NeDF complexes discussed in Section B.

The J independent "baseline" rate may reflect a more direct predissociation pathway, or perhaps simply the overlapping of many or much broader resonance features in the spectrum. Of course, the "doorway" picture of predissociation dynamics becomes a less useful zero-order perspective in the context of such a broad resonance interaction. But there is clearly evidence in Fig. 8 for these special states in ONN-HF, nearly resonant with the HF stretch, that over a relatively narrow range of energies mediate additional coupling to the predissociation continuum.

Our second example is that of ArH_2O complexes,[26] excited in the $\nu_3=1$ asymmetric stretch manifold. Excitation to the internal rotor state correlating with 0_{00} (here J_{KaKc} denotes the asymmetric top labels for free H_2O) leads to instrumentally narrow linewidths. In contrast, excitation to the Π component of the 1_{01} internal rotor state leads to quite evident predissociative broadening in the P/R branch. However, this broadening exhibits a clear resonance structure characteristic of a crossing between the bright e parity states and an intermediate "doorway" state in the complex that couples more strongly with the continuum.[27] (See Fig. 9.) Interestingly, this broadening effect is conspicuously absent in the Q branch which, as in the rare gas triatomic complexes, can access only the f parity levels. However, the Q branch lines are strongly perturbed at the same J values as the P/R branch, indicating that there exists a close pair of "doorway" levels, one of e and f parity, and only the e levels of which can energetically predissociate. A more detailed analysis of the perturbing rotational constants shows that these levels are not excited in the van der Waals stretch. Given the ortho/para symmetry requirements and the sparse level density of the internal H_2O rotor, this information proves sufficient to identity the doorway state as a symmetric stretch excited vibration of H_2O, and in particular to the Δ component of the 3_{13} internal rotor level. This internal rotor level can rotationally predissociate into H_2O ($\nu_1=1$, 0_{00}) + Ar, releasing barely enough energy (120 cm^{-1}) to break the 100 cm^{-1} van der Waals bond. However, the corresponding f levels remain bound because the lowest f parity internal rotor state ($\nu_1=1$, 1_{11}) is 10 cm^{-1} too high in energy to be a predissociation product.

The physical picture of the "doorway" state in this problem is quite clear. An asymmetric stretch $\nu_3=1$ H_2O molecule undergoes intramolecular vibrational transfer to the $\nu_1=1$ manifold, by virtue of the "half collision" with the Ar atom. This asymmetric to symmetric OH stretch transition requires an asymmetric push on one of the OH groups relative to the other. This develops a torque which speeds up the rotor, thereby leading to internal rotor predissociation as in the lighter rare gas-hydrogen halide species. A more detailed discussion and analysis of this intriguing example of resonance enhanced vibrational predissociation will be presented elsewhere.[27] But this brief description serves to illustrate the role of such "doorway" state phenomena in van der Waals predissociation, in perhaps one of the rare cases where the identity of the "doorway" state can be unambiguously determined.

Acknowledgments

Support for this research from the National Science Foundation (grant CHE 86-05970) through the University of Colorado is gratefully acknowledged. Travel support to attend this symposium was obtained from the National Science Foundation, administered through the Weizmann Institute.

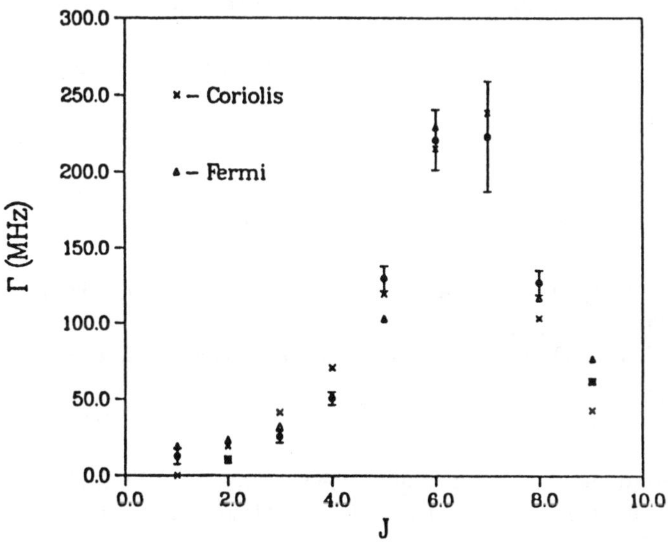

Fig. 9 J dependent linewidths in the Π_{101} $\nu_3=1$ state of ArH_2O. The clear resonance behavior is due to coupling with the Δ_{313} $\nu_1=1$ level. This represents a clear example of intramolecular V-V transfer in a complex, which then acts as a "doorway" state to accelerate the predissociation event.

References

1. R. L. Vander Wal, J. L. Scott, and F. F. Crim, J. Chem. Phys. 92, 803 (1990).
2. A. Sinha, M. C. Hsiao, and F. F. Crim, J. Chem. Phys. 92, 6333 (1990).
3. C. M. Lovejoy and D. J. Nesbitt, J. Chem. Phys. 86, 3151 (1987).
4. A. McIlroy and D. J. Nesbitt, J. Chem. Phys. 91, 104, (1989).
5. A. McIlroy and D. J. Nesbitt, Opt. Lett. (submitted).
6. D. J. Nesbitt, Chem. Rev. 88, 843 (1988).
7. C. M. Lovejoy and D. J. Nesbitt, J. Chem. Phys. 93, 5387 (1990).
8. D. J. Nesbitt, C. M. Lovejoy, T. G. Lindeman, S. V. ONeil, and D. C. Clary, J. Chem. Phys. 91, 722 (1989).
9. C. M. Lovejoy and D. J. Nesbitt, J. Chem. Phys. 94, 208 (1991).
10. C. M. Lovejoy and D. J. Nesbitt, Chem. Phys. Lett. 147, 490 (1988).
11. M. D. Schuder and D. D. Nelson, Jr., and D. J. Nesbitt, J. Chem. Phys. 94, 5796 (1991.
12. C. M. Lovejoy and D. J. Nesbitt, J. Chem. Phys. 91, 2790 (1989).
13. C. M. Lovejoy, J. Dana, and D. J. Nesbitt (manuscript in preparation).
14. D. J. Nesbitt and C. M. Lovejoy, Faraday Disc. 86, 1 (1988).
15. D. C. Clary, C. M. Lovejoy, S. V. ONeil, and D. J. Nesbitt, Phys. Rev. Lett 61, 1576 (1988).
16. C. M. Lovejoy, D. D. Nelson, Jr., and D. J. Nesbitt, J. Chem. Phys. 87 5621 (1987).
17 M. D. Schuder, D. D. Nelson, Jr., E. Kruus, and D. J. Nesbitt (manuscript in preparation).
18. K. W. Jucks and R. E. Miller, J. Chem. Phys. 87, 5629 (1987).
19. C. M. Lovejoy, D. D. Nelson, Jr., and D. J. Nesbitt, J. Chem. Phys. 89, 7180 (1988).
20. M. D. Schuder, D. D. Nelson, Jr., and D. J. Nesbitt, J. Chem. Phys. (submitted).
21. M. D. Schuder, D. D. Nelson, Jr., and D. J. Nesbitt, J. Chem. Phys. (submitted).
22. R. Callaghan, S. Arepalli, and R. J. Gordon, J. Chem. Phys. 86, 5273 (1987).
23. D. J. Nesbitt and C. M. Lovejoy, J. Chem. Phys. 93, 7716 (1990).
24. C. M. Lovejoy, M. D. Schuder, and D. J. Nesbitt, J. Chem. Phys. 86, 5337 (1987).
25. C. M. Lovejoy and D. J. Nesbitt, J. Chem. Phys. 90, 4671 (1989).
26. R. Lascola and D. J. Nesbitt, J. Chem. Phys. (in press).
27. D. J. Nesbitt and R. Lascola (manuscript in preparation).

OBSERVATIONS OF THE LEVEL MIXING ASSOCIATED WITH THE METHYL ROTOR ACCELERATION OF INTRAMOLECULAR VIBRATIONAL REDISTRIBUTION. p-FLUOROTOLUENE

Zhong-Quan Zhao and Charles S. Parmenter
Department of Chemistry
Indiana University
Bloomington, IN 47405, USA

ABSTRACT. S_1-S_0 cold jet fluorescence excitation and dispersed fluorescence spectroscopy have been used to characterize the low-lying (0-410 cm^{-1}) S_1 level structure in p-fluorotoluene, pFT, with either a $-CH_3$ or $-CD_3$ rotor. Special attention has been given to assigning the zero-order character of the states and to discovering interactions between degrees of freedom and between specific states. The appearance of many small forbidden bands in the fluorescence excitation spectrum is accounted for by interaction between internal rotation (i) and overall rotation (r). This i•r interaction also explains the wide variety of rotational contours observed in the spectrum. Abundant specific state-to-state interactions with 1-10 cm^{-1} matrix elements are revealed by the dispersed fluorescence. These state mixings occur by two types of interactions. One is the direct interaction of internal rotation with ring vibrations, i•v. The other is the triple interaction i•v•r. These couplings are very likely the interactions that underlie the acceleration of intramolecular vibrational redistribution when a fluorine in p-difluorobenzene is replaced with a methyl group to make pFT.

1. INTRODUCTION

Intramolecular vibrational redistribution, IVR, has always been a companion of mode selective chemistry because of its role as the Great Frustrater of efforts to achieve mode selectivity. That role is so central to reaction dynamics that a small chemical industry devoted to the explicit study of IVR has arisen. The industry includes those who are attempting to devise strategies to beat the IVR problem. One of those strategies has focussed on chemical modifications that alter IVR characteristics in a way that increases the possibilities of mode selectivity.

Whatever the successes of such strategies, it is clear that chemical modification can indeed alter IVR time scales substantially. A specific example is seen by comparing the IVR time scales in p-difluorobenzene, pDFB, and p-fluorotoluene, pFT. In room temperature experi-

J. Jortner et al. (eds.), Mode Selective Chemistry, 127–137.

ments[1], the replacement of a pDFB fluorine by a methyl group to make pFT increases the IVR rates by roughly a hundred-fold within the region $\epsilon_{vib} \approx 1500\text{-}3000\ cm^{-1}$ of the S_1 manifold.

The IVR change occurs even though the dominant vibrational structure of the S_1-S_0 spectrum and the ring mode frequencies of pDFB are little changed by the internal rotor transplant. Furthermore, the accelerated IVR occurs in energy regions of the S_1 manifold too low for the eight methyl modes to contribute appreciably to the state density. Thus we are seeing an IVR response to the addition of only one more degree of freedom, namely an internal rotation. From this point of view, the big IVR acceleration is perhaps surprising since the internal rotor is nearly free with a rotation barrier[2] of only about 30 cm^{-1}.

In one sense, the IVR change is uninteresting. The IVR goes in the wrong direction for those pursing mode selectivity. It is, however, highly interesting from another point of view. This pronounced IVR response to chemical modification occurs in molecules that offer good opportunity to learn about the reasons for the IVR change. The instruction is provided by the exceptional accessibility of these molecules to S_1-S_0 fluorescence spectroscopy in cold supersonic expansions[2,3].

2. THE SPECTROSCOPIC APPROACH

Attention is focussed on the role of the methyl internal rotation. The pioneering fluorescence spectroscopy of Okuyama, Mikami and Ito[2] (OMI) led to the identification of some low lying S_1 internal rotation levels and to the determination of a 33.7 cm^{-1} S_1 six-fold barrier to internal rotation. This information gives an understanding of the increase in S_1 level density due to internal rotation. The real question, however, concerns how the increased level density is used in coupling S_1 states with different vibrational identities.

One proposition for this coupling was put forward by Moss, Parmenter and Ewing[4]. It is based on the concept of "internal collisions" in which the methyl hydrogens interact during internal rotation with the alpha ring carbons and their hydrogen, due to the overlap of the van der Waals radii. The interaction is highly sensitive to the interatomic distances. These distances are modulated by vibrational motion of the ring so that a natural coupling exists between internal rotation and ring vibrations. The effect of the interaction is to allow coupling between nearby combination states having high vibrational energy and low rotor energy on the one hand, and low vibrational energy and high rotor energy on the other. The theory defines selection rules for such coupling and shows that large matrix elements, ca. 1-10 cm^{-1}, should occur for favorable cases. Another theoretical treatment[5] examines the results of a classical system operating with similar potential interactions. At low energies, extensive and chaotic mixing is found between the internal rotation and vibrational motions.

We have made several preliminary explorations of the interactions by extensions of OMI's S_1-S_0 fluorescence spectroscopy[6]. More recently, an extensive study of the S_1 region $\epsilon_{vib} \approx 0\text{-}410\ cm^{-1}$ has been completed in pFT with both a $-CH_3$ and a $-CD_3$ rotor. Our goal has been to define the positions and zero-order identities of every state in a region where levels are well separated. We then search among those levels for specific cases of mixing. Since the mixing involves states of known zero-order assignments, symmetry arguments can be used to comment on the mixing mechanisms.

If a large enough number of interactions can be catalogued, a generic picture may begin to arise of coupling between molecular degrees of freedom in pFT. These mechanisms surely would persist in regions of higher level densities where IVR occurs. Thus by looking at many individual trees, so to speak, we anticipate that the forest of IVR may be discerned.

The S_1-S_0 spectroscopy we use is fluorescence excitation and dispersed fluorescence. Fluorescence excitation spectra show the level positions. The dispersed fluorescence from single vibrational levels (SVL) in the S_1 state is used to establish their zero-order assignments. These data reveal the rotor levels and the rotor potential energy surface (PES) as well as the low frequency vibrational levels. The information can then be put together to build a picture of the low-lying zero-order S_1 level structure.

OMI led the way in this approach by working out rotor levels and the PES for o-, m-, and pFT. Our additional work has confirmed their basic picture for pFT-h_3.

One of the products from our revisitation of the fluorescence excitation spectra has been a characterization of the rotational contours. Replication of these contours by computer simulation provides information about the interactions of the internal rotation with other degrees of freedom. Much of our interest in the revisitation of the dispersed fluorescence spectroscopy concerns the small extra fluorescence bands that are the hall marks of the mixing among the zero-order levels.

3. RESULTS AND DISCUSSION

3.1 The S_1 Level Structure

Fig. 1 shows fluorescence excitation spectra of pFT-h_3 and pFT-d_3 at about 5 K. The spectra extend about 510 cm^{-1} above the 0^0_0 band and contain the structure from which the rotor levels and the low frequency ring fundamentals can be determined. (Here we are using the term "rotor" for the methyl internal rotation.) As OMI have shown, the structure is exceedingly weak. The pFT-h_3 band marked m^4_1, for example, is about one percent of the 0^0_0 intensity. Analysis of these spectra yields the information from which the S_1 zero-order level structure can be constructed.

Fig. 2 shows the levels in the region $\epsilon_{vib} = 0\text{-}410\ cm^{-1}$ of S_1 pFT-h_3. The levels are constructed from fundamentals, overtones and combinations of low frequency ring vibrations, from rotor states, and from combinations between vibrations and the rotor.

The region encompasses 117 such states. Twelve are pure rotor levels, 20 are pure vibrational levels and the remaining 85 are vibration-rotor combination levels. The presence of the methyl rotor increases the number of levels by about a factor of six relative to those that exist in pDFB.

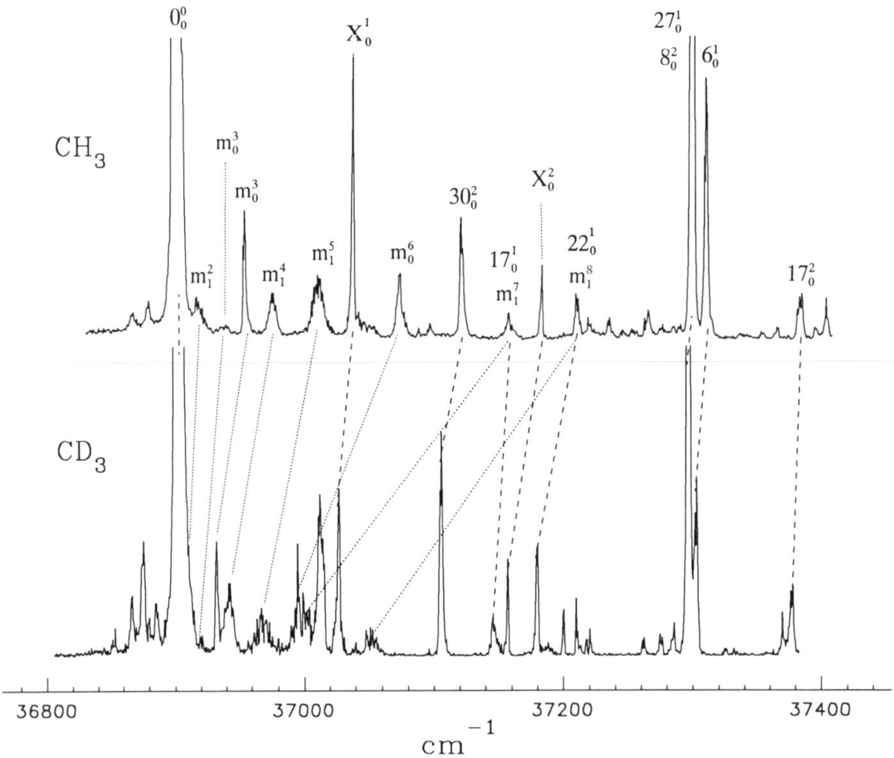

Figure 1. The $S_1 \leftarrow S_0$ fluorescence excitation spectra of jet-cooled p-fluotoluene (pFT) with both a $-CH_3$ and a $-CD_3$ rotor. The rotational temperature of the jet is about 5K. The spectrum of pFT-h_3 (top) has been shifted 27 cm^{-1} to the blue so that its 0_0^0 transition at 36860 cm^{-1} can be aligned with the 0_0^0 transition of pFT-d_3 occurring at 36887 cm^{-1}. The bands connected by dotted lines are pure internal rotor transitions. The pure vibrational transitions are linked by short dashed lines. The assignments in the top spectrum use notation such as m_1^4 to identify transitions between internal rotor states m in a manner analogous to that used for vibrational transitions. Two m_0^3 transitions involve m^3 levels of different symmetry. Some transitions are overlapped.

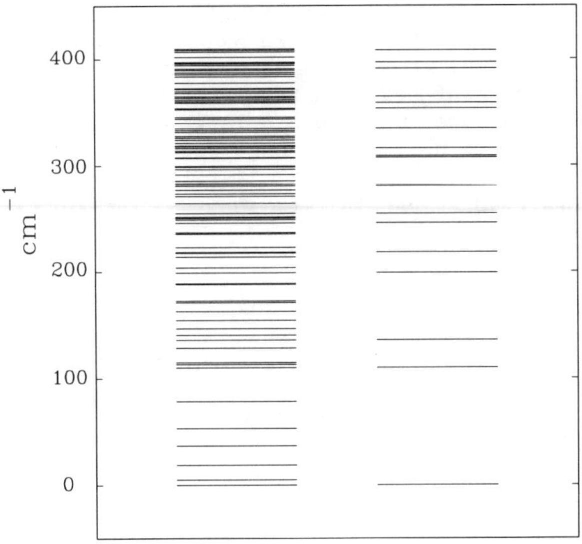

Figure 2. (Left) A display of all S_1 levels in the region 0-410 cm^{-1} of pFT-h_3. The levels are comprised of ring fundamentals, overtones and combinations, plus methyl internal rotation levels and their combinations with the ring modes. (Right) The selection of levels that involve only ring vibrations.

The display in Fig. 2 also shows those states for which the internal rotor quantum number m is zero. These m=0 states are, of course, purely vibrational levels. These levels would be the only ones present in pDFB. The large increase in level density brought about by adding the single degree of freedom of an internal rotation is seen graphically by comparing the two level structures in Fig. 2.

Our calculated S_1 internal rotation barrier in pFT-h_3 is 32 cm^{-1}, in agreement with OMI. The barrier in pFT-d_3 is 25 cm^{-1}. The rotor energy levels are not too different from those of a free methyl rotor for which E_{rot} (cm^{-1}) $\approx 5m^2$ where m is the rotor quantum number 0, 1, 2... The two-fold degeneracy of the free rotor states is split, in principle, in our hindered rotor for the levels m = 3, 6, 9... The splitting is large enough for us to see clearly only for m = 3.

The energy levels in Fig. 2 encompass eight ring fundamentals: ν_{30}' = 110 cm^{-1}, ν_x' = 136 cm^{-1}, ν_8' = 199 cm^{-1}, ν_{17}' = 255 cm^{-1}, ν_{22}' = 308 cm^{-1}, ν_{16}' = 359 cm^{-1}, ν_{27}' = 397 cm^{-1} and ν_6' = 408 cm^{-1}. Activity in all of these modes is observed in the fluorescence excitation spectrum.

Mode X remains a mystery. Its frequency in the ground electronic state is ν_x'' = 183 cm^{-1}. It is not a mode of a van der Waals complex, since its bands do not disappear when the molecular beam temperature is increased. It belongs to pFT and not an impurity because SVL fluorescence obtained by pumping its $S_1 \leftarrow S_0$ bands contain all the pFT frequencies. Comparisons of pFT-h_3 with pFT-d_3 spectra confirm that it is a ring vibration not to be confused with rotor states. Thus some ring mode, presumably out-of-plane, undergoes a large change in S_1 and S_0 when pDFB is transformed to pFT. The change occurs in both pFT-h_3 and pFT-d_3. This mode must lie above 410 cm^{-1} in pDFB since the lower fundamentals have all been accounted for.

3.2 Interactions Among Degrees of Freedom

In Fig. 1, only the 0^0_0, m^5_1, m^7_1, m^6_0, 30^2_0, X^2_0, 8^2_0, 6^1_0 and 17^2_0 bands are zero-order symmetry allowed. Thus, about half of the observed bands in our fluorescence excitation spectra are forbidden. Their presence indicates interaction among different degrees of freedom. We can easily find, by symmetry arguments, the simplest interactions that would make those transitions allowed.

Many of the forbidden transitions involve only the internal rotation states. The simplest interaction that turns on these transitions is one commonly seen in microwave spectroscopy[7], namely that between the internal rotation and overall rotation. We designate it as i•r interaction. This interaction alone makes all of the internal rotation transitions allowed. A second group of transitions involve vibrational levels only. The Herzberg-Teller vibronic interaction (e•v) can account for all but two of the forbidden transitions, i.e. 30^1_0 and 17^1_0.

Fig. 3 shows the rotational contours of the m^3_0 and m^4_1 bands and our attempts to fit these contours with an approximate treatment. Both m^3_0 and m^4_1 band contours should be perpendicular type in a symmetric top approximation (pFT is a near prolate top). The observed contours, however, are markedly different. The differences are typical. The contour of almost every band is unique. Some of this variability can be seen in the lower resolution spectra of Fig. 1.

Crude simulations of the contours have been made with a symmetric top program[8] using calculations that limit J" populations to $0 \leq J" \leq 8$. The essential aspect of the program centers on its inclusion of the i•r interaction in establishing the overall rotation eigenvalues and eigenfunctions.

Trials with such simulations produce the broad range of contours observed in the excitation spectra. Furthermore, the simulated contours generally replicate the actual contours of specific bands, at least to the extent that such approximate calculations could do so. The two examples shown in Fig. 3 are characteristic.

Our contour simulations reveal convincingly that the dominant influence toward establishing the contour is the i•r interaction. This interaction is also the one that gives intensity to the forbidden rotor transitions.

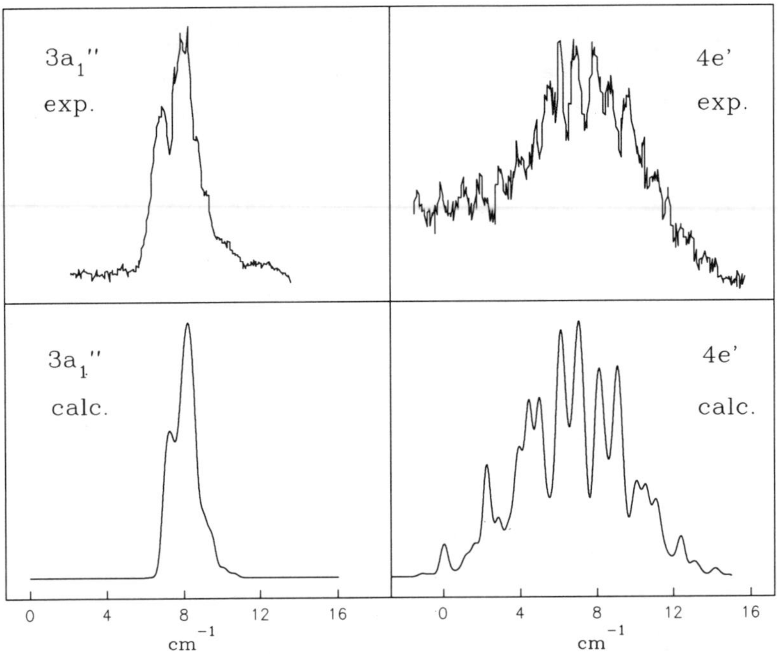

Figure 3. The experimental (top) and calculated (bottom) rotational contours of the m_0^3 and m_1^4 bands in the $S_1 \leftarrow S_0$ fluorescence excitation spectrum of pFT-h_3. The experimental and simulated spectral resolution is 0.5 cm^{-1}.

3.3 State-to-State Level Mixing.

SVL dispersed fluorescence spectra have been obtained by pumping almost every absorption band in the spectra of Fig. 1. The exercise has been completed for both the -CH_3 and -CD_3 molecules. For many bands, several spectra have been obtained with excitation tuned to different parts of the rotational contour.

The most subtle aspect of the dispersed fluorescence is the most interesting. We have searched among the weak structure to find fluorescence bands that reveal mixing among the S_1 levels. The search is only possible when fairly secure zero-order assignments of the S_1 level structure have been made.

Fig. 4 shows an example with pFT-d_3. We have generated dispersed fluorescence by pumping an absorption region where the rotor transitions m_0^6 and m_1^7 overlap. (The nomenclature identifies the rotor level m in the S_1 and S_0 states analogous to the nomenclature used for vibrational transitions.) As we tune excitation within the absorption band contour, we can emphasize pumping of m^6 or m^7. In fact, the emission from m^7 almost disappears when the low frequency part of the absorption band contour is pumped.

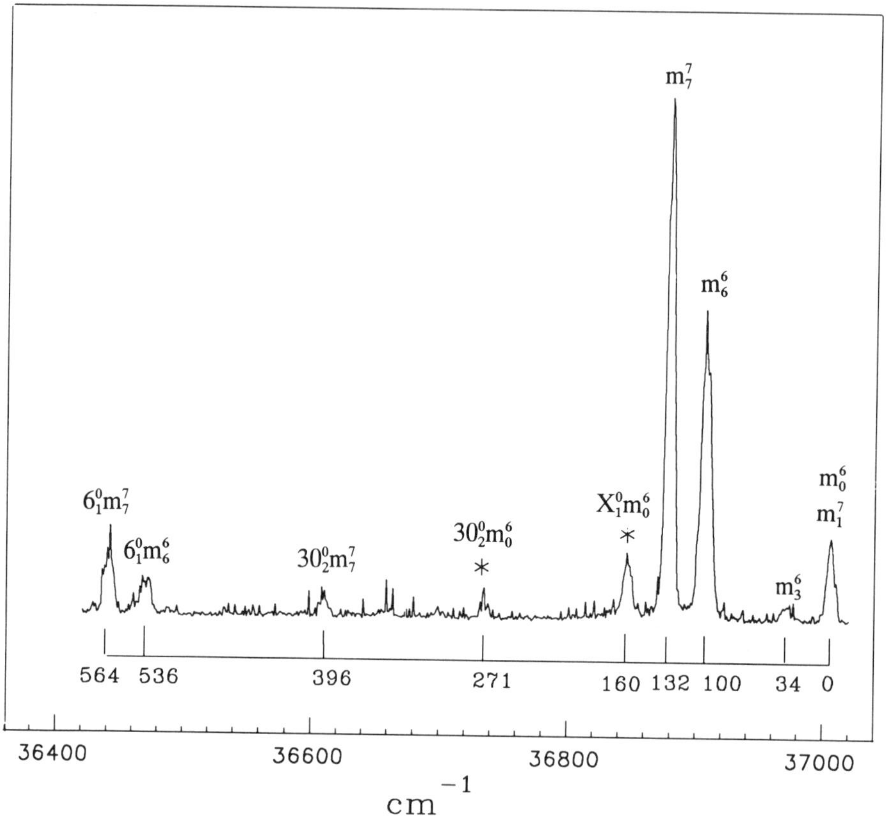

Figure 4. The dispersed fluorescence spectrum from the rotor levels m^6 and m^7 of S_1 pFT-d_3. The spectrum was generated by pumping the overlapping m_0^6 and m_1^7 transitions. The upper cm^{-1} scale shows displacement of bands from the pump position. The lower cm^{-1} scale shows absolute positions. The bands marked with an asterisk occur on account of mixing of the pumped level m^6 with nearby levels. The intensity at the pump position includes scattered exciting light.

The strongest fluorescence bands are the expected Δm=0 transitions, m_7^7 and m_6^6. Other bands are those expected for the less intense structure from m^6 and m^7. Most are combination bands involving some vibrational transition occurring with the Δm=0 components of the rotor.

Two small bands displaced 160 cm^{-1} and 271 cm^{-1} from excitation are special. The displacements fit the energy of the S_0 levels X_1 and 30_2, respectively, indicating emission band assignments $X_1^0 m_0^6$ and $30_2^0 m_0^6$. One could propose that these bands are part of the normal weak structure expected from the zero-order level m^6. If that were true, however, then we should expect to see some related bands that are much brighter. Consider, for example, the $X_1^0 m_0^6$ transition. If it were a normal band, then the band $X_1^0 m_6^6$ with similar activity in ν_x, but with no quantum change in the rotor should occur with much greater intensity. There is, however, no such band at the expected 260 cm^{-1} displacement. Thus we must seek another explanation. The more successful account is an S_1 level mixing between m^6 and X^1 levels. Thus the transition $X_1^0 m_0^6$ gains intensity because it can be considered as being an X_1^1 transition from the mixed component of the pumped state.

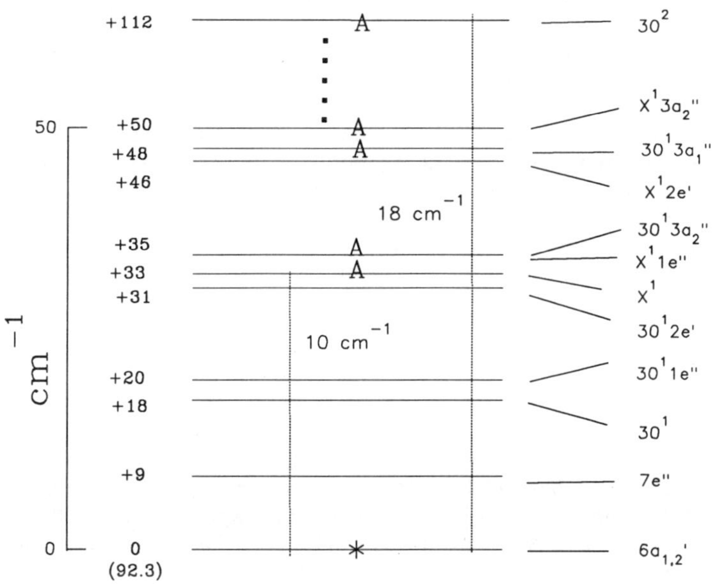

Figure 5. The energy level structure of S_1 pFT-d_3 within 50 cm^{-1} of the m^6 level at 92.3 cm^{-1}. The zero-order level identities are given to the right. The level 30^2 is off-scale, lying 112 cm^{-1} above m^6. The two mixings observed in the spectrum of Fig. 4 are marked with dotted lines along with their coupling matrix elements. The five levels marked "A" are symmetry allowed to mix with m^6 by i•v interaction. Only two levels occur within 50 cm^{-1} below m^6. One is at -25 cm^{-1}; the other is at -50 cm^{-1}.

Fig. 5 shows a segment of the S_1 pFT-d_3 level structure near the level m^6. It shows the mixing that is detected by the spectroscopy. The mixing is of course highly selective among the neighbors. We can inquire using symmetry arguments as to how many of these levels could be coupled with the m^6 level by the i•v coupling scheme proposed by Moss et al.[4] The answer is that five among the 13 levels shown in the diagram could couple. Two of these are the levels actually involved in the mixing.

The i•v interaction occurs naturally in pFT, but as we have observed in the rotational contour analysis, so does the i•r interaction. It is therefore probable that all three degrees of freedom interact to generate the wonderful acronym, i•v•r. Symmetry arguments tell us that six of the 13 levels appearing in the diagram of Fig. 5 could be mixed with the level m^6 by such interaction. In our example the two detected couplings can be simply turned on by i•v interaction. Of course, they can also be turned on by i•v•r interaction. In other examples, we found many state mixings that can occur only by i•v•r interaction.

We can direct our inquires about state mixing in pFT to a simple question. What is the simplest mechanism that can produce the observed state mixing? The answer is always found in i•v and i•v•r interactions. If it is not the former, the latter always fulfills the symmetry requirement.

Another aspect of level interactions concerns magnitudes. Estimates of the coupling matrix elements can be obtained from our fluorescence spectra in the standard way. They generally prove to be large. The 10 cm^{-1} and 18 cm^{-1} coupling matrix elements of the two interactions displayed in Fig. 5 are typical. Of course, we can only detect interactions with matrix elements on the order of the level separations. Since the levels are separated by at least a few cm^{-1} in this low energy region of the S_1 manifold, we can not see the weak interactions.

4. SUMMARY

Evidence for substantial interactions among degrees of freedom and between specific energy levels appear everywhere in the S_1-S_0 spectroscopy that involves the lowest 410 cm^{-1} in the S_1 manifold. The fluorescence excitation spectra contain many forbidden bands whose presence can be explained by interactions between internal rotation and overall rotation, i•r, or by vibronic interactions, e•v. The level interactions revealed by the fluorescence spectroscopy are ubiquitous and strong. They can be accounted for by either the mechanism of internal rotation-vibration interaction, i•v, or by coupling of that interaction with overall rotational motion, i•v•r.

An impression of the ubiquitous nature of level mixing in pFT can be obtained by looking at the fraction of observed levels involved in mixing. In pFT-h_3 we examined by dispersed fluorescence 36 S_1 levels in the range 0-410 cm^{-1}. Fifteen of them are involved in state mixing. The fact that almost half of the levels examined are mixed is even more impressive when it is understood that the coupling matrix elements exceed 1 cm^{-1} in every case.

If we extrapolate this picture to higher S_1 regions, the implications are clear. The methyl internal rotor not only creates a huge increase in level density, but it also provides extremely effective routes for coupling that density.

5. ACKNOWLEDGEMENTS

We greatly appreciate interactions with Lian Yu and Professor T.A. Miller who have provided the symmetric top rotational contour program containing provisions for coupling between internal rotation and overall rotation. This work was supported by the National Science Foundation Grant CHE88-17860.

6. REFERENCES

1. C.S. Parmenter and B.M. Stone, J. Chem. Phys. 84, 4710 (1986).
2. K. Okuyama, N. Mikami, and M. Ito, J. Phys. Chem. 89, 5617 (1985).
3. A.E.W. Knight and S.H. Kable, J. Chem. Phys. 89, 7139 (1988).
4. D.B. Moss, C.S. Parmenter, and G.E. Ewing, J. Chem. Phys. 86, 51 (1987).
5. C.C. Martens and W.P. Reinhardt, J. Chem. Phys. 93, 5621 (1990).
6. R. Longfellow and C.S. Parmenter, Hyperfine Interactions 38, 625 (1987) J. Chem. Soc. Faraday Trans. 2, 84, 1499 (1988).
7. H.D. Rudolph and H. Seiler, Z. Naturforsch. A20, 1682 (1965); P.N. Ghosh, J. Mol. Spectrosc. 138, 505 (1989); *ibid.*, 142, 295 (1990).
8. The program was provided to us by L. Yu and T.A. Miller.

DYNAMICS AND SPECTROSCOPY OF HETEROCLUSTERS

NARDA BEN-HORIN, DAR BAHATT, UZI EVEN AND JOSHUA JORTNER
School of Chemistry, Tel-Aviv University,
Tel-Aviv 69978, Israel

ABSTRACT. Spectroscopic interrogation of the structure, energetics and nuclear dynamics of mass-selected 9,10dichloroanthracene-rare-gas heteroclusters was conducted by resonance two-color two-photon ionization in conjunction with constant energy molecular dynamics (MD) simulations of the spectroscopic observables. MD simulations of the nuclear motion identified a hierarchy of isomerization processes, which involve correlated atomic motion, surface melting, side crossing, wetting-nonwetting and rigid-nonrigid transitions. The isomerization phenomena could not be identified by the temperature dependence of the spectral shifts, homogeneous linewidths and ionization potential shifts, while radiative lifetimes are more sensitive to the nuclear dynamics. The size dependence and isomer specificity of the spectral shifts, homogeneous linebroadening, ionization potential shifts and radiative lifetimes of these heteroclusters could be well accounted for by the combination of microscopic theories for the spectroscopic observables with MD simulations. The intriguing phenomenon of spectral line narrowing with increasing the heterocluster size is attributed to the reduction of inhomogeneous broadening, which originates from the combined effects of isomer specificity of spectral shifts and of homogeneous line-broadening effects.

1. Introduction

The nature of phase transformations in clusters is of considerable interest for the elucidation of the interrelationship between energetic, thermodynamic and dynamic size effects in finite systems [1-3]. It was predicted in 1909 by Pawlow [4] that the melting temperature of droplets increases with increasing size. This prediction has been borne out experimentally in 1976 by Buffat and Borel [5] for metal clusters supported on a substrate. Extensive computer simulations of neat clusters and of heteroclusters encompassing "weakly interacting" van der Waals systems [6,7] and "strongly interacting" ionic [8] and metallic systems in three dimensions as well as in two dimensions, reveal that "melting" of clusters may be synonymous with isomerization between a multitude of isomers, with the underlying processes exhibiting marked size effects. The gross features of the size and temperature dependence of cluster isomerization crucially depend on the following chemical properties:

(1) The intracluster interaction strength and range. "Weak" short-range interactions in rare-gas clusters, e.g., Ar_{13}, are manifested by a frightfully large number ($\geq$988) of isomers [7], while the "strong" long-range interactions in alkali halide clusters, e.g., $Na_4C\ell_4$ involve a molecular isomerization between well-characterized isomers [8].

J. Jortner et al. (eds.), Mode Selective Chemistry, 139–152.

(2) The cluster composition. The characteristics of isomerization are distinct for neat clusters and for heteroclusters. In particular, embedding a probe molecule, e.g., an inorganic (SF_6 or SiF_4) [9,10] or an organic (aromatic) molecule [11-15] in or on a rare-gas cluster modifies the structure, energetics and density of vibrational states of the heterocluster.

This paper is devoted to the spectroscopic interrogation of cluster size effects and isomer-specific effects on spectroscopic observables, i.e., $S_0 \rightarrow S_1$ electronic spectra, ionization potentials and radiative lifetimes of heteroclusters, consisting of a large aromatic molecule solvated by rare-gas atoms. We have studied the heteroclusters of 9,10dichloroanthracene (DCA) with rare-gas atoms. The choice of DCA as the probe-molecule rests on the unique photophysics of its S_1 state. For the electronic origin of S_1 the $S_1 \rightarrow T_1$ intersystem crossing channel is closed in the bare molecule and is not enhanced by the external heavy atom effect [13]. Consequently, (i) no cluster fragmentation is induced by the formation of a vibrationally excited triplet state, and (ii) pure radiative lifetimes of DCA-rare-gas heteroclusters can be explored. The experimental facts were confronted with the results of constant energy molecular dynamics (MD) computer simulations [15], elucidating cluster size effects and isomer-specific features of the structure, energetics, nuclear dynamics and isomerization of large finite systems.

2. Experimental Procedures

The experimental method is based on the interrogation of the $S_0 \rightarrow S_1$ electronic excitation of heteroclusters, which are generated and cooled in a supersonic expansion, utilizing the resonance two-color two-photon ionization (R2C2PI) method in conjunction with mass-spectrometric detection [14]. One laser induces the $S_0 \rightarrow S_1$ excitation of a size-selected cluster, while a second laser ionizes the cluster near threshold to prevent fragmentation. Introducing a controlled time delay between the pulses of the two lasers enabled us to measure mass-selective or even isomer-selective lifetimes. In our experiments [14] the cluster beam was generated in a three-chamber differentially pumped vacuum system. The two-color resonance ionization was achieved by two dye lasers that were simultaneously pumped by an excimer laser. The lasers beams were mildly focused to the center of the ionization volume, and were aligned for counter propagation with the cluster beam, resulting in a relatively long ionization cylinder. The ions masses were analyzed in a reflectron time-of-flight mass-spectrometer [14].

3. Spectroscopy

The $S_0 \rightarrow S_1$ R2P2CI spectra of mass-resolved DCA·Ar_n (n = 1-55) heteroclusters (Figure 1) exhibit the following major features:

(1) Small clusters (up to n ~ 9) exhibit sharp and structured spectra, which can be attributed to the electronic origin of one or more isomers, together with intermolecular vibrational excitations of the nuclear motion of the rare-gas atoms.

(2) At n ≈ 10 the spectral structure is washed out, and the spectra become broad and structureless.

(3) For DCA·Ar_{20} the linewidth reaches a maximum, and for larger clusters, in the range n = 20-35, the linewidth decreases with increasing n.

(4) For clusters with n > 35, the linewidth and the spectral shift exhibit a very weak size dependence.

Heteroclusters of DCA·Kr_n (n = 1-20), portrayed in Figure 2, show a very similar spectral behaviour. In addition:

(5) For DCA·Kr_n a sudden line-narrowing is exhibited at n = 17, due to the appearance of a structured spectrum for n = 17-20.

These patterns are summarized in figure 3, which describes the size dependence of the spectral shift $\delta\nu$ and linewidth Δ of these heteroclusters.

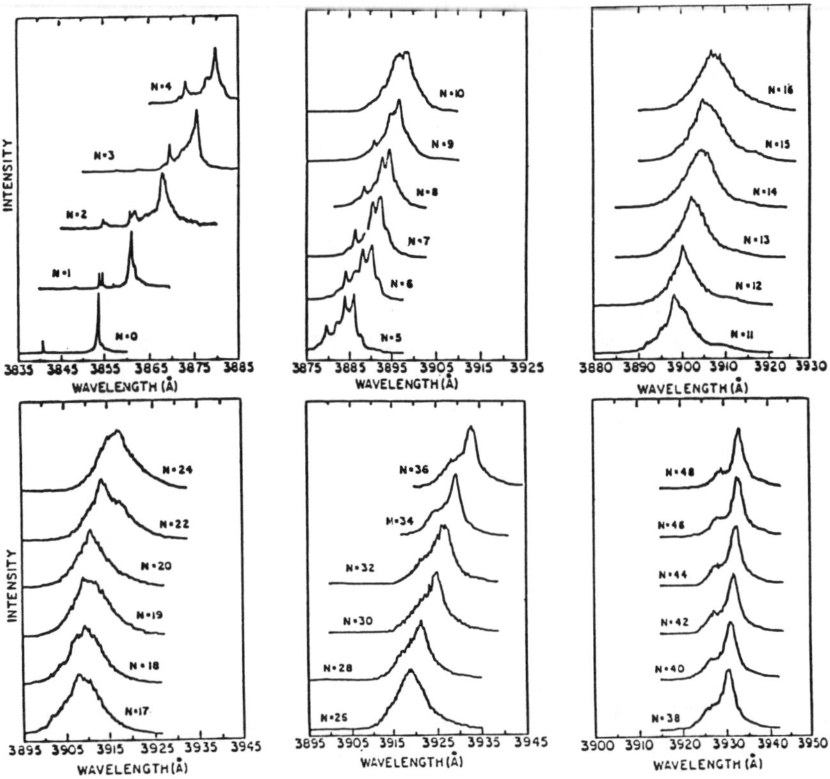

Figure 1. Mass-resolved R2P2CI spectra for the $S_0 \rightarrow S_1$ transition of DCA·Ar_N (N = 1-48) heteroclusters.

4. Simulations of Structure, Energetics, Nuclear Dynamics and Isomerization

Constant energy classical MD simulations [15] utilized a fifth-order predictor-corrector method for the integration of the equations of motion. The potential surfaces were specified using Lennard-Jones pairwise atom-atom potentials. For each of the heteroclusters studied the first stage in the calculation involved an equilibration stage. Subsequently, trajectories were generated in the ground electronic state for a time of 1.0-1.5 nsec. A pictorial presentation of the DCA·Kr_{10} heterocluster with a (7|3) structure (Figure 4) reveals temperature induced structural changes. Here the (n|m) structure denotes n+m atoms in the first solvation layer with n atoms above the molecule microsurface and m atoms below it. With increasing the temperature to 48K, the two-dimensional (2D) structure is converted into a three-dimensional structure, revealing the precursor of a wetting-nonwetting transition. Energetic and structural ground state properties were simulated over a broad temperature range for the energetically most stable isomers.

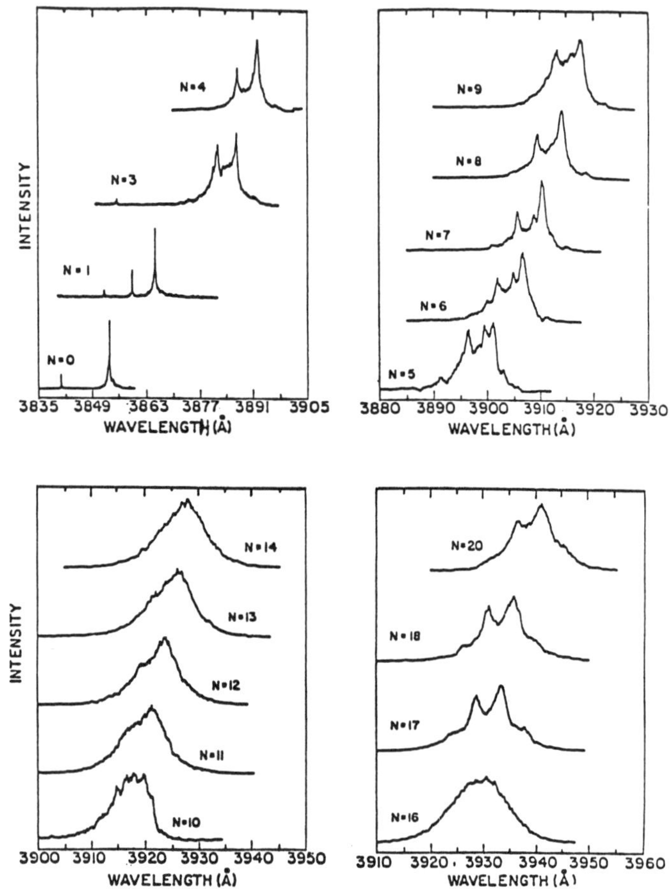

Figure 2. Mass-resolved R2P2CI spectra for the $S_0 \rightarrow S_1$ transition of DCA·Kr_N (N = 1-20) heteroclusters.

Various isomerization processes can be monitored from the temperature dependence of the fluctuations in the structural parameters of the (n|m) structure, e.g., the normalized standard deviation (nsd) σ of the distance between the rare-gas atoms to the centre of mass of the molecule, the cartesian components of σ along the long parallel axis (σ_x), along the short parallel axis (σ_y) and along the perpendicular axis (σ_z), the corresponding nsd $\sigma^{(n)}$ or $\sigma^{(m)}$ for the (n|0) and (0|m) substructures and their cartesian components, and the nsd δ of the rare-gas-rare-gas distance. In Figure 5 we present some examples of the temperature behavior of the fluctuations in the structural parameters. For DCA·Kr_5, the break in σ_y at 15K, which corresponds to the parallel motion of the rare-gases, manifests the onset of correlated free surface motion. The surface melting is manifested by a simultaneous break in σ_y, σ_x and δ. An example for this pattern is exhibited for DCA·Kr_{16} around 30K (Figure 5). For this cluster size this temperature is also the onset of side crossing, as can be seen from the break in σ_z, which reflects the perpendicular motion of the atoms relative to the molecule. The wetting-nonwetting transition was identified by the structural changes monitored along trajectories at different temperatures. A pictorial presentation of this transition is given in Figure 4.

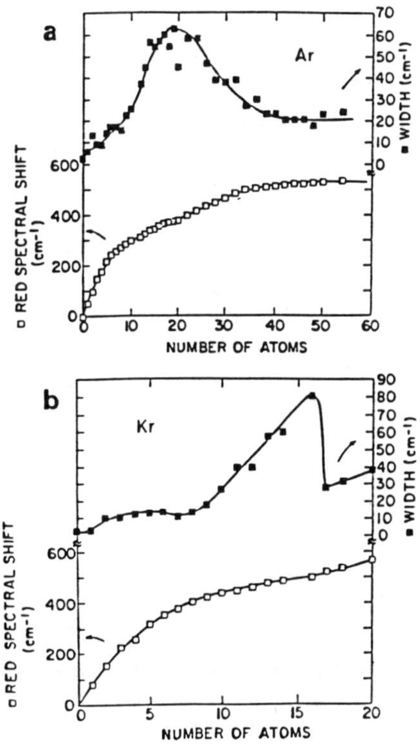

Figure 3. Size dependence of the spectral shifts $\delta\nu$ and linewidth Δ of the prominent feature in the spectra of DCA·Ar_n and DCA·Kr_n heteroclusters.

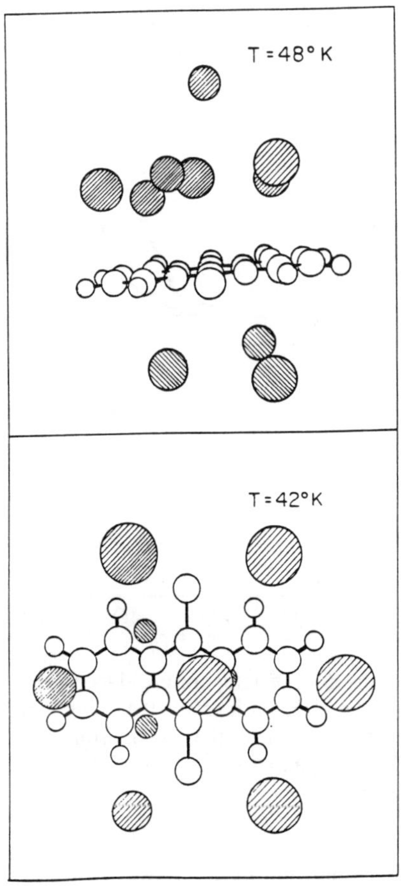

Figure 4. Structure of the (7|3) isomer of the DCA·Kr_{10} heterocluster obtained from MD simulations (short-time averaging of 100 psec). Note the thermally induced 2D → 3D structural change at 48K.

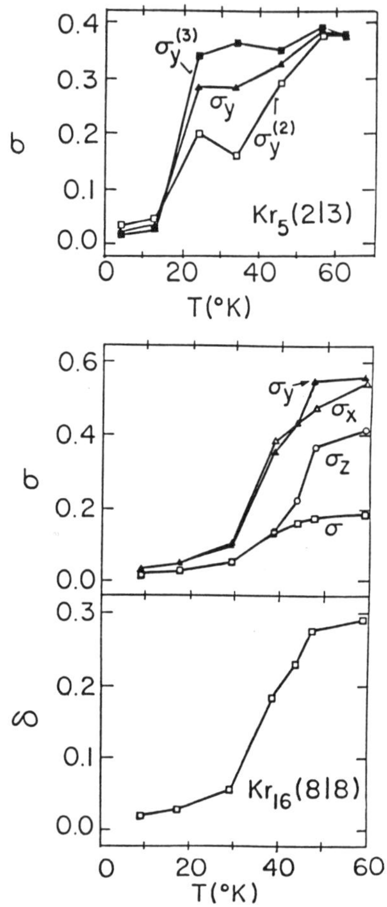

Figure 5. The temperature dependence of the fluctuations in the structural parameters for the (2|3) isomer of $DCA{\cdot}Kr_5$ and for the (8|8) isomer of $DCA{\cdot}Kr_{16}$.

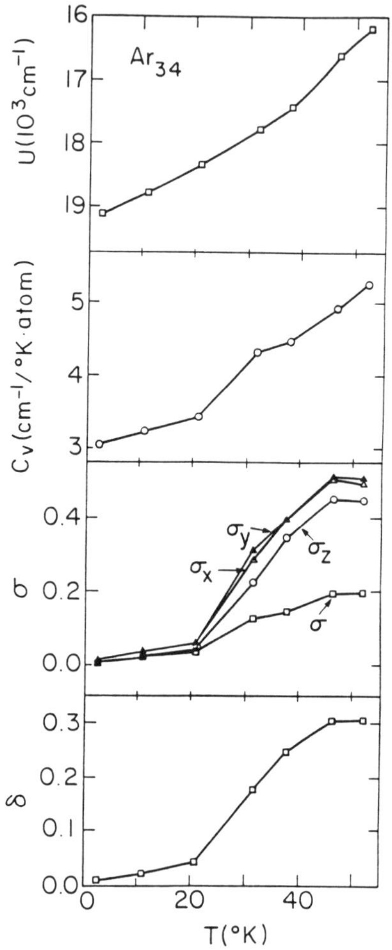

Figure 6. The temperature dependence of the potential energy (U), the specific heat (C_v) and the fluctuations in the structural parameters σ and δ for $DCA{\cdot}Ar_{34}$.

The following heterocluster isomerization phenomena were identified from MD structural and nuclear dynamic data in the $DCA{\cdot}A_n$ (A = Ar,Kr) heteroclusters:

(i) Correlated motion of several atoms sitting together on one side of the molecule.
(ii) Surface melting, which involves uncorrelated 2D motion of rare-gas atoms parallel to the microsurface of the probe molecule.
(iii) Side crossing of atoms sitting above one side of the molecule to the other side of it.
(iv) Wetting → nonwetting transition (or 2D → 3D transition).
(v) Rigid → nonrigid transition.

This identification of isomerization phenomena is in accord with the original work of Leutwyler et al. on carbazole-rare gas heteroclusters [12] and subsequent work on tetracene-rare-gas systems [16]. From the MD simulations we infer that:

(1) Distinct isomerization processes are exhibited for different cluster sizes. Correlated free surface motion was observed for small heterocluster substructures up to a substructure of 4 atoms. Surface melting was observed for small and medium-sized clusters in the range n = 5-16. Side-crossing was observed for all the examined cluster sizes with n = 5-34, 2D → 3D transitions were observed only for n ≥ 10.

(2) Several isomerization processes, with different temperature onsets, can be exhibited for a single heterocluster composition.

(3) For very large clusters, e.g. n = 34, the surface melting, side-crossing and 2D → 3D transitions are indistinguishable, merging into a single "(nearly) rigid → nonrigid" transition (Figure 6). For this cluster size all the structural parameters, as well as the heat capacity, manifest a break in the same temperature. No break is observed for the caloric curve, due to the absence of an energy gap in the potential energy surface.

(4) The temperature onset for a given type of isomerization tends to decrease with increasing cluster size. This is shown in Figure 7, which corresponds to the "phase diagram" for $DCA \cdot Kr_n$ heteroclusters. This pattern indicates that the potential barriers for these isomerization processes decrease with increasing the coverage of the finite molecular microsurfaces. This behaviour is qualitatively different from what was documented for other finite systems, i.e., neat rare-gas clusters [6] and metal clusters supported on an (infinite) substrate [5], which exhibit an overall increase of the melting temperature with increasing cluster size. We expect that for larger heteroclusters the trend of the size dependence of the isomerization temperature onset will reverse, converging to the melting temperature of the corresponding infinite system.

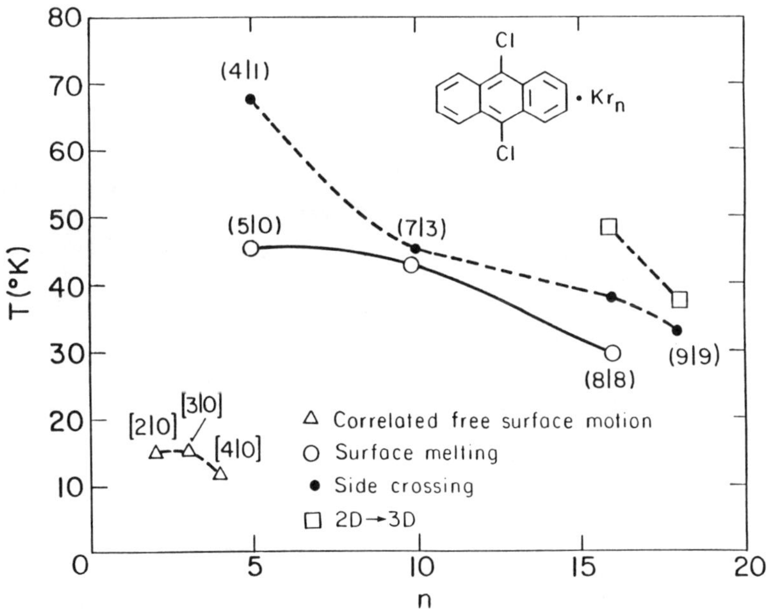

Figure 7. The size dependence of the temperature onsets of isomerization processes of $DCA \cdot Kr_n$ heteroclusters.

5. Simulations of Spectroscopic Observables

Simulations of spectroscopic data for the first spin allowed $S_0 \rightarrow S_1$ electronic transition [17], for the radiative lifetime of the S_1 state [18] and for the ionization of the aromatic molecule [19] in rare-gas heteroclusters were performed using quantum mechanical expressions for the relevant spectroscopic attributes at a fixed nuclear configuration, which were summed over the entire trajectories on the potential surface of the ground electronic state. The spectral shifts, $\delta\nu$, and homogeneous linewidths, Δ, of the absorption spectra of DCA·A_n were calculated from the classical first and second moments [15]. The excited-state potential surface was calculated using the semiempirical theory of Shalev and Jortner [17] for dispersive stabilization. The energetic shifts in the ionization potentials relative to the bare molecule were calculated from the electrostatic stabilization energy of the ionic state relative to the neutral ground state [19]. The change $\Delta\tau_r$ of pure radiative lifetimes of the aromatic molecule in the heterocluster relative to the bare molecule was expressed in terms of electrostatic (transition-monopoles induced-dipoles) interactions formulated by Shalev and Jortner [18]. What can be inferred from the interrelationship between structure, nuclear dynamics, and electronic vibrational excitations of heteroclusters? The simulated spectral shifts and linewidth reveal that [14]:

(i) The temperature dependence of the spectral shift and of the homogeneous linewidth for a given initial structure of a heterocluster does not show pronounced changes, which may mark the onset of isomerization processes. This can be demonstrated by the behaviour of the spectral shift and width with increasing temperature for DCA·Kr_{16} (Fig. 8), which do not reflect the surface melting and side-crossing which start at 32K, neither the 2D → 3D transition at 48K.

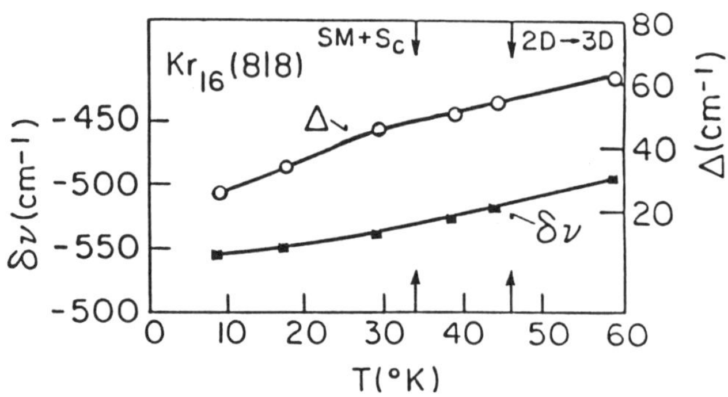

Figure 8. The temperature dependence of the spectral shift and spectral linewidth of the DCA·Kr_{16} heterocluster with the initial low-temperature structure of (8|8), as obtained from MD simulations. The temperature onsets of surface melting (SM), side crossing (SC) and wetting-nonwetting (2D → 3D) isomerization are marked by arrows.

(ii) The spectral shifts exhibit an isomer specificity which is size-dependent. The energetic splitting between isomers of a given composition is larger for Kr than for Ar. This observation can be rationalized in terms of the dependence of the spectral shift on the rare-gas polarizability.

(iii) The homogeneous spectral widths at a constant temperature reveal a weak isomer and size specificity.

The ionization potential shifts (Fig. 9) reveal the following features:
(i) The overall change of the IP shift, δI, over the temperature range 10-60K is small, being in the range of 5%-10%.
(ii) The temperature dependence of the IP shift and width does not manifest any break at the onset of surface melting.
(iii) The values of δI for n = 10 and n = 16 exhibit a weak break at ~45K which mark the onset of side-crossing, while δI for n = 18 reveals a pronounced break at ~40K, marking side-crossing and 2D → 3D isomerization. The temperature dependence of the width does not reveal these isomerization processes.

The pure radiative lifetime exhibits a marked structure sensitivity and isomer-specificity [15,18]. To emphasize further the sensitivity of the lifetime to the geometry of the heterocluster, we note that the dominating terms in $\Delta\tau_r$ can be either negative or positive, depending on the position of the rare-gas atom relative to the molecular transition moment [18]. In particular, the occupation of peripheral sites by the rare-gas atom results in its reduction, while for atom positions above the molecular plane the lifetime is increased. Combining the structural parameter calculations with the lifetime simulations (Fig. 10), the following observations emerge:

(i) For the n = 5 and n = 10 heteroclusters the parallel one-sided correlated free-surface motion and surface melting bring the atoms to peripheral regions, resulting in the shortening of the lifetime with increasing temperature. In contrast, for the n = 16 and n = 18 heteroclusters, the side-crossing and the 2D → 3D transitions cause an increase of the lifetime with increasing temperature.
(ii) For DCA·Ar_{34}, the single rigid → nonrigid transition at 22K is characterized by the descent of atoms from the second to the first solvation layer. The resulting structures have therefore considerably shorter lifetimes. The temperature dependence of the lifetime of DCA·Ar_{34} reveals a sharp break at 22K, which marks this transition.

From this analysis it is apparent that the radiative lifetime data are expected to serve as a useful spectroscopic probe for the identification of structural isomers and for the interrogation of isomerization processes in heteroclusters. However, from the experimental point of view, this approach is limited in some cases by the relatively small changes of this observable at the isomerization onsets.

6. Size Dependence, Isomer Specificity and Inhomogeneous Broadening

From the foregoing analysis we conclude that the spectroscopic observables studied herein are rather limited in the interrogation of heterocluster isomerization processes. On the other hand, we infer from our simulations that these spectroscopic observables do show a marked isomer-specificity. We have measured the spectral shifts, ionization potential thresholds, as well as the radiative lifetimes for the rare-gas clusters of DCA, and confronted the experimental results with the calculated values. Figures 11 and 12 demonstrate the fit between experimental and calculated values for distinct isomers of the spectral shift and the ionization potential shift. The attribution of spectral features in the excitation spectra to different isomers, according to the appropriate calculated spectral shift, was supported by the same procedure for the ionization potential shift and the radiative lifetime of these isomers. In Figure 13 we show the same fit for the pure radiative lifetime, which reveals good matching between the experimental and calculated values for different isomers, corresponding to various cluster sizes. The three observables give the same identification of isomers, so that together they help to achieve a complete and coherent picture of spectroscopic interrogation of size dependence and isomer specificity of heteroclusters.

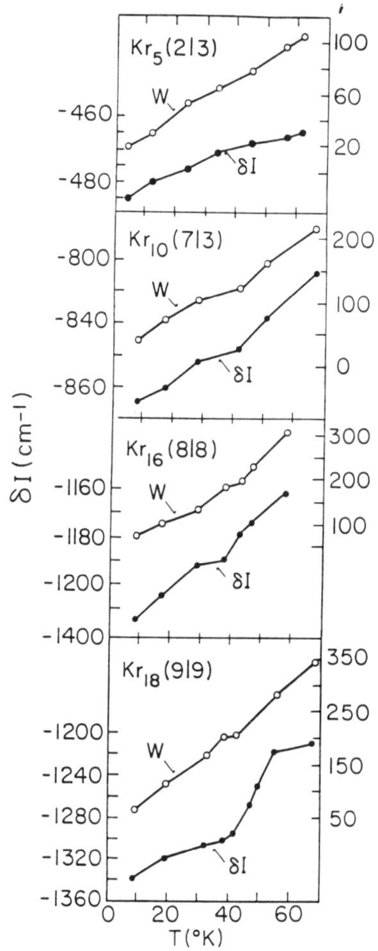

Figure 9. Temperature dependence of ionization potentials δI and the energetic widths of the ionization potential threshold(s) for several DCA·Kr_n (n = 5,10,16,18) heteroclusters.

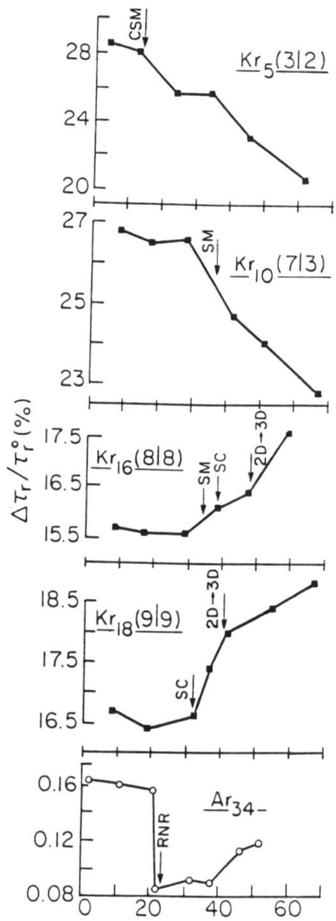

Figure 10. Temperature dependence of radiative lifetimes (relative changes $\Delta\tau_r/\tau_r^0$ of the cluster radiative lifetime relative to the value τ_r^0 for the bare molecule) for several DCA·Kr_n (n = 5,10,16,18) and DCA·Ar_{34} heteroclusters. The temperature onsets of correlated surface motion (CMS), surface melting (SM), side crossing (SC), wetting-nonwetting (2D → 3D) and rigid-nonrigid (RNR) isomerization are marked by arrows.

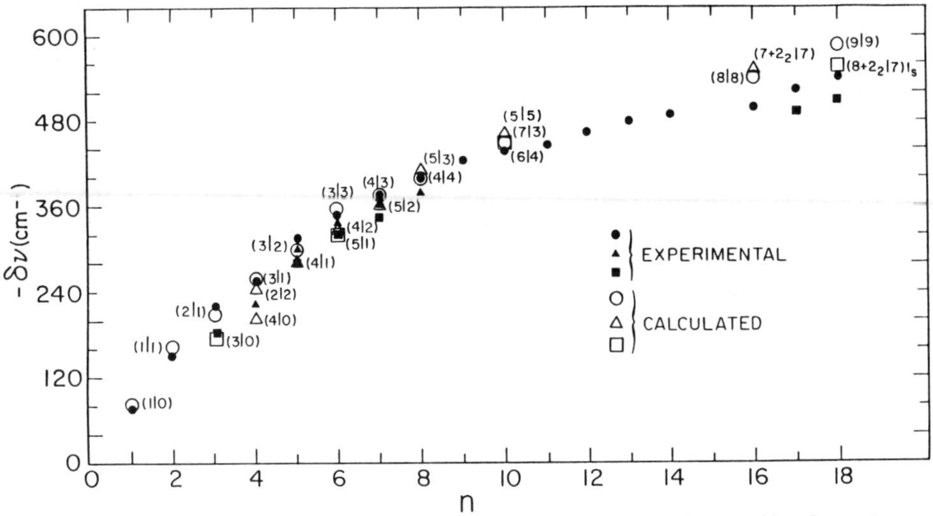

Figure 11. Size dependence and isomer specificity of the spectral shifts for DCA·Kr_n heteroclusters. The experimental $\delta\nu$ data are: $\delta\nu$ for the most prominent spectral feature (•) and $\delta\nu$ for the secondary spectral features (Δ and ■) in the order of decreasing intensity. The calculated $\delta\nu$ data are: $\delta\nu$ for the energetically most stable isomer (O) and $\delta\nu$ for less stable isomers (Δ) in the order of decreasing energetic stability. The isomer structures corresponding to the calculated data are marked by (n|m).

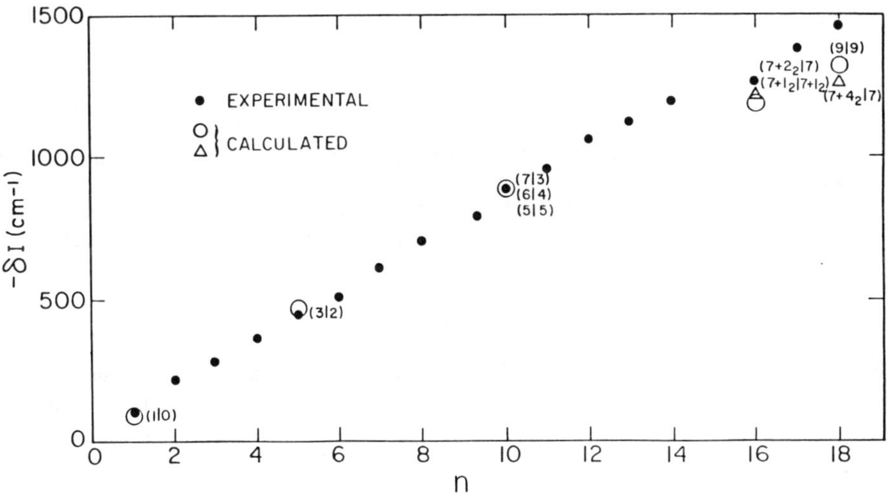

Figure 12. Size dependence and isomer specificity of the ionization potential shifts for the DCA·Kr_n heteroclusters. Notation for experimental and calculated data as in Figure 12.

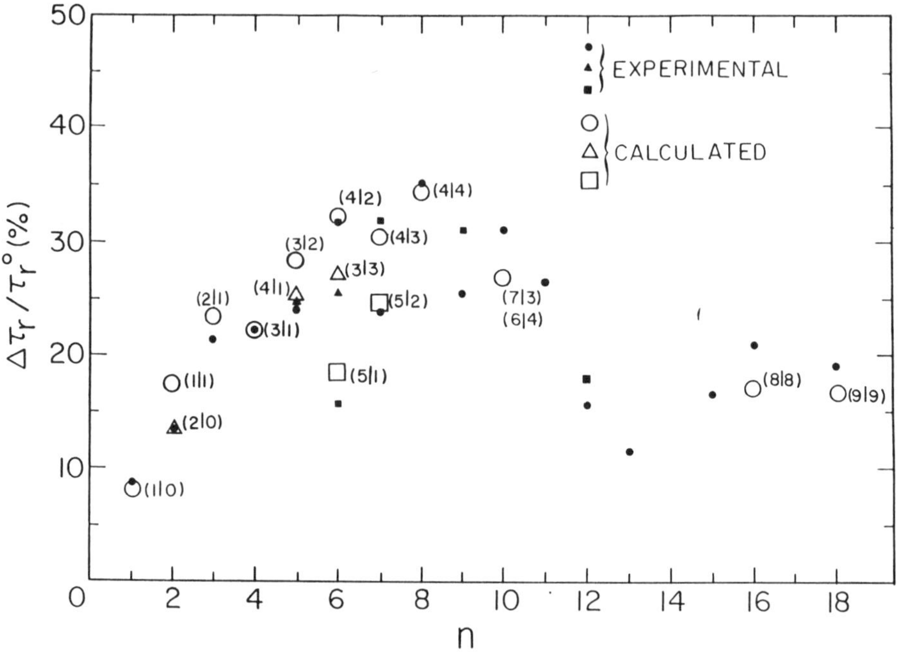

Figure 13. Size dependence and isomer specificity of radiative lifetimes. $\Delta\tau_r/\tau_r^0$ corresponds to the change in the pure radiative lifetime relative to its value τ_r^0 for the bare molecule. Notation for experimental and calculated data as in Figure 12.

These approaches constitute clues for the deconvolution of the complex inhomogeneously broadened spectral features of heteroclusters. We consider the composite inhomogeneously broadened spectral features of a given size, which are determined by the delicate balance between two factors:

(1) The isomer specificity of the spectral shifts, which is size-dependent. The spectral shift has the largest sensitivity for atoms sitting above the centre of the molecule; then for atoms sitting in the periphery, while for atoms sitting in the second layer, there is almost no sensitivity.

(2) The homogeneous spectral width of a distinct isomer.

In Figure 14 we portray the size dependence of the maximum energetic splitting between the stable isomers for DCA·Ar_n heteroclusters, together with the homogeneous spectral width of a distinct isomer. The homogeneous linewidth is taken according to the very rough estimate of the cluster experimental temperature. This plot can qualitatively explain the spectra of these clusters. Small clusters, e.g., n = 5, are cold and thus their experimental homogeneous linewidth is rather narrow (~ 4cm^{-1}). The difference in the spectral shifts between the possible isomers (3|2) and (4|1) is around 14 cm^{-1}. Therefore, a well resolved, structured spectrum is obtained. Increasing the cluster size to n = 10, the spectral shifts of the energetically stable isomers (7|3), (6|4) and (5|5) are almost the same, with a difference of only 2 cm^{-1}, and thus the spectrum is expected to show a homogeneous broadened band. Indeed, the experimental linewidth of about 20 cm^{-1}

agrees with the calculated spectral width of each of these isomers at T = 10K. When the first solvation layer is sufficiently large to support a second layer, i.e., n $\gtrsim$ 15, the spectral shift difference between the isomers is increasing again. This is due to the difference in the spectral shift of one-layered isomers and two-layered isomers. According to the relative magnitude of this energetic splitting vs the homogeneous linewidth, one of the following two spectral appearances can be realized: (A) When the splitting is smaller than the width, we will get an unresolved inhomogeneous spectral feature. This is what probably happened in medium-sized Ar clusters of about 20 atoms, and for Kr heteroclusters for n = 16. (B) When the spectral shift splitting of the abundant isomers is of the same order of magnitude of the homogeneous linewidth, we shall get again a structured spectra. A "transition" from range (A) to range (B) with increasing the cluster size is manifested by line-narrowing. This is our interpretation of the line-narrowing which occurs for DCA·Kr_{17} (Figures 2 and 3). A further increase of the cluster size will bring us to a new domain, where the first layer of atoms is full and atoms occupy the second solvation layer. In this size-range the spectral shift is not sensitive any longer for the various isomeric structures, resulting in an essentially homogeneous linewidth. This size domain is represented in our studies by the Ar_{34} heteroclusters (Fig. 14), for which the experimental linewidth matches the calculated homogeneous spectral width at 21K.

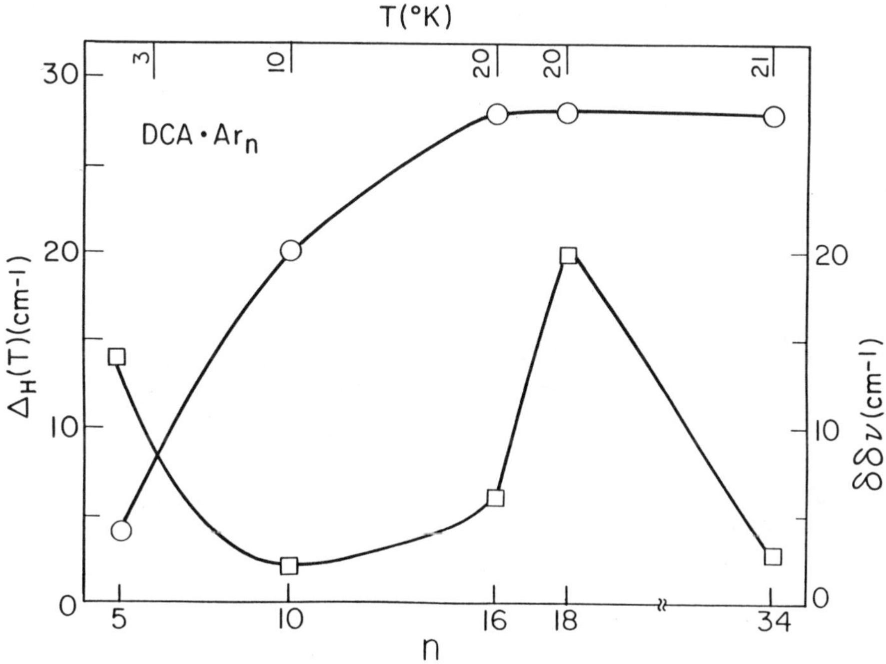

Figure 14. Origins of inhomogeneous broadening in DCA·Kr_n heteroclusters. The size dependence of the difference $\delta\delta\nu$ (□) between the spectral shifts for the energetically stable isomers and the homogeneous linewidths (O) of individual isomers. The temperatures (marked in the figure) constitute rough estimates of the heterocluster temperature.

The confrontation between the computer simulations of spectroscopic attributes of realistic systems and experimental reality elucidates the interrelationship between isomer-specificity and inhomogeneous linebroadening in governing the spectroscopy of mass-selected heteroclusters.

Acknowledgment. We are grateful to Dr. Dafna Scharf for her valuable help with the molecular dynamics simulations. This research was supported by the Binational German-Israeli James Franck research program for laser matter interaction.

References

1. Imry, Y and Bergman, D.J. (1971) Phys. Rev. A3, 1413.
2. G. Scoles (ed.), Proceedings of the International School of Physics "Enrico Fermi", Course CVII - The Chemical Physics of Atomic and Molecular Clusters, North-Holland (1990).
3. Bixon, M. and Jortner, J. (1989) J. Chem. Phys. 91, 1631.
4. Pawlow, P. (1909) Z. Phys. Chem. 65, 1.
5. Buffat, D.A. and Borel, J.P. (1976) Phys. Rev. A13, 2289.
6. Berry, R.S., Beck, T.L., Davis, H.L. and Jellinek, J. (1988) Adv. Chem. Phys. 70, 74.
7. Hoare, M.R. and Mc.Innes, J.A. (1985) Adv. Phys. 32, 791.
8. Jortner, J., Scharf, D. and Landman, U. (1981) in G. Benedek, T.P. Martin and G. Pacchioni (ed.), Elemental and Molecular Clusters, Springer, Berlin, p. 148.
9. Roy, R.J., Shelly, J.C. and Eichenauer, D. (1987) in J. Jortner, B. Pullman and A. Pullman (eds.), Large Finite Systems, Reidel, Dordrecht, p. 145.
10. Gu, X.J., Levandier, D.J., Zhang, B., Scoles, G. and Zhuang, D. (1990) J. Chem. Phys. 93, 4898.
11. Even, U., Amirav, A., Leutwyler, S., Ondrechen, M.O., Berkovitch-Yellin, Z. and Jortner, J. (1982) J. Faraday Discuss. Chem. Soc. 73, 153.
12. Leutwyler, S. and Bösiger, J. (1990) Chem. Rev. 90, 489.
13. Amirav, A., Horovitz, H. and Jortner, J. (1988) J. Chem. Phys. 88, 3092.
14. Even, U., Ben-Horin, N. and Jortner, N. (1989) Chem. Phys. Lett. 156, 138.
15. Ben-Horin, N., Even, U. and Jortner, J. J. Chem. Phys. (submitted).
16. Ben-Horin, N., Even, U., Leutwyler, S. and Jortner, J. (to be published).
17. Shalev, E., Ben-Horin, N., Even, U. and Jortner, J. (1991) J. Chem. Phys. (in press).
18. Shalev, E., Ben-Horin, N. and Jortner, J. (1991) J. Chem. Phys. (in press).
19. Ben-Horin, N., Sanderovitch, H., Kaldor, U., Even, U. and Jortner, J. J. Phys. Chem. (submitted).

MODE SELECTIVITY IN CLUSTER-MOLECULE INTERACTIONS: $Ni_{13} + D_2$*

Julius JELLINEK and Ziya B. GÜVENÇ
Chemistry Division
Argonne National Laboratory
Argonne, Illinois 60439, USA

ABSTRACT. Results of a detailed quasiclassical simulation study of the $Ni_{13} + D_2$ collision system are presented. The dissociative adsorption of the molecule as well as its scattering from the cluster are analyzed as functions of the initial rovibrational molecular state, collision energy and structure of the cluster. Mode-specific features of the reactive and nonreactive channels of the cluster-molecule interaction are displayed and discussed. Evidence for resonances and for a strong cluster structure-reactivity correlation is presented.

1. Introduction

Chemical reactivity of atomic and molecular clusters or, more generally, interactions of clusters with individual atoms and molecules, is a research area "at the beginning of the road". The interest in these interactions grew out of the special conceptual role of clusters as a new "bridging" (between atoms/molecules and solids/liquids) phase of matter, and of the recognition of the importance of cluster-molecule interactions in practical applications as well as in natural phenomena (e.g., in the upper atmosphere and in the soil). Heterogeneous catalysis is one of the processes the control over which may be greatly enhanced through the understanding of the mechanisms of cluster-molecule, especially metal cluster-molecule, interactions.

In recent years a variety of experimental studies was performed [1] to probe the kinetics and energetics of metal cluster-molecule reactions. Most of the theoretical work, however, was focused on interactions of molecules with metal surfaces [2]. Although clusters were used in some of these studies to represent the metal surface, their relevance in the context of cluster-molecule interactions is limited by two factors: (1) the clusters were chosen as slabs of a crystal lattice, (2) the slabs were often assumed to be rigid. Only a few theoretical investigations addressed the problem of metal cluster-molecule interaction as the central subject. These can be divided into electronic structure calculations [3] and dynamical treatments [4,5]. The former identify preferred binding sites and yield binding energies as well as (a usually restricted) mapping of the potential energy surface and of the reaction path. The dynamical treatments use model or

*Work performed under the auspices of the Office of Basic Energy Sciences, Division of Chemical Science, US-DOE under contract number W-31-109-ENG-38.

J. Jortner et al. (eds.), Mode Selective Chemistry, 153–164.

semiempirical potentials to characterize the time-evolution of a cluster-atom or a cluster-molecule collision system; they quantify the different processes in terms of probabilities, cross sections, rate constants, etc.

Reactions of D_2 with Ni_n, n = 4-13, were treated dynamically by Raghavan, Stave and DePristo (RSD) [4] under the assumption that the clusters are rigid [6]. We have performed an extensive quasiclassical simulation study of the $Ni_{13} + D_2$ collision system allowing for all degrees of freedom of the system to participate in the dynamics. A partial account of the results for the reactive channel (dissociative chemisorption of D_2 on an icosahedral Ni_{13}) has been given [7]. Here we present a more complete account which includes effects of cluster structure as well as state-to-state data for inelastic scattering of the molecule on the cluster. In section 2, the interaction potential and the computational details are outlined. The results, including mode-selective features of the interaction, and their discussion are given in section 3. We conclude with a summary in section 4.

2. The Potential and Computational Details

The potential energy surface (PES) V defining the forces acting on each atom in the $Ni_{13}+D_2$ collision system is of the form

$$V = V_{EA} + V_{LEPS}, \tag{1}$$

where V_{EA} is a so-called embedded-atom potential describing the interactions between Ni atoms in the cluster, and V_{LEPS} is a LEPS (London-Eyring-Polanyi-Sato) function mimicking the intramolecular (D-D) and molecule-cluster interactions. We have employed Voter and Chen's [8] parameterization of V_{EA} which, in addition to a number of bulk properties of nickel, also correctly reproduces the equilibrium bond length (2.2 Å) and bond energy (~1.95 eV) of Ni_2. The so-parameterized embedded-atom potential can be expected to be of relevance for the intermediate size range, i.e., clusters, as well. The LEPS function is the PESII of RSD [4], modified by a smoothing function used by Truong et al. [9] (details will be given elsewhere). We have chosen PESII, rather than PESI employed by RSD, because the former was fitted to reproduce the binding energy and the binding height of an H atom on a threefold site of an icosahedral Ni_{13} as calculated in the framework of the corrected effective medium (CEM) approach [4]; the parameters of PESI, on the other hand, were determined from dynamical simulations of the dissociative chemisorption of H_2 on different faces of the bulk (fcc) nickel [10].

The time-evolution of the $Ni_{13} + D_2$ collision system was obtained by solving numerically Hamilton's equations of motion for all 15 atoms. A fourth-order variable step-size predictor-corrector propagator was employed. Choice of the maximal step sizes of 5×10^{-16} - 10^{-15} sec assured conservation of the total energy as well as of the total linear and angular momenta within 0.03-0.15%. The initial conditions were specified as follows. Nontranslating and nonrotating Ni_{13} clusters were prepared in an icosahedral (ico), cubooctahedral (cubo) or hexagonal closed-packed (hcp) structure at temperature T defined as

$$T = \frac{2\langle E_k \rangle}{(3n-6)k}, \tag{2}$$

where E_k and n are the total kinetic energy and the number of atoms in the cluster, respectively, k is the Boltzmann constant, and $< >$ denotes time averaging. The center of mass of the D_2 molecule was placed asymptotically far (8.5 Å) from the center of mass of the cluster, and the initial coordinates and momenta of the D atoms were chosen in accordance with a quasiclassical prescription [11] to define a desired initial quantal vibrational v_i and rotational j_i state of D_2. The molecule was sent towards the cluster with a specified initial relative kinetic (collision) energy E^i_{tr} and impact parameter b. The collision event (trajectory) was qualified as reactive when it resulted in a dissociative adsorption of the molecule on the cluster; the molecule was deemed to be dissociated when the D-D bond length exceeded the value of 2.223 Å (three times the equilibrium D-D bond length). The collision could result in a nonreactive scattering of the molecule from the cluster. Upon reaching the asymptotic region, the scattered D_2 molecule was characterized by its final relative translational energy E^f_{tr}, final vibrational v_f and rotational j_f quantum numbers defined as rounded off to the closest integer

$$v_f = -\frac{1}{2} + \frac{1}{h}\oint p_r dr \qquad (3)$$

and

$$j_f = -\frac{1}{2} + \frac{1}{2}\left(1 + \frac{16\pi^2 J_f^2}{h^2}\right)^{1/2}, \qquad (4)$$

where r and p_r are the D-D bond length and its conjugate momentum, respectively; J_f is the magnitude of the classical final rotational angular momentum of the molecule; and h is Planck's constant. We have found that each of the final channels of the interaction (reactive or nonreactive) could be reached through a direct process or via a molecularly adsorbed precursor state. Some of the precursor states had quite long lifetimes. Care was taken to run each trajectory long enough to allow for the molecule to dissociate or, in the case of scattering, to reach the asymptotic region. In the case of dissociative adsorption the trajectory was terminated as soon as the dissociation criterion was met, thus preventing a possible recombination and subsequent desorption of the molecule.

Five hundred trajectories were run for each fixed set of initial conditions. These trajectories differed in the initial orientation of the molecule and of the cluster, and in the initial phase of the D_2 oscillator. The E^i_{tr}- and b-dependent probabilities $P_{v_i,j_i,T,...}(E^i_{tr},b)$ of the different processes (dissociative chemisorption, scattering of the molecule) labeled by the initial and , if appropriate, final state parameters were calculated as

$$P_{v_i,j_i,T,...}(E^i_{tr},b) = \frac{\tilde{N}_{v_i,j_i,T,...}(E^i_{tr},b)}{N}, \qquad (5)$$

where $\tilde{N}_{...}$ is the number of trajectories resulting in the process of interest, and N = 500. The cross sections $\sigma_{...}$ of the processes were determined in accordance with

$$\sigma_{v_i,j_i,T,...}(E^i_{tr}) = 2\pi \int_0^{b_{max}} bP_{v_i,j_i,T,...}(E^i_{tr},b)db, \qquad (6)$$

where b_{max} is (in the reactive case, for which the cross sections are presented in the next section) the lowest value of b at which $P...(E_{tr}^{i},b) = 0$. The translational temperature T_{tr} dependent state-resolved rate constants $\kappa...(T_{tr})$ were obtained from

$$\kappa_{v_i,j_i,T,...}(T_{tr}) = \pi\mu\left(\frac{2}{\pi\mu kT_{tr}}\right)^{3/2} \int_0^\infty \sigma_{v_i,j_i,T,...}(E_{tr}^{i})E_{tr}^{i}\exp\left(-\frac{E_{tr}^{i}}{kT_{tr}}\right)dE_{tr}^{i}, \tag{7}$$

where μ is the reduced mass of the cluster-molecule system. The low-temperature segments of $\kappa...(T_{tr})$ were fitted to the Arrhenius formula

$$\kappa_{v_i,j_i,T,...}(T_{tr}) = A_{v_i,j_i,T,...}\exp\left(-\frac{E_A^{v_i,j_i,T,...}}{RT_{tr}}\right) \tag{8}$$

to extract state-specific (translational) activation energies $E_A^{...}$; R in Eq. 8 is the universal gas constant.

3. Results and Discussion

Here we display and discuss some representative results for the reactive and nonreactive channels of the interaction of a D_2 molecule with a Ni_{13} cluster. It has been predicted by CEM calculations [4] and also using an embedded-atom potential [12] that the icosahedral packing is the lowest-energy, most stable geometry of Ni_{13}. The less stable hcp and cubooctahedral structures of Ni_{13} are almost degenerate energetically with the cubooctahedron possibly being just a bit higher in energy. In Fig. 1 the cross sections for dissociative chemisorption of the molecule on an icosahedral Ni_{13}, initially at room temperature, are shown as functions of the collision energy for different initial rovibrational states of D_2. The (v_i=0, j_i=3) state was considered because it is the most populated at room temperature. The choice of the (v_i=1, j_i=0) and (v_i=0, j_i=10) states was motivated by the fact that they have almost equal energies and thus are appropriate for testing the mode-selectivity of the cluster-molecule interaction. It is seen in the figure that for all the cases considered the cross sections display a small but nonzero threshold value of E_{tr}^{i} below which they are zero and above which they increase monotonically with E_{tr}^{i}. The higher the initial rovibrational energy of the molecule, the lower the value of the threshold energy and the larger the corresponding cross sections of the dissociative adsorption. While an increase of the probability of breaking the D-D bond with an increase of total energy (higher E_{tr}^{i} and/or higher initial rovibrational energy of D_2) is quite a natural result, a decrease in the threshold value of E_{tr}^{i} is a more intricate phenomenon. It can be a consequence of (a) a decrease, with an increase of v_i or j_i, of an adiabatic-type barrier encountered by the impinging molecule; (b) an enhanced dynamical coupling which channels part of the rovibrational energy into the "reaction coordinate" and thereby helps the molecule to climb the activation barrier; (c) both. Comparison of the cases (v_i=1, j_i=0) and (v_i=0, j_i=10) points clearly to the mode-selectivity of the reaction: an initial vibrational excitation of D_2 results in substantially larger cross sections for dissociative adsorption than an energetically equal (in fact, by 0.023 eV even higher) initial rotational excitation. Although the threshold values of the collision energy are very close for these two cases, the reactive cross section grows initially much faster for (v_i=1, j_i=0). At large E_{tr}^{i}'s the cross sections grow almost linearly and with the same rate for both cases.

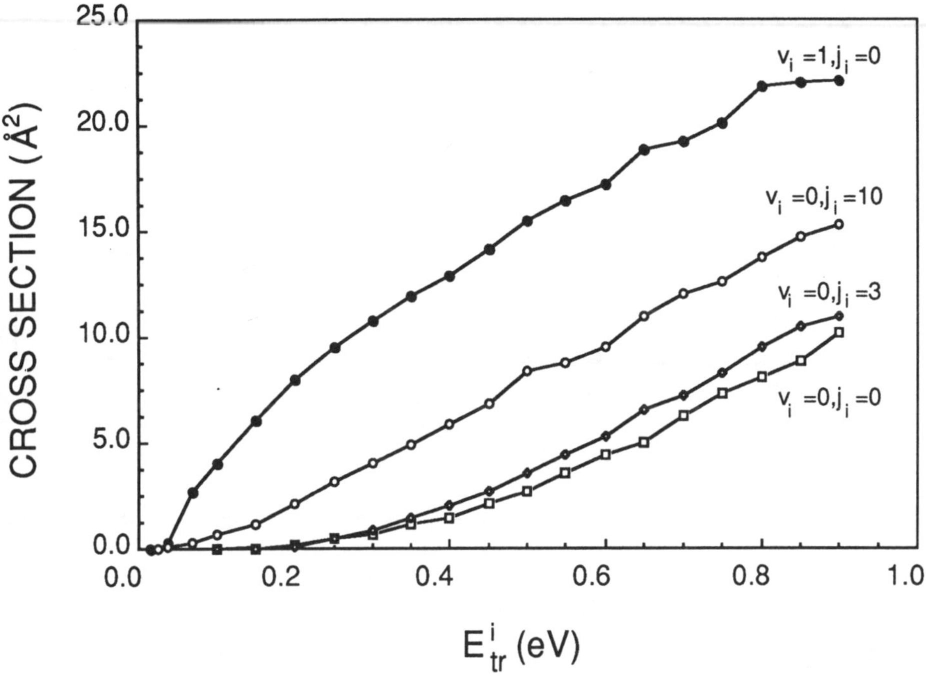

Figure 1. Cross sections of dissociative adsorption of $D_2(v_i, j_i)$ on a room-temperature icosahedral Ni_{13} as functions of the collision energy.

Fig. 2 displays the calculated rate constants (circles) for the dissociative adsorption of $D_2(v_i=1, j_i=0)$ and $D_2(v_i=0, j_i=10)$ on a room-temperature icosahedral Ni_{13}. The rate constants clearly behave in a non-Arrhenius fashion. Their low-translational-temperature segments can, however, be fitted to straight lines. The activation energies extracted from the fits are 0.027 eV for the (v_i =1, j_i =0) case and 0.031 eV for the (v_i =0, j_i =10) case. These values are in excellent agreement with the corresponding threshold energies (cf. Fig. 1). An agreement between the threshold energies and the activation energies extracted from the fits of the low-temperature segments of the corresponding rate constants has been found for all the (v_i, j_i) cases considered. This suggests that the state-specified threshold energies can be interpreted as measures of the heights of the state-dependent *dynamical* barriers for dissociative adsorption of the molecule on the cluster.

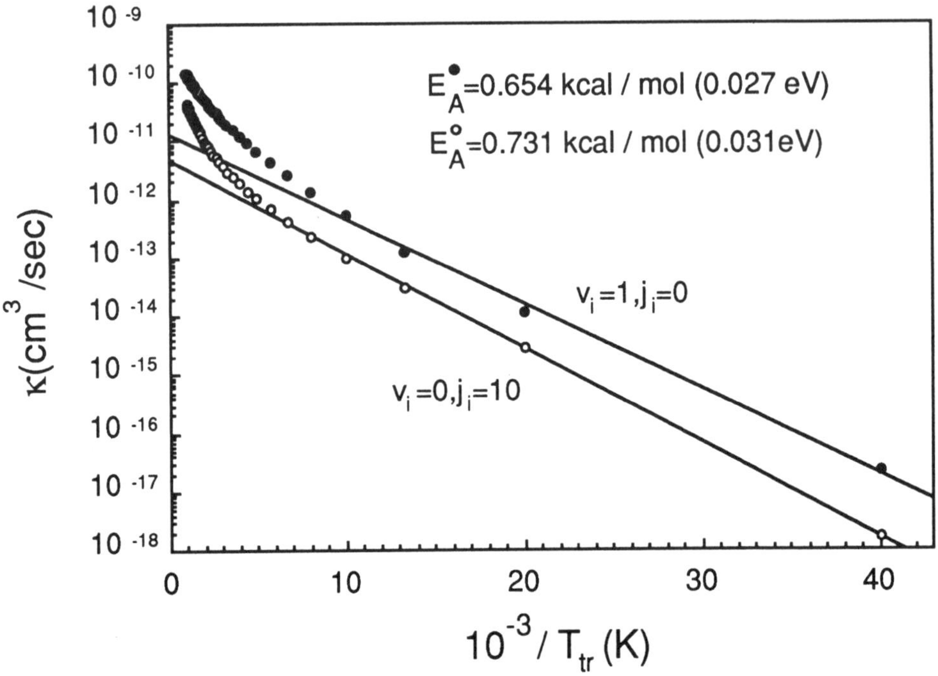

Figure 2. Rate constants of dissociative adsorption of $D_2(v_i, j_i)$ on a room-temperature icosahedral Ni_{13} as functions of the translational ("collision") temperature. The circles are the calculated results, the lines are the Arrhenius fits in the low-temperature regime. E_A are the activation energies extracted from the fits.

In Fig. 3 the cross sections of dissociative adsorption on a zero-temperature cubooctahedral Ni_{13} are shown for different (v_i, j_i) states of D_2. The large differences in the cross sections for the (v_i=1, j_i=0) and (v_i=0, j_i=10) cases indicate again a pronounced mode-selectivity of the reaction. However, the most striking feature of Fig. 3 is the nonmonotonic behavior of the cross sections. Particularly intriguing are the peaks for the (v_i=1, j_i=0) and (v_i=0, j_i=10) cases at very low collision energies. We have found that at these energies the predominant pathway for the reaction is the one involving molecularly adsorbed precursor states. The peaks thus can be identified with reaction resonances. Other distinctive features of reactivity of the cubooctahedral Ni_{13} are that the cross sections do not indicate any thresholds (barriers) for the reaction and they are larger than the corresponding cross sections for the icosahedral Ni_{13}.

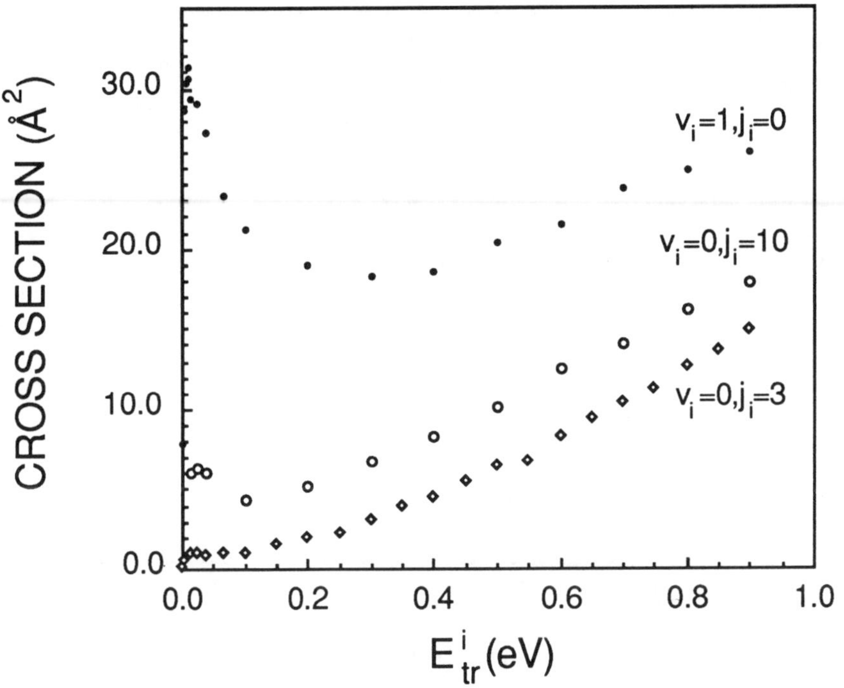

Figure 3. Cross sections of dissociative adsorption of $D_2(v_i, j_i)$ on a zero-temperature cubooctahedral Ni_{13} as functions of the collision energy.

The features of dissociative adsorption of D_2 on a zero-temperature hcp Ni_{13} are essentially the same as those for the cubooctahedral Ni_{13}. The only difference is that for most E^i_{tr}'s the magnitudes of the cross sections for the hcp cluster are slightly lower than the corresponding values for the cubooctahedral cluster. Figs. 4 and 5 show the reactive cross sections for the (v_i=0, j_i=3) and (v_i=1, j_i=0) initial molecular states, respectively, and each of the three isomers of the cluster. The cluster structure-reactivity correlation is apparent. Its features can be summarized as follows. The reactivity of Ni_{13} changes inversely with the stability of its structure; this is in agreement with what has been found by RSD [4] for rigid isomers of Ni_{13} interacting with a D_2(v_i=0, j_i=0). The qualitative features of the E^i_{tr}-dependence of the reaction cross sections for the hcp and cubooctahedral geometries are drastically different from those for the icosahedral structure. The qualitative dissimilarities (e.g., presence or absence of thresholds and of resonance peaks) are particularly pronounced at low collision energies, at which the molecule stays a longer time in the vicinity of the cluster surface. Taking, in addition, into account that the reactivity properties of the hcp and of the cubooctahedral structures closely resemble each other even in details and that the surface topologies of these two structures are very similar (for example, both have threefold and fourfold faces, while the icosahedron has only threefold faces), one is led to the conclusion that the topology of the cluster surface (or its structure) is another "initial condition" or "mode" which affects its

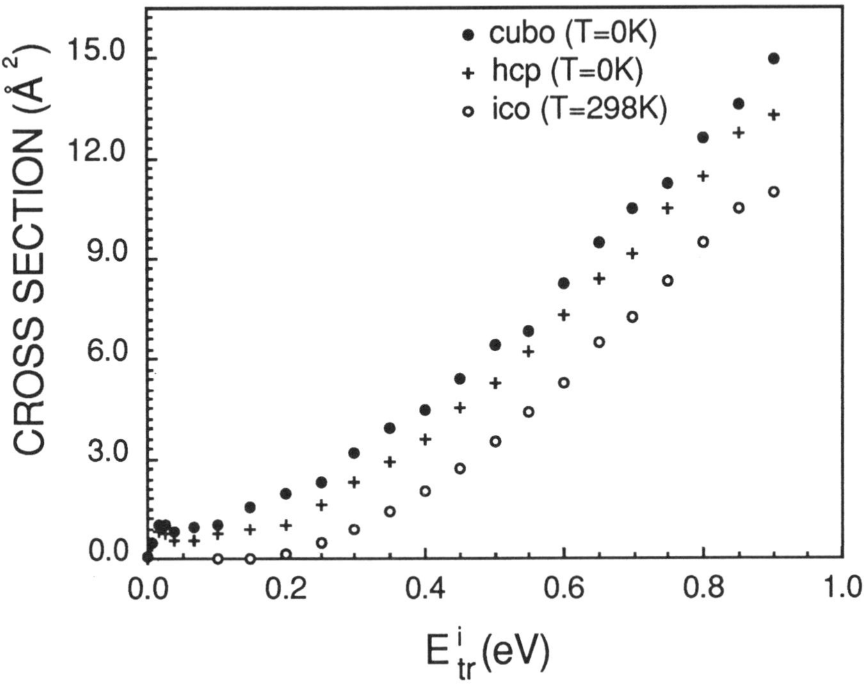

Figure 4. Cross sections of dissociative adsorption of $D_2(v_i=0, j_i=3)$ on three isomers (ico, hcp and cubo) of Ni_{13} as functions of the collision energy.

reactivity. As seen in the figures, the effect of this "mode" diminishes at high collision energies. (It should be remarked that the hcp and cubooctahedral geometries of Ni_{13}, as defined by the embedded-atom potential, are metastable minimum energy structures in that they are not separated by a computationally detectable barrier from the icosahedral structure. This was the reason for "preparing" them initially at T=0K. Interaction with a D_2 molecule, either reactive or nonreactive, triggers a structural "fluctuation" to the icosahedral geometry. The induction time for the structural change is, however, much longer than the collision or reaction time. The fact that the results are compared to those for an icosahedral Ni_{13} at room temperature is of no consequence since the reactivity features of the latter are insensitive to temperature changes in the range 0-298K [7]). The dependence of the reactivity of a cluster on its structure has been utilized recently [13] to measure the ionization potentials for different isomers of niobium clusters.

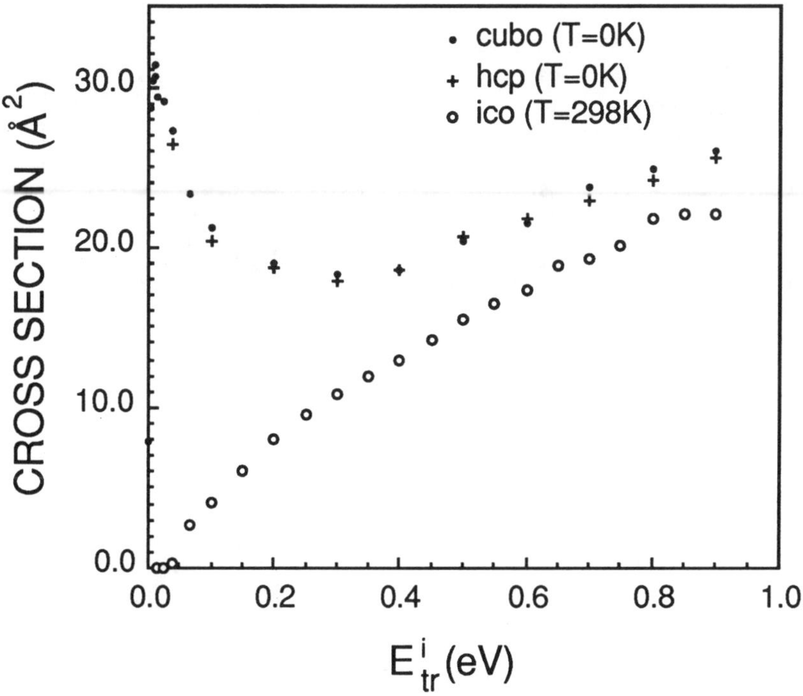

Figure 5. The same as Figure 4 but for $D_2(v_i=1, j_i=0)$.

We turn now to a brief discussion of the nonreactive interaction of D_2 with Ni_{13}. In Fig. 6 the probability distributions of the final vibrationally (v_f) and rotationally (j_f) resolved states of D_2 scattered from a room-temperature icosahedral Ni_{13} are shown. The collision energy is 0.9eV. Three panels, corresponding to different impact parameters, are displayed for each of the (v_i=0, j_i=0), (v_i=1, j_i=0) and (v_i=0, j_i=10) cases. Each P_{v_f} is the (summed over the final rotational states) total probability of a specified final vibrational state. By adding up all the P_{v_f}'s, one arrives at the initial-state-resolved total probability of the nonreactive scattering. One easily verifies that this latter is an increasing function of the impact parameter b for all three (v_i, j_i) cases (cf. Ref. [7]). At large impact parameters the molecule samples only the peripheral parts of the potential where the weak dynamical couplings channel less energy into the D-D bond. This makes the probability of the dissociative adsorption lower and, consequently, of the nonreactive scattering higher. The same reasoning applies to the b-dependence of the probability distributions of the final rovibrational states of the molecule for the (v_i=0, j_i=0) and (v_i=0, j_i=10) cases. At low impact parameters and high collision energies, as in this case, the molecule penetrates deeply into the core part of the potential where the strong translational-vibrational and rotational-vibrational couplings result not only in bond breaking but also in populating the initially unoccupied v_f = 1,2 and possibly even v_f = 3 states. As the impact parameter increases, v_f = 0 becomes by far the most probable state. The pattern of the probability distributions of the final rovibrational states is very different for the

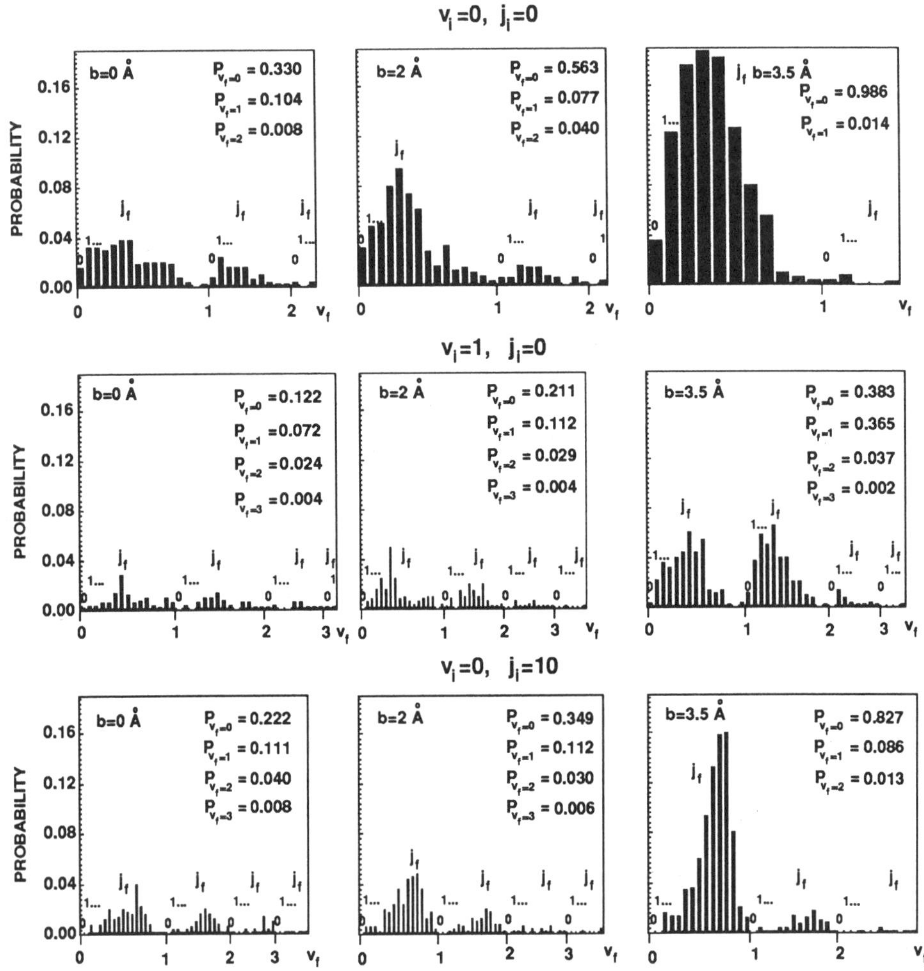

Figure 6. Probability distributions of the final rovibrational states of D_2 scattered from a room-temperature icosahedral Ni_{13}. Each bar is the probability of a specified (v_f, j_f) state. The P_{v_f}'s are the probabilities of the final vibrational states v_f. Three panels, corresponding to different values of the impact parameter b, are presented for each of the ($v_i = 0$, $j_i = 0$), (v_i=1, j_i=0) and (v_i=0, j_i=10) initial states of the molecule.

(v_i=1, j_i=0) case, where it appears to be, at least qualitatively, independent of b. The b-dependent population of the different v_f states by the translational-vibrational coupling is masked by the partial depopulation of the v_i=1 state primarily into the v_f=0 state. The difference in the distribution patterns of the final rovibrational states for the v_i=0 and v_i=1 cases, respectively, indicates and illustrates a (vibrational) mode-selectivity (or specificity) in a nonreactive cluster-molecule interaction. One notices the non-Boltzmann character of the distribution of the final rotational states for all the v_f's. Work is in progress on investigation of the scattering of D_2 from the hcp and cubooctahedral Ni_{13}. Based on reactivity properties of these isomers, one may expect interesting structural effects also in the nonreactive interaction of the molecule with the cluster, especially at low collision energies.

4. Summary

We have reported here results of a quasiclassical simulation study of the interaction of a D_2 molecule with a Ni_{13} cluster. Both channels, the reactive (dissociative adsorption of the molecule on the cluster) and the nonreactive (scattering of the molecule from the cluster), were considered. The processes were examined as functions of the initial quantal rovibrational state of the molecule, collision energy and structure of the cluster.

The main findings can be summarized as follows. Energizing the system (increasing either the collision energy or the rovibrational energy of the molecule) results in larger cross section of the dissociative adsorption. The reaction is mode-selective: initial vibrational excitation of the molecule leads to a more efficient bond breaking than an energetically equal initial rotational excitation. The structure of the cluster is an important factor in the specification of its reactivity features. It appears that the topology of the cluster surface plays a crucial role, especially at low collision energies. The topologically similar hcp and cubooctahedral structures of Ni_{13} are not only more reactive than the icosahedral Ni_{13}, but the dynamical details of the interaction of the molecule with these isomers are also very different. While the reaction cross sections for the icosahedral geometry display thresholds (dynamical activation barriers) and increase monotonically with the collision energy, there are no barriers for the reaction with the hcp and cubooctahedral isomers and the reactive cross sections change nonmonotonically, exhibiting possible resonance features. The resonance features (heights and widths of the peaks) depend on the initial vibrational and rotational state of the molecule.

The distribution of the final vibrational states v_f of the D_2 scattered from a room-temperature icosahedral Ni_{13} depends strongly on the initial vibrational state v_i of the molecule. With increase of the impact parameter the distribution shifts almost exclusively towards v_f=0 for the v_i =0 cases. On the contrary, the probabilities for the v_f=0 and v_f=1 final states of the D_2 remain of the same order of magnitude at all impact parameters when the molecule impinges with v_i=1. We expect to find particularly interesting features in the nonreactive interaction of D_2 with the hcp and cubooctahedral isomers of Ni_{13}. Work in this direction is in progress.

Acknowledgments

We are grateful to Dr. A. F. Voter for providing us the embedded-atom potential for nickel. J. J. thanks Professors A. E. DePristo and P. E. M. Siegbahn for useful discussions.

References

1. See, for example, J. L. Elkind, F. D. Weiss, J. M. Alford, R. T. Laaksonen, and R. E. Smalley, J. Chem. Phys. 88, 5215 (1988); W. D. Reents, Jr. and M. L. Mandich, J. Phys. Chem. 92, 2908 (1988); W. F. Hoffman, E. K. Parks, and S. J. Riley, J. Chem. Phys. 90, 1526 (1989); Y. M. Hamrick, and M. D. Morse, J. Phys. Chem. 93, 6494 (1989); S. K. Loh, L. Lian, and P. B. Armentrout, J. Chem. Phys. 91, 6148 (1989); S. Nonose, Y. Sone, K. Onodera, S. Sudo, and K. Kaya, Chem. Phys. Lett. 164, 427 (1989); P. Fayet, A. Kaldor, and D. M. Cox, J. Chem. Phys. 92, 254 (1990); K. M. Creegan, and M. F. Jarrold, J. Am. Chem. Soc. 112, 3768 (1990); R. E. Leuchtner, A. C. Harms, and A. W. Castleman, Jr., J. Chem. Phys. 92, 6527 (1990); and references therein.
2. See, for example, J. Tully, J. Chem. Phys. 73, 633 (1980); P. E. M. Siegbahn, M. R. A. Blomberg, and C. W. Bauschlicher, J. Chem. Phys. 81, 2103 (1984); C.-Y. Lee and A. E. DePristo, J. Chem. Phys. 85, 4161 (1986); I. Panas, J. Schüle, P. Siegbahn, and U. Wahlgren, Chem. Phys. Lett. 149, 265 (1988); G.-Q. Xu, S. L. Bernasek, and J. Tully, J. Chem. Phys. 88, 3376 (1988); J. Harris, Surf. Sci. 221, 335 (1989); U. Nielsen, D. Halstead, S. Holloway, and J. K. Nørskov, J. Chem. Phys. 93, 2879 (1990); A. Chattopadhyay, H. Yang, and J. L. Whitten; J. Phys. Chem. 94, 6379 (1990); E. A. Colbourn, in *Catalysis*, vol. 8, Royal Society of Chemistry, Cambridge, 1989, p. 42; A. J. Cruz and B. Jackson, J. Chem. Phys. 94, 5715 (1991); and references therein.
3. See, for example, T. H. Upton, Phys. Rev. Lett. 56, 2168 (1986); T. H. Upton, D. M. Cox, and A. Kaldor, in *Physics and Chemistry of Small Clusters*, P. Jena, B. K. Rao and S. N. Khanna (Eds.), Plenum Press, NY, 1987, p. 755; K. Przybylski, J. Koutecký, V. Bonacić-Koutecký, P. von Ragué-Schleyer, and M. F. Guest, J. Chem. Phys. 94, 5533 (1991); and references therein.
4. K. Raghavan, M. S. Stave, and A. E. DePristo, Chem. Phys. Lett. 149, 89 (1988); J. Chem. Phys. 91, 1904 (1989).
5. J. E. Adams, J. Chem. Phys. 92, 1849 (1990).
6. Recently DePristo and coworkers performed simulation studies of the interaction of D_2 with nonrigid clusters (private communication).
7. J. Jellinek and Z. B. Güvenç, Z. Phys. D, 18, 000 (1991).
8. A. F. Voter and S. F. Chen, Mater. Res. Soc. Symp. 82, 175 (1987).
9. T. N. Truong, D. G. Truhlar, and B. C. Garett, J. Phys. Chem. 93, 8227 (1989).
10. C. Y. Lee and A. E. DePristo, J. Chem. Phys. 85, 4161 (1986); ibid. 87, 1401 (1987).
11. R. N. Porter, L. M. Raff, and W. H. Miller, J. Chem. Phys. 63, 2214 (1975).
12. J. Jellinek, unpublished.
13. M. B. Knickelbein and S. Yang, J. Chem. Phys. 93, 1476 (1990); ibid 93, 5760 (1990).

Cluster Photoexcitation and Evaporation: Two Regimes?

F. G. Amar† and S. Weerasinghe
Department of Chemistry
University of Maine
Orono, Maine 04469
USA

ABSTRACT: We discuss simulation studies of cluster evaporation dynamics for several systems which can be considered typical of two limiting regimes: a statistical regime and a mode-specific regime. At the statistical limit, we compare the results of molecular dynamics (MD) simulations of the evaporation of small neutral argon clusters with both approximate statistical theories and exact classical phase space theory. In the limit of mode-specific excitation we discuss the photoexcitation and caging of diatomic molecules and ions in rare-gas and molecular clusters. Special attention is given to the $I_2^-(CO_2)_n$ system. We compare simulation results with existing experimental data and discuss the transition from non-statistical to statistical evaporation as a function of cluster size.

1. INTRODUCTION

In this paper we shall discuss the evaporation of highly excited van der Waals clusters (both neutrals and ions) and the way in which the dynamics of this evaporation depends on the mode of excitation of the clusters. One can imagine two limiting regimes: a) one in which the excitation is initially localized on one molecule of the cluster or in one bond or vibronic state; and b) the case in which the cluster is initially excited statistically or thermally.

Clearly, the first case is the appropriate limit to consider in a spectroscopic experiment in which a chromophore in the cluster absorbs a photon, initiating some dynamics. The spectroscopic experiment focusses attention on the chromophore--the pattern of spectral shifts and changes in line shape give information on how the chromophore is perturbed by its cluster environment. It is, of course, also possible to turn the question around and ask how the chromophore disrupts its microsolvation environment? In this case, it is ultimately of interest to ask how fast the excitation relaxes and to consider the kind of mechanism at work in the dissociation of the cluster. Is there a direct coupling between the excitation and the dissociation channel to yield a direct fragmentation of the system or is there an intermediate vibrational relaxation step to produce a thermally excited cluster which can then evaporate?

The statistically excited or thermal cluster limit, on the other hand, is the appropriate

†Chercheur associé au CNRS while on sabbatical leave, academic year 1990-91 at Laboratoire de Photophysique Moléculaire, Université de Paris-Sud, 91405 Orsay, France.

J. Jortner et al. (eds.), Mode Selective Chemistry, 165–188.

starting point to understand the evaporation of clusters produced in a beam source, collisionally excited clusters, or even evaporation of clusters prepared by photoexcitation but which undergo rapid internal redistribution of the energy before dissociation can take place. This last case will be of great interest to us here.

The study of the dynamics of cluster evaporation processes is of cardinal importance for the understanding of the appearance of "magic numbers" in cluster mass spectra and as a tool for inferring cluster structure. The mass spectra of a wide variety of cluster systems display peaks with enhanced intensity at particular cluster sizes. In rare gases the pattern of these special stabilities (13, 55, 147, ...) is associated with geometric shell closings in the icosahedral growth sequence for the packing of spheres [1,2]. For the alkalis, the magic numbers (8, 20, ...) are associated with electronic shell closings of the jellium model [3]. In both cases, the clusters associated with shell closings are especially stable with respect to their neighbors. The expression of this stability is mediated by the kinetics of the evaporation processes which transform the cluster distribution arising at the cluster source into that observed at the detector.

In effect, a relatively small enhancement of stability can affect the evaporation rate of a cluster by many orders of magnitude, leading to a pooling of population at more stable cluster sizes and a depletion of population at less stable cluster sizes, as for example, in the mass spectra of carbon clusters [4]. On the other hand, an overall reduction of the importance of evaporation in the cluster beam, e.g., by cooling the beam, will tend to suppress the intensity variations in the mass spectrum. This has been nicely demonstrated by Whetten and coworkers [5] for sodium clusters.

Klots has introduced the notion of the evaporative ensemble [6] to describe the distribution of clusters in a beam which have a finite time in which to evaporate before being detected. The population distribution observed at the detector will then depend on rate constants which are larger than the inverse transit time of a cluster in the apparatus. Since these evaporative rate constants are sensitive functions of the relative stability of the clusters, it is possible to estimate binding energies of clusters by modelling the kinetics of the evaporation cascade using statistical rate theory. Several groups have used this approach: Bréchignac and coworkers [7] have extracted the binding energies of sodium cluster ions from their mass spectra and similar calculations for Hg_n^+ have been performed by Haberland and coworkers [8]. Engelking [9] has proposed a modified version of RRK theory to model the evaporation data of $(CO_2)_n^+$ obtained by Lineberger and his group [10].

In this paper we address several questions related to the issue of modelling the kinetics of evaporation. In Section 2, we shall review the results of our earlier MD simulations [11] of the size dependent cage effect in Br_2Ar_n and at the same time discuss the experimental results of the Lineberger group on the related $Br_2^-(CO_2)_n$ and $I_2^-(CO_2)_n$ systems [12-14]. We shall examine these results [15] for hints about the relevant time scales for relaxation of the diatomic chromophore as compared to the timescale for evaporation.

The central part of the paper is a comparison of our recent MD simulations [16,17] of the evaporation rates of Ar_n ($12 \leq n \leq 14$) with accurate phase space theory predictions. In this regard we shall briefly describe new approaches to calculating the density-of-states of highly anharmonic systems (Sections 3.3 and 3.5). Besides evaporation rates we also focus on kinetic energy release distributions. Finally, in Section 4, we shall take up the issue of statistical *vs* nonstatistical evaporation in clusters subject to localized excitation ("caging" systems) and then discuss some of the future directions in this field.

2. Evaporation and the Size-dependent Cage Effect

2.1. BACKGROUND

The photodissociation of a diatomic molecule may be regarded as the simplest chemical reaction, proceeding, in the gas phase, via excitation to a repulsive excited state or surface on which the molecule dissociates in 10^{-14} sec. or less. In the condensed phase, the situation is much more complicated due to solvation and especially due to the effect of the solvent *cage* which may inhibit the dissociation. This kinetic cage effect and related processes such as geminate recombination have been extensively studied both experimentally and theoretically both in condensed phases and in high pressure gases. We shall briefly highlight some work of particular relevance to clusters but because the amount of work in this field is so large we refer the interested reader to the review article of Hynes with its extensive bibliography [18].

The classic picosecond pump-probe experiments of Eisenthal and coworkers [19] showed that when I_2 (in CCl_4) is photoexcited to the repulsive Π state, the population relaxes back down to the ground state with a time constant of about 100 ps. This "bleaching" experiment essentially measured the recovery of the absorption signal as a function of the time delay between the excitation laser and the probe pulse. The modelling of Nesbitt and Hynes [20] showed that this time was characteristic of the vibrational relaxation of the I_2 by energy transfer to the solvent); the recombination time must be very much shorter. This work also raised a number of fascinating questions about the role of internal modes of the solvent (effectiveness of V-V vs V-T transfer) in effecting the relaxation of the diatomic.

Given the obvious importance of the first solvation shell in caging and relaxation of the photoexcited diatomic, some interesting questions from the point of view of cluster science are: How big does a solvent cluster need to be to observe caging? What are the systematics of caging and relaxation as a function of the size? As the solvent medium changes from an infinite solvent to a finite cluster, the mechanism of energy loss from the diatomic "hotspot" must change from heat dissipation to the medium to evaporation: how and at what size does this transition take place?

A partial answer (and still a rather tantalizing one) to the first question came with the "one-atom cage effect' experiment of Valentini and Cross [21]. In that experiment, a van der Waals complex of I_2Ar was excited above the dissociation limit of the (attractive) B state. Somewhat surprisingly, no dissociation was detected but rather fluorescence was observed from a series of vibrational levels of the B state. The results are consistent, at least qualitatively, with the presumption of a linear complex in which the kinetic energy of a dissociating I atom is transferred to the Ar in a kind of internal collision [22,23]. This effect has been extensively investigated for other solvent partners [24] but it appears that this explanation is complicated, however, by the fact that the dissociative Π state crosses the B state and may contribute to the caging.

We shall now take up the cage effect in clusters *per se* in order to begin to answer some of the above questions.

2.2. SIMULATING THE CAGE EFFECT IN CLUSTERS

Our simulation study of the cage effect in Br_2Ar_n clusters was perhaps the first systematic attempt to assess the effect of cluster size on the caging process [11]. For the details of the

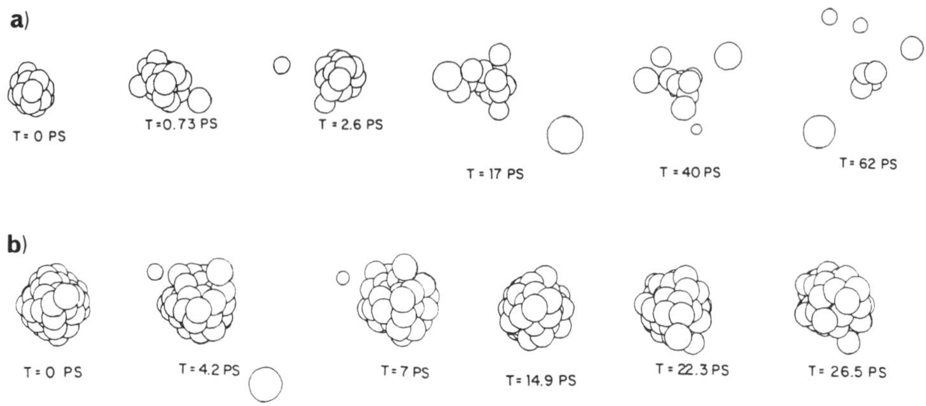

Figure 1. **a)** Snapshots of the dissociation of a Br_2Ar_{20} cluster following excitation to the $^1\Pi_{1u}$ state. **b)** Snapshots of a Br_2Ar_{70} cluster. From reference [11].

simulation, the reader is referred to the original paper; here we reproduce two figures which show snapshots of cluster trajectories following simulated photoexcitation of Br_2 to the repulsive $^1\Pi_{1u}$ excited state. Two sizes are considered: first a cluster in which the first solvation shell is filled (20 argons) and then a cluster consisting of two shells (70 argons).

Figure 1a shows how, in a typical trajectory, Br_2 is caged inside an Ar_{20} cluster and then, as the Br_2 relaxes, the cluster is completely fragmented in less than 100 ps. For an ensemble of such trajectories, no loss of Br atoms was observed--Br_2 is caged with unit efficiency by a single solvation layer. Analysis of the trajectories for Br_2Ar_{20} shows that the relaxation of the diatomic and the evaporation occur on the same timescale--in fact it is possible to correlate the evaporation of a specific Ar atom with a particular collision of Br_2 with that atom! This seems to be telling us that the evaporation process is strongly dependent on the fact that the excitation is localized in a particular mode in this system.

Turning now to Figure 1b, we see that in the case of a two shell cluster, two Ar atoms are lost at very early times, essentially by transfer of momentum during the Br_2's first violent interaction with the solvent cage. However, on the timescale in which Br_2Ar_{20} is being totally dissociated, the two shell cluster heats up slightly but does not further dissociate. Again an analysis of trajectories indicates unit efficiency for the caging process and shows that Br_2 relaxes just as fast (if not faster) as in the 20 atom cluster. Clearly, the larger cluster is able to receive and store energy from the diatomic without immediately evaporating. Presumably if we had been able to run longer trajectories we would have observed further evaporations from the totally thermalized clusters.

We should note that for clusters with less than one solvation shell, the caging is not 100% efficient. We shall not dwell on this interesting size dependence effect since our main interest here is on the evaporation processes following photoexcitation. It is fair to note, however, that important mass effects and strong dependence on interaction potentials may change the effiency of caging even for a one-shell cluster as in the $HIXe_n$ clusters considered by Alimi and Gerber [25]. Scharf, Landman, and Jortner [26] have also observed very similar caging and evaporation effects in their simulations of rare gas excimer systems like $Xe_2{}^*Ar_n$.

To summarize this Section, we emphasize that there is an apparent transition with increasing cluster size from a regime in which solvent evaporation is strongly coupled to the relaxation of the localized excitation (the diatomic molecule) to a regime in which the energy relaxation and evaporation occur on different time scales. There are two obvious explanations: a) while the Br_2 can interact strongly with the first solvent shell in a two-shell cluster, the first-shell atoms cannot easily evaporate because they are themselves 'caged' by the second shell and energy transfer to the second shell is not efficient enough to cause immediate evaporation; b) the larger thermal mass of the two shell cluster distributes the excitation among more modes, reducing the (statistical) evaporation rate. This second argument implies that we really know how to describe statistical evaporation processes in clusters. We argue in Section 3 that statistical rate theory for evaporation should include the effects of cluster anharmonicity.

2.3. EXPERIMENTS ON THE CAGE EFFECT IN CLUSTERS

Carl Lineberger and his group in Colorado have pioneered the investigation of the cage effect in cluster ions [12-14]. Focussing on ions like Br_2^- and I_2^- in clusters of CO_2 they have been able to investigate the efficiency of caging as a function of cluster size and photon energy. First cold cluster ions are created in a supersonic expansion, and then a particular cluster size is selected by a mass spectrometer. After photoexcitation of this given parent ion, the fragment ions resulting from caging and evaporation processes similar to those described in Section 2.2 are detected in a second mass spectrometer. The Colorado group first focussed on $Br_2^-(CO_2)_n$ clusters [12]: by analyzing the proportion of ion fragments containing Br^- as opposed to Br_2^-, the caging fraction could be determined for each cluster size. This principal result is that, when excited with 700 nm photons, Br_2^- is more and more efficiently caged as the number of solvent CO_2s rises from 2 to 12 at which point no Br^- based fragments are observed. Our simulations [15] of the $Br_2^-(CO_2)_n$ and $Br_2^-Ar_n$ systems are in qualitative agreement with these results.

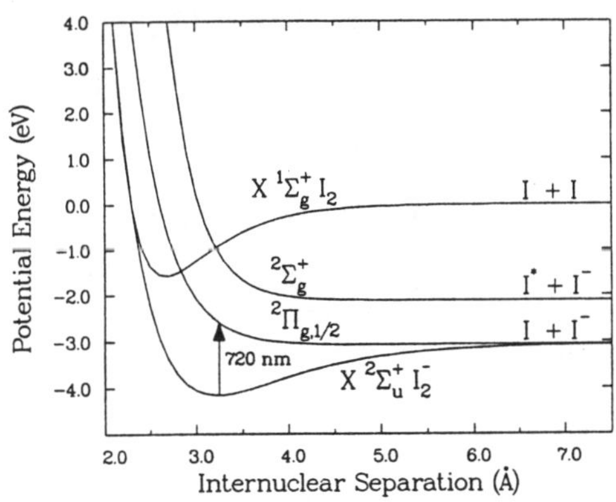

Figure 2. Potential energy curves for I_2^-. From reference [13].

We turn now to the more recent results on $I_2^-(CO_2)_n$ for two reasons: first, the Colorado group [13,14] has now performed picosecond pump-probe experiments very similar in principle to the experiments of Eisenthal referred to above; and second, we have now adopted an improved potential model for the CO_2-CO_2 interaction [27] so that simulations may now be a more trustworthy guide to understanding the experiments. Figure 2 shows the relevant potential energy curves [28] for the I_2^- diatomic; the 720 nm photon corresponds to an energy of ~13900 cm^{-1} or 1.72 eV. Figure 3a shows the recombination probability for $I_2^-(CO_2)_n$ as a function of cluster size. Note that all clusters below n=6 dissociate and that caging is completely efficient for n=16 and above; it is also interesting that the recombination probability is almost perfectly linear between the two extremes. The larger threshold for strong caging in I_2^- based clusters as compared to Br_2^- is most likely the result of the larger size of I_2^- leading to an expanded first solvent shell.

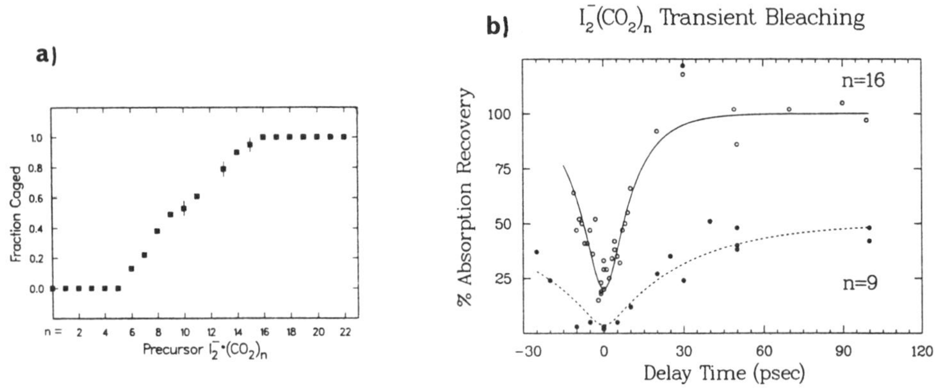

Figure 3. a) Fraction of $I_2^-(CO_2)_n$ clusters caged as a function of n. Unpublished results of Lineberger *et al.* as quoted by Stace [14]; **b)** Absorption recovery signal for $I_2^-(CO_2)_{16}$ and $I_2^-(CO_2)_9$. From[13].

The results of the picosecond experiments on $I_2^-(CO_2)_n$ are shown in Figure 3b. The recovery of the absorption signal is shown as a function of the delay between the pump and the probe laser. The signal is symmetrical about t=0 since both laser pulses are equivalent [13]. For the n=16 cluster, the recovery of the signal is complete, reflecting the efficiency of caging in this system. For the n=9 cluster the recovery is only partial since some clusters (about half) dissociate and cannot reabsorb. The larger cluster has a exponential recovery time of about 10 ps while the smaller cluster recovers in about 30 ps. This relaxation time is extremely fast compared with the bulk experiments on I_2/CCl_4. Whether this is a cluster effect, or is due to the nature of the solvent, or simply to the fact that an ion is involved instead of a neutral is still not completely clear.

We have recently used a realistic potential for the CO_2-CO_2 interaction [27] together with a self-consistent treatment of ion-induced dipole interactions [15] to find the minimum energy structures of the $I_2^-(CO_2)_n$ clusters ($2 \leq n \leq 17$) [29]. The structures for several clusters are displayed in Figure 4. The $I_2^-(CO_2)_n$ cluster structures are characterized by

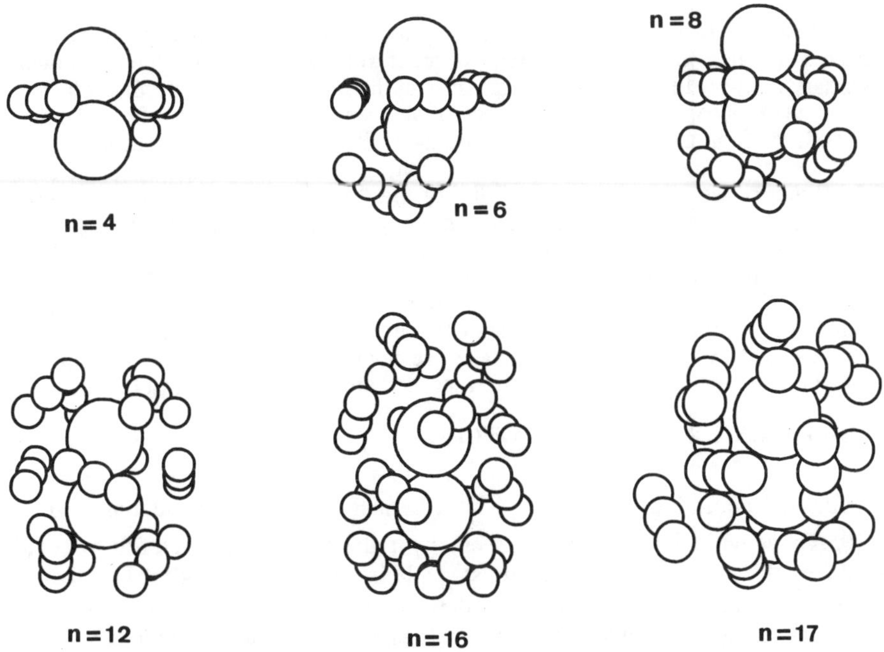

Figure 4. Minimum energy structures for selected $I_2^-(CO_2)_n$ clusters. At n=6, we observe the first singly-capped structure. The first solvation shell closes at n=16. See reference [29] for more details on the energetics of these clusters.

the development of 4-membered rings of CO_2s around the axis of I_2^-. The relative angle of the CO_2s within a ring is reminiscent of the canted parallel arrangement of the CO_2 dimer and is due to the large quadrupole moment of CO_2. Most striking is the fact that the first capped structure is observed at n=6 just where the Colorado group observes the beginning of caging; at the same time the first solvent shell closes at n=16 just where caging becomes perfectly efficient. Simulations of the dynamics of $I_2^-(CO_2)_n$ clusters are ongoing [30]. Some of the issues being addressed will be discussed in Section 4. In order to understand or even identify nonstatistical effects, one must first understand statistical behavior. Thus we now present a brief account of some of our recent work [16,17] on understanding evaporation in the statistical limit.

3. Simulation and Theory of Evaporation Compared

The basic idea of this section is to use MD simulations of evaporation of a system with a realistic intermolecular potential to test statistical rate theories of such a unimolecular process. For the evaporation reaction:

$$Ar_n \dashrightarrow Ar_{n-1} + Ar \tag{1}$$

we would expect the correct statistical theory to be phase space theory (PST) since the reverse reaction has no activation barrier. In order to implement such a theory, however, it is necessary to know the density-of-states of both the reactant and the product cluster. We calculate the classical anharmonic densities-of-states exactly in order to check the range of validity of the often-used harmonic approximation. By simulating evaporation rates at high internal energies (and short times) and comparing to exact formulations of classical statistical theory we can verify the validity of the theory and then, with some degree of confidence, extrapolate to lower energies and longer timescales where direct simulations are not feasible even if accurate potentials are known.

This section is divided into several parts: first, we describe our MD simulations of evaporation which serve as computer experiments against which we test theory. Then we summarize the formulations of several statistical rate theories, followed by a brief account of how exact anharmonic densities-of-states are calculated. Finally, a critical comparison of simulation and theory leads us to a different method for calculating absolute densities

3.1. MD SIMULATIONS OF Ar_n EVAPORATION

We model the neutral argon clusters with a Lennard-Jones 12-6 potential (ε=119.8K, σ=3.405Å); we will take ε as our unit of energy in the discussion which follows. Table 1 gives the total binding energy of the most stable isomers of Ar_n ($14 \geq n \geq 11$); the second column shows the differences between these numbers--to be interpreted as adiabatic dissociation energies and which we shall call E_0 in what follows; obviously, a cluster must have energy greater than E_0 in order to evaporate. The exceptionally large E_0 of Ar_{13} is a well-known consequence of its "perfect" icosahedral structure.

Table 1. Binding energies and dissociation energies of argon clusters (ε_{Ar} units).

n	E_{min}	E_0
14	-47.845	3.518
13	-44.327	6.359
12	-37.968	5.202
11	-32.766	

MD simulations of the evaporation rate of Ar_n ($12 \leq n \leq 14$) for an ensemble of trajectories at precisely defined internal energies and angular momenta yield evaporation rate constants, k(E,J), and kinetic energy release distributions. At each energy considered, a set of 1400 random phase points, consistent with the required energy and J=0 were used to initiate phase space trajectories. A fifth order predictor-corrector algorithm was used to integrate each trajectory, with a timestep of 10^{-14} sec for a total time of 200 ps. Energy was conserved to 1 part in 5 x 10^4.

From simple unimolecular rate theory we obtain the familiar expression:

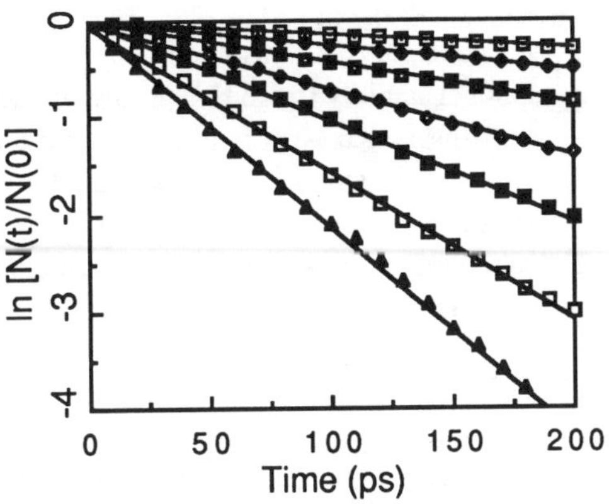

Figure 5. Plots for Ar_{13} unimolecular evaporation at energies: $20\varepsilon \leq E \leq 26\varepsilon$. Straight lines are least-squares fits to the MD data.

$$\ln[N(t)/N(0)] = -kt, \tag{2}$$

where N(0) is the total number of trajectories, N(t) is the number of clusters surviving (not yet evaporated) at time t, and k is the unimolecular rate constant. Thus, the rate constant, k(E,J), is given by the negative slope of the straight line plot of ln[N(t)/N(0)] *vs* time. Figure 5 shows a series of such plots for the reaction $Ar_{13} \rightarrow Ar_{12} + Ar$ with $20\varepsilon \leq E \leq 26\varepsilon$ and J=0. We use a distance criterion to determine whether an atom has left the cluster but then use a dynamical test to fix the time of evaporation in order remove delay effects associated with the variable velocity of the leaving particle. Additional details are given in [16] and [17].

In Figure 6, we summarize the evaporation rates obtained from MD for Ar_{12}, Ar_{13} and Ar_{14} with J=0. The rates are plotted against the total energy per atom in order to

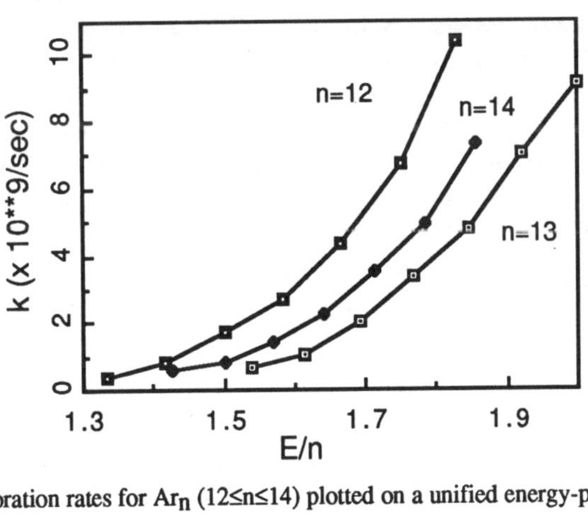

Figure 6. MD evaporation rates for Ar_n ($12 \leq n \leq 14$) plotted on a unified energy-per-atom energy scale (in ε_{Ar}). Note that the rate scale is linear and spans about a decade.

compare the relative evaporation rates of the three clusters: on an energy-per-atom basis, we observe that $k_{12}>k_{14}>k_{13}$. The same trend has been observed by Smith [31]. Not surprisingly, Ar_{13}, with its exceptional stability (large E_0) has the smallest rate constant for a given E/n value. What is surprising is that Ar_{14} is found to be more stable than Ar_{12} despite the fact that the E_0 for Ar_{14} is very much smaller than that for Ar_{12} (see Table 1). Simple (harmonic) RRK theory (see section 3.2.1) would predict, on that basis, that $k_{14}>>k_{12}$. We shall see later that the inclusion of the cluster anharmonicity effects into the densities-of-states will resolve this puzzle.

3.2. STATISTICAL THEORIES OF EVAPORATION

Statistical rate theory in its various forms requires the computation of the ratio of the phase space volume of a reaction bottleneck or transition state to the volume of the reactant system [32,33]. We now present rate expressions obtained from three different theories representing different assumptions about the transition state and different levels of approximation: RRK theory, phase space theory, and the Engelking model.

3.2.1. *RRK Theory* . One of the simplest such theories, the so-called RRK model, considers the dissociating molecule as an m-dimensional oscillator. At the transition state, one oscillator is singled out as the dissociation coordinate and the density of states for the transition state is thus taken as that of an (m-1)-dimensional oscillator. The harmonic density of states of an m-dimensional harmonic oscillator is [34]:

$$\rho_{H.O.}(E)=E^{m-1}/[(m-1)!\prod_{i=1}^{m}(h\nu_i)] \tag{5}$$

The following RRK expression:

$$R(E,\varepsilon_t) = D(m-1)(E-E_0-\varepsilon_t)^{m-2}/E^{m-1}, \tag{6}$$

is the differential rate of dissociation for a given total energy E, and energy ε_t in the dissociation coordinate; E_0 is the energy of the transition state. Equation 6 is usually obtained by using the harmonic density formula of equation 5 for both the transition state and the reactant, assuming the same isotropic oscillator model for both (a single Einstein frequency). D is the remaining frequency factor after cancellation of the rest of the terms in the frequency products.

Integrating equation 6 over all possible values of ε_t between 0 and $E-E_0$, we get the following standard RRK expressions for the microcanonical rate, the kinetic energy release distribution, and the average kinetic energy release:

$$k(E) = D[(E-E_0)/E]^{m-1} \tag{7a}$$

$$P(E,\varepsilon_t) = (m-1)(E-E_0-\varepsilon_t)^{m-2}/(E-E_0)^{m-1} \tag{7b}$$

$$\langle\varepsilon_t\rangle = (E-E_0)/m. \tag{7c}$$

3.2.2. *Phase Space Theory.* PST rigorously incorporates angular momentum constraints assuming a loose transition state (i.e. the products are the transition state) [35,36]. We loosely follow the discussion of Jarrold [33] in what follows, letting primed quantities stand for products. For reaction (1), the differential rate is given by:

$$R(E,J,\varepsilon_{tr}) = \frac{gS_r' \rho'(E-E_0-\varepsilon_{tr})\Gamma'(J,\varepsilon_{tr})}{S_r \rho(E-E_r)}, \tag{8}$$

where E is the total energy and E_r the rotational energy of the reactant, J is the total angular momentum, and ε_{tr} is the total kinetic energy release (translational plus rotational); ρ and ρ' are the vibrational densities-of-states of the reactant and product, respectively; Γ' is the rotational density-of-states of the products; g is the channel degeneracy factor (ratio of symmetry numbers); and S_r and S_r' are rotational degeneracies. PST expressions analagous to Equations 7 are:

$$k(E,J) = \int_0^{E-E_0} R(E,J,\varepsilon_{tr})d\varepsilon_{tr} \tag{9a}$$

$$P(E,J,\varepsilon_{tr}) = R(E,J,\varepsilon_{tr})/\int_0^{E-E_0} R(E,J,\varepsilon_{tr})d\varepsilon_{tr} \tag{9b}$$

$$<\varepsilon_{tr}> = \int_0^{E-E_0} \varepsilon_{tr} P(E,J,\varepsilon_{tr})d\varepsilon_{tr}. \tag{9c}$$

Clearly the difficulty in applying PST exactly even in its classical version is the calculation of the exact vibrational densities-of-states of reactant and products; the rotational density of states, Γ' must also be obtained. In this work we have calculated all of these quantities extremely accurately for J=0. We shall not dwell, here, on our approach to obtaining the product rotational density, except to mention that the radial potential on which the centrifugal barriers (which determine Γ') are built is calculated numerically and then fit to a function of the form $-C/(r-r_0)^6$. This provides a much better representation of the radial dissociation potential for clusters than the $1/r^6$ form usually used for neutral molecule dissociation (essentially because the radius of the cluster, r_0 is taken into account). We refer the interested reader to references [16] and [17] for details.

It turns out that the anharmonic contribution to the density-of-states is of much greater significance to the calculation of rates than the fine points of the rotational density, at least for the clusters considered here. Before turning to the vibrational density-of-states problem in Section 3.3, we shall consider one additional approximate statistical theory.

3.2.3. *Engelking model.* This model was first proposed [9] to study the evaporation cascade of $(CO_2)_n^+$ clusters [10]. As pointed out by Jarrold [33], the model of Engelking is related to the Weisskopf model of nuclear decay [37]. The differential dissociation rate in this model is given by (m=3n-6):

$$R(E,\varepsilon_t) = \frac{8\pi Sg\mu(m-1)(m-2)(m-3) \prod_{i=1}^{m} (h\nu_i)(E-E_0-\varepsilon_t)^{m-4}\varepsilon_t}{h^3 \prod_{j=1}^{m-3} (h\nu_j') E^{m-1}} \tag{10}$$

where S is the collision cross section for the product cluster, μ is the monomer mass, and

ε_t replaces ε_{tr} since the Engelking model does not explicitly account for angular momentum. This model uses the harmonic density of states. Engelking expressions analagous to Equations 7 and 9 are:

$$k(E) = \prod_{i=1}^{m}(h\nu_i)8\pi Sg\mu(m-1)(E-E_0)^{m-2}/\prod_{j=1}^{m-3}(h\nu_j')E^{m-1} \tag{11a}$$

$$P(E,\varepsilon_t) = (m-2)(m-3)(E-E_0-\varepsilon_t)^{m-4}/E^{m-2} \tag{11b}$$

$$<\varepsilon_t> = 2(E-E_0)/(m-1). \tag{11c}$$

It is interesting to note that $\Gamma'(J=0,\varepsilon_{tr})$ of equation 8 is nearly a linear function of the kinetic energy release--the Engelking theory gives a strictly linear term, ε_t, in equation 10. Also note that the simple RRK expression for the average kinetic release (equation 7c) differs by a factor of about 2 from the Engelking prediction (equation 11c).

In our applications of the various theories briefly presented here, we permit no adjustable parameters: the dissociation energies of these clusters are known from simulations (Table 1) and the frequency factors present in the rate formula (equation 7a) of conventional RRK and in the original formulation of the Engelking model are eliminated by diagonalizing the Hessian matrix at the minimum energy of both reactants and products to obtain explicit values for all vibrational frequencies within our potential model. For the RRK expression we take D to be the cube root of the ratio of frequency products obtained from this procedure. One final note concerning the expressions for the kinetic energy release distributions and the average kinetic energy release: because of the normalization, most obviously displayed in equation 9b, all of the troublesome constants such as frequency factors, degeneracy factors, and the like, have no influence on the calculation of the quantities associated with the kinetic energy release.

3.3 THE VIBRATIONAL DENSITY-OF-STATES OF VERY ANHARMONIC CLUSTERS

3.3.1. *The multiple histogram method.* We outline here our use of the multiple histogram method to calculate vibrational densities-of-states. Our approach is similar to that of Labastie and Whetten [38], who used a Monte Carlo procedure to generate the densities-of-states of Ar_{13}, Ar_{55}, and Ar_{147} in order to investigate the melting of these clusters from a statistical thermodynamic point of view, comparing their results with the trajectory calculations of Jellinek, Beck, and Berry [39].

Figure 7 schematically shows a set of overlapping energy distributions or histograms which give the probability of Ar_{13} having energy E during a canonical simulation at different temperatures. Each distribution is indexed by temperature ($\beta=1/kT$) and can be expressed as:

$$P_\beta(E) = N_\beta(E)/n_\beta = \Omega(E)e^{-\beta E}/Q(\beta), \tag{12}$$

where $N_\beta(E)$ is number of samples of energy E during the simulation at β, n_β is the total number of samples during the run, and $\Omega(E)$ and $Q(\beta)$ are the microcanonical density and the canonical partition function, respectively. Taking logarithms of both sides:

$$\ln N_\beta(E) - \ln n_\beta = \ln \Omega(E) - \beta E - \ln Q(\beta). \tag{13}$$

We then discretize over N energy bins, E_i (i=1,N) and M simulations at different

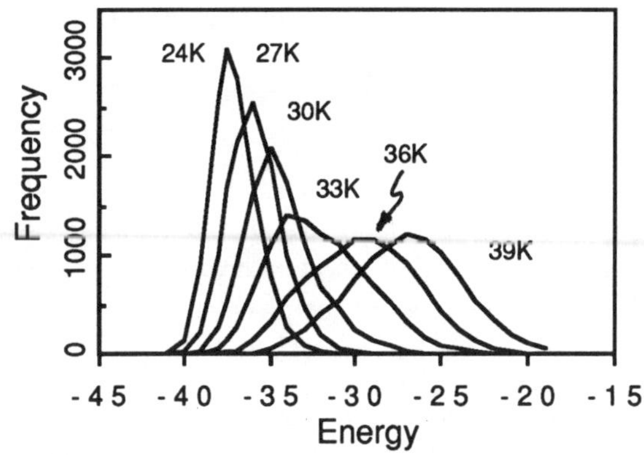

Figure 7. Energy histograms for Ar_{13} from constant temperature Nosé dynamics. Each distribution consists of an equal number of samples. The asymmetry and broadening in the distributions at 33K and 36K reflects the melting transition in Ar_{13} that occurs between 34 and 35 K. In this plot, the energy zero is taken at the atomization limit.

temperatures, β_j (j=1,M), to obtain:

$$\ln N_j(E_i) - \ln n_j = \ln \Omega(E_i) - \beta_j E_i - \ln Q(\beta_j) \qquad (i=1,N;\ j=1,M) \qquad (14)$$

This is an overdetermined system of (NxM) equations in (N+M) unknowns for which a least-squares fitting procedure must be used to find the unknowns, $\ln\Omega_i$ and $\ln Q_j$. Rearranging, we have (with an obvious shift in notation):

$$\ln N_{ij} + \beta_j E_i = \ln \Omega_i - \ln Q_j + \ln n_j \qquad (i=1,N; j=1,M) \qquad (15)$$

The left hand side of this expression contains all the (computer) experimental data from our constant temperature simulations. Calling the l.h.s. W_{ij}, we then form the residuals:

$$R_{ij} = [W_{ij} - (\ln \Omega_i - \ln Q_j + \ln n_j)] \qquad (i=1,N; j=1,M) \qquad (16)$$

and then the χ^2 is:

$$\chi^2 = \sum_{i=1}^{N} \sum_{j=1}^{M} N_{ij} R_{ij}^2. \qquad (17)$$

Taking derivatives and setting = 0,

$$\begin{aligned} \partial(\chi^2)/\partial(\ln \Omega_i) &= 0, \qquad i = 1,N \\ \partial(\chi^2)/\partial(\ln Q_j) &= 0, \qquad j = 1,M. \end{aligned} \qquad (18)$$

yields a set of (N+M) equations in (N+M) unknowns! This procedure solves the problem of deconvoluting the intrinsic density-of-states from the Boltzmann factor in equation 13. Note that several other authors [40,41] have independently developed similar techniques.

In contrast to Labastie and Whetten, we use a thermostatted MD procedure rather than Monte Carlo to obtain the energy histograms that enter into the procedure outlined above. Among other advantages [17], the Nosé dynamics discussed below conserves angular momentum and thus yields appropriate densities for comparison with the J=constant evaporation dynamics whose results are given in Section 2.

3.3.2. *Nosé dynamics*. In 1984, Nosé [42] proposed an extended dynamics in which a fictitious degree of freedom, s, serves as a heat bath for the physical system of interest. The energy of the extended system is given by:

$$H = \sum_{i=1}^{3n} p_i^2/2m_i + \phi(\{q_i\}) + p_s^2 s^2/2Q + gkT_{ext}\ln(s) \tag{19}$$

where $\{q_i,p_i\}$ are the coordinates and momenta of the physical system and s and p_s represent the coordinate and momentum of a heat bath coordinate coupled to the system of interest. Here, ϕ is the potential energy of the physical system; Q is a parameter with dimensions of *energy*x*time*^{2} which acts as a mass for the bath coordinate and modulates the energy flow between the physical degrees of freedom and the heat bath; and g is the effective number of degrees of freedom in the system (g=3n-7 for an n atom cluster). The temperature control parameter, T_{ext}, is held constant and the value of the parameter, Q, affects the magnitude of kinetic and potential energy fluctuations. The H of equation 19 is technically not a Hamiltonian since the appropriate equations of motion (see reference 17 and 42 for details) are not of the conventional form.

For Nosé dynamics, the average value of the kinetic temperature,

$$kT_{kin} = 2\langle\sum_{i=1}^{3n} p_i^2/2m_i\rangle/g \tag{20}$$

of the physical system and the total H of the extended system remain constant (with $T_{kin}=T_{ext}$), while the energy, E, of the physical system varies as in the canonical ensemble. Additional details about the procedure are given elsewhere [17], but it is important to remark that at high T, the Nosé trajectories can dissociate. We use a dynamical criterion similar to that used for the microcanonical MD study of Section 3.1 to stop such trajectories in order to avoid counting part of the product phase volume in with the bound or quasibound states.

To determine the vibrational density-of-states of Ar_{13}, for example, we bin the energies from Nosé trajectories at 42 temperatures between 0.5 K and 40 K with between 80000 and 360000 energy samples depending on the temperature. Using the fitting procedure outlined in Section 3.3.1, the anharmonic density-of-states is determined up to a multiplicative constant. Figure 8 shows the natural logarithm of the vibrational density of Ar_{13} and Ar_{12} as a function of energy (we note that $\Omega(E)$ has units of ε_{Ar}^{-1}). Also shown are the harmonic reference densities for collections of oscillators whose frequencies are those of the Ar_{13} and Ar_{12} normal modes at their respective minima. We remind the reader that only the *shape* of the product density curve is used in predicting the kinetic energy release distribution because of the normalization condition. However, calculating the rate from PST requires a knowledge of the *absolute* densities of product and reactant clusters.

We have attempted to determine absolute classical densities by matching the slope of the anharmonic density to that of the harmonic density at the origin (E-->0) as can be seen in Figure 8. This fitting procedure implicitly assumes that the anharmonic density is always higher than the harmonic density. As will be seen in Section 3.4.2, this assumption is not necessarily valid and can lead to poor results, so we will turn to the adiabatic switching or Hertz invariant method of Reinhardt [43,44] to attempt to determine the absolute densities-of-states.

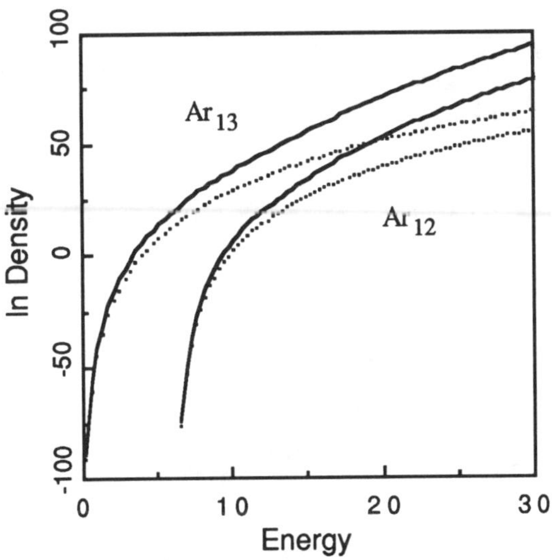

Figure 8. Anharmonic densities of Ar_{13} and Ar_{12}. The anharmonic curves are fit to the slope of the harmonic reference densities at the origin(s). The anharmonic contribution is very large at high energy. Energy in ε_{Ar}.

3.4. RESULTS

3.4.1. *Kinetic Energy Release Distributions.* We show in Figure 9 a plot of the distribution of the kinetic energy release, ε_{tr}, for the evaporation: Ar_{13} ---> Ar_{12} + Ar from an ensemble of MD trajectories at E=23 ε_{Ar}. Also shown are the predictions of RRK, PST, and the Engelking model. PST is seen to give perfect agreement with simulation; the Engelking model has a maximum away from the origin but is too broad with too long a tail; and RRK theory has the wrong behavior at the origin since the reaction coordinate is assumed to involve only one mode of the cluster.

In Section 3.2.3, we noted that the rotational density of the product, $\Gamma'(0,\varepsilon_{tr})$ is very nearly a linear function, making the probability distribution of the kinetic energy release a test of the influence of the *shape* of the vibrational density-of-states function, ρ', of the product. Indeed, if we replace the harmonic density used in the Engelking expression by the anharmonic density determined above, there is much closer, though not perfect, agreement between Engelking and simulation (and thus with PST). The results of Figure 9 are quite typical and plots of the average kinetic energy release *vs* ε_{tr} show that PST works perfectly in comparison with simulation results. It is quite interesting that the harmonic Engelking model tends to grossly overestimate (by about a factor of 2) the average kinetic energy release while RRK underestimates this quantity but comes close to the PST value. Despite the wrong shape of the RRK distribution, the tail of the RRK curve of Figure 9 follows the accurate PST curve quite closely. Results for Ar_{14} and Ar_{12} are qualitatively the same as those presented here for Ar_{13}.

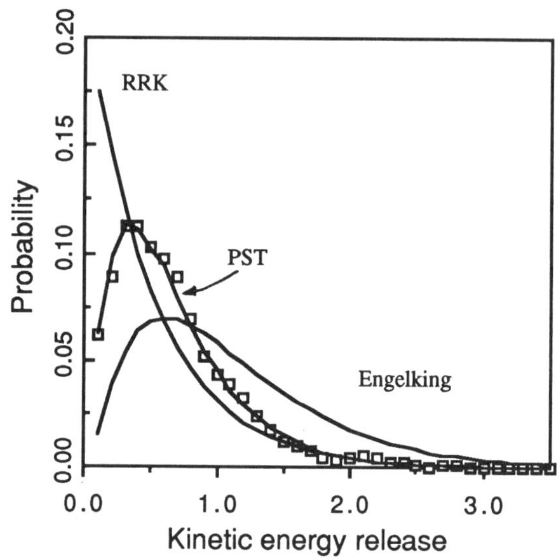

Figure 9. Kinetic energy release distributions for Ar_{13} at 23 ε. Squares are MD data; PST curve is **not** fit to the MD results.

3.4.2. *Rates.* We now turn to a consideration of the evaporation rates predicted by the various theories and their comparison with the simulation results. Recall that the application of equation 7a requires that the absolute densities for both reactant and product be known. The Engelking model and RRK are treated in the harmonic approximation while the anharmonic densities shown in Figure 8 are used for PST. Table 2 gives the results for one energy taken from the middle of the range considered for each cluster. A more complete account is given elsewhere [17]; the data presented here is typical of the whole body of results. Let us first look at the results for Ar_{13} since this is the first case that we investigated and the relatively good agreement between the MD result and the prediction of PST (less than a factor of 2 difference) is really very good for a rate theory with no adjustable parameters. For the same cluster, the Engelking model is way too high amd RRK is a factor of 3 too high. Before congratulating ourselves, however, we must

Table 2. Evaporation rate constants for Ar_n clusters in ns^{-1}.

n	E	k_{MD}	k_{PST}	$k_{Eng.}$	k_{RRK}
14	23	2.26	0.028	16000	25.7
13	23	3.29	2.12	1475	9.35
12	19	2.93	18.8	1200	0.96

look at the data for Ar_{12} and Ar_{14}, calculated in the same way as for Ar_{13}. Here the situation looks much worse. PST predicts an evaporation rate that is 100 times too low for Ar_{14} while the same prediction for Ar_{12} is a factor of about 6 too high! The Engelking model gives results which are uniformly high. RRK, interestingly enough gives results which are in the right range (within an order of magnitude in every case.

Careful checking of rotational constants, fitting procedures, and the like reveals that errors in these pieces of the overall calculation may contribute an error of the order of a factor of 2 to the rates. This is, in some sense, encouraging since it tells us that for Ar_{13} we've done about as well as possible. We are still left with the problem of the PST rates for Ar_{14} and Ar_{12} evaporation. In looking for a possible source of systematic error we must remember that the deviation is found to be positive for one cluster and negative for another. A major assumption of our calculation is that we have determined the absolute densities-of-states by fitting the curvatures of the anharmonic densities to their harmonic counterparts at the origin. If this assumption is incorrect we can account for the deviations in Table 2. We have thus used a new method to determine absolute densities to try to eliminate the remaining error.

3.5 ABSOLUTE DENSITIES BY ADIABATIC SWITCHING

3.5.1. *The Hertz Invariant Method.* Reinhardt [43] has recently proposed an adiabatic switching method to determine the densities-of-states and other thermodynamic properties of realistic systems and Watanabe and Reinhardt [44] have presented results of this technique for liquid water. Briefly, the method is based on the observation by Hertz [45] that the energy shell is an invariant with respect to adiabatic modification of the Hamiltonian; this idea coupled with Liouville theorem implies that the phase volume is conserved during adiabatic switching. The Hertz invariance principle is strictly true only for a small class of Hamiltonians; the idea of Reinhardt is to use this technique as an averaging method even in situations where the Hertz invariance principle is not strictly obeyed. The hope, which seems to be fulfilled for the cases studied, is that final energies of the trajectories which are switched into the reference Hamiltonian will not have too large a dispersion and that the density-of-states for the reference system at the average energy will be close to the true density of the physical system of interest.

We transform the Hamiltonian of our argon clusters into a harmonic Hamiltonian

$$H(\{p,r\}) = \sum_{i=1}^{3n} p_i^2/2m_i + a(t)\sum_{i<j} {}^1\!/_2 k(r_{ij} - r_{ij}^0)^2 + [1 - a(t)]\sum_{i<j} V_{L\text{-}J}(r_{ij}) \qquad (21)$$

where the switching function is of the C^2 form given by Reinhardt,

$$a(t) = t - \sin(2\pi t)/2\pi \qquad (0 \le t/T \le 1) \qquad (22)$$

and where T is the total switching time. In the second term of equation 21, the harmonic potential is defined by the structure of the minimum energy isomer of the Lennard-Jones potential; only nearest neighbors are allowed to interact and the harmonic force constant is adjusted so that the product of normal mode frequencies in the harmonic limit is identical to that of the Lennard-Jones minimum. Angular momentum is conserved (and kept equal to zero) during the switching.

For argon clusters, switching times of 70-100 ps give converged results, reliable enough for our subsequent modelling of evaporation kinetics. This is in accord with the findings of Watanabe and Reinhardt [44] who showed that when a sample of liquid water is switched into an ideal gas, switching times greater than 2-3 ps are adequate for convergence. The longer switching time needed for argon as compared to water is due to the longer period of a typical argon vibration. The beauty of the method is that, practically speaking, switching times of the order of a few vibrational periods are sufficient rather than, as might have been supposed, the much longer time more required to achieve ergodicity. The adiabatic switching estimate of the density of states of a cluster of Lennard-Jones particles is then given by:

$$\Omega_{L\text{-}J}(E) = \Omega_{H.O.}(\mathfrak{E}) \qquad (23)$$

where $\mathfrak{E}$ is the average final energy of an ensemble of trajectories initiated in the Lennard-Jones Hamiltonian at energy E.

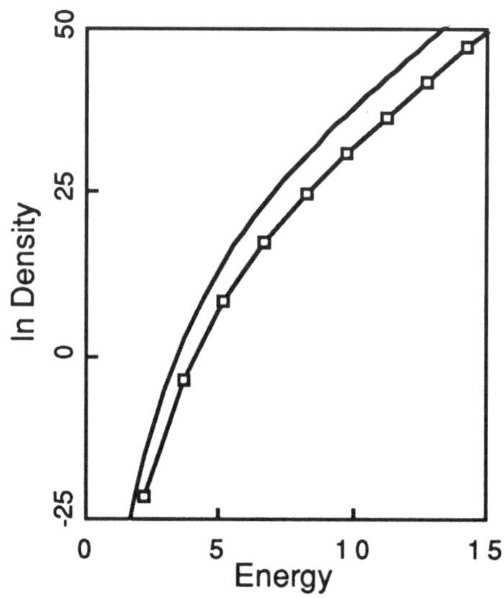

Figure 10. Comparison of vibrational densities-of-states for Ar_{13} determined by the multiple histogram method (upper curve) and by adiabatic switching (squares). The difference between the two curves converges to a constant value as E-->0. Energy in ε_{Ar} units.

A quick look at the results of adiabatic switching is offered in Figure 10. The upper, smooth curve is the same anharmonic density curve ($\ln\Omega_{13}$) as shown in Figure 8. Its position on the vertical scale is determined by the curvature at the origin. The lower curve, on the other hand, is the density-of-states determined by adiabatic switching at the points marked by squares and connected by straight lines. Careful inspection shows that the difference between the two curves is nearly constant and that the shift converges as one approaches the origin [17].

Since the multiple histogram method has permitted us to determine the shape of the density-of-states curve extremely accurately, we use the results of adiabatic switching to determine the absolute density at one or a few points away from E=0, in order to shift the whole curve into the right position. The second column of Table 3 gives the converged difference in the logarithms of the densities determined by the multiple histogram method fit to the harmonic reference (denoted HH) and by the adiabatic switching (AS) technique. In all cases, the AS density was found to be lower than the HH density. These numbers are thus the values that must be subtracted from the $\ln\Omega$ curves previously fit to the harmonic origin as in Figure 8, in order to shift the densities into agreement with the results from switching.

Table 3. Results from adiabatic switching. Δ_n is the amount the HH density curve for Ar_n must be shifted *down* in order to match the AS results. The last column gives the corrected rates (in ns^{-1}).

n	$\Delta_n \equiv \ln\frac{\Omega_{HH}}{\Omega_{AS}}$	k_{MD}	k_{PST}^{HH}	k_{PST}^{AS}
14	11.66 ± 0.44	2.26	0.028	7.42
13	6.08 ± 0.54	3.29	2.12	9.79
12	4.55 ± 0.42	2.93	18.8	3.10
11	6.35 ± 0.23			

3.5.2. *Rates Revisited.* Table 3 also gives (for the same energies as in Table 2) the PST rates calculated using densities adjusted as indicated above. The effect of the density shifts on the rates is very simple to calculate. Since the reactant density appears in the denominator of equation 8, a downwards shift of this density results in an increase in the rate. Thus by multiplying the PST/HH rates by the simple expression: $\exp[\Delta_n - \Delta_{n-1}]$, one obtains the PST/AS rates.

In order to facilitate comparison, the MD rates and the PST/HH rates are also shown. The results are very encouraging. The Ar_{13} rate which was about a factor of 2 too low is now found to be about a factor of 3 too high while k_{12} shifts from too high a value into almost perfect agreement with the MD rate. Finally, k_{14} which was about two orders of magnitude too small is now brought to within a factor of 3 (too high) of the MD value. Thus the adiabatic switching adjustment to the density has in no case reduced the agreement between PST and MD while in the one case where a serious discrepancy existed (for Ar_{14}), the adjustment resolves the problem. In evaluating these results it should be stressed that for all three clusters, the PST/AS rates and the "experimental" MD rates agree to within the uncertainty imposed jointly by the MD statistics and the residual uncertainties in the "absolute" densities as determined by switching (see Table 3).

We have shown that, in addition to getting the kinetic energy release distribution right, rigorous PST (with no adjustable parameters) is capable of predicting the absolute rates of evaporation of argon clusters. We can thus proceed, with some degree of confidence to extrapolate the PST rate predictions to lower energies and longer timescales where MD trajectory calculations become too expensive to perform. Figure 11 is a plot of the PST rates determined for Ar_{12}, Ar_{13} and Ar_{14} *vs* E/n. Since MD and PST agree to within the statistical uncertainties of both calculations, we have taken the liberty, in this plot, of shifting the rate curves for Ar_{13} and Ar_{14} into agreement with the MD values. In contrast to Figure 6, the rate scale is now logarithmic. Both the k_{12} and k_{13} curves show relatively smooth behavior with k_{12} higher than k_{13} over the whole energy range. It is the k_{14} curve which is most interesting for its dramatic change in slope. At low energies, the evaporation rate of Ar_{14} is much higher than that of Ar_{13} as would be expected from the RRK model. In fact, at the reduced energy at which Ar_{14} evaporates on the microsecond timescale, Ar_{13} lives on the order of seconds! At high energy, on the other hand, the effect

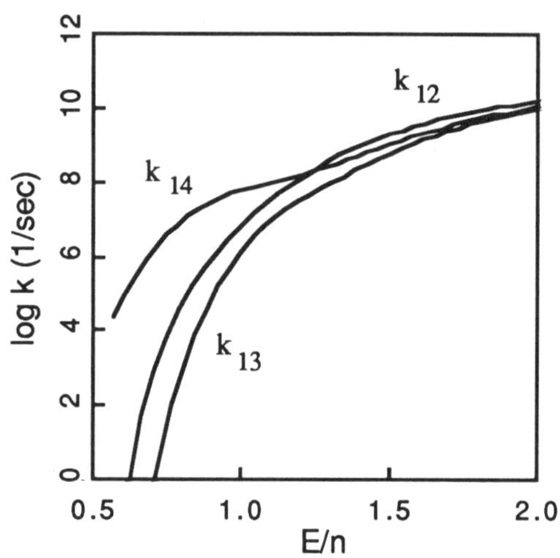

Figure 11. PST rates for Ar_n evaporation extrapolated to low energies. The shape of the k_{14} curve is due to cluster anharmonicity. Energy in units of ε_{Ar}.

of cluster anharmonicity is clearly noticeable in the reduction of the Ar_{14} rate.

These results, their interpretation, and their consequences are discussed in greater detail elsewhere [17]. It is most interesting to note here that the accurate PST calculations confirm the *ordering* of the cluster evaporation rates predicted by the simple RRK picture for low energies! At high energies, both MD and PST show that the effects of abnormal cluster stability or instability ("magic") tends to be washed out due to cluster anharmonicity effects. The k_{14} and k_{12} rate curves are essentially parallel in the RRK results though, of course, at even higher energies, the RRK evaporation rates for the three clusters must come together as they converge to the limits imposed by the frequency factors. It will be interesting to see if experiments can be devised to probe the 1-10 nanosecond time window to verify whether the picture presented here is correct. It is also possible [46] to extend the MD calculations to longer times than those reported here in order to check the change in slope of the Ar_{14} rate curve.

4. Caging, Evaporation, and the Statistical Limit

Evaporation processes are ubiquitous in the field of clusters: they are seen to play an important role in the cooling of clusters prepared in supersonic expansions; in amplifying the effect of slightly abnormal stabilities of certain cluster sizes, yielding "magic numbers" from which structural properties can be inferred; and finally, these processes can even be a diagnostic of photochemistry as in the caging experiments of Lineberger *et al.* [10,12,13]. The central question raised here is: How do the observed evaporation phenomena depend on the mode of excitation of the cluster?

Statistical theories such as PST or RRKM (for a "tight" transition state [32]) are applicable if the energy deposited in the system is redistributed more rapidly than the characteristic time for evaporation to occur. This criterion appears to be satisfied for large clusters given the success of approximate evaporation cascade calculations in explaining and fitting results for a variety of systems [7,8,47]. The snapshots of Figure 1b also illustrate the fact that while the Br_2Ar_{70} cluster receives a localized excitation, evaporation is much slower than the relaxation of the excitation of the diatomic [11].

On the experimental side, the observation by Lineberger *et al.* [10,12,13] that the average number of CO_2s lost is a linear function of the photon energy implies that all the energy deposited in the diatomic is equally available for evaporation, supporting the notion that the excitation energy is rapidly redistributed. As mentioned above, the picosecond experiments on $I_2^-(CO_2)_n$ demonstrate that the relaxation of the diatomic is very rapid: ~10-30 ps depending on the size of the cluster. What is not known, however, are the rates of evaporation of the CO_2s which leave (presumably sequentially) from the cluster. One could attempt to shorten the time window of the experiment in order check whether fewer CO_2s leave the cluster (on average) if less time is given for the last, slowest evaporation.

Alternatively, one could simulate the evaporation process directly in order to check a) whether simulation gives the same rapid I_2^- relaxation observed in experiment; b) how fast the CO_2s leave the cluster; and c) whether this dissociation is statistical or nonstatistical. Very preliminary results of calculations of this sort [30] indicate that both the the I_2^- relaxation and the evaporation processes are very rapid (10-100 ps timescale). To decide the question of whether the evaporation processes deviate from a statistical limit even using simulation with a given potential model may require the use of the methods of Section 3 in order determine the statistical result first. A simpler but perhaps less sure method is to compare the results of cluster trajectories at the same energy but prepared both with localized and nonlocalized excitation [30].

We turn finally to the question of the kinetic energy release distributions resulting from cluster evaporation. This distribution can be measured experimentally and of course can be determined by simulation or from statistical theory as shown in Section 3.4. Substantial progress in this area has already been made for the Ar_n^+ cluster system. On the basis of spectroscopic studies [48,49] and theoretical calculations [50-52], it is now well known that Ar_n^+ is consists of small molecular ion core (most likely linear Ar_3^+) surrounded by essentially neutral argons. There is also evidence for the existence of rare gas cluster ions containing a tetramer ion core [50b,53]. Several groups [54-56] have measured the kinetic energy release of Ar^+ ions following the photoexcitation of Ar_3^+ using both unpolarized light and light polarized along the beam direction to enhance the dissociation in the direction of the detector (a time-of-flight mass spectrometer). The results summarized here are for photoexcitation at wavelengths shorter than 600 nm to exclude the Ar_2^+ channel shown to exist [48] at longer wavelengths.

The results show that the Ar^+ signal is split into a triplet of peaks. The central peak corresponds to argon ions moving with essentially zero kinetic energy release while the outer components correspond to fast ions with center-of-mass velocities towards and away from the detector. The interpretation of the results is somewhat controversial; the recent hemiquantal dynamics calculations of Gadea *et al.* [57,58] indicate that the fast ions correspond to the outer atoms of the trimer ion taking the charge during the dissociation, while an internal conversion process, leaving the charge on the central atom, gives rise to the slow component in the TOF spectrum. Nagata *et al.* [59] have also extended these measurements to larger clusters and find that an Ar_2^+ signal is detected from Ar_4^+ and that this signal gets larger for the larger clusters. This result indicates that some kind of caging is taking place, especially since this channel is closed for Ar_3^+ itself at the short wavelengths (higher photon energies) of this experiment. It will be quite interesting to see what measurements of the kinetic energy release distributions of the neutral atoms [60] reveal about this kind of caging. Hemiquantal dynamics calculations are also underway for Ar_n^+ (n≥4) [61].

For larger argon ion clusters, of course, MD simulations such as the ones described here offer the most reasonable route to understanding the dynamics of evaporation. The result of Nagata *et al.* [59] is most encouraging since it implies that the photon energy is more and more efficiently transferred to a diatomic species with increasing size, offering the hope that statistical methods will be valid even for very small sizes. The methods of Section 3, supplemented perhaps by quantum corrections, offer the possibility of accurate calculations of the thermodynamic properties of both ionic and neutral atomic and molecular clusters. These equilibrium properties then form the basis for statistical reaction rate theories against which the results of experiment and MD simulations of non-equilibrium systems can compared.

Acknowledgements

We acknowledge the University of Maine computer center for a generous grant of computer time on its IBM 3090 computer. FGA thanks the Laboratoire de Photophysique Moléculaire and the CNRS for their hospitality and support while this article was prepared. We also thank Dr. L. Perera for stimulating discussions.

5. References

1. M.R. Hoare and P. Pal, Adv. Phys. **20**, 161 (1971). See also, M.R. Hoare, Adv. Chem. Phys., Vol. XL, pp. 49-135, edited by I. Prigogine and S.A. Rice (Wiley, New York), 1979.
2. A. Mackay, Acta. Crystallogr. **15**, 916 (1962).
3. W.D. Knight, K. Clemenger, W.A. de Heer, W.A. Saunders, M.Y. Chou, and M. Cohen, Phys. Rev. Lett. **52**, 2141 (1984).
4. H.W. Kroto, J.R. Heath, S.C. O'Brien, R.F. Curl, and R.E. Smalley, Nature **318**, 162 (1985).
5. E.C. Honea, M.L. Homer, J.L. Persson, and R.L. Whetten, Chem. Phys. Lett. **171**, 147 (1990).
6. C.E. Klots, J. Chem. Phys. **83**, 5854 (1985).
7. C. Bréchignac, Ph. Cahuzac, J. Leygnier, and J. Weiner, J. Chem. Phys. **90**, 1492 (1989).
8. H. Haberland, H. Kornmaier, H. Langosch, M. Oschwald, and G. Tanner, in J. Chem. Soc. Faraday Trans. **86**, 2473 (1990).
9. P.C. Engelking, J. Chem. Phys. **85**, 3103 (1986); *ibid.* **87**, 936 (1987).
10. M.L. Alexander, M.A. Johnson, and W.C. Lineberger, J. Chem. Phys. **82**, 5288 (1985).
11. F.G. Amar and B.J. Berne, J. Phys. Chem. **88**, 6720 (1984).
12. M.L. Alexander, N.E. Levinger, M.A.Johnson, D.Ray, and W.C. Lineberger, J. Chem. Phys. **88**, 6200 (1988).

13. D. Ray, N.E. Levinger, J.M. Papanikolas, and W.C. Lineberger, J. Chem. Phys. **91**, 6533 (1989).
14. A. Stace, Nature **342**, 346 (1989)--a discussion of the Lineberger results on the picosecond dynamics of $I_2^-(CO_2)_n$ clusters.
15. L. Perera and F.G. Amar, J. Chem. Phys. **90**, 7354 (1989).
16. S. Weerasinghe and F.G. Amar, Z. Phys. D., in press, 1991.
17. S. Weerasinghe and F.G. Amar, manuscript in preparation.
18. J.T. Hynes, Ann. Rev. Phys. Chem., Vol. 36 (Academic Press, New York, 1985), pp. 574-578.
19. T.J. Chuang, G.W. Hoffman, and K.B. Eisenthal, Chem. Phys. Lett. **25**, 201 (1974).
20. D.J. Nesbitt and J.T. Hynes, J. Chem. Phys. **77**, 2130 (1982).
21. J.J. Valentini and J.B. Cross, J. Chem. Phys. **77**, 572 (1982).
22. I. NoorBatcha, L.M. Raff, and D.L. Thompson, J. Chem. Phys. **81**, 3715 (1984).
23. F.G. Amar, unpublished work.
24. J.A. Beswick, R. Monot, J.-M. Philippoz and H. van den Bergh, J. Chem. Phys. **86**, 3965 (1987) and references contained therein.
25. R. Alimi and R.B. Gerber, Phys. Rev. Lett. **64**, 1453 (1990).
26. D. Scharf, J. Jortner, and U. Landman, J. Chem. Phys. **88**, 4273 (1988).
27. C.S. Murthy, K. Singer, and I.R. McDonald, Mol. Phys. **44**, 135 (1981).
28. E.C.M. Chen and W.E. Wentworth, J. Phys. Chem. **89**, 4099 (1985).
29. F.G. Amar and L. Perera, Z. Phys. D. in press, 1991. See also, L. Perera, Ph.D. thesis, University of Maine, 1990.
30. F.G. Amar and L. Perera, work in progress.
31. R. W. Smith, Z. Phys. D, in press, 1991.
32. W. Forst, *Theory of Unimolecular Reactions*, Acadamic Press, New York, 1973.
33. M. F. Jarrold, in *Clusters of Atoms and Molecules*, H. Haberland, ed., Springer-Verlag, Berlin, in press, 1991. This review also contains references to the development of RRK theory and PST.
34. R.K. Pathria, *Statistical Mechanics*, Pergamon Press, Oxford, 1972.
35. C.E. Klots, J. Phys. Chem. **75**, 1526 (1971).
36. W.J. Chesnavich and M.T. Bowers, J. Chem. Phys. **66**, 2306 (1977); J. Am. Chem. Soc. **99**, 1705 (1977).
37. V. Weisskopf, Phys. Rev. **52**, 295 (1937).
38. P. Labastie and R.L. Whetten, Phys. Rev. Lett. **65**, 1567 (1990).
39. J. Jellinek, T.L. Beck, and R.S. Berry, J.Chem. Phys. **84**, 2783 (1986).
40. A.M. Ferrenberg and R.H. Swendsen, Phys. Rev. Lett. **63**, 1195 (1989).
41. C. Bichara, J.-P. Gaspard, and J.-C. Mathieu, J. Chem. Phys. **89**, 4339 (1988); Phys. Lett. A **119**, 462 (1987).
42. S. Nosé, Mol. Phys. **52**, 255 (1984); J. Chem. Phys. **81**, 511 (1984).
43. W.P. Reinhardt, J. Mol. Struct. **223**, 157 (1990). See also W.P. Reinhardt, M. Watanabe, and R.L. Waterland, in *Proceedings of the First International Conference on Classical Mechanics in Atomic and Molecular Physics*, edited by P. Gruijic, T. Grozdanov, and P. Krstic (World Scientific, Singapore, 1989), p. 26.
44. M. Watanabe and W.P. Reinhardt, Phys. Rev. Lett. **65**, 3301 (1990).
45. P. Hertz, Ann. der Physik **33**, 537 (1910).
46. C.E. Román and I.L. Garzón, Z. Phys. D, in press, 1991.
47. Extensive calculations of the evaporation cascade of photoexcited rare gas ion clusters have been carried out by H. Haberland and M. Schmidt, private communication.

48. N.E. Levinger, D. Ray, M.L. Alexander, and W.C. Lineberger, J. Chem. Phys. **89**, 5654 (1988).
49. Z.Y. Chen, C.R. Albertoni, M. Hasegawa, R. Kuhn, and A.W. Castleman, Jr., J. Chem. Phys. **91**, 4019 (1989).
50. J. Hesslich and P.J. Kuntz, Z. Phys. D **2**, 251 (1986); P.J. Kuntz and J. Valldorf, Z. Phys. D **8**, 195 (1988).
51. H.-U. Böhmer and S.D. Peyerimhoff, Z. Phys. D. **11**, 239 (1989).
52. M. Amarouche, F.X. Gadea, and J. Durup, Chem. Phys. **130**, 145 (1989).
53. M. Amarouche, G. Durand, and J.P. Malrieu, J. Chem. Phys. **88**, 1010 (1988).
54. C.A. Woodward, J.E Upham, A.J. Stace, and J.N. Murrell, J. Chem. Phys. **91**, 7612 (1989); C.A. Woodward, B.J. Whitaker, and A.J. Stace, J. Chem. Soc. Faraday Trans. **86**, 2069 (1990).
55. J.T. Snodgrass, C.M. Roehl and M.T. Bowers, Chem. Phys. Lett. **159**, 10 (1989).
56. T. Nagata, J. Hirokawa, T. Ikegami, T. Kondow, and S. Iwata, Chem. Phys. Lett. **171**, 433 (1990).
57. F.X. Gadea and M. Amarouche, Chem. Phys. **140**, 385 (1990).
58. F.X. Gadea, Z. Phys. D., in press, 1991; F.X. Gadea and J. Durup, presented at the 6th European Workshop on Molecular Spectroscopy and Photon-Induced Dynamics, Sant Feliu de Guixols, Spain , May 28-31, 1991.
59. T. Nagata, J. Hirokawa, and T. Kondow, Chem. Phys. Lett. **176**, 526 (1991).
60. A. Stace, private communication: Direct measurement of neutral fragment velocities should yield rather different information than that obtained by the previous measurements [A. Stace, J. Chem. Phys. **85**, 5774 (1985)] of the average kinetic energy release for Ar_n^+ evaporation based on the analysis of metastable peaks in the TOF mass spectra.
61. F.X. Gadea, private communication.

SEARCH FOR MODE SELECTIVE CLUSTER ISOMERIZATION

A. HEIDENREICH, I. SCHEK and J. JORTNER
School of Chemistry, Tel-Aviv University,
69978 Tel-Aviv, Israel.

ABSTRACT. The time-resolved dynamics of diffusionless cube → ring isomerization of the $(NaC\ell)_4$ ionic cluster was interrogated by constant energy molecular dynamics simulations utilizing the first passage time method. The isomerization was induced by several excitation modes of the nuclear motion, i.e., nonselective, bond selective, ion selective and normal mode selective vibrational excitations. The dependence of the isomerization rates on the excess vibrational energy was successfully analysed in terms of the statistical RRK theory. Concurrently, we have calculated the intracluster vibrational energy redistribution times, their dependence on the excitation mode and on the excitation energy. At high excess vibrational energies traces of the onset of mode selectivity are manifested in the nonstatistical temporal distribution of the isomerization events.

1. Introduction

Structural, thermodynamic, dynamic and transport properties of finite systems depend on the nature of the intermolecular interactions, the cluster size and the characteristics (canonical or microcanonical) of the ensemble [1-4]. Extensive information is currently available regarding the dynamics of cluster isomerization pertaining to "long-time" observables [1-7]. Significant supplementary and complementary information will emerge from time-resolved dynamics of cluster isomerization [8]. Using classical molecular dynamics (MD) simulations, we have recently explored [8] the time-resolved dynamics of the $(NaC\ell)_4$ cluster isomerization induced by an "instantaneous" temperature jump. The stable low-temperature configuration of the $(NaC\ell)_4$ cluster corresponds to a cube (the precursor of the bulk crystal structure), while the high temperature structure corresponds to a ring [5,6]. In this note we explore the time-resolved dynamics of this cube → ring isomerization reaction, which is induced by different modes of vibrational excitation. We focus on the following issues:

(1) Establishing the time scales for isomerization.
(2) Determination of the temporal distribution of the isomerization events and the (averaged) reaction rates.
(3) Investigating the consequences of distinct modes of vibrational excitation on the dynamics.
(4) Examining the applicability of statistical kinetic theories for the analysis of the isomerization rates.
(5) Establishing the time scales for intracluster vibrational energy redistribution (IVR).

J. Jortner et al. (eds.), Mode Selective Chemistry, 189–199.

(6) Exploring the competition between the IVR and the isomerization process.
(7) Searching for mode selectivity in cluster isomerization.

2. Methodology

We have performed constant energy classical molecular dynamics (MD) calculations with interionic potentials consisting of a point Coulomb term and a Born-Mayer repulsion term [9]

$$V = \sum_{i} \sum_{< j} \left[\frac{Q_i Q_j}{R_{ij}} + A_{ij} \exp\left(- \frac{R_{ij}}{\rho_{ij}} \right) \right] \tag{1}$$

(Q_i,Q_j: charges of ion i and j, respectively; R_{ij}: distance between ions i and j; A_{ij},ρ_{ij}: parameters of the Born-Mayer potential). A 5th order predictor-corrector algorithm [10] has been used to integrate the equations of motion. The integration time step has been chosen to be 0.1 fsec up to 1800K and 0.05 fsec at higher temperatures. The simulation time of a single trajectory was up to 1.2 nsec. To prepare an equilibrated cluster for further MD simulations, an initial MD calculation was performed with a simulation time of around 0.1 nsec, where the velocities were scaled several times according to the particular desired temperature. The calculations were carried out on a Silicon-Graphics 4D-80 GTX and on an IBM RISC RS/6000 workstation.

3. The Caloric Curve and Excitation of Nuclear Motion

The caloric curve (the total energy per ion vs the time-averaged kinetic energy per ion) for $(NaC\ell)_4$ was first simulated by Luo et al. [6]. The caloric curve (fig. 1) consists of two nearly parallel branches corresponding to the low-temperature cube and to the high-temperature ring structure. Each of the data points in fig. 1 was obtained by averaging the kinetic energy over a simulation time of 1 nsec for a given constant total energy. The linear branch of the cube terminates at ~740K, as at higher temperatures the cube is not stable anymore on the time scale of 1 nsec. The ring can be supercooled to low temperatures.

The differences in the intersects (at the U axes) of the two parallel branches give the isomerization energy, which is 310 cm^{-1} per ion. This is in agreement with Martin's static calculation result of 270 cm^{-1} per ion [5], which is based on geometry optimizations of the cube and the ring. The isomerization energy is relevant for the dynamics, because it defines a lower limit for the cube-ring reaction barrier.

Proceeding from thermodynamics to time-resolved dynamics we excited the nuclear motion of the cube at 600K, exploring the dynamic consequences of different (non-selective, bond selective, ion selective and normal mode selective) vibrational excitations. The excitation process can be represented in the caloric curve (fig. 1) as follows: We raise the total energy to a new value U_{new}. This is denoted by the vertical arrow in fig. 1. At the higher new total energy U_{new} the metastable cube undergoes isomerization after the vibrational excitation, so that the temperature T_{new} cannot be obtained by long-time averaging. Instead, the new average kinetic energy (and the new temperature T_{new}) after excitation was estimated by extrapolating the linear branch of the cube. This is denoted by the horizontal arrow in fig. 1. As a common feature of all excitation modes the nuclear configuration and therefore the potential energy was unchanged at the instant of nuclear excitation. Accordingly, the new total energy U_{new} was achieved only by rising the kinetic energy. Thus, directly after the excitation the cube contains a high excess of kinetic energy, which is then partly converted into potential energy. This implies that at the excitation instant the system is at the right side of the extrapolated branch of the caloric curve. After this the relaxation takes place along the U_{new} horizontal line, towards the new average kinetic energy value. We note that for the application of the Arrhenius and statistical unimolecular theories to describe the time-resolved dynamics, only the total energy U_{new} is of interest, rather than the temperature T_{new}. Nevertheless, the temperature T_{new} derived from the caloric curve is useful for the characterization of the system after excitation.

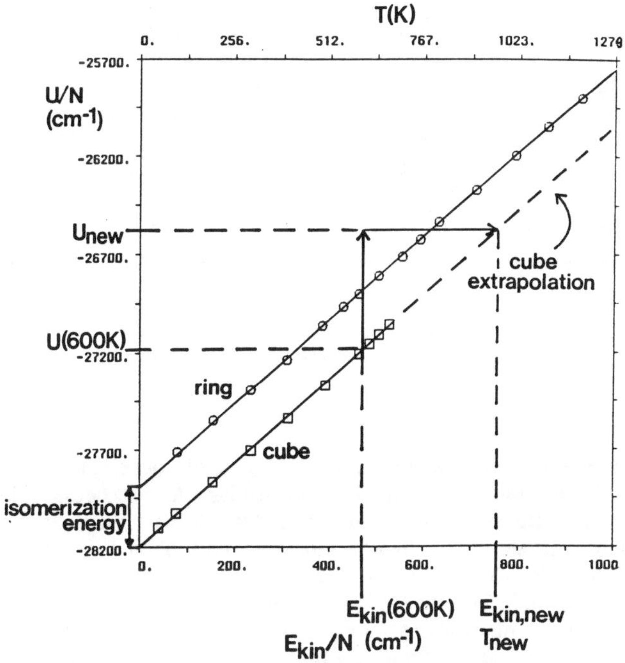

Figure 1. The caloric curve for $(NaC\ell)_4$. Data points (□ cube structure and o ring structure) are obtained from 1 nsec MD trajectories.

4. Excitation Modes

We turn to the specification of particular excitation modes of the nuclear motion:

(i) Nonselective excitation. The momental velocities of all particles of the 600K trajectory were scaled up by a uniform factor.

(ii) Bond selective excitation. The excitation energy is put into a single Na-Cℓ bond. Both bond stretching and bond compression have been carried out.

(iii) Ion selective excitation. A sodium ion gets velocity towards the center of mass, and the other ions get the appropriate velocities in the opposite direction to keep the total linear momentum zero.

(iv) Normal mode excitation. The particle velocities are a superposition of the velocities of the 600K trajectory and an appropriate amount along the particular normal coordinate. The normal mode coordinates and frequencies (Table 1) were obtained by diagonalizing the force constant matrix. The frequencies agree with earlier results of Martin [5]. From the data of Table 1 we note that the frequencies (characterized by vibrational periods in the range 150-300 fsec) lie quite close together.

TABLE 1. Normal mode energies (E) and harmonic vibrational period lengths (τ_H) for $(NaC\ell)_4$

normal mode number	normal mode	E/cm^{-1}	τ_H/fsec
1	$1A_1$	227	147
2,3,4	$1T_2$	213	157
5,6,7	$2T_2$	205	162
8	$2A_1$	196	170
9,10	1E	171	195
11,12,13	$1T_1$	156	213
14,15	2E	117	285
16,17,18	$3T_2$	110	294

All the excitations of the nuclear motion have been carried out so that the total linear momentum remains zero. This applies also to the total angular momentum except the normal mode excitation, where a small component of the angular momentum has not been quenched.

In real life the bond selective and ion selective excitations can possibly occur in cluster-atom or cluster-molecule collision processes. Nonselective excitation can perhaps be achieved by the collision of the cluster with a surface. Alternatively, nonselective excitation can perhaps be accomplished by high-order infrared multiphoton excitation, provided that intracluster intramolecular vibrational energy redistribution is fast, a problem which deserves further study. For the normal mode excitations we do not see an experimental possibility to put so many quanta into a single normal mode as high-order vibrational overtones are very weak in their spectral intensity. Nevertheless, the results of normal mode excitation are of methodological interest, as the normal modes are used for the analysis of the other excitation modes. As far as the interrogation of the time-resolved dynamics is concerned, we note that the cube and ring configurations can be experimentally distinguished by their infrared, Raman or electronic-vibrational spectra.

5. Reaction Rates

In our MD study the reaction rates for isomerization were obtained from the first passage times, τ, that means the time between excitation and the first occurrence of the ring isomer. The reaction mechanism was already reported by Luo et al. [6]. In the first step the cube opens to a ladder, whose central rungs open then into a ring in the second step. Several criteria were advanced to detect the ring isomer and to distinguish it from other configurations. One of the criteria rests on the analysis of the coordination number, which must drop from 3 in the cube to 2 in the ring.

We have calculated first passage times for excitations from 600K to temperatures between 900K and 2,400K (corresponding to excitation energies between 4,000cm^{-1} and 24,000cm^{-1} for the cluster). For different excitation modes at each temperature we collected a large number n (=200-400) of τ values, using several distinct times of the 600K MD trajectory as starting points for the excitation. The τ values for a given excitation mode and temperature can be represented in terms of a histogram of the temporal distribution of isomerization events (fig. 2). Such histograms give us a measure for the probability density p(τ), that an isomerization event happens at time τ after the excitation. From these τ values we have calculated the arithmetic mean value

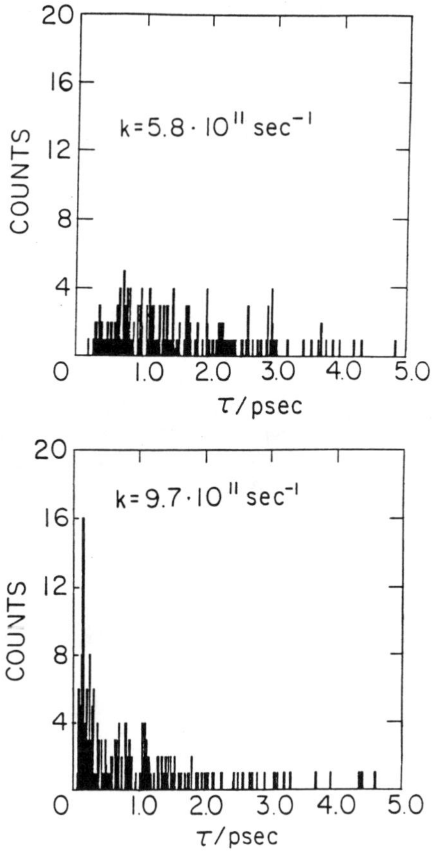

Figure 2. Histograms of the distribution of the first passage times, τ, for two mode selective excitations from 600K to 2400K. Upper histogram 1A_1 normal mode excitation and lower histogram 1T_1 normal mode excitation.

$$\langle \tau \rangle = \frac{1}{n} \sum_{i=1}^{n} \tau_i \tag{2}$$

and the isomerization rates, k, were defined by

$$k = \frac{1}{\langle \tau \rangle} \quad . \tag{3}$$

In fig. 3 we present the dependence of the isomerization rates on the excitation temperature T_{new}, or, equivalently, the excitation energy, U_{new}, for the distinct excitation modes. The reaction rates for all excitation modes show a similar general behaviour. Up to about 1400K the k values increase exponentially with increasing temperature and subsequently decrease slightly after passing a maximum between 1700 and 2200K. This "saturation" of the reaction rates at high temperatures is due to proliferation of configurations. With increasing the excitation energy, bond breaking takes place, whereupon more nuclear configurations become accessible and are passed, e.g., the opening to a chain as an extreme case. This causes a delay in reaching the ring configuration.

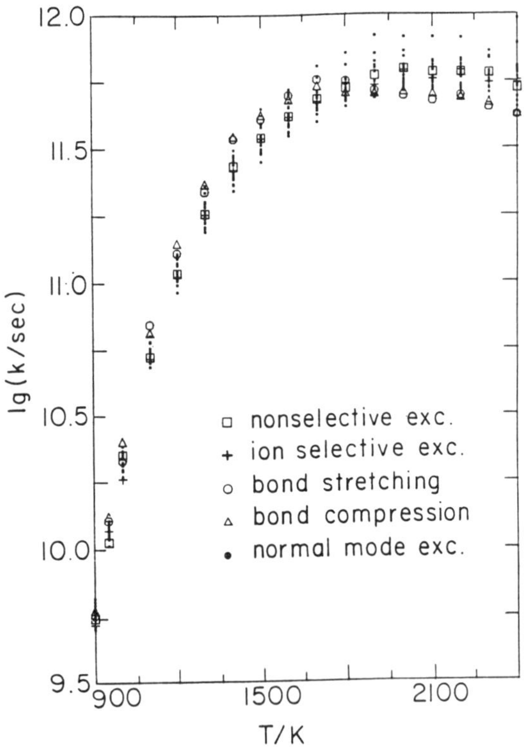

Figure 3. The temperature dependence of the isomerization rates, k, for all excitation modes.

6. Analysis

Quantitative information on the isomerization dynamics emerges from the dependence of k on the excess vibrational energy E_v per ion

$$E_v = [U_{new}(T) - U(0)]/N \tag{4}$$

where U(0) is the energy extrapolated to T = 0 and N = 8 is the number of ions. We have analysed the reaction rates in terms of the Arrhenius equation

$$k = A \exp(-E_A/E_V) \tag{5}$$

where E_A is an activation energy and A a frequency factor, as well as in terms of the RRK theory [11,12]

$$k = A(1 - E_B/E_V)^{s-1} \tag{6}$$

where E_B is the barrier height and s the number of vibrational degrees of freedom.

Both the Arrhenius and the RRK equations give good fits for the reaction rates within the temperature range 900K to 1400K. As a typical example we present in figure 4 the RRK plot for the nonselective excitation. The RRK plots for the other excitation modes look very similar. In the RRK fit we have always kept the number of vibrational degrees of freedom fixed at s = 3N-6 = 18. For the RRK barrier height for nonselective excitation we obtained E_B = 630cm^{-1} per ion or a total value of 5000cm^{-1} per cluster. This value of E_B is lower than Martin's estimate of 840cm^{-1} per ion from potential energy hypersurfaces [5], which provides an upper limit for the reaction barrier. The frequency factor was found to be A = 6.2·10^{13} sec^{-1}. The parameters derived for the other excitation modes are similar to those of the nonselective excitation. We found for the other excitation modes, E_B = 610-650 cm^{-1} and A = 4.8·10^{13}-9.8·10^{13} sec^{-1}. We conclude that the overall averaged rate constant does not show mode selectivity.

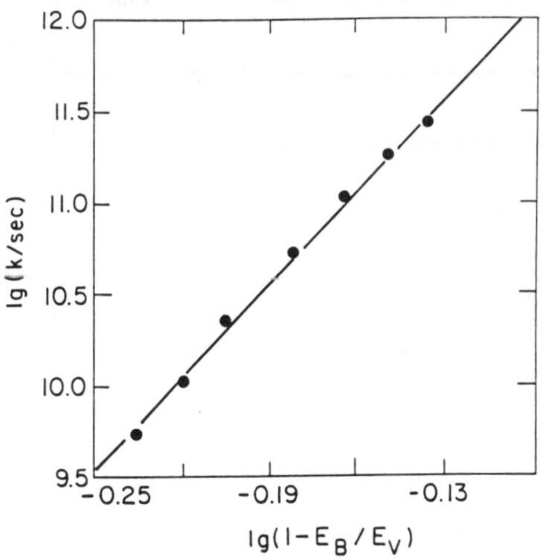

Figure 4. RRK plot of the isomerization rates for nonselective excitation. The points represent the rates for excitations from 600K to 900K-1400K. The solid line is the RRK fit with s = 18 (fixed), E_B = 630cm^{-1} per ion and A = 6.2·10^{13}sec^{-1}.

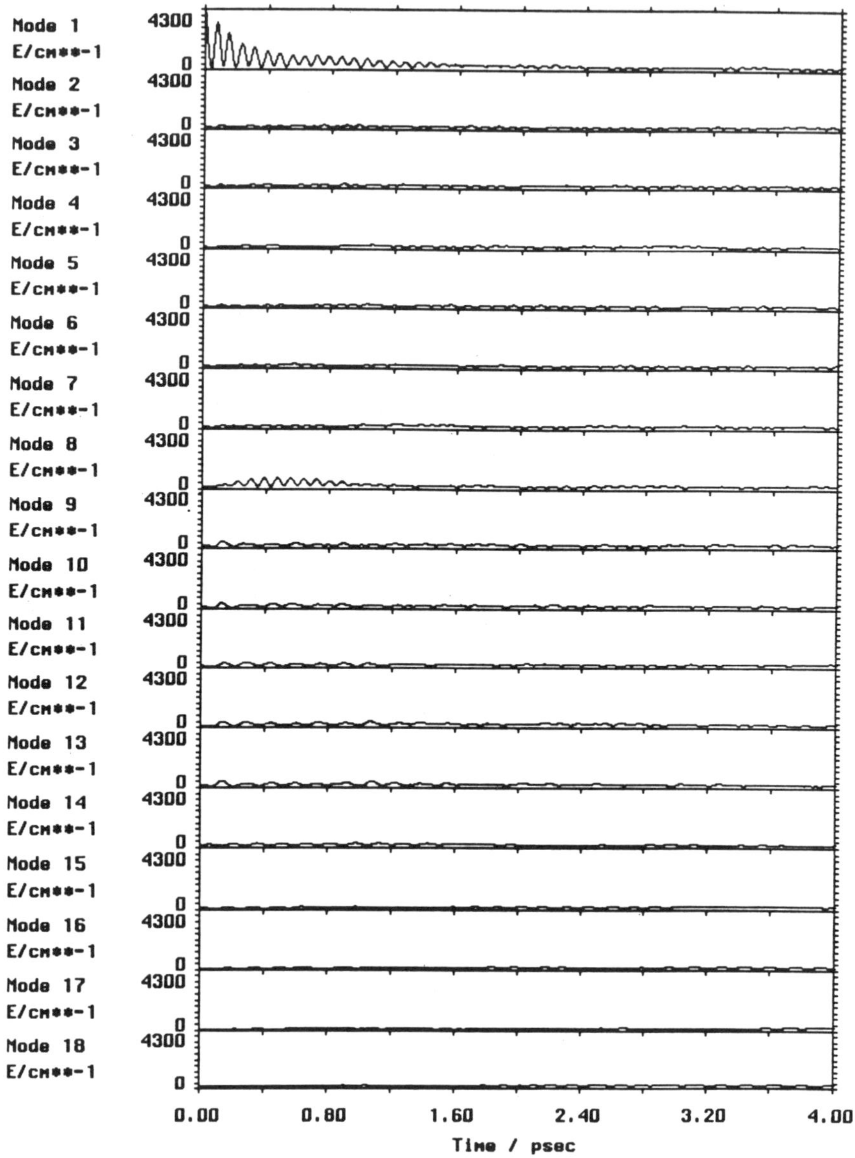

Figure 5. Intracluster vibrational energy redistribution following the 600K → 900K excitation of the $1A_1$ normal mode (mode 1). The kinetic energy (E) is decomposed into normal mode contributions. The numbering of the normal modes is given in Table 1.

7. Intracluster Vibrational Energy Redistribution

We have investigated the intracluster vibrational energy redistribution (IVR) for all excitation modes by decomposing the total kinetic energy into normal mode contributions. As a typical example we present in figure 5 the IVR induced by the $1A_1$ excitation from 600K to 900K, demonstrating the decay of the energy in the initially excited normal mode and its redistribution among the other normal modes. For each excitation mode we have considered excitations from 600K to 900K, 1500K and 2300K, in order to study also the influence of the excitation energy on the IVR process. The redistribution times τ_{IVR} were determined from the decay of the kinetic energy amplitudes to 5% of their initial value (which corresponds to 3 lifetimes for exponential decay). The τ_{IVR} values are presented in figure 6. These results provide some of the first computational evidence regarding vibrational energy flow in large systems at high excitation energies. The energy redistribution times of the various excitation modes are quite distinct (fig. 6). The largest redistribution time we observed is almost 4 psec for the $3T_2$ mode at 900K. The general trend is that the redistribution times become shorter with increasing the excitation energy, as expected.

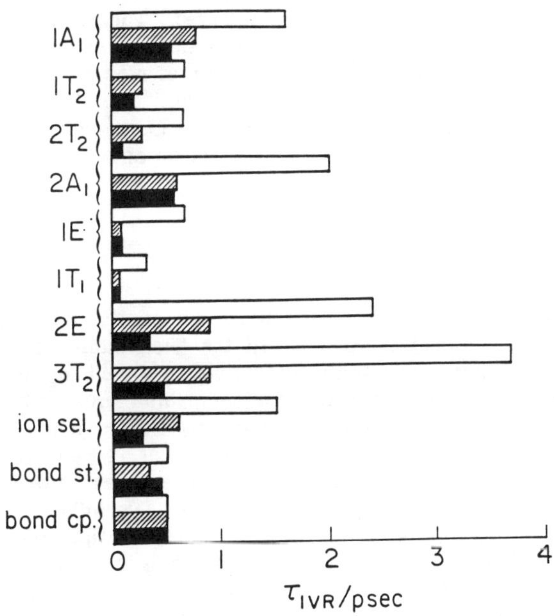

Figure 6. IVR times for different excitation modes. The τ_{IVR} values for the normal mode selective excitations and for the ion selective excitation were determined from the decay of the kinetic energy amplitudes to 5% of their initial value, while τ_{IVR} for the bond selective excitations were determined at the onset of appearance of smooth time dependence of the kinetic energy.

We proceed to consider the competition between the IVR and the isomerization reaction. At 900K the longest redistribution time is roughly 4 psec. Also at 900K only 2% of all isomerization events occur within this time interval. Consequently, we cannot expect mode selectivity at this temperature. From the RRK fit an estimate of $E_B \simeq 5000\ cm^{-1}$ per cluster for the reaction barrier emerges. This energy is necessary to break open the cube. Only for excitation temperatures greater than 1000K the energy amount of $\geq 5000 cm^{-1}$ is deposited in an excitation mode. This conclusion is in agreement with our observations of the system at 900K, where within the energy redistribution time the cube configuration is almost always conserved. At higher temperatures, 1500K and 2300K, there is sufficient excitation energy to overcome the barrier before the IVR is over. Indeed, it is observed that the cube often breaks within the IVR time. There are high-temperature bond breaking processes, which compete with IVR. After the opening of the cube to a ladder a wide distribution of nuclear configurations can be generated. The ring configuration is subsequently reached, after a delay, which is manifested in the "saturation" of the (average) rate k at high temperatures T > 1400K (fig. 3). In addition, at 2300K the cube can break without passing the ladder. In most of the excitation modes this bond breaking process also does not lead directly to a ring but to a wide distribution of nuclear configurations. Exceptions are special linear combinations of the T_1 normal coordinates, where the cube opening occurs at high temperatures (> 1800K) in a way which is favourable to achieve the ring configuration. This pattern is evident from the histograms of the temporal distribution of isomerization events (fig. 2). Mostly, such histograms are very broad as in the case of $1A_1$ mode excitation, meaning that the isomerization events are nearly statistical. In the T_1 mode excitation to high temperatures, a deviation from the pure statistical behaviour is observed. The maximum at small τ values is considerably enhanced, meaning that these short-time isomerizations are directly induced by the excitation. In the extreme case, which is presented here, this short-time contribution covers 40 per cent of all the isomerization events. The effect on the reaction rates is an enlargement by a numerical factor of 2 compared with the statistical case. From these results we conclude that some first signs of an onset of mode selectivity of cluster isomerization are exhibited.

8. Concluding Remarks

We have investigated the cube → ring isomerization dynamics of the $(NaC\ell)_4$ cluster induced by several modes of vibrational excitation. In the temperature range 900K-1400K the reaction rates for isomerization were successfully analysed by the statistical RRK theory, indicating that the averaged rate does not exhibit gross mode selectivity. We have investigated the intracluster vibrational energy redistribution times, their dependence on the excitation mode and on the excitation energy. For all excitation modes at 900K, $\tau_{IVR} <<$ (isomerization time) and the separation of time scales between IVR and isomerization erodes mode selectivity. At higher temperatures T > 1400K the cube breaks within the IVR time achieving a wide distribution of nuclear configurations (differing from the ring). For some specific excitation modes a nonstatistical temporal distribution of the isomerization events is observed. Traces of the onset of mode selectivity of cluster isomerization are exhibited at high excitation energies.

Acknowledgments

One of us (A.H.) thanks the Minerva Foundation for granting a fellowship. We are grateful to Professor J. Troe for stimulating discussions.

References

1. Jortner, J., Scharf, D. and Landman, U. (1988) 'Elemental and Molecular Clusters', Benedek, G., Martin, T.P. and Pacchinoni, G. (eds.), Springer Verlag, Berlin, p. 148.
2. Berry, R. S., Beck, T. L., Davis, H. L. and Jellinek, J. (1988), Adv. Chem. Phys. 70, 75.
3. Jortner, J., Even, U., Ben-Horin, N., Scharf, D., Barnett, R.N. and Landman, U. (1989), Z. Phys. D12, 167.
4. Jortner, J. and Bixon, M. (1989), J. Chem. Phys. 91, 1631.
5. Martin, T. P. (1983), Phys. Reports 95, 167.
6. Luo, J., Landman, U. and Jortner, J. (1987) 'Physics and Chemistry of Small Clusters', P. Jena, B. K. Rao and S. N. Khanna (eds.), Plenum Publishing Corporation, New York, p. 201.
7. Hoare, M. R. and McInnes, J. A. (1983) Adv. Phys., 32, 791.
8. Heidenreich, A., Schek, I., Scharf, D. and Jortner, J. (1991), Z. Phys. D (in press).
9. Fumi, F. G. and Tosi, M. P. (1964), J. Phys. Chem. Solids, 25, 31, 45.
10. Nordsieck, A. (1962), Math. Comp. 16, 22.
11. Rice, O. K. and Ramsperger, H. C. (1927), J. Am. Chem. Soc. 49, 1616.
12. Rice, O. K. and Ramsperger, H. C. (1928), J. Am. Chem. Soc. 50, 617.

Dynamics of Photoinduced Reactions in Clusters

R.B. Gerber* [(a),(b)], R. Alimi[(a),(b)],
A. Garcia-Vela[(a),(b)] and Y. Hurwitz[(a)]

[a] *Department of Physical Chemistry and*
The Fritz Haber Research center for Molecular Dynamics,
The Hebrew University, Jerusalem 91904, Israel.

[b] *Department of Chemistry,*
University of California, Irvine, CA 92717,USA. USA

ABSTRACT. A theoretical study is presented on the dynamics of elementary photochemical reactions in van der Waals and hydrogen-bonded clusters. The processes treated are the photolysis of HI in the clusters $Xe_N(HI)$ (for $N = 1,2,...,12$); the photodissociation of HCl in ArHCl; and photoinduced reactions following UV excitation of $(HCl)_2$. The studies were pursued in part by classical trajectory calculations, and in part by a mixed quantum/classical method that treats light atoms (H) by time-dependent wavepackets, and heavier atoms classically. The studies focus on the existence of a cage effect in clusters and to dependence upon cluster size; on the role of collisions between photofragments and "solvent" atoms in the cluster, and on reactions between photofragments and another molecule in the cluster. It is found that a cage effect exists even for small clusters such as XeHI and ArHCl, that have only a single "solvent atom". The cage effect manifests itself as a resonance that prolongs the photodissociation lifetime. For clusters having about half or more of the solvation shell (e.g., $Xe_N(HI)$, with $N \geq 5$), the resonances dominate the process, and lead to long lifetimes. Collisions between nascent photofragments and "solvent" atoms, e.g., in ArHCl are shown to result in much broader energy distribution of the H product than found for photolysis of isolated HCl. The study of $(HCl)_2$ shows that a photoinduced reac-

J. Jortner et al. (eds.), Mode Selective Chemistry, 201–215.

tion of the type $H+ClH' \rightarrow HCl+H'$ occurs with high probability following photoexcitation, and that resonances contribute to the reaction. On the whole, solvation is found to have very large effects on photochemistry in clusters.

I. Introduction

Photodissociation of molecules in weakly bound cluster has been the topic of vigorous research efforts in the last few years.[(1)–(8)] One major motivation for this novel research discipline is that it makes it possible to explore, in a very advantageous framework, the role of weak "solvation" bonds upon chemical reactivity. This, of course, is a central question in solvation chemistry, but clusters make it possible to explore this issue in simple systems, having few degrees of freedom. Also experimentally, clusters offer important advantages over liquids as a framework for pursuing this important subject. Another merit to the study of reactions in clusters is that by varying cluster size one can gain understanding of how does reaction dynamics change from the gas-phase to the framework of condensed matter. The general role of clusters as a natural bridge between the isolated molecule and the extended system limits is well known, but it appears that almost no attention was given until very recently to this aspect for the specific issue of chemical reaction dynamics. Finally, an important motivation for many researchers in this field is that cluster reactions may provide insight on the role of molecular excitations, that we restricted in many clusters, in chemical processes.[(3)–(5)]

This paper presents an overview of several theoretical studies by the authors on photoinduced chemical processes in van der Waals and hydrogen-bonded clusters. While the results of one published study are reviewed in this framework,[(8)] most of the material presented pertains to very recent unpublished results. It is hoped that the accumulating results will point to some coherent principles, to several general features and effects that appear to play a role in the dynamics of photochemistry in clusters. Emphasizing these characteristic effects is our main aim in this paper. We thus focus on the physical mechanisms of the processes involved and comment only briefly on the theoretical methods, the details of the calculations, etc.

II. Methods and Systems

Classical trajectory calculations were employed in several of the studies reported here. On physical grounds, the method should be adequate for most systems and processes discussed here, at least on a semiquantitative basis. In these calculations, one first calculated the vibrational motions of the cluster system in its ground electronic state. In most systems studied, the cluster was assumed initially to be vibrationally completely cold, so the energy of the system was taken to correspond to the zero-point energy. Configurations of the system in the ground state motion are sampled at different, random time intervals, and employed as initial configurations for the photodissociation dynamics. That is, each configuration on the ground state surface is promoted vertically to the excited state potential surface. For a monochromatic excitation (as is usually the case), an acceptable configuration in the initial state sampling must correspond to the specified excitation energy. Also, each acceptable configuration on the excited state surface must be (in the spirit of the Franck-Condon principle) in the classically allowed regions and at (or very near to) the classical turning points on that surface. We note that these two conditions may imply non-trivial restrictions on initial configurations otherwise sampled randomly from the ground-state trajectories.[9]

II 1. QUANTUM VS. CLASSICAL INITIAL STATE SAMPLING:

In the above standard algorithm each mode is assigned the zero point energy (if the cluster is completely cold before excitation), but the probability density distribution for the initial configurations is, of course, completely classical. It is desirable to explore the effects of using a quantum distribution of initial configurations. This can be done simply, although very approximately, by using the quantum distribution $|\psi(q_n)|^2$ for each vibration mode q_n (assuming the modes are separable). Comparison of this semiclassical configuration with the pure classical sample of initial configurations, made in the case of ArHCl and ArHCl[(9)] showed some quantitative differences in the subsequent photodissociation results but, on the whole, the effects were not large.

II 2. MIXED QUANTUM/CLASSICAL DYNAMICS:

In cases where, e.g. a light atom is involved, the possibility of quantum effects associated with that particle should be examined. This is the case, e.g., for the clusters Xe_NHI discussed below. In such cases, we employed a hybrid dynamics that treats the light atom (H) by time-dependent quantum mechanics, and the heavier atoms by classical dynamics.[(10),(11)] The approach we used samples the quantum and the classical sub-systems by the Time-Dependent Self Consistent Field (TDSCF) method.[(10),(11)] In this approach, there is an effective time-dependent potential that governs the quantum motions (and is a function of the quantum degrees of freedom only), and another effective potential, which is a function of the "classical" modes only and governs their motions. However, the potential for the quantum modes is determined from the motions of the classical modes, and the potential for the classical motion depends on the wavepacket of the quantum subsystem. This approach leads therefore to a self-consistent treatment of the propagation of the wavepacket for the quantum modes, with the propagation of the trajectories for the "classical" modes. This approach, as any other available algorithm for combining quantum and classical dynamics, involves dynamical approximations in addition to the treatment of the heavy atoms as classical. Its validity regime is thus far from obvious. It is therefore important to note that the validity of the mixed quantum/classical TDSCF was tested against exact calculations for photodissociation in a collinear model of Xe···HI, a system closely related to the realistic clusters pursued here.[12] The results obtained are very encouraging, and are the basis for our confidence in the applications of the hybrid method in the present study.

For the ground electronic states of some of the clusters treated here (Ar/HCl, $(HCl)_2$), fairly reliable potential surfaces are available in the literature, based both on *ab initio* calculations and on IR spectroscopy. For other systems ($Xe_N(HI)$), such potential were developed from data in the literature. For the excited electronic states, the saturation is far more uncertain. We developed and used a pairwise potential from H/Xe, Xe/Xe and H/I interactions for the photodissociation potential surface of $Xe_N(HI)$. The details are given in (8), and in references therein. Likewise, we employed a pairwise potential surface for the excited (repulsive) state of ArHCl.[9] The situation for $(HCl)_2$ is far more complicated. A semi-empirical excited state potential surface was developed for this system,[13] and it includes a repulsive interaction between H and Cl in one of the monomers, dispersion forces between the Cl atoms, and a (reactive) three-body H-Cl-H LEPS potential surface. This potential surface seems to have at least reasonable semi-quantitative properties.

III. Results

III-A. CAGE EFFECT IN PHOTOLYSIS OF $Xe_N(HI)$

The calculated classical equilibrium geometry of the clusters XeHI, Xe_2HI, Xe_5HI and Xe_5HI is shown in Fig. 1.

Fig. 1: Structure of Xe_NHI clusters

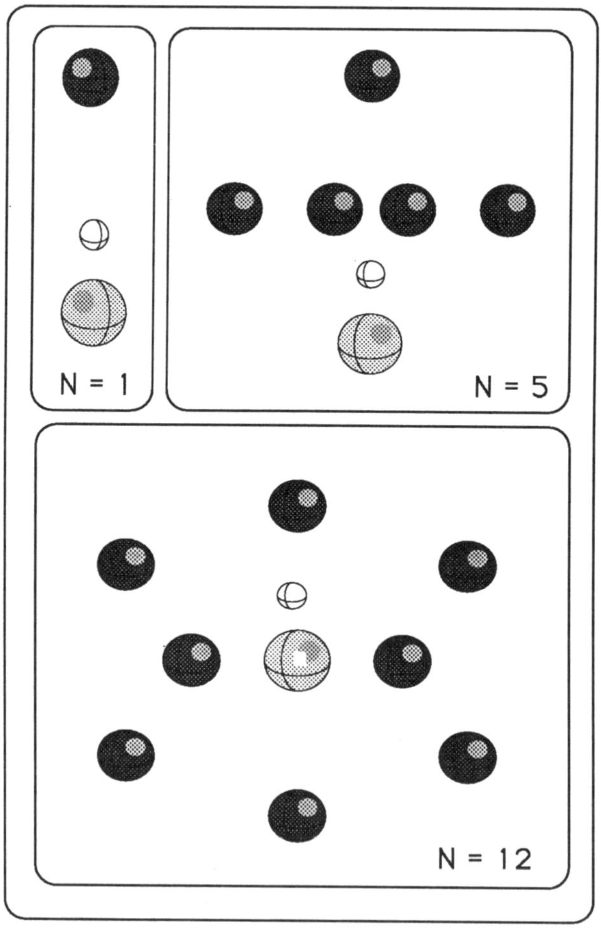

The geometry plays an important role in the subsequent photodissociation dynamics The calculations of Ref. (8), using the mixed wavepacket/classical trajectory approach, showed that *a cage effect is found in all the clusters studied, including* XeHI. The cage effect manifests itself as a resonance, that prolongs the photodissociation lifetime. The resonance involves bouncing of the H atom among the heavy atoms, until the hydrogen atom escapes. Fig. 2 shows the effect at the level of a completely classical calculation, for the cluster XeHI.

Fig. 2: Position of H atom *vs.* time, in a trajectory

desorbing a resonance in photolysis of XeHI

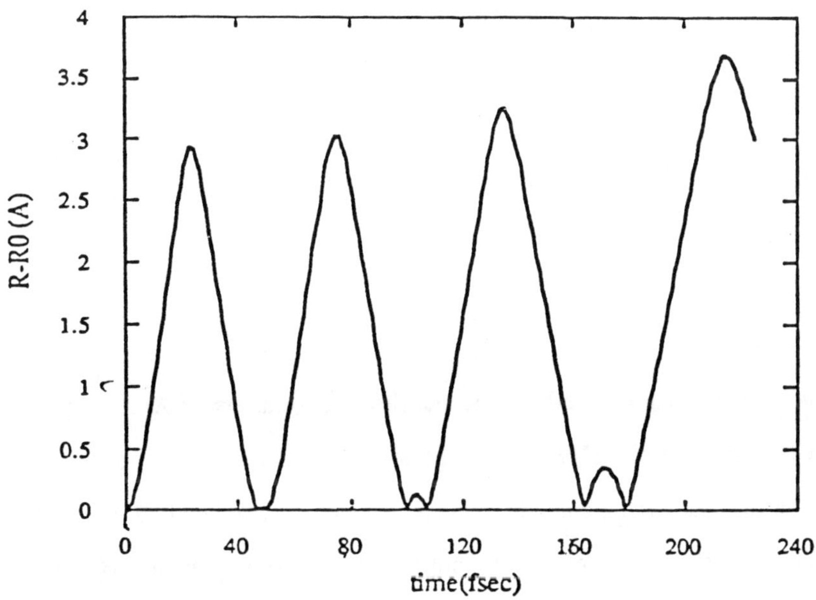

For the small clusters the resonances contribute relatively little compared with the direct process (i.e., there are four such events), and the resulting lifetimes are still short (although longer than for the direct process). A sharp transition occurs for clusters with

$N \geq 5$. The resonances dominate in importance, and are also much longer lived. As Fig. 3 shows, the photodissociation lifetimes of $Xe_N(HI)$ clusters with $N \geq 5$ are of the order of 0.5 picoseconds, much longer than found for the smaller clusters.

Fig. 3: Photodissociation lifetime distributions for the clusters $Xe_N(HI)$

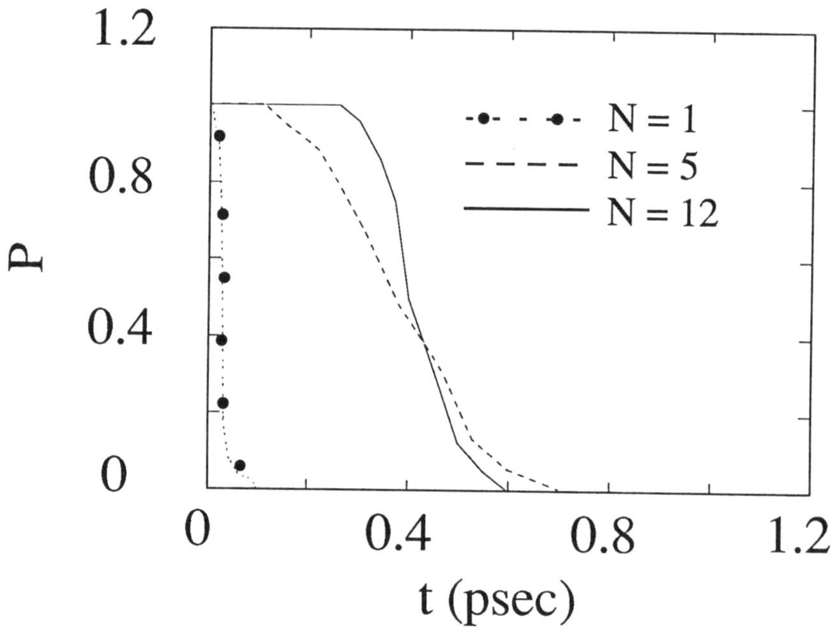

This behavior is due to the ground state geometries of the clusters, shown in Fig. 1. In the clusters with $N \geq 5$, the photolyzed H atom finds itself locked by heavy atoms blocking its exit. It carried out large amplitude motions in the cluster, and collides with the heavy atoms many times (in the order of ~ 50) until the cluster opens up and the H exits. Fig. 4 shows the correlation function of the wavefunction $|\langle\psi(t)|\psi(0)\rangle|^2$ for the photodissociation process in Xe_NHI.

Fig. 4: The correlation function $|\langle\psi(t)|\psi(0)\rangle|^2$

vs. time for photodissociation of $Xe_N(HI)$.

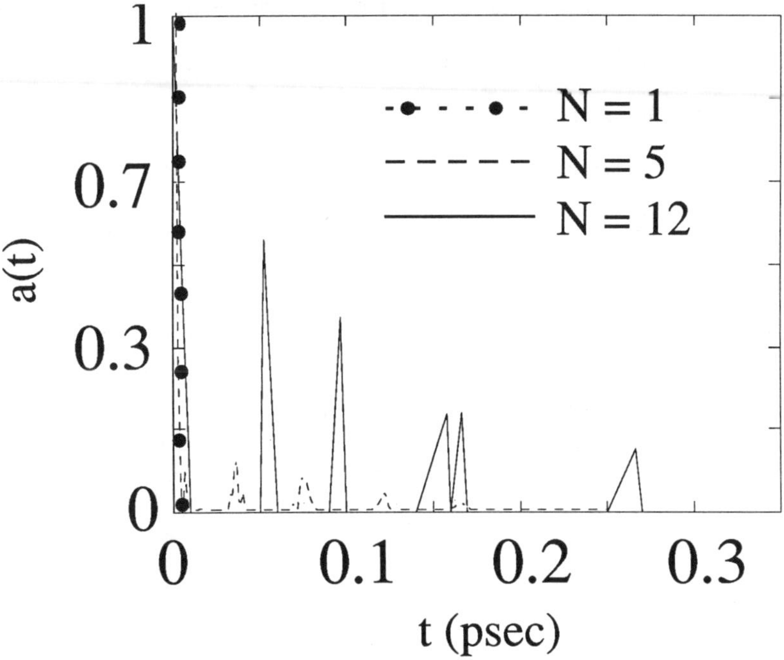

This throws light on the quantum manifestations of the photodissociation resonances. A very interesting feature is the manifestation of coherent large amplitude vibrations of the H in the large clusters, although the clusters are slowly disintegrating in the process. The coherent vibrations cause the sharp maxima of the correlation functions, whenever the wavepacket returns to the initial position. The effect should be observable with adequate ultrafast spectroscopy.

III-B. INTRACLUSTER COLLISIONS IN PHOTOLYSIS OF Ar-HCl

In Ar···HCl, as in Xe···HI, once the H photofragment is formed, there is a considerable possibility for it to undergo a "hard" collision with the rare gas atom, followed in some cases by a sequence of additional collisions with the heavy atoms (up to 4-5 collisions may occur). This is the case despite the large bending amplitude in the floppy ground state, a behavior which greatly deviates from the collinear configuration favorable for collisions. The collisions have extremely important consequences. Fig. 5 shows the final energy distribution of the H atom in photolysis of ArHCl.

Fig. 5: Final energy distribution of H from photolysis of ArHCl

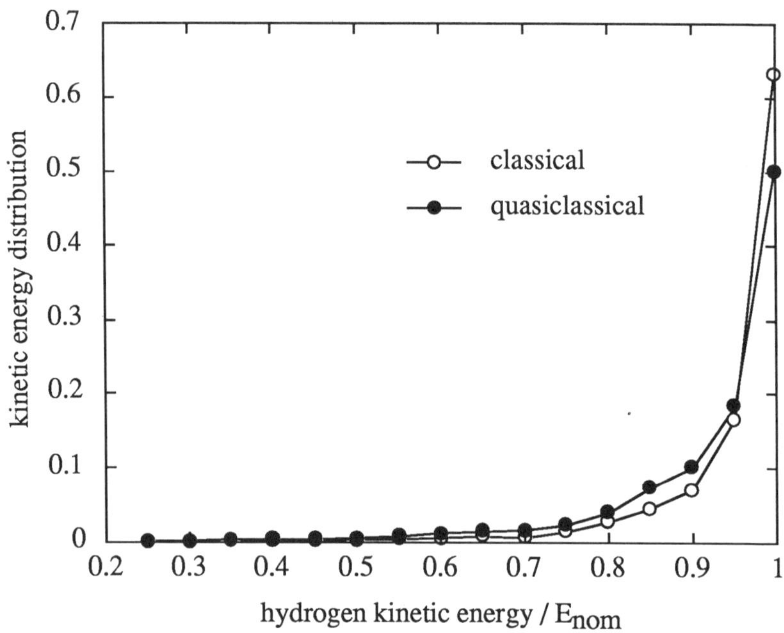

The kinetic energy is divided by E_{norm}, the maximal energy of H in the photolysis of HCl at the same wavelength. The results show a long tail, in sharp contrast to the situation for isolated HCl, representing events of energy loss in collision between the H photofragment and one (or both) of the heavy atoms. Note that the result is insensitive to whether the classical or quantum (semiclassical) distribution is used for the initial state. Fig. 6 represents an even more dramatic effect of the intracluster collisions: The effect on the final angular distribution of the H atom.

Fig. 6: Angular distribution of H atoms from photolysis of ArHCl.

The angle is measured with respect to the initial orientation of the HCl dipole

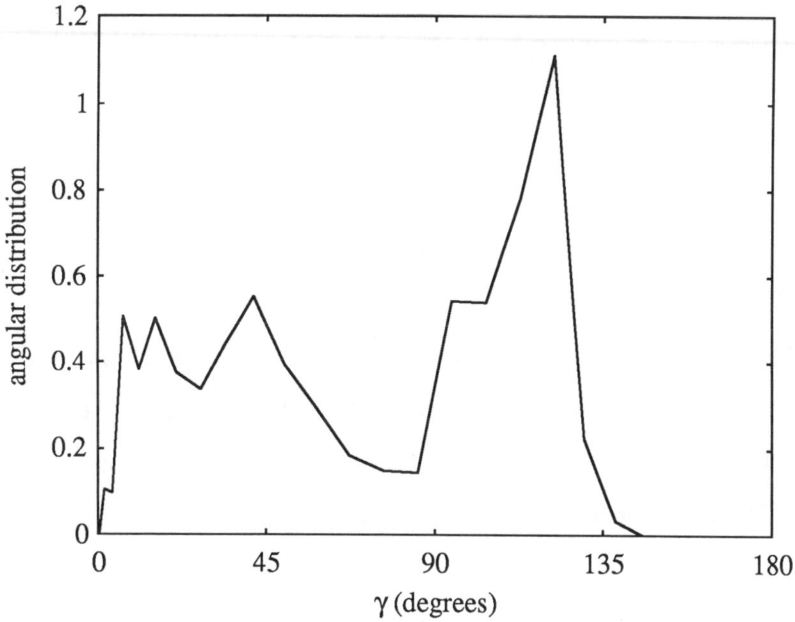

Shown here is the classically computed angular distribution (for the angular distribution the effect of using an initial quantal distribution, as discussed in Sec. II is large). For isolated HCl, the angular distribution is a δ-function spike at $\gamma = 0$. The effect of the clustering is thus enormous. Basically, the effect of the collisions is to make the angular distribution much less anisotropic than for the isolated molecule. There are two regions of peaks in Fig. 6, that can be attributed to H atoms that underwent all and even numbers of collisions, respectively. The angular distribution cannot be measured in practice. However, the "anisotropy parameter" that is easily measured in polarization experiments is related to the average of the above distribution, and we calculated that it should be greatly shifted from the isolated molecular value. In conclusion, the effects of intracluster collisions on the angular distribution of the fragments is very large, and measurements of the anisotropy parameter should offer a sensitive experimental probe of such collisions.

III-C. PHOTOINDUCED REACTIONS IN $(HCl)_2$

The production of a hot H atom by photoabsorption leads in this molecular cluster to reaction processes as well as to photofragmentation. The fact that the ground state of the $(HCl)_2$ cluster is very floppy gives rise to a wide range of possible collision geometries between the H photofragment and the second HCl molecule bound in the cluster. While the orientations of the two HCl axes in the cluster cover a wide range in the electronic ground state, the range of Cl-Cl distances is more limited, and this sets important restrictions on the impact parameters that contribute to the processes. An important and surprising result is that the substitution reaction caused by the hot H atom $H + ClH' \rightarrow HCl + H'$ occurs with very high probability. The collision path from a typical trajectory is shown in Fig. 7.

Fig. 7: Path of trajectory leading to a substitution reaction following photoexcitation of $(HCl)_2$

H' + ClH(v=0,J=0) -------> H'Cl(v=6,J=30) + H

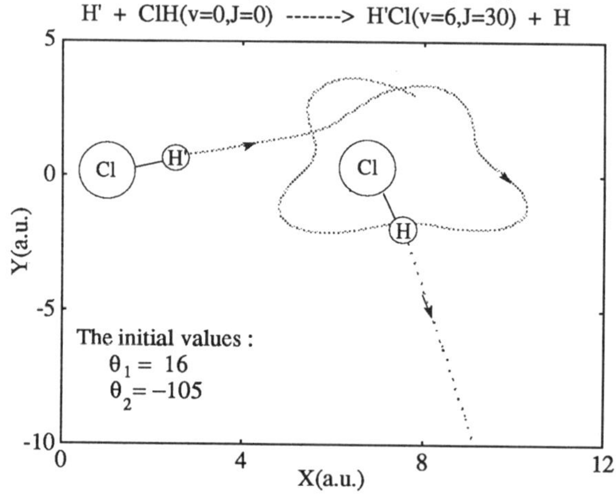

As dictated by the geometry, the "new" HCl produced is vibrationally and rotationally very hot, which suggests an important experimental signature for the occurrence of the reaction. The reaction path itself combines elements of an S_N2 mechanism, with an orbiting collision, or part thereof. This appears to be a new type of a reactive orbiting collision. Another interesting possible collision, although much less probable than the H-substitution process (the yield for it is of the order of 2%), leads to the formation of H_2. The path of the trajectory is shown in Fig. 8.

Fig. 8: Path of trajectory leading to formation of H_2 upon photolysis of $(HCl)_2$.

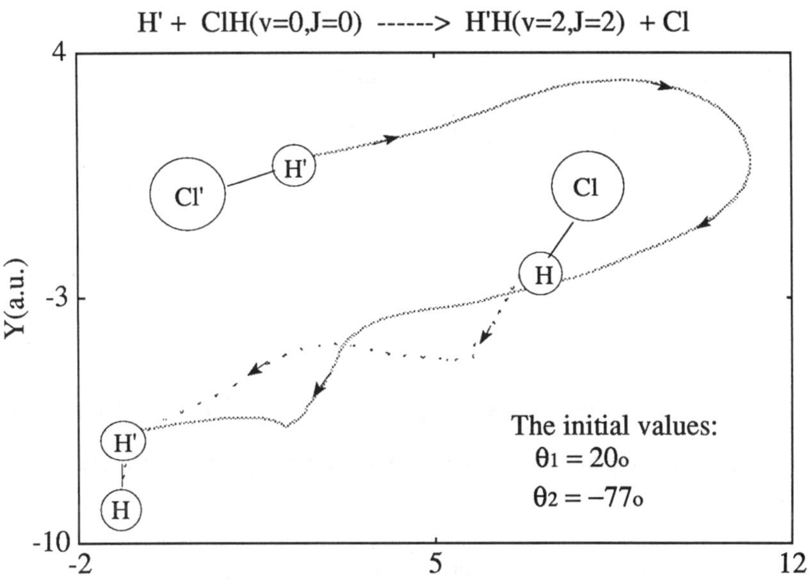

It should be noted that trajectories involving hard H-H collisions do not result in H_2 formation, but in substitution. It is an orbiting type of path that results in the formation of H_2. The trajectory is, however, of short duration in time, and not a resonance. $(HCl)_2$

photolysis does involve several interesting resonances, but these will not be discussed here.

IV. Conclusions

A general conclusion that emerges from these results is that cluster formation has a dramatic effect on the photodissociation behavior of molecules that are subject to the weak bond. This is true also when the solvation is by inert atoms. One of the important effects is the onset of resonances, which enter photolysis processes directly into time delayed events for the isolated molecule. The lifetime and role of these resonances can vary critically with cluster size. These resonances are manifestations of cage effects, and when the cluster corresponds to about half a solvation layer, the cage effect on the reaction can be quite similar, e.g. for $Xe_5(HI)$, to that in condensed matter. Collisions between photofragments on "solvent" atoms or molecules occur with high probability. Their effect on the energy and angular distribution of the product is a major one even on the qualitative level.

These and other findings obtained here reinforce the suggestion that chemistry in clusters is profoundly related to chemistry in condensed phases, and may provide important insights to mechanisms and dynamics of condensed matter processes.

Acknowledgement: The Fritz Haber Center is supported by the Minerva Gesellschaft für die Forschung, mbH, Munich, BRD. A.G.-V. thanks the Ministry of Education and Science of the Spanish Government for a Postdoctoral Fellowship.

References

(1) C. Jouvet and B. Soep, Chem. Phys. **96**, 426 (1983).

(2) W.H. Breckenridge, C. Jouvet and B. Soep, J. Chem. Phys. **84**, 1443 (1986).

(3) S. Buclow, G. Radhakrishnan, J. Catanzante and C. Wittig, J. Chem. Phys. **83**, 444 (1985).

(4) C. Wittig, S. Sharpe and R.A. Beaudet, Accts. Chem. res. **21**, 341 (1988).

(5) N.F. Scherer, C. Sipes, R.B. Bernstein and A.H. Zewail, J. Chem. Phys. **92**, 5239 (1990).

(6) J.A. Beswick, R. Manot, J.M. Philippoz and H. van den Bergh, J. Chem. Phys. **86**, 3965 (1987).

(7) J.M. Philippoz, R. Manot and H. van den Bergh, J. Chem. Phys. **93**, 8676 (1990).

(8) R. Alimi and R.B. Gerber, Phys. Rev. Lett. **64**, 1453 (1990).

(9) A. Garcia-Vela, R.B. Gerber and J.J. Valentini, Chem. Phys. Lett. (submitted).

(10) R.B. Gerber, V. Buch and M.A. Ratner, J. Chem. Phys. **77**, 3022 (1982).

(11) R.B. Gerber and M.A. Ratner, Adv. Chem. Phys. **70**, 97 (1988).

(12) R. Alimi, R.B. Gerber, A.D. Hammerich, R. Kosloff and M.A. Ratner, J. Chem. Phys. **93**, 6484 (1990).

(13) Y. Hurwitz and R.B. Gerber, to be published.

MODE- AND BOND-SELECTED BIMOLECULAR REACTION OF WATER

F. F. CRIM, A. SINHA, M. C. HSIAO, and J. D. THOEMKE
Department of Chemistry
University of Wisconsin-Madison
Madison, Wisconsin 53706 USA

ABSTRACT. Selective excitation of overtone vibrations permits the control of the bimolecular reaction of water with either H atoms or $C\ell$ atoms, in a demonstration of both mode-selected and bond-selected chemistry. The reaction rate depends strongly on the identity of the initially excited mode, even for states having comparable energies, and the distribution of the products among their quantum states also strongly reflects the distribution of energy in the excited reactant. In isotopically substituted water, HOD, initial excitation of the O-H stretching vibration causes preferential cleavage of that bond in the bimolecular reaction.

1. Introduction

Mode selected chemistry is the alteration of the outcome of a chemical process, such as dissociation, isomerization, or bimolecular reaction, by initial deposition of energy in a particular molecular degree-of-freedom. For example, one can test the ability of translational, vibrational, or rotational energy to accelerate the rate of a bimolecular reaction or to control the partitioning of energy among the degrees-of-freedom of reaction products.[1] An even more detailed test of mode selectivity is reaction of different vibrational states with comparable energies in a polyatomic molecule. We have used selective excitation of water to control photodissociation from different vibrational states[2,3] and to alter rates and product state distributions in bimolecular reactions.[2,4,5]

There are several possible indicators of vibrational mode selected chemistry. For example, a dependence of the reaction rate or the product internal energy on the identity of the initially excited state signals mode selective chemistry. Controlling the cleavage of a bond by selective vibrational excitation (bond selected chemistry) is an intriguing example of mode selected chemistry and has been a long standing goal of chemical dynamics. The approach we use to achieve mode- and bond-selected chemistry is to prepare a particular state in a molecule by vibrational overtone excitation and to react it with an atom. Here we describe mode- and bond- selected bimolecular reactions of water with hydrogen or chlorine atoms.

The stretching vibrations of water differ in the distribution of the excitation between the two identical bonds, and the local mode picture of molecular vibrations is particularly convenient for describing these different states.[6] Local mode notation designates a vibrational overtone state containing "a" quanta of vibration in one O-H bond and "b" in the other as $|ab\rangle$, but only the symmetric and antisymmetric linear combinations of the states $|ab\rangle$ and $|ba\rangle$, designated $|ab\rangle^+$ and $|ab\rangle^-$, respectively, have the symmetry of the Hamiltonian for the molecule. In the experiments described here, we excite the antisymmetric states, for example, $|04\rangle^- = (1/2)^{1/2}\,[\,|04\rangle - |40\rangle]$ and $|13\rangle^- = (1/2)^{1/2}\,[\,|13\rangle - |31\rangle]$. We denote states containing bending excitation as $|ab\rangle^-|c\rangle$ where "c" is the number of bending quanta.

J. Jortner et al. (eds.), Mode Selective Chemistry, 217–225.

The bimolecular abstraction reactions of water with H and $C\ell$,

$$H_2O + H \rightarrow OH + H_2 \qquad \Delta H_0^0 = 61.6 \text{ kJ/mol}$$
$$H_2O + C\ell \rightarrow OH + HC\ell \qquad \Delta H_0^0 = 65.5 \text{ kJ/mol},$$

produce the readily detected OH radical and have the proper chemistry for vibrational enhancement. They are very endothermic, and the bound O-H stretching vibration should be an important part of the reaction coordinate for an abstraction reaction. Figure 1 is an energy level diagram for the reaction of H_2O with H atoms that includes a schematic illustration of the energy along the reaction coordinate. (The energy level diagram for the reaction with $C\ell$ is quite

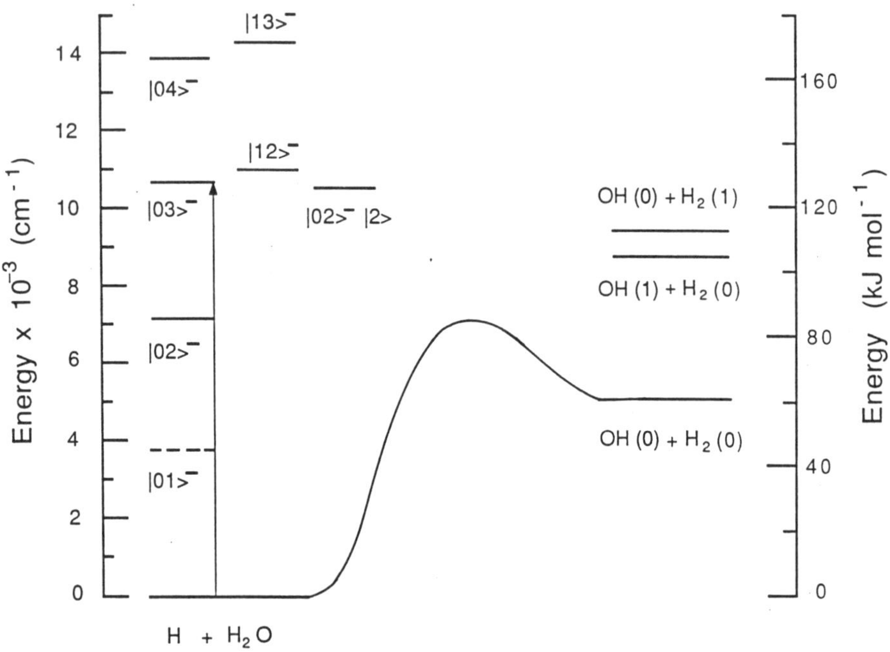

Figure 1. Schematic energy level diagram for the H + H_2O reaction. The energy levels on the left show the vibrational states in H_2O. The horizontal lines on the right mark the energies of the H_2 and OH products.

similar but with a lower barrier. The activation energy is $E_a \sim 72$ kJ/mol[7].) We excite all the vibrations indicated as solid lines on the diagram and observe the OH product with single-quantum state resolution by laser induced fluorescence (LIF). Because we measure the internal states of the OH product and can estimate limits to the relative translational energy from the widths of the transitions in LIF excitation spectra, these experiments also provide a means of assessing the energy disposal[5], but here we only discuss the selectivity of the reaction.

2. Experimental Approach

The experimental apparatus is similar to the one used in our earlier studies.[4,5] A conventional discharge of flowing gas generates hydrogen atoms from H_2 or a photolysis laser generates chlorine atoms from $C\ell_2$. Vibrational overtone excitation energizes the reactant molecules, and laser induced fluorescence detects the products. The reaction chamber is a glass cell equipped with a viewing window for monitoring laser induced fluorescence, two Brewster angle side arms for introducing laser light, and several inlet ports for introducing reagents. The reagents flow continuously through the cell at a total pressure between 80-90 mTorr and into a partially throttled mechanical pump. Water vapor enters the cell through a separate port at a partial pressure regulated between 15-20 mTorr. For the HOD experiments, we use the vapor from a mixture containing equal parts of H_2O and D_2O. The partial pressures of H_2 and He in the cell are 20-30 mTorr and 40-50 mTorr, respectively.

A Nd:YAG laser pumped dye laser provides the energy for vibrational overtone excitation. For excitation of vibrational levels below 11,000 cm^{-1}, we shift the output of the dye laser by stimulated Raman scattering in H_2. To identify the various rovibrational transitions of water, we direct a small portion ($\approx 4\%$) of the vibrational overtone excitation beam into a photoacoustic cell

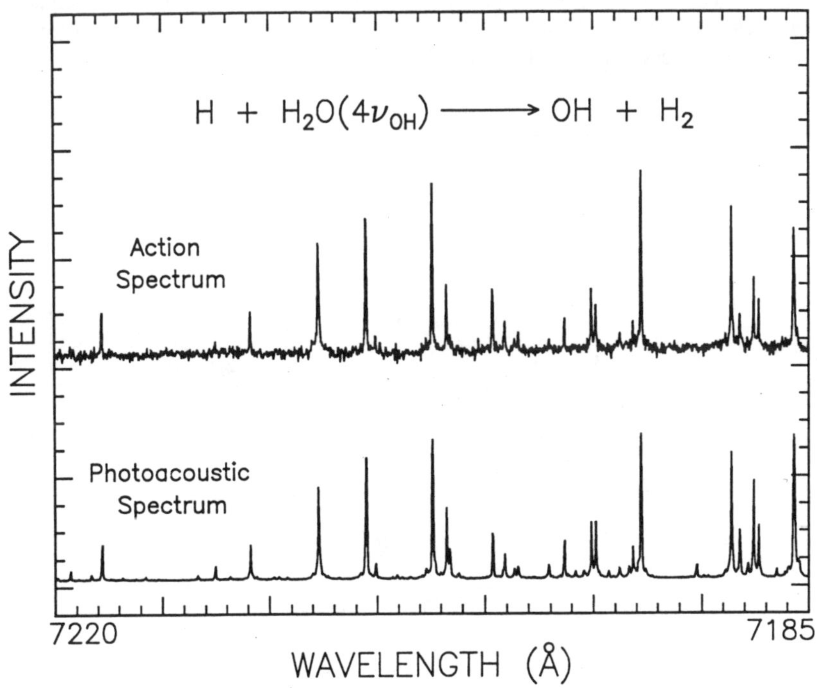

Figure 2. Photoacoustic spectrum (lower trace) and action spectrum (upper trace) for the $|04\rangle^-$ state of water. The intensities in the action spectrum reflect both the excitation cross section and the reaction probability. Those in the photoacoustic spectrum reflect only the excitation cross section.

containing water vapor. The continuously monitored photoacoustic signal provides a convenient diagnostic of the vibrational excitation process. Some of the measurements use an intracavity etalon in the vibrational excitation laser to reduce its bandwidth to ≈ 0.05 cm^{-1} for more efficient excitation of the narrow water transitions. We probe the OH and OD product state distributions by partially saturated laser induced fluorescence using frequency doubled light around 308 nm from a XeCl excimer laser pumped dye laser. The probe laser light propagates counter to that from the vibrational overtone excitation laser, and the delay between the two laser pulses is between 50 and 1000 ns, depending on the particular measurement. Figure 2 illustrates the ability of vibrational overtone excitation to enhance the rate of reaction between water and hydrogen atoms. The upper trace is the photoacoustic spectrum of water in the region of the $|04\rangle^-$ transition and the lower trace is an action spectrum showing the dependence of the OH product LIF signal on the vibrational overtone excitation wavelength. The action spectrum reproduces the photoacoustic spectrum almost exactly. There is no reaction of the unexcited molecule but rather efficient reaction of the vibrationally excited species.

3. Reaction Rate

The reaction rates for the $|0b\rangle^-$ states of water having b=2,3,4 increase sharply with vibrational excitation. Figure 3 illustrates this behavior for the reaction of water with hydrogen atoms. The $|03\rangle^-$ state reacts at about half the rate of the $|04\rangle^-$ state, and the $|02\rangle^-$ state reacts 16 times more slowly. This very slow rate for $|02\rangle^-$ is consistent with the calculated barrier being close

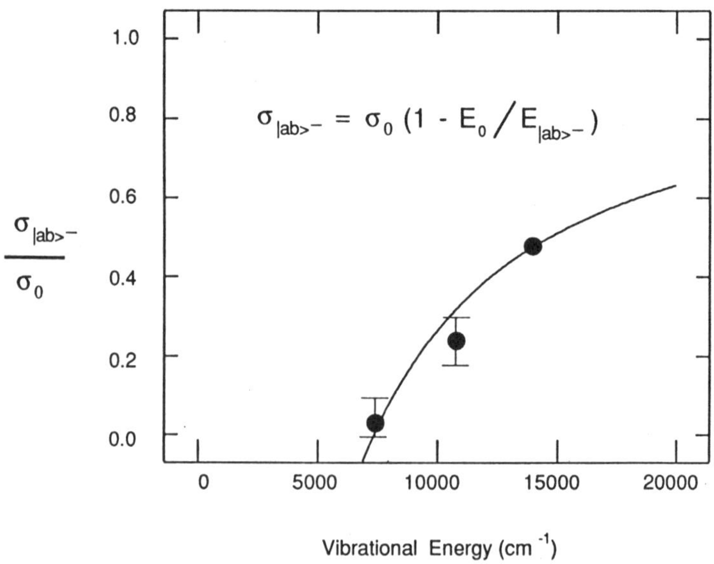

Figure 3. Relative rate for the H_2O + H reaction as a function of vibrational energy. The data are results for the $|04\rangle^-$, $|03\rangle^-$, and $|02\rangle^-$ vibrational states.

to the threshold energy for reaction. A very approximate estimate of the rate constant for $H_2O(|04\rangle^-)$ suggests that the reaction of the vibrationally excited molecule proceeds at about the gas kinetic collision rate. The vibrational state dependence of the reaction rate for H_2O with $C\ell$

is similar, but we can estimate the size of the rate constant more precisely by comparison to the oxygen atom abstraction of H from methanol to make OH. In the calibration reaction, we produce O atoms photolytically from NO_2, in the same way as we produce $C\ell$ atoms from $C\ell_2$. Knowing the relative photolysis cross-sections for the two precursor molecules and the rate constant for the thermal reaction allows us to extract an absolute rate constant for $H_2O(|04\rangle^-)$ + $C\ell$ of $k(|04\rangle^-) = 2 \times 10^{-10}$ cm^3 molecule^{-1} s^{-1}. The sharp increase of the rate constant with vibrational energy and the agreement of the observed threshold with the calculated barrier[8] indicate that all of the vibrational excitation in the $|0b\rangle^-$ states is available for overcoming the barrier.

The variety of states accessible by vibrational overtone excitation allows us to determine the effect of different vibrational modes on the rate of reaction. As Figure 1 illustrates, there are three states, $|03\rangle^-$, $|12\rangle^-$, and $|02\rangle^-|2\rangle$, lying within 500 cm^{-1} of the same energy but corresponding to rather different nuclear motions. These three, nearly isoenergetic, states react very differently. Both the $|12\rangle^-$ state and the $|02\rangle^-|2\rangle$ state react with H and $C\ell$ much more slowly than the $|03\rangle^-$ state. Thus, energy not in the vibration that becomes the reaction coordinate does not help overcome the barrier to the reaction. The nonreacting bond is apparently a spectator in the reaction.

4. Product States

Observing the population of product states for different initial vibrational states is a means of testing the suggestion that the nonreacting bond is a spectator. Figure 4 shows the vibrational overtone excitation spectra obtained by monitoring the OH product from the reaction $H_2O(|ab\rangle^-)$ + H → OH(v) + H_2. The left hand portion of the figure shows spectra taken by monitoring

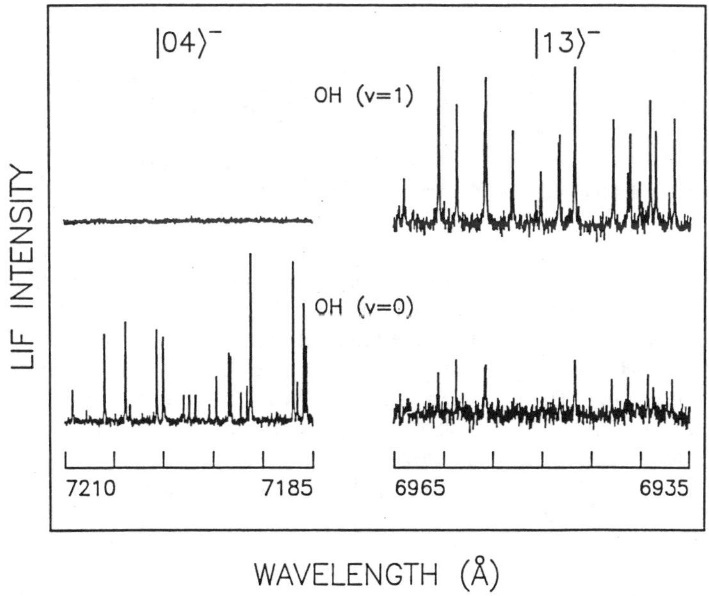

Figure 4. Vibrational overtone excitation spectra for transitions to the $|04^-$ and $|13\rangle^-$ states of water obtained by probing OH(v=0) and OH(v=1) products resulting from the $H_2O|ab\rangle^-$ + H reaction.

OH(v=1) (top) or OH(v=0) (bottom) from the reaction of H_2O ($|04\rangle^-$). This state, which has no excitation in the nonreacting bond, forms products without any vibrational energy. The reaction of H_2O ($|13\rangle^-$) leads to a very different outcome, as the right side of the figure illustrates. This state, which has one quantum of vibrational excitation in the nonreacting bond, produces *mostly OH (v=1)*. The reaction with chlorine atoms is very similar. It also preserves the initial excitation in the nonreacting bond as vibration of the OH product. Table 1 summarizes these results and shows one subtle difference between the H-atom and Cℓ-atom reactions. The reaction of the heavier and more translationally energetic Cℓ atom produces a smaller difference in the OH (v=1) and OH (v=0) product populations. This may reflect a greater mixing of the vibrational states in collisions with chlorine atoms compared to those with hydrogen atoms.

TABLE 1. Relative populations of OH(v=0) and OH(v=1) product states from the reaction of H_2O with H atoms or with Cℓ atoms.

	N(v=1)/N(v=0)	
	$\|04\rangle^-$	$\|13\rangle^-$
$H_2O(\|ab\rangle^-)$ + H	0.01 ± 0.01	6 ± 1
$H_2O(\|ab\rangle^-)$ + Cℓ	0.03 ± 0.01	3 ± 1

5. Product Identity

The same approach that allows us to control the rate and product states in the reaction of H_2O with H or Cℓ is a means of performing *bond-selected bimolecular chemistry*. Replacing a hydrogen atom in water with deuterium allows us to distinguish two bonds and their reactivity,

$$\mathrm{HOD}(|40\rangle) + \mathrm{X} \longrightarrow \begin{cases} \mathrm{OH} + \mathrm{DX} \\ \mathrm{OD} + \mathrm{HX} \end{cases}$$

where X is either an H atom or a Cℓ atom. (The initially excited vibrational state has four quanta of O-H stretching excitation.) Figure 5, which is a portion of the LIF excitation spectrum of the hydroxyl products, illustrates the selective cleavage of the vibrationally excited bonds. Exciting the O-H stretching vibration in HOD causes the reaction of only that bond, which produces solely OD. The very high selectivity favors the reaction of the vibrationally excited bond by at least a factor of 100, and this bond selectivity is comparable for the reaction with thermal H atoms or translationally energetic Cℓ atoms. The keys to the bond selectivity in the reaction of water are the isolated, local mode character of the stretching vibrations and the direct abstraction reactions with H and Cℓ.

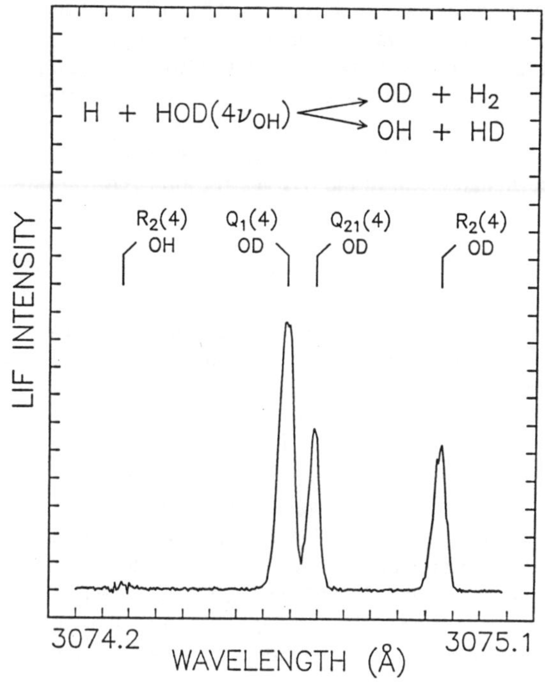

Figure 5. Laser induced fluorescence excitation spectrum of OH and OD resulting from the reaction of HOD(|04⟩) with H atoms.

6. Conclusion

The measurements described here clearly illustrate both mode- and bond-selected bimolecular chemistry of vibrationally excited water. By preparing a molecule in a single vibrational eigenstate, we can control the rate of reaction and the disposal of energy into the vibrations of the OH product. Extending this approach to an asymmetric molecule, HOD, produces bond selected chemistry, in which the initial excitation controls the identity of the bond that breaks in the bimolecular reaction.

These ideas extend beyond the reactions of water, although its direct hydrogen atom abstraction chemistry and the extreme local mode character of its vibrations make it a particularly good example. A rudimentary model provides the criteria for mode-selected reaction in general. The molecular eigenstate prepared by vibrational overtone excitation is a linear combination of zero-order states. Typically, the states corresponding to excitation of the internal coordinates, such as bond lengths and angles, are particularly convenient zero-order states that appeal to chemical intuition. As usual, we can express the highly vibrationally excited molecular eigenstate $|n\rangle$ as a linear combination of the zero-order states $|s\rangle$ and $\{|\ell\rangle\}$,[9]

$$|n\rangle = C_s^n\ |s\rangle + \sum C_\ell^n\ |\ell\rangle.$$

The state $|s\rangle$ is the bright zero-order state, in our case the O-H stretching state that carries the oscillator strength for the vibrational overtone transition. The states $\{|\ell\rangle\}$ are the dark zero-order states such as the bend and O-D stretch that have no oscillator strength. The transition probability in the vibrational overtone excitation step depends on the coefficient of the bright state, but, more important in the present context, the coefficient influences the bond selective reaction as well. The outcome of the subsequent interaction of eigenstate $|n\rangle$ with an H atom in the bimolecular abstraction reaction involving HOD($|40\rangle$) depends on the amount of zero-order O-H stretching character in the molecular eigenstate. We can write the cross section for this interaction as $\sigma \propto |\langle f|P|n\rangle|^2$ where P is the operator for reaction of a molecule in state $|n\rangle$ to form products in the final state $|f\rangle$. If the reaction probability depends *only* on the amount of pure O-H stretching excitation, all of the $\langle f|P|\ell\rangle = P_{f\leftarrow\ell}$ are zero. In this case, the cross section is

$$\sigma \propto |\langle f|P|n\rangle|^2 = |C_s^n \langle f|P|s\rangle + \sum C_\ell^n \langle f|P|\ell\rangle|^2 = |C_s^n|^2 \; |P_{f\leftarrow s}|^2.$$

This expression emphasizes the requirement for large bond selectivity or rate enhancement, such as we observe in the bimolecular reaction of water. The coefficient C_s^n and the reaction probability $P_{f\leftarrow s}$ of the zero-order state $|s\rangle$ must be large. We know this is the case for water since its stretching vibrational levels are good local mode states and the O-H stretching excitation promotes the abstraction reaction. In the language of intramolecular dynamics, a large C_s^n corresponds to an isolated state that closely resembles the zero-order state. For such a molecule, the zero-order state, if it were prepared, would largely retain its excitation and not transfer it to other zero-order states extensively. The key to predicting this selectivity for reactions of other molecules is to assess the size of C_s^n. Vibrational overtone spectroscopy can provide this information and should be a guide to identifying likely candidates for bond selective reaction. The reaction involving water is probably a limiting case, and other species should show varying degrees of enhancement depending on the extent of delocalization of the initial excitation.

Acknowledgment

We gratefully acknowledge the support of this research by the Division of Chemical Sciences, Office of Basic Energy Sciences, of the United States Department of Energy.

References

1. Levine, R. D. and Bernstein, R. B. (1987) *Molecular Reaction Dynamics and Chemical Reactivity*, Oxford University Press, New York.

2. Crim, F. F., Hsiao, M. C., Scott, J. L., Sinha, A., and Vander Wal, R. L. (1990) Phil. Trans. Roy. Soc. **A332**, 259.

3. Vander Wal, R. L. and Crim, F. F. (1989) J. Phys. Chem. **93**, 5331; Vander Wal, R. L., Scott, J. L., and Crim, F. F. (1990) J. Chem. Phys. **92**, 803; Vander Wal, R. L., Scott, J. L., and Crim, F. F. (1991) J. Chem. Phys. **94**, 1859; Vander Wal, R. L., Scott, J. L., and Crim, F. F., Weide, K., and Schinke, R. (1991) J. Chem. Phys. **94**, 3548; Schinke, R., Vander Wal, R. L., Scott, J. L., and Crim, F. F. (1991) J. Chem. Phys. **94**, 283.

4. Sinha, A., Hsiao, M. C., and Crim, F. F. (1990) J. Chem. Phys. **92**, 6333; Sinha, A., Hsiao, M. C., and Crim, F. F. (1991) J. Chem. Phys. **94**, 4928.

5. Hsiao, M. C., Sinha, A., and Crim, F. F. (1991) J. Phys. Chem. (in press).

6. Child, M. S. (1985) Acc. Chem. Res. **18**, 45; Child, M. S. and Halonen, L. (1984) Adv. Chem. Phys. **57**, 1.

7. Westley, F., Herron, J. T., Cvetanović, R. J., Hampson, R. F., and Mallard, W. G. (1990) NIST Chemical Kinetics Database.

8. Schatz, G. C., Colton, M. C., and Grant, J. L. (1984) J. Phys. Chem. **88**, 2971.

9. For example, Stannard, P. R. and Gelbart, W. M. (1981) J. Phys. Chem. **85**, 3592 and references cited.

MODE-SELECTIVE BOND FISSION IN FUNDAMENTAL STRETCHINGS OF HOD

Y. COHEN, D. DAVID, T. ARUSI-PARPAR, I. BAR and S. ROSENWAKS
Department of Physics
Ben-Gurion University of the Negev
Beer-Sheva 84105
Israel

and

J.J. VALENTINI
Department of Chemistry
Columbia University
New York
NY 10027

ABSTRACT. Stimulated Raman excitation and coherent anti-Stokes Raman scattering prepare and detect, respectively, HOD molecules with *one* quantum of vibrational excitation in the O-H or O-D stretch vibrations. This selective vibrational excitation prepares densities of the excited molecules in particular rotational levels that are sufficient for molecular reaction dynamics studies. An ArF excimer laser at 193 nm promotes vibrationally excited molecules to the first electronic surface $A(^1B_1)$ where they dissociate to produce OH + D or OD + H fragments. The photodissociation products OD and OH are detected via laser induced fluorescence. The photodissociation cross section of HOD (0,0,1) vibrationally excited molecules with one quantum of vibration in the O-H stretch is greatly enhanced over that for HOD (0,0,0). The dissociation of this vibrationally excited molecule is also very selective, the OD + H yield being 2.5 ± 0.5 times greater than the OH + D yield. In the photodissociation of the O-D stretch, excited HOD (1,0,0), no enhancement of the yield of the fragments is obtained. These results demonstrate that bond cleavage does not necessarily occur on the weakened bond, and they agree with theoretical calculations indicating that the yield of OD and OH fragments depends on the Franck-Condon overlap of the ground state vibrational wavefunction with the continuum wavefunction on the repulsive surface of the upper state. Also, our results show that even the very lowest possible level of vibrational excitation can be "leveraged" to effect selective bond breaking.

1. Introduction

Although the important role of vibrational energy in chemical reactions has long been recognized, until the advent of powerful lasers and nonlinear optical techniques, research lacked a tool for efficiently studying this particular mode of excitation. Most studies investigate photodissociation from ground vibrational states, although during the last few years there are examples involving vibrationally excited systems [1-7]. Vibrational excitation gives a molecule access to regions of the ground electronic state potential surface that are unavailable to a vibrationless molecule and it may influence the absorption of a second photon. The change in the initial geometry from that of vibrationless molecule can substantially affect the wavelength dependence of the continuous photoabsorption spectrum, change the magnitude of the absorption cross section and alter the nascent product quantum yield.

The effects of initial vibrational excitation on the intensity and shape of the absorption spectra

J. Jortner et al. (eds.), Mode Selective Chemistry, 227–240.

and on the distribution of the photofragments states are very sensitive to the shape and symmetry of the potential surfaces of the connected electronic states. Therefore it is expected that each vibrational mode of a polyatomic molecule has a significantly different effect on the photoabsorption spectrum depending on the potential functions and vibrational mode symmetries, and not simply on the amount of vibrational energy. In fact, by dissociating vibrationally excited molecules we study different processes as compared to those occuring from vibrationless molecules. However, the fragmentation still proceeds along the same excited-state potential surface and the vibrational excitation just changes the initial conditions.

Vibrationally excited molecules are of considerable importance in many electronic photodissociation and photochemical processes and they present rich vibrational photodissociation dynamics of their own. They are important reactive species in thermal kinetics, combustion processes, and atmospheric chemistry. Vibrational excitation can influence the reaction pathway, and so understanding the dynamical behavior of vibrationally excited molecules is essential to designing schemes to selectively direct the course of a chemical reaction. Moreover, photodissociation through vibrational states allows us to study the spectroscopy of vibrationally excited molecules.

Of particular interest in unimolecular reactions is mode-selective control of the reaction pathway by selective excitation of particular reagent vibrational states which correspond to particular motions in the molecule. The initiation of a bond-selected chemical reaction can be achieved by special energy investment in the reaction coordinate, either by its direct excitation or by preferential coupling of energy into it from another vibration [1]. Selective excitation of the parent molecule with *one* quantum of vibrational excitation is the cleanest possible way to get such specific energy deposition in a particular bond, since coupling to other vibrational modes is minimized for such excitation and often even specific rotational levels can be selectively pumped. This energy can then induce the molecule to dissociate selectively along the vibrationally excited coordinate in a photodissociation process.

Triatomic molecules are particularly appealing objects of both theoretical and experimental studies, since they are small enough to allow *ab initio* calculations of potential surfaces and photodynamics and yet retain the complexity of different vibrational degrees of freedom.

The HOD molecule is particularly interesting for investigating the influence of reagent fundamental vibrational excitation in bond-selective photodissociation, since theoretical predictions on this molecule are available. The first fully three-dimensional quantum evaluation of mode-selective bond fissions has been carried out by Segev and Shapiro [8] for HOD in the second absorption band ($B \leftarrow X$) and predicted enhanced yields for the OH + D channel. Studies on HOD photodissociation conducted by Engel and Schinke [9] and independently by Imre and coworkers [10] using *ab initio* potential energy surfaces [11] indicate that this molecule potentially can also exhibit mode-specific behavior in the first absorption band ($A \leftarrow X$). These authors have predicted the isotopic branching ratio for the photodissociation HOD $\rightarrow$ OD + H (OH + D), and showed that the branching ratio is a strong function of the frequency of the dissociating photon and the vibrational state of the HOD. The theoretical studies predict OD/OH ratios that agree with the results for photodissociation of HOD (4νOH) [12], for 157 nm photolysis of HOD (0,0,0) [13], and with our preliminary results for 193 nm photodissociation of HOD (0,0,1) [7]. Similarly, good agreement with theory [9,10,14] is obtained for the OH product quantum state distribution in 157 and 193 nm photodissociation of H_2O (0,0,0) and in 193 nm photolysis of H_2O (0,0,1) prepared by infrared excitation [3,15,16].

The major motivation of this study is to control the reaction dynamics of molecules from particular rovibrational states, which are selected in the vibrational excitation step. The most important aspect of this is the preparation of different, distinct initial vibrational states by selective investment of minimal amounts of energy. This paper describes the photodissociation dynamics of HOD excited with one quantum of vibrational energy in the pure O-H or O-D stretches. The selective excitation of HOD with just one quantum of vibrational excitation in the O-H or O-D stretches leads to different photodissociation channels. When HOD (0,0,1) molecules are photodissociated the OH + D and especially the OD + H yields are appreciably enhanced relative to

HOD (0,0,0), while when HOD (1,0,0) molecules are photodissociated no enhancement of the yields of either OH + D or OD + H is observed. These experiments that prepare and subsequently photodissociate a particular low lying initial state provide an important test of theoretical predictions [9,10].

2. Experiment

The experimental protocol is described in Fig. 1. It involves excitation of a specific

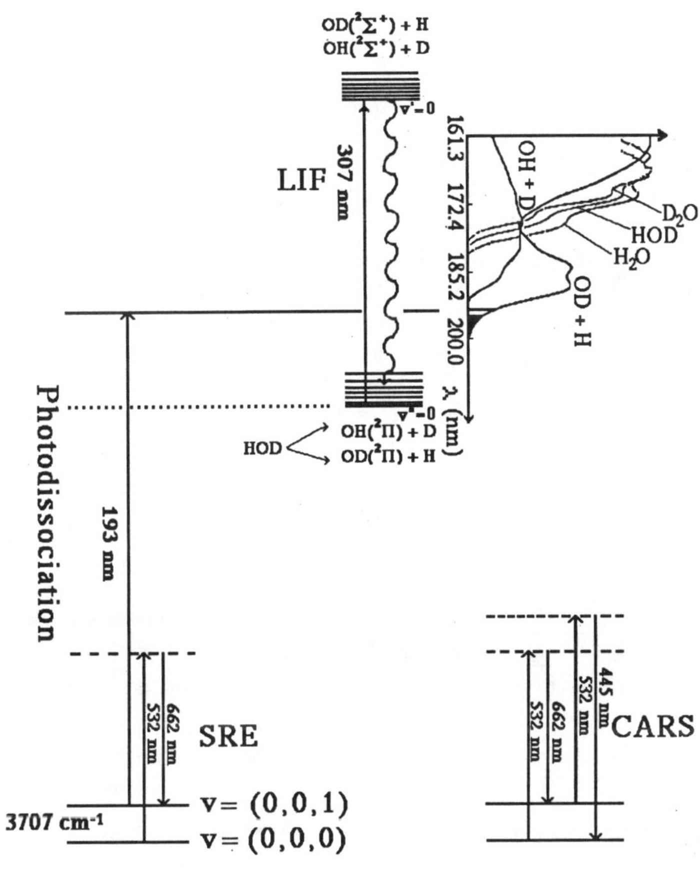

Fig. 1. Scheme of the experiment for photodissociation of HOD (0,0,1). 3707 cm-1 corresponds to excitation of HOD (0,0,0) to HOD (0,0,1). From the vibrational excited state the molecule is photodissociated using 193 nm light and the OH and OD products are detected by LIF. At the upper right side the calculated absorption curves of H_2O, HOD and D_2O starting from ground state are shown. The curves for OD + H and OH + D are partial absorption spectra for the two dissociation channels starting from HOD (0,0,1), and the shaded region indicates where the selective bond breaking is possible (adapted from Imre and coworkers [10]).

fundamental vibrational mode of HOD, photolysis of the vibrationally excited molecule, and laser induced fluorescence (LIF) [17] detection of the hydroxyl species OH and OD. Vibrational excitation of HOD (0,0,1) and HOD (1,0,0) is effected by stimulated Raman excitation (SRE), and the excited molecules are monitored by coherent anti-Stokes Raman scattering (CARS) [18]. The Raman transition of the SRE is [19]:

$$\text{HOD (0,0,0)}\ J_{KaKc} \rightarrow \text{HOD (v1,v2,v3)}\ J'_{Ka'Kc'},$$

where (v1,v2,v3) are the quantum numbers for vibration and J, K_a and K_c are the quantum numbers for the total angular momentum and its projection on the a and c axes in the prolate and oblate symmetric limits. The cross section for absorption at 193 nm is much larger for vibrationally excited HOD (0,0,1) than for ground state species [10] (see upper right corner of Fig. 1).

The experimental setup is depicted in Fig. 2. In the SRE technique the frequency difference of two visible lasers is tuned to the transition of a specific fundamental vibrational mode. Here we use the second harmonic of a Nd:YAG laser at 532 nm (ω_p) and a tunable dye laser (ω_s) pumped by it (Quantel, YG580, TDL50) for the two visible beams. A mixture of DCM and LDS 698 allows generation of ~ 662 nm for HOD (0,0,1) preparation and Rhodamine 640 provides ~ 622 nm for HOD (1,0,0) production. Typical ω_p energy is 30 mJ and ω_s is 20 mJ. Excitation of the molecule is monitored and optimized by measuring the CARS signal (ω_{as}) generated by the Raman excitation beams. Following the SRE pulse, after a delay of 50 ns, the vibrationally excited molecules are photodissociated by a ~ 2 mJ, 193 nm beam (ω_d) from an ArF excimer laser (Lambda Physik, EMG 101MSC). After a delay of another 50 ns the OH and OD photofragments are probed by LIF on the A, v' = 0 ← X, v" = 0 transition [20-22], with < 10 μJ, ~ 307 nm pulse (ω_{pr}) from the frequency-doubled output of a Nd:YAG-pumped dye laser (Quanta-Ray, DCR-2A, Lambda Physik, FL3002). The probe laser is kept at this low value of energy to minimize saturation.

The ω_p, ω_s and ω_{pr} beams propagate collinearly through the cell, after passing through a 12" focal length lens, which brings them to a common focus at the center of the cell. An optical delay line provides the temporal control necessary to keep the ω_p, and ω_s pulses temporally overlaped. In order to check the spatial overlap of the three beams a small portion (~ 4 %) of the beams is split off by a window and imaged onto a viewing screen, after passing through a pinhole located at the focus. At the exit of the gas cell ω_p, ω_s, ω_{pr} and ω_{as} are transmitted by a dichroic mirror that reflects ω_d. The beams are recollimated by a 20" focal length lens, then ω_{as} is separated from the input beams by a Pellin-Broca prism and color filters. A lens focuses ω_{as} onto a slit in front of a photomultiplier tube (PMT). The photodissociation beam ω_d counterpropagates, relative to the SRE and LIF beams, through a 40" focal length lens which brings it to the focus of the other beams.

The LIF from the hydroxyl fragments is observed at a right angle to the laser beams and is imaged by a 2" focal length f/1 lens onto a 5 mm slit in front of a PMT and an interference filter (310 ± 5 nm). A boxcar integrator (SRS, SR250 and SR280) captures the resulting LIF and CARS signals. A PC controls the scanning of the dye laser. It also accumulates, pulse by pulse, the LIF and CARS signals along with signals from photodiodes that monitor the energy of the photodissociation and probe lasers. Each data point in the ensuing spectra is obtained from these signals as an average of ten pulses. The LIF is normalized, point by point, to the photolysis and probe laser energies. Since the CARS signal is saturated with respect to each of the exciting beams the LIF signal is not normalized to it.

The experiments are carried out in a gas cell through which the sample flows at a pressure of 0.35 Torr. The sample is drawn from a liquid reservoir containing a statistical mixture of HOD, H_2O and D_2O with H:D ratio of 1. Because of the differences in the vibrational frequencies of

HOD, H_2O and D_2O [19], the SRE produces no vibrationally excited H_2O or D_2O when tuned for HOD excitation.

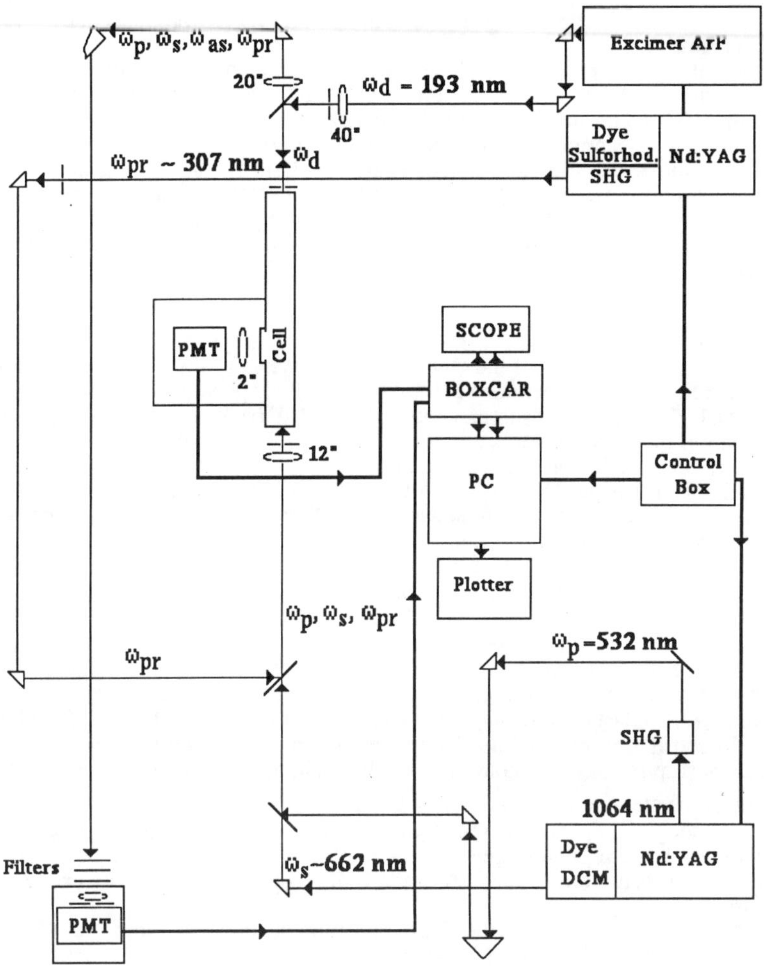

Fig. 2. The experimental setup. ω_p, ω_s, ω_{as}, ω_d and ω_{pr} are the pump, Stokes, anti-Stokes, dissociation and probe wavelengths, respectively. ω_s ~ 662 nm corresponds to HOD (0,0,1) excitation. For HOD (1,0,0) ω_s ~ 622 nm.

Low pressures and short temporal delays are used to minimize collisioal perturbation of the initially prepared HOD state and the rotational state of the OH/OD fragments. Indeed, decreasing the delay between the SRE and photolysis lasers to ~ 15 ns does not affect the results, showing that collisional relaxation of the rovibrationally excited HOD molecules is unimportant. Using measured rotational relaxation rate constant [23] for OH by H_2O, $2.2 \pm 0.5 \times 10^{-10}$ cm^3s^{-1}, we calculate that only about 13 % of the OH products are relaxed rotationally at the pressure and delay of our experiment. Assuming that the relaxation efficiency of OD is comparable to that of OH, it seems that the collisional relaxation of the rotations of the OD/OH products does not affect the observed branching ratio.

Two types of experiments are conducted with this system. In the first, the probe laser wavelength is varied and the SRE laser kept fixed on a particular rovibrational transition of HOD. This procedure yields OH and OD LIF product excitation spectra from which the distribution of products among their quantum states can be extracted. In this case the CARS signal is used to identify and to optimize the SRE excitation of a particular Raman transition. In the second type of experiment we monitor particular OH or OD rovibrational states at fixed LIF wavelength while varying the SRE laser wavelength, to generate a reactant state yield spectrum, that is a spectrum of those HOD molecules that photodissociate to produce OH and OD fragments in the specific quantum state.

3. Results

The LIF excitation spectrum of the OD and OH fragments as measured between 307.1 and 307.7 nm in the first type of experiment is shown in Fig. 3 for photodissociation of HOD (0,0,1). The "net" signal, Fig. 3(c), is a result of subtracting the LIF background signal (no SRE), due to photodissociation of ground state isotopomers of water (Fig. 3(b)), from the LIF signal obtained when SRE excitation occurs (Fig. 3(a)).

It is clearly seen that the OD and, to a lesser extent, the OH signals are enhanced when vibrationally excited molecules are photodissociated. Since the $R_2(4)$ transition for each of the fragments is not overlapped by other transitions it is used for calculating the OD/OH branching ratio. The ratio between the peaks of the OD and OH net LIF signals measures their relative populations when the $R_2(4)$ absorption transition probabilities [22] are accounted for. When HOD (0,0,1) is photodissociated, the branching ratio of the OD/OH product is found to be 4.9 ± 2.5. This result is based on an average of nine measurements. The error in this branching ratio obtained from the LIF measurements is large, because the OH net signal is very small.

The presence of appreciable background LIF signals, due to photodissociation of vibrational ground state H_2O, HOD and D_2O may seem surprising in view of the fact that Fig. 1 shows the 193 nm excitation is well outside the absorption band of the vibrationless water isotopomers. However, the absorption of these molecules is only small, not zero at 193 nm. Based on a minor extrapolation of the measured [24,25] water absorption band we estimate that $\sigma = 5 \times 10^{-22}$ cm^2 at 193 nm. Although it appears that the absorption cross section of HOD (0,0,1) is about 1000 times larger than this, the photofragment yield from the HOD (0,0,1) does not overwhelm the background yield because only about 1% of the HOD molecules are actually excited to (0,0,1) (see Discussion).

Shown in panel (a) of Fig. 4A is the CARS spectrum obtained by tuning the SRE excitation through the Q-branch transitions of (0,0,1), which are assigned in Table I. Panels (b) and (c) present the reactant yield spectra for OD and OH. The OD (v" = 0) and OH (v" = 0) LIF signals are shown as a function of the SRE scan with the probe laser fixed on the $R_2(4)$ transitions for OD and OH, respectively. The position of the rotational lines in panels (b) and (c) correspond to those seen in the CARS spectrum (a). These peaks are due to the photodissociation of molecules in particular rotational states of the O-H stretch vibration. It is clearly seen that the OH and OD photofragment yields are enhanced when the (0,0,1) state is prepared by SRE. Comparison of the OD fluorescence to that of OH demonstrates that the photodissociation of HOD (0,0,1) always produces more OD than OH and the branching ratio of the OD/OH product, obtained from an average of four OD and four OH scans, is 2.5 ± 0.5. This result represents the ratio averaged

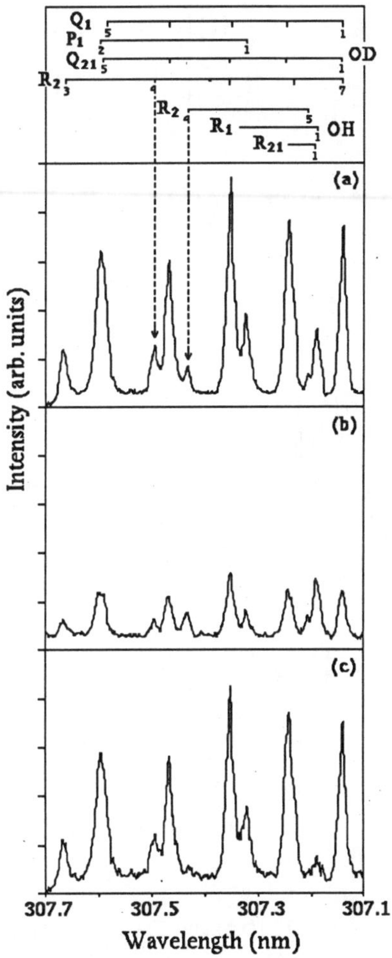

Fig. 3. LIF excitation spectra of the OH and OD products of the photodissociation. The "net" signal (c) comes from subtracting the LIF background signal (b) (which is present even without vibrational excitation) from the LIF signal, (a), obtained when SRE excitation occurs at a frequency that corresponds to transition No. 11 in Fig. 4A(a). All three panels have the same intensity scale.

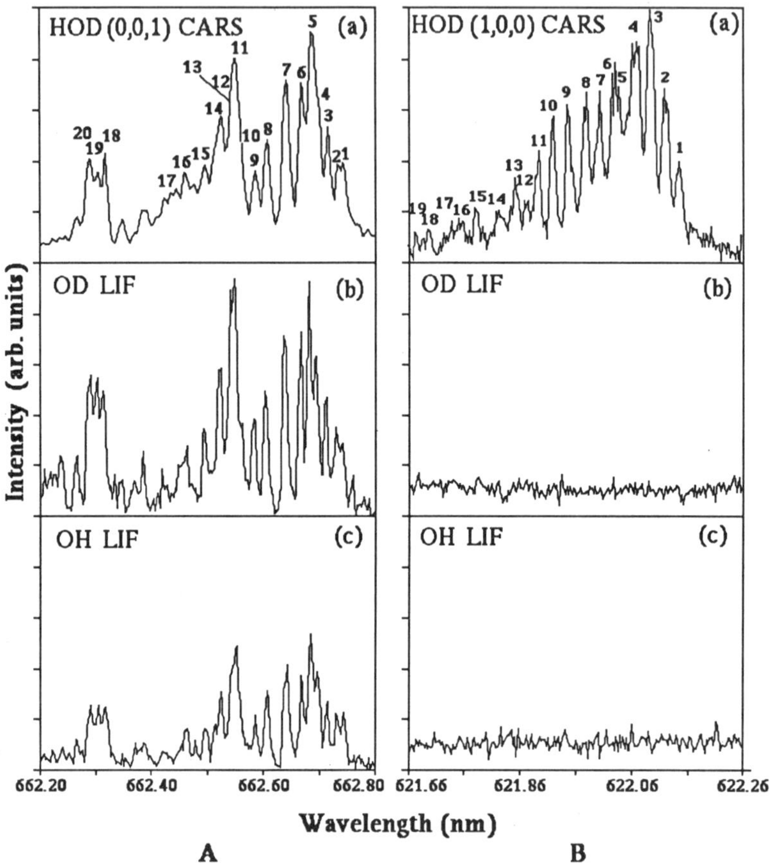

Fig. 4. Direct excitation (A) of HOD (0,0,1) and (B) of HOD (1,0,0) obtained by scanning the SRE excitation over the rotational levels of the vibrational states: (a) saturated CARS signal, (b) and (c) LIF of the R2(4) transition of the OD and OH fragments, respectively. The vertical scale in (b) and (c) is the same. The background signal due to photodissociation of vibrationless ground state molecules, which is comparable to the signal resulting from photodissociation of HOD (0,0,1) molecules, has been subtracted. The peaks in the CARS spectra correspond to Q-branch rotational transitions assigned in Table I.

over the rotational states of (0,0,1) and therefore it is adopted rather than the result obtained from the product excitation spectra, which represents the ratio for a specific rotational state of HOD (0,0,1).

The top panel of Fig. 4B illustrates the CARS signal obtained by tuning the SRE excitation through the Q-branch transitions of (1,0,0) which are assigned in Table I. Panels (b) and (c) show the $R_2(4)$ LIF signals of the OD and OH photofragments when the SRE is tuned through rotational levels of (1,0,0), the OD stretch excitation in HOD. No enhancement of the OH and OD fluorescence is observed, in contrast to the strong enhancement of OD and OH yields for photodissociation of HOD (0,0,1).

Table I. Assignment of the HOD (0,0,1) and HOD (1,0,0) Q-branch rotational transitions [26-28] monitored via CARS.

No. in Fig 4A, 4B	J_{KaKc} (0,0,1)	J_{KaKc} (1,0,0)
1	0_{00}, 1_{01}	0_{00}, 1_{11}, 1_{10}, 1_{01}
2	2_{02}	2_{20}, 2_{21}, 2_{12}, 2_{02}, 2_{11}
3	3_{03}	3_{13}, 3_{21}, 3_{02}, 3_{22}, 3_{12}
4	1_{10}, 1_{11}, 2_{11}	4_{14}
5	3_{12}, 2_{12}, 4_{04}	4_{13}, 4_{23}, 4_{22}, 4_{31}, 4_{32}
6	4_{13}, 3_{13}	4_{41}
7	5_{14}, 5_{05}, 4_{14}	5_{24}
8	6_{15}, 5_{15}	5_{14}, 5_{33}, 5_{23}, 5_{32}
9	6_{06}	6_{06}, 6_{16}, 5_{42}
10	6_{16}	6_{34}, 6_{33}, 5_{51}, 6_{15}
11	2_{21}, 2_{20}, 7_{16}, 3_{21}, 4_{22}	7_{07}, 7_{17}, 6_{42}, 6_{43}
12	3_{22}, 5_{23}	7_{26}
13	7_{25}, 6_{24}	7_{16}, 6_{52}, 7_{35}
14	5_{24}, 4_{23}, 7_{07}, 7_{17}	7_{34}, 8_{18}, 8_{08}
15	8_{26}, 8_{17}	8_{27}, 9_{19}, 9_{09}
16	6_{25}, 8_{08}	8_{36}
17	8_{18}	8_{26}, 8_{44}, 8_{45}
18	3_{30}, 3_{31}, 4_{32}, 4_{31}, 9_{28}	9_{18}, 8_{54}, 8_{53}
19	5_{32}, 5_{33}, 10_{19}	9_{37}
20	6_{33}, 7_{34}, 6_{34}	

4. Discussion

Our measurements show that the lowest possible level of vibrational excitation of HOD has a dramatic effect on the photodissociation cross section, and that the effect is vibrational-mode-specific. These results also indicate that the vibrational excitation leads to bond selective dissociation. The enhancement of the photodissociation cross section is very large. If we knew exactly the 193 nm photodissociation cross section for each water isotopomer in the (0,0,0) vibrational state and the exact fraction of the HOD molecules that we promote to (0,0,1) via SRE we could determine exactly the quantitative enhancement in the photodissociation cross section due to vibrational excitation. But neither of these quantities is known precisely and neither is measured in our experiments, so an exact calculation is not possible. Nonetheless, we can make reasonable estimates that allow an approximate calculation.

As mentioned above, extrapolation of the measured [24,25] water absorption band leads to an estimate of 5×10^{-22} cm^2 for the absorption cross section of the (0,0,0) vibrational state water molecules at 193 nm. We estimate that 30 % of the HOD molecules having Raman transitions within the bandwidth of the SRE lasers are excited to (0,0,1). This estimate is based on an analysis of our SRE conditions in terms of the extensive SRE vibrational pumping efficiency measurements made by Tabor [29], and estimates made by others [30,31]. Note that $v = 0$ to $v = 1$ conversion efficiencies of up to 40 % have been demonstrated with SRE [29]. The Boltzmann distribution over rotational states determines the percentage of the molecules having Raman transitions within the SRE bandwidth. On an average for our experimental conditions ~ 2 % of the HOD molecules fall within the SRE bandwidth. Therefore, about 0.6 % of the HOD molecules (0.3 % of all the water isotopomers) are vibrationally excited. Fig. 3 implies that the yield of OD from HOD (0,0,1) is about 3 times larger than that from the (0,0,0) isotopomers. Fig. 4 shows that this yield is comparable to that from (0,0,0) isotopomers when the subtraction of the background signal is accounted for. This implies that at 193 nm the photodissociation cross section for HOD (0,0,1) is ~ 300 times that of the (0,0,0) molecules. Based on theoretical predictions of the cross section for HOD (0,0,1) [10] and our estimate of the 193 nm absorption cross section for the vibrationless ground state molecules we would expect an enhancement of ~ 1000.

In contrast to the behavior of HOD (0,0,1), for HOD (1,0,0) the theoretical work predicts no enhancement of the photodissociation cross section. Our results are consistent with this prediction; as shown in Figs. 3 and 4 we see no increase in the yield of OH or OD fragments when we prepare HOD (1,0,0). Based on our signal-to-noise ratio we estimate that the 193 nm photodissociation cross section for HOD (1,0,0) is no more than six times that of the (0,0,0) molecules. Thus, the cross section for HOD (0,0,1) at 193 nm is at least 50 times greater than that of HOD (1,0,0), even though these two states differ in energy by only about 2 % of the bond energy.

The magnitude of the bond selectivity we observe in the fragmentation of the HOD (0,0,1) is also consistent with the theoretical predictions of Imre and co-workers [10], at least qualitatively. We find that breaking the O-H bond to form OD + H is favored over braking the O-D bond to form OH + D by a factor of 2.5 ± 0.5. The theoretically predicted ratio, ~ 10, is actually considerably greater than we measure. However, the predicted branching ratio is estimated from the thoretical OH + D and OD + H partial absorption cross section for HOD (0,0,1), and since the absorption cross section for the OH + D channel is very small the estimate is only approximate.

The theoretical calculations reveal that Franck-Condon (FC) effects are the source of both the difference in photodissociation behavior of HOD (0,0,1) and HOD (1,0,0) and the selectivity in the bond breaking when HOD (0,0,1) is dissociated. The enhancement of the photodissociation cross section of HOD (0,0,1) is the result of much better FC overlap of the dissociative continuum with the (0,0,1) vibrational wave function than with either the (0,0,0) or the (1,0,0) vibrational wave functions. Even though the vibrational excitation of the O-D stretch in HOD (1,0,0) gives the O-D bond an initial "push" in the direction of the dissociation coordinate to OH + D, the unfavorable FC factors keep this from enhancing the dissociation. In the case of HOD (0,0,1) the favorable FC factors enable the additional energy along the OD + H dissociation coordinate to be very effective at enhancing the dissociation.

Fig. 5, shows a cut along the dissociation coordinate for the ground and first electronically excited potential surfaces of HOD, which explains and illustrates the photodissociation of the vibrationally excited species. The photolysis photon (193 nm) dissociates the vibrational fundamentals by their promotion to the repulsive electronic state (A^1B_1). In the FC approximation the dissociation samples the two channels, corresponding to breaking of the O-H or the O-D bond, based on the overlap between the ground electronic state vibrationally excited wave function and the dissociative excited state continuum wave function. For a steeply repulsive excited state surface, like that encountered in direct photodissociation of HOD, the continuum function has large amplitude only at the classical turning point. The FC overlap of the vibrational wave function with the continuum excited state function is influenced by the location of the turning point. Due to difference in the reflection of the two lower vibrational states, (0,0,1) and (1,0,0), at the repulsive dissociative surface a different FC overlap is obtained. For 193 nm excitation of HOD (0,01) the overlap of the ground state wavefunction with the continuum function at the excited state classical turning point is good, and dissociation is promoted. In contrast, for HOD (1,0,0) this FC overlap is poor, and dissociation is disfavored.

This bond selectivity is also isotope selectivity, so vibrational excitation and subsequent photodissociation of HOD can lead to isotopic fractionation. This is a direct consequence of the large enhancement of the photodissociation cross section of HOD (0,0,1) and the selective dissociation of this species to give OD + H, coupled with the lack of enhancement of the dissociation cross section for HOD (1,0,0). In fact, it is clear that isotopic fractionation could even be achieved without mode-specific vibrational excitation. Photodissociation of thermally excited, i.e. heated HOD, having appreciable populations of both (1,0,0) and (0,0,1), would still result in preferential production of OD + H over OH + D. Even photodissociation by broad-band source could do this. The resulting isotopic fractionation would show an unusual mass dependence.

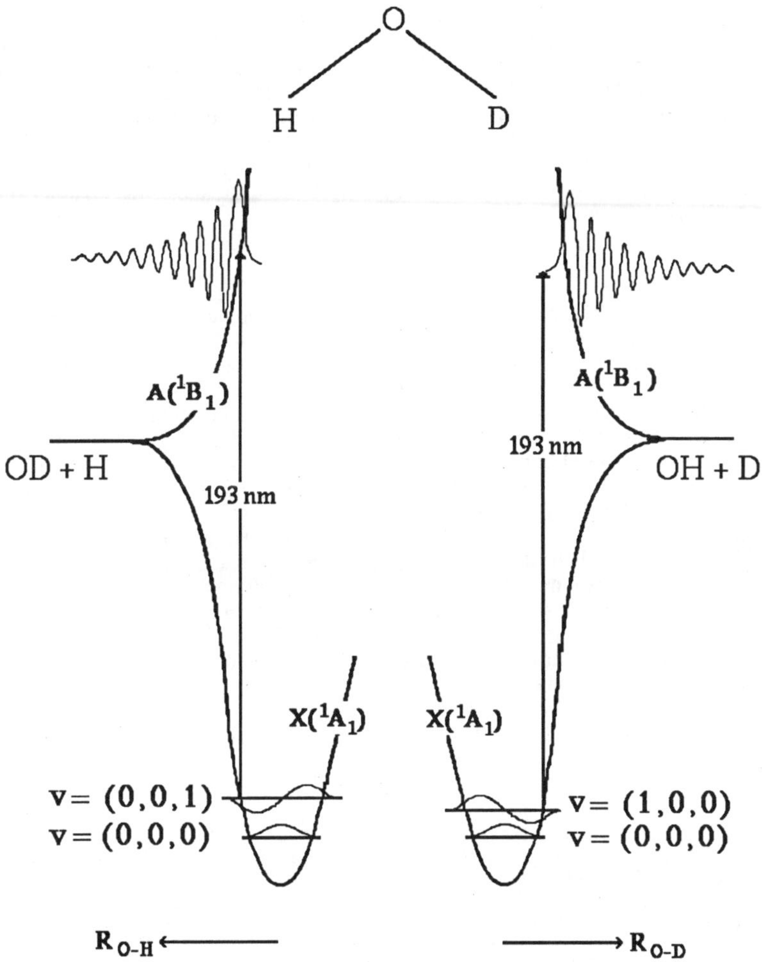

Fig. 5. Schematic description of the transition from the different stretches of HOD in the ground electronic state to the repulsive A state showing a better FC overlap for the transition from (0,0,1) than from (1,0,0).

Comparison of the 193 nm photodissociation of HOD (0,0,1) and HOD (1,0,0) molecules to the vibrationally mediated photodissociation of HOD (4vOH) at 239.5 nm shows that although the excess energy for photodissociation over bond dissociation energy are comparable [32], the relative yields of the hydroxyl photofragments are different, OD/OH ~ 2.5, OD/OH ~ 1 and OD/OH > 15, respectively. In the experiments of Crim and coworkers [12] the vibrational excitation is quite high, about 1.7 eV or 33 % of the 5.118 eV HOD bond dissociation energy, and about 24 % of the 6.9 eV energy of the photons that effect dissociation. However, in our experiments the vibrational excitation energy is very small, only about 9 % and 7 % of the dissociation energy and only about 7 % and 5 % of the energy of the photons that effect dissociation, for HOD (0,0,1) and HOD (1,0,0), respectively. These results and the studies of single photon photodissociation at 157 nm [13] support theoretical calculations [9,10]. The present study, where two distinct vibrational modes are prepared before the photodissociation, confirms that it is not enough to weaken the bond to be cleaved in order to get bond selectivity, but a good FC overlap of the vibrational wave function with the corresponding channel on the dissociative excited state is essential.

5. Summary

The photodissociation of HOD prepared in particular rovibratational states in the fundamental vibrational stretching modes, photodissociated with a 193 nm beam and detected by LIF of the OH and OD photofragments, has been investigated. The most striking result of these experiments is the observation that it is possible to control the branching ratio of the OH and OD fragments by specific selection of the fundamental vibrational wavefunctions. Excitation of the O-H stretching, HOD (0,0,1), leads to a preference in the O-H bond fission, while that of O-D stretching, HOD (1,0,0), does not enhance the O-D bond cleavage. Our results demonstrate that bond cleavage does not necessarily occur on the weakened bond, but rather depends on the FC overlap between the initial and final states. This agrees with theoretical studies and shows that even at the very lowest level of vibrational excitation, selective bond breaking can be effected.

The photodissociation of HOD (0,0,1) gives a bond selectivity that is probably far lower than ultimately achievable. In order to increase the selectivity in photodissociation of HOD (0,0,1) and to obtain only OD + H products one should photodissociate the molecule at wavelengths > 195 nm [9]. Therefore, to refine our measurements of the yield of OD/OH from photodissociation of HOD (0,0,1) we intend to extend our studies to longer wavelengths at which the absorption from ground state molecules will be minimal.

Our experiments on mode-selective chemistry in the photodissociation of water and related studies by other groups [1,12,13,34,35] represent an attempt to understand and ultimately control the course of unimolecular reactions.

Acknowledgement. This work has been supported by a grant from the United States-Israel Binational Science Foundation (BSF), Jerusalem, Israel. JJV acknowledges support of this work from the National Science Foundation.

References

1. Vander Wal, R.L., Scott, J.L. and Crim, F.F. (1991) 'State resolved photodissociation of vibrationally excited water: Rotations, stretching, vibrations, and relative cross sections', J. Chemical Physics, 94, 1859 - 1867; Crim, F.F. (1990) 'State- and bond-selected unimolecular reactions', Science 249, 1387 - 1392; Crim, F.F., Hsiao, M.M., Scott, J.L., Sinha, A. and Vander Wal, R.L. (1990) 'State- and bond-selected photodissociation and bimolecular reaction of water', Phil. Trans. Royal Society London A 332, 259 -272.
2. Zittel, P.F. and Masturzo, D.F. (1986) 'Photodissociation of vibrationally excited OCS', J. Chemical Physics 85, 4362 - 4372 ; Zittel, P.F., Darnton, L.A. and Little, D.D. , (1983) ' Separation of O, C, and S isotopes by two-step, laser photodissociation of OCS', J. Chemical Physics, 79, 5991 - 6005.
3. Hausler, D., Andresen, P. and Schinke R. (1987) 'State to state photodissociation of H_2O in the first absorption band', J. Chemical Physics 87, 3949 - 3965.
4. Brouard, M., Martinez, M.T., O'Mahony, J. and Simons, J.P. (1990) 'Energy and angular momentum disposals in the vibrationally mediated photodissociation of HOOH and HOOD intramolecular properties via photofragment mapping', Molecular Physics 69, 65 - 84.

5. Luo, X. and Rizzo, T. R. (1991) 'Unimolecular dissociation of hydrogen peroxide from single rovibrational states near threshold', J. Chemical Physics 94, 889 - 898 ; Luo, X. and Rizzo, T. R. (1990) 'Rotationally resolved vibrational overtone spectroscopy of hydrogen peroxide at chemically significant energies', J. Chemical Physics 93, 8620 - 8633.
6. Foy, B.R., Casassa, M.P., Stephenson, J.C. and King, D.S. (1990) ' Overtone-excited $HN_3(X^1A')$: Anharmonic resonance, homogeneous linewidths, and dissociation rates', J. Chemical Physics, 92, 2782 - 2789; Foy, B.R., Casassa, M.P., Stephenson, J.C. and King, D.S. (1988) 'Unimolecular dynamics following vibrational overtone excitation of HN_3 v_1=5 and v_1=6: $HN_3(X;v,J,K) \rightarrow HN(X^3\Sigma^- v,J,\Omega,+ N_2(X^1\Sigma_g^+)$', J. Chemical Physics 89, 608 - 609.
7. Bar, I., Cohen, Y., David, D., Arusi-Parpar, T., Rosenwaks, S. and Valentini, J.J. (1991) 'Mode-selective bond fission: Comparison between the photodissociation of HOD (0,0,1) and HOD (1,0,0)', J. Chemical Physics, submitted; Bar, I., Cohen, Y., David, D., Rosenwaks, S. and Valentini, J.J. (1990) 'Direct observation of preferential bond fission by excitation of a vibrational fundamental: Photodissociation of HOD (0,0,1)', J. Chemical Physics 93, 2146 -2148.
8. Segev, E. and Shapiro, M. (1982) 'Three-dimensional quantum dynamics of H_2O and HOD photodissociation', J. Chemical Physics 77, 5604 - 5623.
9. Engel, V. and Schinke, R. (1988) 'Isotope effects in the fragmentation of water: The photodissociation of HOD in the first absorption band', J. Chemical Physics 88, 6831 - 6837.
10. Imre, D.G. and Zhang, J. (1989) 'Dynamics and selective bond breaking in photodissociation', Chemical Physics 139, 89-121; Zhang, J., Imre, D.G and Frederick, J.H. (1989) 'HOD spectroscopy and photodissociation dynamics: Selectivity in OH/OD bond breaking', J. Physical Chemistry 93, 1840 - 1851; Zhang, J. and Imre, D.G. (1988) 'OH/OD bond breaking selectivity in HOD photodissociation', Chemical Physics Lett., 149, 233 - 238.
11. Staemmler, V. and Palma, A. (1985) 'CEPA calculations of potential energy surfaces for open-shell systems. IV Photodissociation of H_2O in the A^1B_1 state', Chemical Physics 93, 63 - 69.
12. Vander Wal, R.L., Scott, J.L. and Crim, F.F. (1990) 'Selectivity breaking the O-H bond in HOD', J. Chemical Physics 92, 803 - 804.
13. Shafer, N., Satyapal, S. and Bersohn, R. (1989) 'Isotope effect in the photodissociation of HDO at 157.5 nm', J. Chemical Physics 90, 6807 - 6808.
14. Weide, K., Hennig, S. and Schinke, R. (1989) 'Photodissociation of vibrationally excited water in the first absorption band', J. Chemical Physics 91, 7630 - 7637; Engel, V., Schinke, R. and Staemmler, V. (1988) ' Photodissociation dynamics of H_2O and in the first absorption band: A complete *ab initio* treatment', J. Chemical Physics 88, 129 - 147; Schinke, R. (1988) 'Rotational distributions in direct molecular photodissociation', Annual Review of Physical Chemistry 39, 39 - 68; Schinke, R., Engel, V. and Staemmler, V. (1985) 'Rotational state distribution in the photolysis of water: Influence of the potential anisotropy', J. Chemical Physics 83, 4522 - 4533.
15. Andresen, P. and Schinke, R. (1987) 'Dissociation of water in the first absorption band: A model system for direct photodissociation', in Ashfold, M.N.R. and Baggott, J.E. (eds.) Molecular Photodissociation Dynamics, Royal Society of Chemistry, London, pp. 61 - 113; Andresen, P., Ondrey, G.S. ,Titze, B. and Rothe, E.W. (1984) 'Nuclear and electron dynamics in the photodissociation of water', J. Chemical Physics, 80, 2548 - 2569.
16. Grunewald, A.U., Gericke, K.H. and Comes, F.J. (1987) 'Photodissociation of room-temperature and jet-cooled water at 193 nm', Chemical Physics Lett. 133, 501 - 506.
17. Kinsey, J.L. (1977) 'Laser-induced fluorescence', Annual Review of Physical Chemistry, 28, 349 - 372.

18. Valentini, J.J. (1987) 'Laser Raman Techniques', in Radziemsky, L.J., Solarz, R.W. and Paisner, J.A. (eds.) Laser Spectroscopy and its Applications, Marcel Dekker, New York, Optical Engineering, Vol. 11, p. 507 - 564; Valentini, J.J. (1985) 'Coherent anti-stokes Raman spectroscopy', in Vanasse , G.A. (ed.) Spectrometric Techniques, Academic Press, New York, Vol. 4, pp. 1 - 62.

19. Herzberg, G. (1945) Molecular Spectra and Molecular Structure II. Infrared and Raman Spectra, D. Van Nostrand, New York.

20. Dieke, G.H.and Crosswhite, H.M. (1962) 'The ultraviolet bands of OH fundamental data', J. of Quantitative Spectroscopy & Radiative Transfer, 2, 97 - 199.

21. Coxon, J.A. (1975) 'The $A^2\Sigma^+$ - $X^2\Pi_i$ system of OD determination of molecular constants by the direct two-state fit approach', J. Molecular Spectroscopy 58, 1- 28; Clyne, M.A.A.,Coxon, J.A. and Woon Fat, A.R. (1973) 'The $A^2\Sigma^+$ - $X^2\Pi_i$ electronic band system of the OD free radical spectroscopic data for the 0-0 sequence and rotational term values for $A^2\Sigma^+$ and $X^2\Pi_i$', J. Molecular Spectroscopy 46, 146 - 170.

22. Dimpfl, W.L. and Kinsey, J.L. (1979) 'Radiative lifetimes of OH $A^2\Sigma$ and Einstein coefficients for the A-X system of OH and OD', J. of Quantitative Spectroscopy & Radiative Transfer 21, 233 - 241.

23. Gericke, K.H. and Comes, F.J. (1982) 'Energy partitioning in the reaction $O(^1D) + H_2O \rightarrow$ OH + OH. V. Rotational relaxation of OH($X^2\Pi$,v",J")' Chemical Physics 65, 113 - 121.

24. Watanabe, K. and Zelikoff, M. (1953) 'Absorption coefficients of water vapor in the vacuum ultraviolet', J. Optical Society of America 43, 753 - 755.

25. Laufer, A.H. and McNesby, J.R. (1965) 'Deuterium isotope effect in vacuum-ultraviolet absorption coefficients of water and methane', Canadian J. Chemistry 43, 3487 - 3490.

26. Benedict, W.S., Gailar, N. and Plyler, E.K. (1956) 'Rotation-Vibration Spectra of deuterated water vapor', J. Chemical Physics 24, 1139 - 1165.

27. Perrin, A., Flaud, J.-M. and Camy-Peyret, C. (1985) 'Calculated energy levels and intensities for the ν_1 and $2\nu_2$ bands of HDO' J. Molecular Spectroscopy 112, 153 - 162.

28. David, D., Strugano, A., Bar, I., Rosenwaks, S. and Valentini , J.J. (unpublished).

29. Tabor, K.D. (1991) 'Energy transfer kinetics and reaction dynamics as probes of molecular interaction potentials and transition states.' Ph. D. Thesis, University of California, Irvine.

30. Gladkov, S.M., Karimov, M.G. and Koroteev, N.I. (1983) ' Two-photon Raman excitation and coherent anti-stokes Raman spectroscopy probing of population changes in polyatomic molecules: a novel nonlinear optical technique for vibrational-relaxation studies', Optics Lett. 8, 298 - 300.

31. Brodnikovsky, A.M., Gladkov, S.M. and Koroteev, N.I. (1982) 'Strong selective excitation of Raman-active vibrations and energy transfer between low-lying vibrational states of a CO_2 molecule studies by CARS and PARS' Optics Comm. 40, 312 - 316.

32. The excess energy over bond dissociation energy is $E = E_0 + h\nu - \Delta H_0$ where E_0 is the energy of the vibrational transition, hn the energy of the photolysis photon and ΔH_0 the bond dissociation energy of 5.118 eV [33]. The excess energy for photodissociation of HOD (0,0,1) at 193 nm is 14236 cm^{-1}, for photodissociation of HOD (1,0,0) at 193 nm 13252 cm^{-1}, and for photodissociation of HOD (4vOH) at 239.5 nm is 14310 cm^{-1}.

33. H. Okabe, (1978) Photochemistry of Small Molecules, J. Wiley and Sons, New York.

34. Brum, J.L., Deshmukh, S. and Koplitz, B. (1990), 'Iodoethane photolysis: Which C-H bond leads to H-atom formation ?', J. Chemical Physics 93, 7504 -7505.

35. Cool, T.A., Goodwin, P.M. and Otis, C.E. (1990) 'H/D isotope effect in the predissociation of C_2HD', J. Chemical Physics 93, 3714 - 3715.

ADIABATIC AND NON-ADIABATIC DYNAMICS IN UNIMOLECULAR REACTIONS

J. Troe
Institute of Physical Chemistry
University of Göttingen
Tammannstraße 6
D-3400 Göttingen
Germany

ABSTRACT. Potential energy surfaces for simple bond fission and the reverse association reactions are considered with respect to the attraction and the anisotropy of the potential. Rovibrationally adiabatic channel potential curves are constructed on these surfaces. Simplified relationships, leading from these curves to rate constants, are given. The extent of adiabatic dynamics and non-adiabatic transitions between adiabatic channel curves is discussed. Experimental results confirm the dominance of adiabatic behavior during the rapid bond dissociation or bond formation stage of the reaction.

1. Introduction

The question of mode selectivity in unimolecular reactions is intimately related to the question of intramolecular rovibrational adiabaticity. The present article considers this phenomenon in simple unimolecular bond fissions and in the reverse associations which proceed on bonding potential energy surfaces.

After a potentially mode selective excitation process, in unimolecular reactions generally there is a time lag, due to intramolecular vibrational dynamics, during which the system randomizes and destroys the memory to the initial excitation. Experimental evidence for nonrandomization and survival of the initial mode selection during this period so far is only very spurious. However, during the much shorter final separation of the fragments of a unimolecular bond fission, because of diminishing interfragment couplings, largely rovibrationally adiabatic behavior of the system has been postulated. In this spirit, statisctical models of unimolecular bond fissions generally assume a statistical, random populatioan of dissociation channels generated during the initial longer intramolecular coupling period. The subsequent, shorter separation of the fragments, however, is believed to adiabatically conserve this population to a large extent. Non-adiabatic transitions are not excluded, but the frequency of their occurrence requires further investigations.

J. Jortner et al. (eds.), Mode Selective Chemistry, 241–259.

The following article discusses properties of potential energy surfaces for bond fission reactions which are crucial for the dissociation event: the attraction between the separating fragments and the anisotropy of the potential related to the separation process. The article describes properties of rovibrationally adiabatic, state-selected, channel potential curves which are constructed on these potential energy surfaces. In statistical theories, rate parameters are directly related to the properties of these channel potential curves, in particular to their individual threshold energies. Simplified rate expressions are presented. Finally, the occurrence of non-adiabatic transitions between adiabatic channel potential curves is estimated on the basis of theories of non-adiabatic transitions. The article can only provide a very limited scope of the field which, because of its general importance, is receiving much attention in many applications of gas phase reaction kinetics.

2. Simple Models for Stretch-Bend Relationships on Bond Fission Potentials

Bond fission processes are accompanied by the transformation of deformation vibrations of the disintegrating reactant into rotations of the fragments. These transformations occur during increasing stretching of the interfragment distance. As a consequence, the stretch-bend interaction of the potential in relation with the breaking bond is of crucial importance in unimolecular bond fission reactions. The corresponding modes will be termed "transitional modes", in contrast to other, "conserved" modes which change their properties only to a lesser extent. In the dissociation of methane $CH_4 \rightarrow CH_3 + H$, for instance, the C-H bond stretch leading to dissociation is accompanied by the transformation of two H-C-H deformation vibrations into CH_3 rotational modes. At the same time, other C-H stretches are essentially conserved on the way from CH_4 to CH_3.

By the analysis of thermal dissociation and recombination experiments on the basis of simple rate models such as described later on [1], a simple stretch-bend relation was proposed in [2] and [3]. This can be expressed by the bending force constants, normal mode frequencies or hindered rotation barrier heights as a function of the distance r between the centers of mass of the fragments. We here choose a representation of harmonic oscillator normal mode quanta $\varepsilon(r)$ in terms of the maximum attraction V, radial potential V(r) given by

$$\varepsilon(r) \approx \varepsilon(r_e)(1-(1-V(r)/V(r_e))^{1/2})^{1/2} \qquad (1)$$

At larger distances, eq. (1) approaches

$$\varepsilon(r) \approx \varepsilon(r_e)(|V(r)|/2D_e)^{1/2} \quad (2)$$

with $-V(r_e) = D_e$. However, we emphasize that the model of eqs. (1) and (2) is only proposed to hold for short range potentials. At large distances, electrostatic multipole expansions of the potential apply, see below. It is of central importance to identify which part of the potential dominates the kinetics under the conditions considered.

Although being derived empirically from thermal dissociation and recombination experiments [1] - [3], a more direct approach towards eqs. (1) and (2) would go through ab initio calculations of potential energy surfaces. Unfortunately, such calculations are still rare for the range of r-values governing the dissociation flux and often are not yet of sufficient precision. One of the few exceptions are the ab initio calculations for H_2O_2 from [4] and [5]. The normal mode analysis of this surface leads to the results shown in Table 1. Indeed the empirical relationship of eq. (1) provides a satisfactory approximate characterisation of the stretch-bend relation of three transitional modes, i.e. the HOOH symmetric and antisymmetric bends and the HO-OH torsion. Similar tests of the empirical stretch-bend relationship, e. g. for the bond fissions H-OOH and H-HCO are possible with the ab initio results from [6] and have led to similarly satisfactory results. However, more systematic studies of the stretch-bend relationship with high precision ab initio calculations are required.

At larger distances, switches of the character of the potential occur. E. g., in the HOOH --> HO + OH reaction, the OH-dipoles have to swing around from an HO-OH orientation into an HO-HO orientation corresponding to a long range dipole-dipole interaction. The very complicated transition range of the potential was well illustrated in [5]. In molecular ion fragmentation, likewise, a relatively abrupt change from a short range valence to a long range charge-multipole potential was observed [7]. Again, it is of central importance to identify which part of the potential governs the dissociation (or association) flux. In order to answer this question, adiabatic channel potential curves have to be determined on the potential energy surface of the reaction

3. Adiabatic Channel Potential Curves

Channel i-specific adiabatic channel potential curves are defined as the eigenvalues $V_i(r)$ of the reaction system at fixed interfragment distance r. To a first approximation,

Table 1

Frequencies of transitional modes in the dissociation H_2O_2 --> 2OH (a = ab initio calculations [5], b = empirical approximation of eq. (1) [2], frequencies ω in cm^{-1}, radial potential V(r) in kcal mol^{-1}, O-O distance r in a. u.).

r	-V(r)	$\omega_{sym-bend}$		$\omega_{antisym-bend}$		$\bar{\omega}_{tors}$	
		a	b	a	b	a	b
2.76	52.0	1429	-	1328	-	383	-
3.50	32.7	1063	893	923	830	253	239
4.50	9.5	560	442	438	411	126	119
7.53	2.06	306	202	228	188	67	54
9.06	1.06	151	145	145	128	25	39

in adiabatic channel models one assumes that the dissociating system with increasing r adiabatically remains in channel i. Non-adiabatic transitions, i. e. "channel-hopping", may be induced by neglected terms in the Hamiltonian such as Coriolis and kinetic energy couplings, see below.

Adiabatic channel potential curves today have been determined for quite a series of reaction systems. For a charge-permanent linear dipole potential, e. g., potential curves such as shown in figures 1 and 2 are obtained (at small total angular momenta, J --> 0, of the system). The channel potentials may be labeled by the rotational quantum number j of the dipole and projection quantum number Ω of the dipole angular momentum (projection of j on the line connecting the centers of mass of the charge and the dipole). The reduced form of figure 1 and 2, which is based on the rotational constant B of the dipole, the dipole moment μ_D and the charge q, allows for a universal representation of the channel potentials $V_i(r)$. In order to facilitate an orientation on the energy-distance scale, we indicate that a λ-value of 0.06 for a C^+-HCl system (with $B = 10.6\ cm^{-1}$ and $\mu_D = 1.08$ D) corresponds to a distance of 14 Å where the maximum of the (j = 2, Ω = 0)-channel is located. The maximum of the (j = 30, Ω = 0)-channel in figure 2 then is located at about 1.3 Å, i. e. clearly in a range where a short range potential would have taken over and the charge-dipole potential is inadequate.

For the interaction between two identical linear dipoles, adiabatic channel potential curves have been calculated in [8]. Figure 3 shows examples. At large distances (i. e. small values of the parameter a), free rotor states are approached; at short distances, the energy levels of two doubly degenerate harmonic oscillators, "sitting on" the maximum attraction of the dipole-dipole potential, are approached. At intermediate distances, complicated curve crossing behavior occurs. It appears worthwhile to mention that, outside this intermediate complex crossing region, a "clumping" of the channels occurs: the channel curves of the oscillators as well as of the rotors, which become degenerate at a --> ∞ and a --> 0 respectively, approach each other and no more curve crossings occur in the limiting ranges.

A comparison of the adiabatic channel potential curves of figure 3 with the ab initio potential of H_2O_2 from [5] shows that the dipole-dipole potential is adequate only at HO-HO distances larger than about 10 Å such that most channel maxima are located in the short range potential. Only for the lowest channels, the harmonic oscillator range of figure 3 is relevant, the complicated curve crossing being of nearly no relevance for the channel maxima. As a consequence, the energy pattern of the adiabatic channel

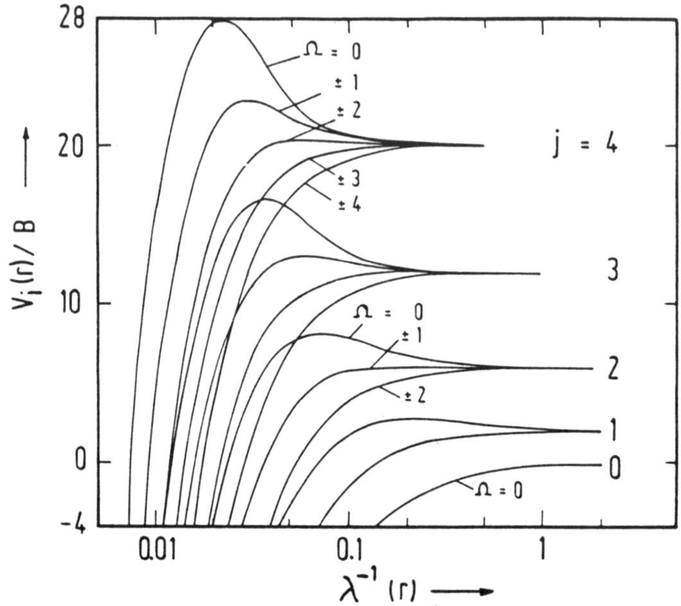

Figure 1 Adiabatic channel potential curves for charge-linear permanent dipole interaction (j = rotational quantum number of the linear dipole, Ω = angular momentum projection quantum number, B = rotational constant of the linear dipole, $\lambda^{-1} = r^2B/q\mu_D$, μ_D = dipole moment, q = charge, r = distance between centers of mass).

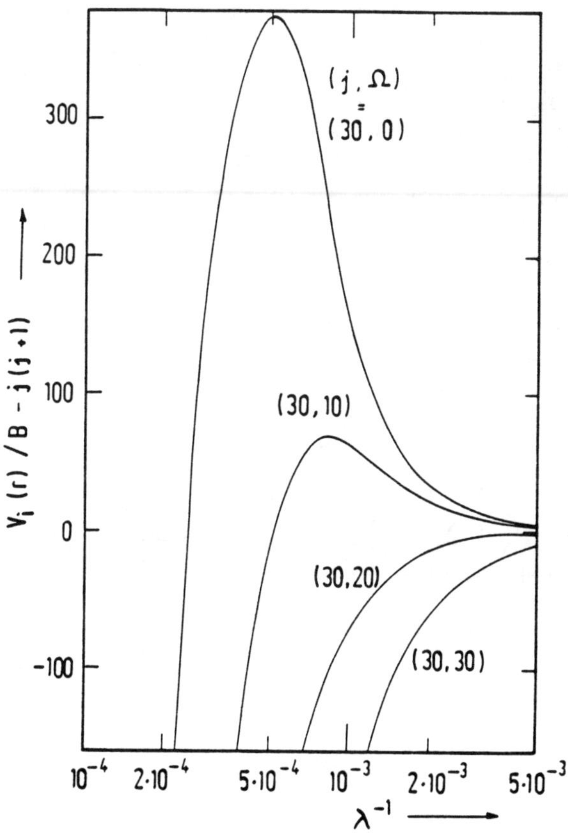

Figure 2 As figure 1, for j = 30

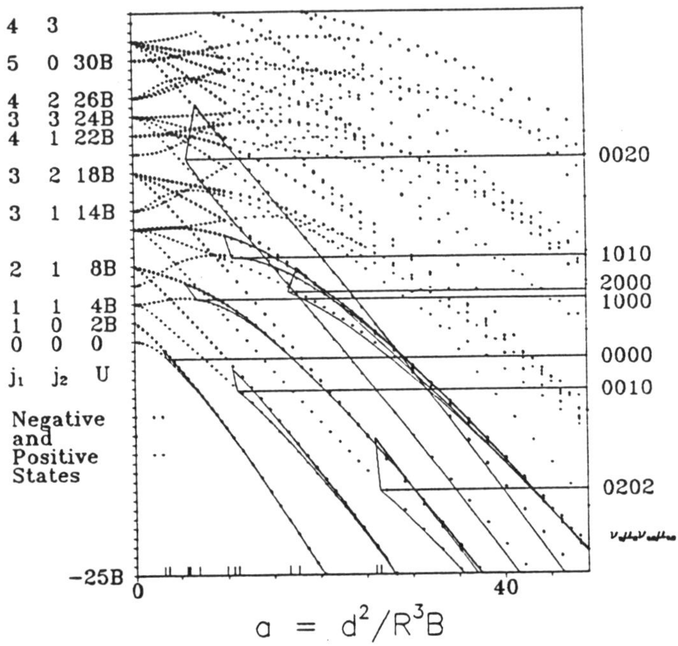

Figure 3 Adiabatic channel potential curves for interaction between two identical dipoles (U = energy, d = dipole moment, B = rotational constant, R = distance between centers of mass, $\nu_s\mu_s\,\nu_{as}\mu_{as}$ = oscillator quantum numbers for large a; j_1, j_2 = rotor quantum numbers for small a, from ref. [8]).

threshold energies, except for the lowest very few channels, is governed by the short range potential where the model of eqs. (1) and (2) becomes nearly valid. However, non-adiabatic "channel-hopping" may result in a by-passing of the individual channel threshold energies such that other ranges of the potential can also become important, see below.

4. Simplified Expressions for Adiabatic Channel Threshold Energies

The basic concept of statistical adiabatic channel models (SACM) [9], [10] is the assumption that individual adiabatic channels i are open if the energy E of the system exceeds the individual threshold energies E_{oi}, i. e. the maxima of the channel potential curves $V_i(r)$. Furthermore, a statistical population of all open channels is postulated which is assured by the dynamics of the system taking place before the threshold region of the potential is approached. Non-adiabatic transitions, by-passing the channel maxima, and tunneling are neglected in first approximation. The determination of threshold energies $E_{oi}(J)$ is, therefore, an essential part of an SACM treatment. As an example, in figure 4 we show threshold energies $E_o(G)$ for some low channels ($j = |\Omega| = 0, 1, 2$) in a charge-linear dipole system. The curves a denote approximate, analytical results based on a long-range perturbation treatment of the channel potential curves; the curves c denote analytical results based on a short-range oscillator approximation of the potential curves; the curves b are exact. G essentially represents the orbital angular momentum in reduced form with $G = J(J+1)h^2/8\pi^2\mu q\mu_D$ (μ = reduced mass, q = ionic charge, μ_D = dipole moment). One notices that the results of an oscillator model are approached for higher G, i. e. for higher angular momenta. It should also be emphasized that channels with decreasing $|\Omega|$ are governed to an increasing extent by the oscillator part of the channel potential. Figure 5 demonstrates this for $|\Omega| = j-1$ (a = oscillator approximation, b = exact). For this reason, it is not surprising that pure oscillator models of channel threshold energies work very well (except at very low collision energies or, in thermal systems, at very low temperatures).

Because of the mostly dominant contribution of the oscillator range of the adiabatic channel potential curves to the channel threshold energies E_{oi}, in the following we consider this limit only. The channel potential curves, e. g., then take the form

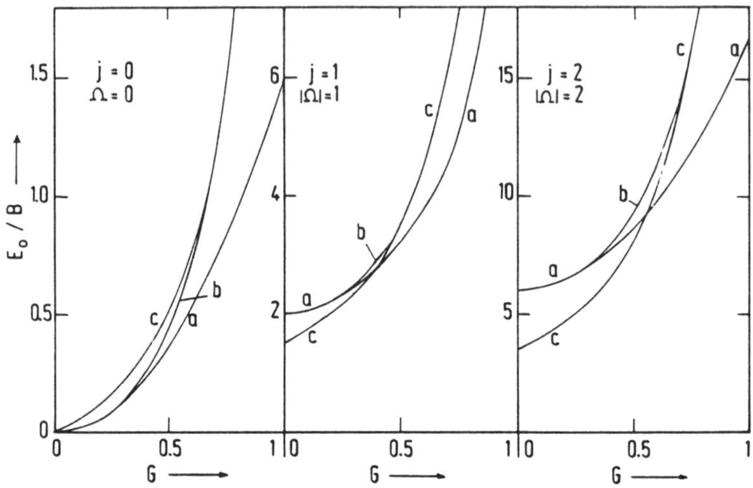

Figure 4 Adiabatic channel threshold energies E_o for charge-linear permanent dipole interactions (B = rotational constant, $G = J(J+1)h^2/8\pi^2\mu q\mu_D$, μ = reduced mass, q = charge, μ_D = dipole moment; a = large-range hindered rotor perturbation limit, b = exact, c = short-range oscillator limit, see ref. [11]).

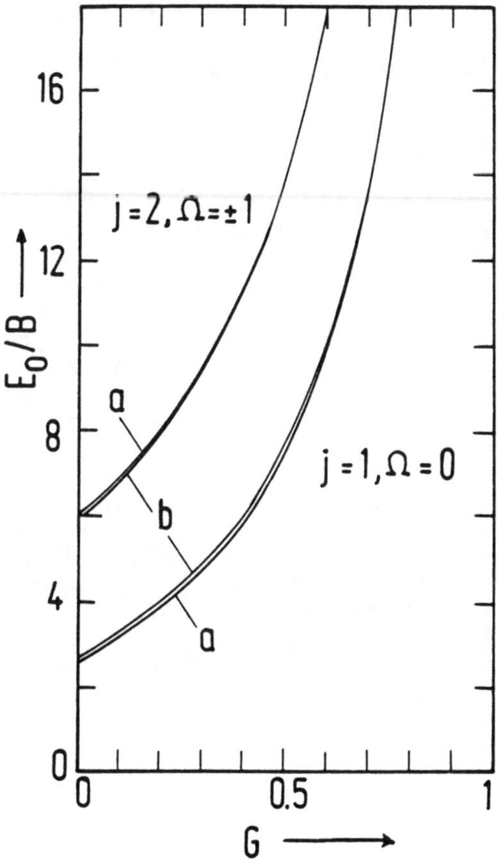

Figure 5 As figure 4, for $|\Omega| < j$ (a = short-range oscillator limit, b = exact, see ref. [11]).

$$V_i(r) \approx -|V(r)| + (i_1+1/2)\varepsilon_1(r_e)(|V(r)|/2D_e)^{1/2} + (i_2+1/2)\varepsilon_2(r_e)(|V(r)|/2D_e)^{1/2} + E_{cent,J}(r) \quad (3)$$

if the case of two transitional modes is considered (approximation of eq. (2), E_{cent} = centrifugal energy, conserved modes neglected). The maxima of eq. (3) give the threshold energies $E_{oi}(i_1,i_2,J)$. They can very easily be obtained numerically on the basis of the actual attraction potential V(r). In some simplified cases, even analytical results are obtained, see [2], [3]. It should be emphasized that eq. (3) applies to short range potentials only which govern most reactions at moderate or high energy or temperature conditions. For this reason, we do not elaborate here the corresponding expressions for long range potentials which were discussed, e. g. in [8] and [11].

In order to illustrate the expressions for the channel threshold energies E_{oi}, we consider the simplified case of a doubly degenerate transitional mode (quantum number v), a radial potential of the r^{-4} type ($V(r) = -A/r^4$) and a centrifugal potential of the r^{-2} type ($E_{cent} = J(J+1)h^2/8\pi^2\mu B_\infty r^2$). In this case, the threshold energies in the oscillator model are approached [2] by

$$E_{oi} \approx C[v + 1 + G\, C^{-1/2}]^2/4 \quad (4)$$

with

$$C = \varepsilon^2(r_e)/2D_eB_\infty \quad (5)$$

and $G = J(J+1)h^2/8\pi^2\mu B_\infty{}^{1/2}A^{1/2}$. Eq. (4) clearly indicates that transitional modes and centrifugal motion in the channel threshold energies cannot be separated. After thermal averaging, acticated complex pseudo-partition functions of centrifugal motion and of transitional modes, therefore, cannot be factored, see below.

5. Simplified Rate Expressions

In statistical rate theories, numbers of open channels W(E,J) at a given energy are counted or pseudo-partition functions at a given temperature are calculated. As an example, we consider the latter quantity in thermal capture or high pressure thermal recombination reactions B + C --> A. The rate constant $k_{rec,\infty}$ here is given by

$$k_{rec,\infty} = \frac{kT}{h} \left(\frac{h^2}{2\pi\mu kT}\right)^{3/2} \frac{Q^*}{Q(B)Q(C)} \tag{6}$$

where

$$Q^* = \sum_i \exp(-E_{oi}/kT) \tag{7}$$

denotes a rovibronic "activated complex pseudo-partition function", and Q(B) and Q(C) are rovibronic partition functions of the associating species B and C. In phase space theory (PST), where the anisotropy of the potential is neglected, one has $Q^* = Q(B)Q(C)Q^*_{cent}$ such that $Q^*/Q(B)Q(C)$ becomes equal to the centrifugal pseudo-partition function Q^*_{cent}. I. e.,

$$k^{PST}_{rec,\infty} = \frac{kT}{h} \left(\frac{h^2}{2\pi\mu kT}\right)^{3/2} Q^*_{cent} \tag{8}$$

The ratio

$$k_{rec,\infty} / k^{PST}_{rec,\infty} = f_{rigid} \tag{9}$$

which we call the "rigidity factor" of the reaction, is smaller than unity. It accounts for the anisotropy of the potential which increases the threshold enregies E_{oi} of the adiabatic channel potential curves beyond the sum of the centrifugal energies and the eigenvalues of B and C.

In order to illustrate the foregoing treatment, we calculate partition functions for the example of eq. (4). Replacing the summations over J and v by integrals (with v >> 1), we have

$$Q^* \approx \int_0^\infty v dv \int_0^\infty 2J dJ \exp[-C(v+GC^{-1/2})^2/4kT] \tag{10}$$

$$= (4kT/CB_\infty)^{3/2}(\pi C)^{1/2}/8 \tag{11}$$

whereas PST would have given

$$Q^{*PST} = (kT/B_\infty)^{3/2}\pi^{1/2} \tag{12}$$

As a consequence, a rigidity factor

$$f_{rigid} \approx 1/C \tag{13}$$

is obtained. The parameter C from eq. (5), therefore, well characterizes the anisotropy character of the potential, at least for conditions governed by the short range part of the potential.

More general investigations of the energy patterns of the channel maxima E_{0i} following from eq. (3) indicate that, in some cases, centrifugal maxima and transient mode energies are better separable than in eq. (4), in other cases they are not separable (see, e. g. [3]). Also, the anisotropy character C may change dramatically from the short range to the long range part of the potential. Depending on the reduced mass μ of the pair B + C, this may or may not appear in the rate constants. More systematic studies in this direction are underway.

In order to illustrate the contribution of the low energy threshold behavior,where the complicated transition from hindered rotor to oscillator channel potential curves appears, we show thermal capture rate constants for charge + linear permanent dipole association. Numerical determination of all E_{0i} and calculation of the pseudo-partition function leads to rigidity factors shown in figure 6. The oscillator limit

$$f_{rigid} = 1/2 \tag{14}$$

is approached rapidly when kT exceeds B_∞. This illustrates the statements, made earlier in this article, that oscillator models govern the channel threshold energy pattern. The comparison with classical trajectory calculations on the same potential is also illuminating. There is agreement within 20 % at kT > B_∞. At lower temperatures, the classical trajectory calculations fail. The 20 % discrepancy at higher temperatures is surprising.

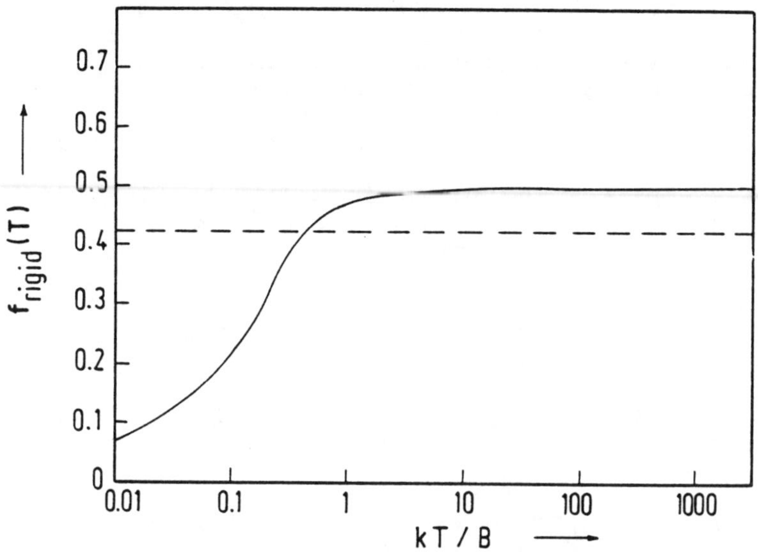

Figure 6 Thermal rigidity factors f_{rigid} from eq. (9) for charge-permanent linear dipole capture processes (full line: SACM, dashed line = classical trajectory calculations).

Its origin is so far unknown. Whether non-adiabatic effects, recrossing of trajectories or numerical artifacts are responsible for this discrepancy, remains to be clarified. One should also mention that variational transition state theory on the same potential, i. e. with the same E_{0i}, differs from SACM for state-resolved capture rate constants [12] but approaches SACM when averaging over all channels is applied. The agreement with approximate quantum scattering calculations was good at high temperatures, but numerical artifacts became apparent for low temperature conditions [11], [14] - [16].

Similar to the treatment of eqs. (10) - (13), the short range anisotropy of the potential and its relationship to the rigidity factors of the rate contants $k_{rec,\infty}$ were analyzed in [2]. The agreement between measured f_{rigid} (on the basis of a Morse potential for the attraction potential V(r)) and calculated f_{rigid} (with very simple relations between f_{rigid} and C, analogous to eq. 13), was surprising. E. G., the measured f_{rigid} for the reactions $CH_3 + CH_3 \rightarrow C_2H_6$, $NO_2 + NO_3 \rightarrow N_2O_5$, and $NO_2 + NO_2 \rightarrow N_2O_4$ were f_{rigid} = 0.15, 0.0062, 0.0020 resp., whereas the calculated f_{rigid} were f_{rigid} = 0.17, 0.013, 0.0014 resp. This type of agreement indicates that, independent of the remaining uncertainties in the potential, the adiabatic channel treatment with the short range part of the potential energy surface provides a useful simplified description of the dominant features of the reaction.

6. Non-adiabatic Transitions between Adiabatic Channel Potential Curves

The success of SACM treatments of thermal recombination or capture process suggests that channel threshold energies are located in a range whre adiabatic dynamics is dominant. In this case non-adiabatic transitions should be tractable by perturbation methods. There are various effects which should be considered. There is, first, Coriolis coupling which is neglected at the simplest level of calculations of adiabatic channel potential curcves. Examples of Coriolis effects have been treated analytically in [17]. These effects lead to marked interactions between adjacent and approaching channel curves at larger distances to such an extent that a new basis of adiabatic curves is required in the hindered rotor range where multiple avoided crossings occur. Thermal rate constants for capture are nearly not influenced by these effects. However, SACM distributions over Ω-states, i. e. product alignments, in part are destroyed [18]. Adiabatic channel potential curves, which account for these effects, were calculated numerically in [8] for the case of two identical dipoles approaching each other. Figure 7 shows examples of potentials curves for channels of identical symmetry. The extent of avoided crossing increases with increasing strength a of the interaction, i. e. with decreasing interdipole distance. These symmetry effects where neglected in the simpler calculation of figure 3.

Non-adiabatic transitions between the channel potential curves shwon in figure 7 are induced by the relative motion of the two dipoles. They can occur at the positions of avoided crossings (Landau-Zener type transitions) or there can be "channel-hopping" between adjacent curves at arbitrary positions ("global"non-adiabaticity). Estimates of the probabilities of non-adiabatic transitions on the basis of the general theory of non-adiabatic transitions were made in [8]. Probabilities P for global "channel-hopping", distant from avoided crossing places, are given by

$$P \approx \exp(-\xi) = \exp\left[-\frac{2\pi^2 \Delta\varepsilon R}{3\, h\, v}\right] \tag{15}$$

where $\Delta\varepsilon$ is the energy gap between the curves, R the dipole-diple distance at which the transition may occur, and v is the relative velocity. The Massey parameter ξ is easily estimated, e. g. for the curves shown in figure 7. Typically ξ is much larger than 3 for total energies not larger than a few hundred cm^{-1} above the channel threshold energies. I. e. global channel-hopping will set in only at high relative velocities of approach. Curve-crossing of the Landau-Zener type at places of avoided crossings is by far more probable. It can be calculated explicitly on the basis of the curve splittings shown in figure 7. As long as no by-passing of channel maxima is created by these crossings, they have only little importance for E_{0i}. Also, the "clumping of channels" in the oscillator range and the dominance of anisotropies of the potential at short ranges, which are generally stronger than in the dipole-dipole potential, suggest only minor contributions of non-adiabatic transitions of the Landau-Zener and the global type. More systematic investigations of non-adiabatic transitions between adiabatic channel potential curves are desirable. However, the analysis of experimental results, such as mentioned in section 5, suggests that adiabatic behavior is dominant under many situations.

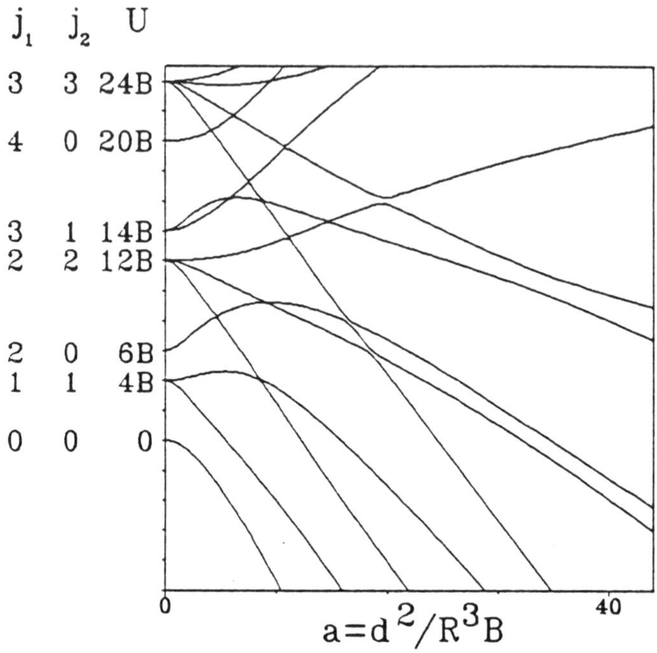

Figure 7 Adiabatic channel potential curves for interaction between two identical dipoles (as figure 3 but accounting for avoided crossings in channels of the same symmetry, see ref. [8]).

References

[1] Cobos, C. J. and Troe, J. (1985), J. Chem. Phys. 83, 1010.
[2] Troe, J. (1989), Z. Phys. Chem. NF 161, 209.
[3] Troe, J. (1991), J. Chem. Soc. Far. Trans, in press.
[4] Harding, L. B. (1989), J. Phys. Chem. 93, 8004.
[5] Harding, L. B. (1991), J. Phy.s Chem., in press.
[6] Harding, L. B., unpublished results.
[7] Troe, J. (1986), J. Phys. Chem. 90, 3485; Lester, W. A. (1970), J. Chem. Phys. 53, 1511.
[8] Maergoiz, A. I., Nikitin, E. E. and Troe, J. (1991), J. Chem. Phys., in press.
[9] Quack, M. and Troe, J. (1974), Ber. Bunsenges. Phys. Chem. 78, 240.
[10] Quack, M. and Troe, J. (1975), Ber. Bunsenges. Phys. Chem. 79, 170, 469.
[11] Troe, J. (1987), J. Chem. Phys. 87, 2773.
[12] Markovic, N. and Nordholm, S. (1989), J. Chem. Phys. 91, 6813; Markovic, N. (1989), Chem. Phys. 135, 109.
[13] Smith, S. C., McEwan, M. J. and Gilbert, R. G. (1989), J. Chem. Phys. 90, 1630.
[14] Troe, J. (1988), Ber. Bunsenges. Phys. Chem. 92, 242.
[15] Clary, D. C. (1984), Mol. Phys. 53, 3; (1985), Mol. Phys. 54, 605.
[16] Dubernet, M. L. and Mac Carroll, R. (1989), Z. Phys. D 13, 255.
[17] Nikitin, E. E. and Troe, J. (1990), J. Chem. Phys. 92, 6594.
[18] Dashevskaya, E. I., Nikitin, E. E. and Troe, J. (1990), J. Chem. Phys. 93, 7803.

MODE SPECIFICITY IN INTRAMOLECULAR CONVERSIONS

DONALD L. THOMPSON
Department of Chemistry
Oklahoma State University
Stillwater, Oklahoma 74078
U.S.A.

ABSTRACT: HONO has proven to be an interesting molecule from the standpoint of chemical dynamics. Extensive classical trajectory calculations have shown that the rates of energy transfer and *cis-trans* isomerization are strongly dependent on initial vibrational energy distribution, rotation, and geometry. Calculations of tunneling rates as a function of the initial mode excitations reveal significant mode selectivity. Also, studies of intramolecular conversion in methyl nitrite and aziridine show mode selectivity.

1. Introduction

For the past decade we have been studying intramolecular vibrational energy redistribution (IVR) and unimolecular reactions in "large" (four or more atoms) molecules by using classical trajectories on "realistic" potential energy surfaces. Of particular interest in these studies has been 1) non-statistical and mode-selective behavior and 2) the practical aspects of doing calculations for realistic, large molecules -- what approximations can be made in the dynamics and potentials? These studies have involved systems ranging from simple van der Waals molecules (e.g., collinear I-I ··· He)[1] to hexahydro-1,3,5-trinitro-1,3,5-triazine (RDX),[2] a 21-atom molecule. We have, on the basis of our calculations and the work of others, tentatively reached some general conclusions that provide some guidelines for identifying systems and conditions for which mode selective chemistry might be observed in the laboratory. In this paper, we focus attention on one kind of "reaction," namely intramolecular inversion, which display mode selectivity.

It is well-established that most experimental data are adequately described by statistical theory -- in fact, RRKM is one of the more successful theories in chemistry because of its agreement with most experiments. On the other hand, it is clear that nonstatistical behavior can occur in many systems, even though it is not manifested in most experiments. It is well-known that when excitation energy is placed in a particular mode or combination of modes, it flows out along specific pathways. Thus, the redistribution of the excitation energy occurs such that there exists for a finite time, a nonstatistical partitioning of the energy within the molecule. If a nonstatisitical distribution of energy persists for times comparable or long relative to that required for reaction, mode selectivity in the reaction can result. In some systems, nonstatistical

J. Jortner et al. (eds.), Mode Selective Chemistry, 261–272.

energy distributions can exist for very long times. Why is it then that mode specific behavior is not more commonly observed in experiments in which mode selective excitation techniques are used? There are several reasons. Some of the reasons stem from the nature of the experiments and choice of systems in experimental studies and others are due to the nature of the dynamics of highly excited molecules.

First, let us consider vibrational overtone excitation. Usually an X-H local bond-mode is excited in a molecule in which unimolecular dissociation is the lowest energy reaction. Thus, the excitation energy is initially located in a high frequency mode while the chemical reaction involves the breaking of bonds between heavy atoms. Thus, the vibrational modes associated with the reaction coordinate motion have frequencies that are low relative to the excitation mode. For mode selective chemistry to occur in these cases, a significant portion of the excitation energy must preferentially transfer from the high frequency excitation mode to the low frequency reaction-coordinate modes. It is unlikely that sufficient energy to affect the reaction will transfer into the reaction coordinate before randomization of the excitation energy throughout the molecule, particularly since the level of initial excitation is often on the order of the energy required for reaction. Stated simply, mode selective chemistry would require that a large portion of the energy in the X-H bond flow more or less directly into the reaction coordinate modes.

We know that the relaxation of excited states of X-H local modes is determined by nonlinear resonances between the stretch and associated bends, benzene being the prime illustration of this.[3] The bending modes then transfer energy to other modes in the molecule. Thus, the IVR between the X-H mode and the reaction coordinate modes is usually indirect. Many of the "bath" modes of the molecule compete effectively with the reaction coordinate modes for the excitation energy and statistical behavior in the dissociation process results.

Anharmonicity causes sufficient coupling (nonlinear resonances) among the various modes of a molecule at high levels of excitation to lead to relatively rapid energy randomization. However, since the initial energy transfer is nonstatistical, mode specific reaction is possible. The question is, under what conditions? The excitation mode and reaction coordinate modes must be sufficiently coupled for direct energy transfer to compete with energy randomization. Also, the level of excitation needs to be large relative to that needed to cause reaction. There are several situations where these criteria can be met with existing experimental limitations.

Since simple bond-rupture reactions usually require energies on the order of that which can, using current experimental techniques, be deposited selectively in molecules, one can turn to reactions which have much lower energy requirements. van der Waals molecules are extreme examples of this. Of more interest from the chemist's point of view are molecules with somewhat higher barriers. Intramolecular conversions such as *cis-trans* isomerizations and weak complexes involving some chemical forces are good examples. These processes involve barriers on the order of 5 to 15 kcal/mol and thus the

the energy range accessible with single-photon excitation is such that it is possible to deposit much more energy than is needed for reaction.

Another situation in which mode selective reaction may be observed is if the energy is deposited in a vibrational mode that is associated with the reaction coordinate. For example, it should be possible (although to our knowledge no experiments have been reported to date) to enhance the rates of certain concerted molecular elimination reactions. For example, elimination reactions involving the transfer of a hydrogen atom allow the possibility of exciting an overtone of the X-H bond mode and thus promoting the migration of the H-atom, the first stage of molecular elimination. We have shown the possibility of this in HONO elimination in dimethylnitramine.[4] Also, we have illustrated this in the nitrogen inversion in aziridine.[5]

Mode selective behavior may also be observed in molecules in which the excitation modes can be isolated from the "bath" modes. A good example of the isolation of energy within a portion of a molecule is "heavy-atom blocking," as has been thoroughly studied by both experimentalists and theorists.[6] If the reaction coordinate, such as the bond that is to be broken, serves as the "block" between the excitation and bath modes, then there is the possibility of mode selective bond rupture even if the excited mode and the reaction coordinate are not strongly coupled. Alkyl hydroperoxides are examples of molecules in which this situation exists. The O-O bond energy is well within the range accessible experimentally and the O-O bond "separates" the OH stretch from most of the molecular modes. There have been some experimental studies of t-butyl hydroperoxide in which overtones of the OH local stretching mode are excited.[7] We have studied the effects of C-H and O-H local stretching overtones on the O-O bond rupture in methyl hydroperoxide.[8] We found that (for a total energy of 104 kcal/mol) the excitation of the OH stretch gives a rate that is twice that for a comparable CH stretch excitation.

In the present paper, we review earlier work on intramolecular conversion in HONO, methyl nitrite, and aziridine, and present results of some recent studies of *cis*→*trans* isomerization in HONO (including mode selectivity effects in tunneling).

2. Methods

2.1 TRAJECTORY CALCULATIONS

The classical trajectories were computed using a general computer program that we have developed.[9] The trajectories are integrated using laboratory-fixed Cartesian coordinates. The equations of motion are integrated by using a fourth-order Runge-Kutta-Gill routine with a fixed stepsize. Integration accuracy is checked by back integration and energy conservation.

Initial conditions are selected by first assigning zero-point energy to each of the normal modes, with random vibrational phases. Excitation energy is then added to a local X-H mode or to one or more of the normal modes. Finally, the momenta and coordinates are scaled to give the desired total enegy, usually to within 10^{-5} kcal/mol.

Lifetimes are computed by defining criteria for separating the reactants and products. Rate constants are determined by least-squares fitting the computed distribution of lifetimes to

$$-kt = \ln(N_t/N_0), \tag{1}$$

where N_t and N_0 are the number of trajectories that have not undergone reaction in time t and the total number of trajectories in the ensemble.

3. Results

3.1 INTRAMOLECULAR CONVERSIONS

Processes, such as *cis-trans* isomerization, involving low potential-energy barriers offer conditions that are favorable for mode selective behavior. Unlike most chemical reactions where a large fraction of the excitation energy (usually < 60 kcal/mol) must flow nonstatistically into the reaction coordinate to surmount the barrier (usually 40 kcal/mol or more), the barriers to intramolecular conversions are quite small (usually 5 to 15 kcal/mol) relative to the levels of excitation common in infrared single-photon experiments. Bauer and coworkers,[10] in a series of papers, have explored nonstatistical behavior in intramolecular conversions. Those studies focus on the applicability/nonapplicability of RRKM theory to intramolecular conversion in the second-order regime in a variety of systems. Bauer and True[10a] reasoned that non-RRKM behavior might be expected in systems where the density of states at the critical energy for reaction is on the order of 10 to $10^2/\text{cm}^{-1}$, and where the relaxation times are long (on the order of nanoseconds rather than picoseconds as predicted by RRKM theory).

We have investigated the influence of mode selective excitation on intramolecular conversion in HONO,[11] methyl nitrite,[12] and aziridine.[5] HONO has been of particular interest to us.

3.1.1 *Classical Dynamics:HONO*. We have carried out extensive classical trajectory calculations for *cis-trans* isomerization in HONO. We have examined the influence of initial vibrational mode excitations, IVR, potential coupling, molecular rotation, and barrier tunneling.

This molecule provides an interesting set of conditions for investigating mode selective behavior. The energy required for the *cis*→*trans* and *trans*→*cis* conversions is about the same (the ΔE is only about 0.6 kcal/mol). The *cis*→*trans* barrier is 9.7 ± 0.7 kcal/mol. Thus, there is the possibility of influencing the isomerization by excitation of low overtones. For example, even the v = 1 state of the OH stretching mode provides sufficient energy to cross the barrier. Experimental studies show that the system displays some interesting behavior. Pimentel and co-workers,[13] in the early 1960's, measured the rate of infrared-induced isomerization of HONO in a nitrogen matrix. They suggested that tunneling may play a role in the conversion. Shirk and co-workers[14] studied the isomerization in nitrogen and argon matrices by selectively exciting various modes with a laser; their results indicate mode specific behavior.

Classical trajectory calculations were carried out at a total energy of 1.7 eV for various initial conditions, with and without molecular rotation, and for two different potential-energy surfaces. The results for nonrotating HONO are summarized in Table 1.

Table 1. Calculated rate constants (in ps^{-1}) of intramolecular conversion in nonrotating HONO.[a]

Excited mode	PES-I[b]		PES-II[c]	
	cis→*trans*	*trans*→*cis*	*cis*→*trans*	*trans*→*cis*
Equipartitioned	0.90	0.04	0.50	0.05
v_1(OH) = 3	0.26	----[d]	0.25	0.02
v_2(N=O) = 6	0.06	0.03	0.10	0.05
v_3(HON) = 8	1.50	0.10	1.81	0.14
v_4(O-N) = 10	0.40	0.06	0.21	0.07
v_5(ONO) = 16	0.37	0.06	0.26	0.10

[a] Unexcited modes are assigned zero-point energy.

[b] From Guan *et al.*, Ref. 11. PES-I is a sum of Morse functions for the stretches, harmonic functions for the bends, and a truncated cosine series for the torsion.

[c] From Guan *et al.*, Ref. 11. PES-II consists of the same functions for the various internal coordinates, but switching functions are used to account for changes in the geometry and potential with large amplitude motion. This introduces potential coupling.

[d] No isomerization was observed for an ensemble of 50 trajectories.

The rate of *cis*→*trans* is significantly faster than *trans*→*cis* for the various initial conditions even though the energy barriers for the two processes differ by only about 0.6 kcal/mol. Note that large differences in the rates exist for both potentials. The kinetic coupling in the *cis* is obviously much greater than in the *trans* conformer.

The results of the trajectory calculations show that there is significant mode selective behavior in this system. The largest rate is obtained when the HON bend is excited and the smallest rate for excitation of the OH stretch. The influence of mode selective excitations on isomerization is determined by the IVR rates and mechanism. It is obvious that the OH stretch relaxes very slowly and is only weakly coupled with the torsion.

The influence of the torsional and bending modes on the dynamics can be illustrated by constraining them. We do this by incorporating Lagrangian multipliers into Hamilton's equations of motion. We have calculated the isomerization rates for excitations of various modes with the HON and ONO bends fixed at their equilibrium values and the rate of relaxation of excited OH stretching by freezing the torsional motion.

The role of the torsional mode is illustrated by the results shown in Fig. 1.

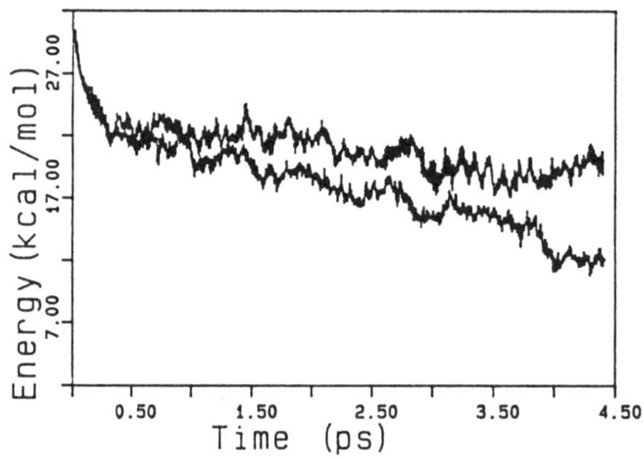

Figure 1. Illustration of the effect of constraining the torsional motion in HONO on the relaxation of the v=3 state of the OH stretching mode.

Figure 1 shows plots of the energy in the OH stretching mode, initially excited to the v=3 level, as a function of time for ensembles of trajectories in which the torsional motion was constrained and in which the motion was allowed. The upper curve is for constrained torsional motion. During an initial period, up to about 0.5 ps, the rate of energy transfer is about the same for the two calculations -- about 1/3 of the excitation energy transfers out of the OH in that time. However, following this initial period, the rate of energy transfer is faster in the case of unconstrained torsional motion. A striking result is obtained when the HON bending motion is constrained. Essentially no energy flows out of the excited OH stretch over the entire period of integration (4.5 ps). These results suggest that the IVR is initially controlled by OH-to-HON energy transfer. The torsion plays a role after other modes have become excited due to IVR. Freezing the ONO mode has little effect on the rate of transfer out of the OH stretch.

The role of the bending angles in the transfer of energy can be studied indirectly by calculating the isomerization rate for constrained ONO and HON motion. Table 2 gives the ratios of the rates for unconstrained to constrained trajectories.

Table 2. Effects of constraining bending angles on the rate constants for *cis*→*trans* isomerization in HONO.[a]

	$k_{Unconstrained}/k_{constrained}$			
Constrained Angle:	HON		ONO	
Excited mode	*cis*→*trans*	*trans*→*cis*	*cis*→*trans*	*trans*→*cis*
Equipartitioned	33	2.4	4.1	2.1
$v_{HON} = 8$			1.8	1.3
$v_{ONO} = 16$	7.3	2.5		

[a] Unexcited modes are assigned zero-point energy.
[b] Isomerization was not observed for the constrained trajectories.

Constraining the HON bend causes the greatest effect on the isomerization rate and it is most pronounced for *cis*→*trans*.

The influence of molecular rotation has also been studied.[11b] The calculations are for the same total energy, 1.7 eV, and for equipartitioning of the vibrational energy among all the modes except torsion (thus, the results can be compared with value given in Table 1, 0.05 ps^{-1}.) A significant increase in the isomerization rate results when only a small amount of energy (0.04 eV) is place in rotation. The rate is 0.28 ps^{-1} when 0.04 eV of rotational energy is equally distributed among the three directions. The rate is also strongly dependent on the orientation of the angular momentum. When the molecule rotates about the torsional axis the rate is 0.34 ps^{-1}, more than a factor of 6 increase over the nonrotating case. When the molecule rotates perpendicular to the torsional axis the rate is only 0.03 ps^{-1}, essentially the same as for no rotation.

This behavior is contrary to what is predicted by statistical theory. RRKM theory predicts a lowering of the rate when energy is placed in rotation rather than vibration. However, the trajectory results are in agreement with RRKM when the rotational energy is somewhat higher (~0.1 eV). The observed increase in the rate for low rotational energy is due to active vibrational-rotational interactions. The energy transfer to the reaction coordinate (the torsion) is significantly increased by the rotational-vibrational coupling. Calculations have also been carried out in which we constrained the torsional motion and calculated the rate of energy transfer from the OH stretch. We find that rotation has essentially no effect on the energy transfer when the torsional motion is constrained to its equilibrium value.

3.1.2 *Tunneling in HONO Cis-Trans Isomerization.* It seems reasonable that the initial mode excitations will influence tunneling rates. There is, of course, the effect due to IVR which determines the rate at which energy is transferred into the "tunneling mode," which determines the frequency of encountering the barrier through which tunneling occurs. Also, in some systems the tunneling mode, like the reaction coordinate mode, may involve vibrational modes which can be selectively excited. Thus, excitation of certain modes could lead to an increase in the frequency of encountering the barrier through which tunneling occurs.

Table 3. Calculated classical plus tunneling rate constants (in ps^{-1}) for intramolecular conversion in HONO.[a,b]

	k (ps^{-1})	
Excited mode	*cis*→*trans*	*trans*→*cis*
$v_1(OH) = 0.5$	1.4×10^{-4} (0.0)	1.0×10^{-6} (0.0)
$v_2(N{=}O) = 1$	4.5×10^{-3} (0.0)	7.8×10^{-3} (0.0)
$v_3(HON) = 1.5$	1.4×10^{-1} (7.6×10^{-2})	1.4×10^{-3} (0.0)
$v_4(O{-}N) = 2$	3.2×10^{-1} (1.8×10^{-2})	2.0×10^{-2} (4.6×10^{-2})
$v_5(ONO) = 3$	2.8×10^{-1} (0.0)	Not calculated

[a] Qin and Thompson, Ref. 11(c).
[b] Unexcited modes are assigned zero-point energy and the torsional mode is initially excited to $v = 3$. The classical rates are given in parentheses.

We have used the Waite and Miller[15] classical-plus-tunneling model to study the *cis-trans* isomerization in HONO. In order to diminish the classical isomerization due to aphysical flow of the zero-point energy, we reduced the ZPE to 1/4 its actual value. Non-integer quantum numbers are used to assign the excitation energy. The total energy was fixed at 13 kcal/mol in each case. Some results are shown in Table 3.

There are significant mode selective effects in the *cis–trans* tunneling rates. The rates range from 10^{-6} to about $3x10^{-1}$ ps^{-1}. The results suggest that, although there may be an effect due to IVR, the tunneling rates are strongly influenced by increasing the amplitudes of certain vibrations by exciting them. For example, consider the results for excited HON and N-O. The rate is much greater in the case of N-O excitation. This is the opposite of the specificity in the purely classical results discussed above. In fact, there is, in general, no correlation between the tunneling and classical rates.

3.1.3 *CH_3ONO Cis-Trans Isomerization.* Experimental studies have shown non-RRKM behavior in the intramolecular conversion of methyl nitrite.[10a] Thus, we have investigated mode selectivity in it. The barrier to internal rotation is about the same as in HONO -- reported values range from 9.5 to 10.5 The *cis* conformer is about a kcal/mol more stable than the *trans*.

The *cis→trans* and *trans→cis* conversion rates in methyl nitrite have been calculated as functions of initial excitations of C-H local stretching modes and the ONO bending normal mode.[12] Three different potential-energy surfaces were used in the trajectory calculations; two purely harmonic (one with coupling terms and another without) and one with anharmonic stretches. The rates of energy transfer and isomerization are faster for the anharmonic potential, but the qualitative behavior is the same for all three potentials. The results show no mode selective effects for excitations of the C-H stretches, however, excitation of the ONO bending mode gives rates that are about twice that for initial conditions in which the energy is equally distributed among all the modes or localized in the C-H stretching modes. The energy transfer from a C-H local stretch is essentially complete in less than a half of a picosecond. As in the case of HONO, the rate of *cis→trans* conversion is much faster than the *trans→cis*.

These results show mode selective behavior, however, not in the case of C-H stretch excitations. The fact that excitation of the ONO bending mode gives faster rates than other energy distributions suggests that there is not energy randomization between the high and low frequency modes on the timescale of the inversion.

3.1.4 *Nitrogen Inversion in Aziridine.* Borchardt and Bauer[10e] have reported the rates for nitrogen inversion in aziridine in the gas phase for a range of temperatures and pressures. There is a significant difference between the behavior of the experimental rates as a function of pressure and RRKM predictions. They concluded that the energy transfer in the molecule is slow relative to the rate of inversion.

We have used classical trajectories to study the influence of initial energy partitioning on the rate of inversion.[12] We constructed a simple anharmonic (stretches) potential energy surface based on an experimentally determined force field. The barrier to inversion is 15.8 kcal/mol for this potential. Rates were calculated for various levels of excitations of the C-H and N-H local stretching modes (see Table 4). At a given total energy, the rate for N-H excitation is about 2.5 times that for C-H excitation or for equipartitioning of the energy. As the N-H relaxes energy flows into the bending mode which is involved in the inversion. The proximity of the N-H mode to the inversion coordinate is an important factor in causing the rate enhancement. The trajectory results also show that the rate of energy transfer is fast relative to the rate of inversion. Nevertheless, mode selective behavior is observed.

Table 4. Rate constants for nitrogen inversion in aziridine.[a]

Excitation	Total Energy	k (ps^{-1})
$v_{NH} = 3$	68.66	0.26
$v_{CH} = 3$	68.66[b]	0.11
$v_{NH} = 4$	76.35	0.48
$v_{CH} = 4$	76.35[b]	0.21
Equipartitioned[c]	76.35	0.24

[a] From Gai and Thompson, Ref. 5.

[b] The energy was adjusted to correspond to that for the comparable NH mode excitation. The actual values for the CH mode excitations are 66.82 and 74.12 kcal/mol for $v_{CH} = 3$ and 4, respectively.

[c] The total energy is equally partitioned among the vibrational modes except for the N-H deformation modes.

4. Concluding Remarks

The dynamics of HONO are interesting. There are pronouced mode selective effects for the isomerization. The difference in the rates of *cis→trans* and *trans→cis* is as much as an order of magnitude even though the barriers for the two processes differ by only 0.6 kcal/mol. It is clear that the kinetic couplings are very different in the two conformers. Rotational effects are large in HONO. Rotation about the torsional axis greatly increases the IVR rate and hence the rate of isomerization. Tunneling in HONO is strongly influenced by mode selective excitation.

Experimental results are sparse for this system, however, those available also show that this molecule displays some interesting behavior. The experiments by Shirk and coworkers[14] indicate mode specific effects. Pimentel and coworkers[13] have interrupted their results to indicate that tunneling plays a role in the isomerization. The calculated and experimental results suggest that additional experimental studies would be worthwhile.

Acknowledgements

This research was supported by the U.S. Army Research Office.

References

[1] (a) S. B. Woodruff and D. L. Thompson, J. Chem. Phys. **71**, 376 (1979); (b) R. Viswanathan, L. M. Raff, and D. L. Thompson, J. Chem. Phys. **77**, 1286 (1982).

[2] T. D. Sewell and D. L. Thompson, J. Phys. Chem., in press.

[3] Y. Guan and D. L. Thompson, J. Chem. Phys. **88**, 2355 (1988); and references therein.

[4] B. G. Sumpter and D. L. Thompson, J. Chem. Phys. **88**, 6889 (1988).

[5] H. Gai and D. L. Thompson, Chem. Phys. Letters **168**, 119 (1990).

[6] See, for example: (a) V. Lopez, V. Fairen, S. M. Lederman, and R. A. Marcus, J. Chem. Phys. **84**, 5494 (1986); (b) V. Lopez and R. A. Marcus, Chem. Phys. Letters **93**, 232 (1982); (c) K. N. Swamy and W. H. Hase, J. Chem. Phys. **82**, 123 (1985).

[7] (a) T. R. Rizzo and F. F. Crim, J. Chem. Phys. **76**, 2754 (1982); (b) D. W. Chandler, W. E. Farneth, and R. N. Zare, J. Chem. Phys. **77**, 4447 (1982); (c) M.-C. Chuang, J. E. Baggott, D. W. Chandler, W. E. Farneth, and R. N. Zare, Faraday Discuss. Chem. Soc. **75**, 301 (1983); (d) J. H. Gutow, D. Klenerman, and R. N. Zare, J. Phys. Chem. **92**, 172 (1988).

[8] H. Gai, D. L. Thompson, and G. A. Fisk, J. Chem. Phys. **90**, 7055 (1989).

[9] K. L. Bintz, M. S. Thesis, Oklahoma State University, 1986.

[10] (a) S. H. Bauer and N. S. True, J. Phys. Chem. **84**, 2507 (1980); (b) S. H. Bauer and K. I. Lazaar, J. Chem. Phys. **79**, 2808 (1983); (c) K. I. Lazaar and S. H. Bauer, J. Phys. Chem. **83**, 85 (1988); (d) S. H. Bauer, J. Chem. Kinetics **17**, 367 (1985); (e) D. B. Borchardt and S. H. Bauer, J. Chem. Phys. **85**, 4980 (1986).

[11] (a) Y. Guan, G. C. Lynch, and D. L. Thompson, J. Chem. Phys. **87**, 6957 (1987); (b) Y. Guan and D. L. Thompson, Chem. Phys. **139**, 147 (1989). (c) Y. Qin and D. L. Thompson, in preparation.

[12] A. Preiskorn and D. L. Thompson, J. Chem. Phys. **91**, 2299 (1989).

[13] (a) J. D. Baldeschwieler and G. C. Pimentel, J. Chem. Phys. **33**, 1008 (1960); (b) R. T. Hall and G. C. Pimentel, J. Chem. Phys. **38**, 1889 (1963).

[14] (a) P. A. McDonald and J. S. Shirk, Chem. Phys. Letters **77**, 2355 (1982); (b) A. E. Shirk and J. S. Shirk, Chem. Phys. Letters **97**, 549 (1983).

[15] B. A. Waite and W. H. Miller, J. Chem. Phys. 73, 3713 (1980).

TEMPERATURES OF FRAGMENTS IN UNIMOLECULAR DISSOCIATIONS

RICHARD BERSOHN
Department of Chemistry
Columbia University
New York, NY 10027

ABSTRACT. A set of unsaturated molecules ranging from ethylene to tetramethylbenzene were irradiated with an excimer laser. The rate of formation of hydrogen (H) atoms and their average kinetic energy were determined by laser induced fluorescence. With increasing molecular size, these observables decrease in accord with theories of unimolecular dissociation. The vibrational temperature of the excited parent molecule before dissociation is equal to the translational temperature of the H atoms after dissociation. This result implies that H atoms relax on a longer time scale than heavy atoms.

Introduction

Let us assume that a set of molecules A in equilibrium with a bath at temperature T decompose according to the equation A --> B + C. At very low pressures of A the rate of decomposition of A is written d(A)/dt = -k(T)(A) where k(T) is the first order rate constant. The rate constant k(T) is an average over rate coefficients k(E) where E is the instantaneous internal energy of the molecule. The rate coefficients are therefore the more fundamental quantities, because knowing the potential surface of molcule A they can in principle be calculated from mechanics, classical or quantum. However until recently it has been difficult to prepare a sufficiently large number of molecules with definite energy E and to measure their rate of decay.

The laser has provided a solution to this problem for certain classes of molecules. We consider a set of unsaturated molecules in a low pressure gas which absorb uv light at energies conveniently above a C-H bond

J. Jortner et al. (eds.), Mode Selective Chemistry, 273–281.

dissociation energy and subsequently dissociate into a radical and a hydrogen atom. By convenient we mean not so close to the dissociation energy that the molecule collides with its container wall before decomposing and not so high that the molecule dissociates in a time too short to measure. The molecule which was originally at temperature T will now have an average energy

$$E = E_{photon} + \sum_i \hbar\omega_i/(\exp(\hbar\omega_i/kT) - 1) \qquad (1)$$

where the sum is over the vibrational frequencies, ω_i. The second term on the right hand side of this equation is the average internal thermal energy of the molecule. This is normally only 1% or less of the first term which is the photon energy. Thus the molecules are prepared in an initial state which is nearly monoenergetic at energy E.

Some molecules fluoresce after absorption of a uv photon. However because of increased densities of states at higher energies the rate of internal conversion increases with increasing photon energy. At the high energies required for rapid dissociation there is no observable fluorescence. In other words the rate of internal conversion is far greater than the rate of radiative decay. Moreover the rate of internal vibrational redistribution (IVR) from the initially populated ground states is assumed to be very fast. The justification for this assumption is experimental, namely that measured rates of IVR in other molecules are in tens of picoseconds, orders of magnitudes faster than dissociation rates of these molecules at uv or vuv energies.

The usual targets for studies of photodissociation dynamics are small molecules such as H_2O, CH_2O and CH_3I which contain three to five atoms and which dissociate from excited electronic states. The relatively large molecules irradiated in the experiments described here dissociate from their ground electronic states. Molecular beam studies often show that the kinetic energy distribution peaks away from zero for dissociation on an excited surface and peaks at zero for dissociation from the ground state.

Experimental

The experiments were carried out by a standard pump and probe technique. The "pump" was an excimer laser usually at 193.3 nm (ArF) but sometimes at 157 nm (F_2) or 222 nm (KrCl). The "probe" was a Lyman alpha source at 121.6 nm obtained by frequency tripling the 364.8 nm output of a dye laser. The tripling is accomplished albeit in very low yield by focusing the laser light into a tube of Kr gas at pressures around 150 Torr. A fluorescence excitation curve is obtained by sweeping the frequency of the dye laser through the Doppler broadened absorption of the H atoms. The Doppler shift is given by the expression $\nu_0 v_z/c$ where ν_0 is the absorption frequency of a stationary H atom and v_z is the component of the H atom velocity along the direction of the probe laser light. Consequently the second moment of the fluorescence excitation curve yields the average kinetic energy associated with the z component of the velocity. For an isotropic velocity distribution the total average kinetic energy is just three times that associated with a single direction.

Results and Discussion

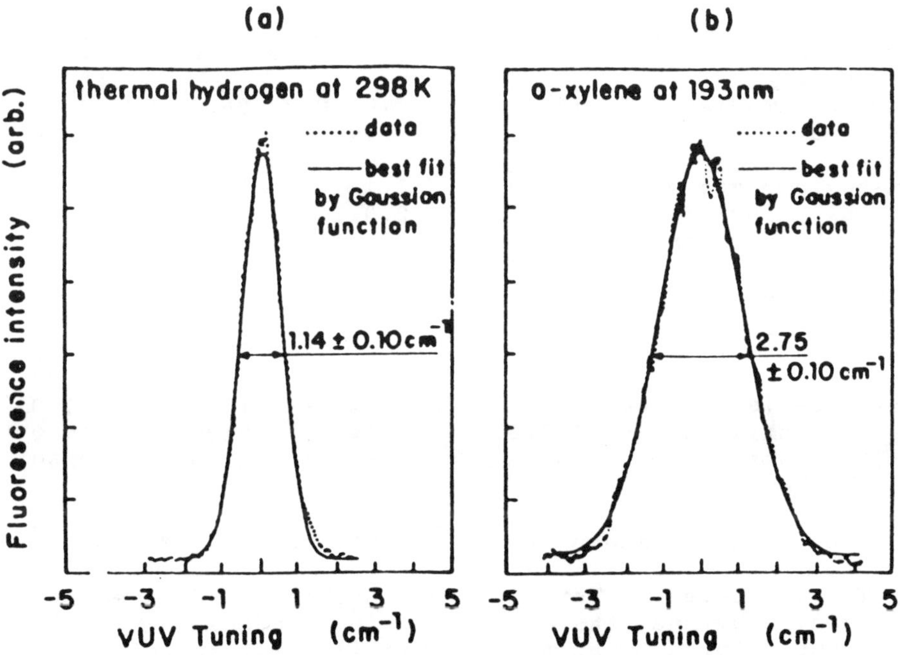

Figure 1. Fluorescence excitation curves of H atoms a) thermalized at 298 K and b) photodissociated from o-xylene at 193.3 nm. The solid curves are the best fits to a Gaussian function.

Table 1 lists the data which we have obtained on the rate of dissociation k(E) and the average kinetic energy of the released H atom.[1-3] In every case the fluorescence excitation curve has a Gaussian shape which shows that the velocity has a Maxwell distribution parametrized by a translational temperature, T_T.(Fig.1) The average kinetic energy of the H atoms is just $3kT_T/2$. On the other hand a vibrational temperature of the molecule before dissociation, T_V may be defined by the transcendental equation

$$E = \sum_i \hbar\omega_i /[\exp(\hbar\omega_i /kT_V)-1] \qquad (2)$$

Table 1 Rates of Dissociation, k(E) of Unsaturated Molecules "Heated" to Temperature T_V and Translational Temperatures, T_T of H Atoms

Molecule	E(Kcal/mol)	k(E)(s^{-1})	T_V(K)	T_T(K)
ethylene	149	($3x10^{11}$)	7410	3650
cyclopentadiene	149	($6x10^{9}$)	3690	4080
indene	150	$7.4x10^{6}$	2480	2510
methylpyrazine	131.2	$1.1x10^{6}$	2810	.
methylpyrazine	150.0	$2.0x10^{7}$	3100	2500
dimethylpyrazine	151.0	$1.9x10^{6}$	2610	2350
trimethylpyrazine	151.9	$1.0x10^{5}$	2275	2150
tetramethylpyrazine	186.8	$1.6x10^{6}$	2350	2310
toluene	131.2	$8x10^{4}$	2630	2520
toluene	150.2	$3.1x10^{6}$	2780	2630
xylenes (all three)	151.3	$3.5x10^{5}$	2390	2180
s-trimethylbenzene	152.3	$1.4x10^{4}$	2130	2200
1,2,4,5-tetra-methylbenzene	187.1	$2.5x10^{5}$	2230	2260

Values in parentheses are calculated using RRKM theory.

The molecules in Table 1 are conveniently divided into four sets, ethylene, the cyclopentadiene-indene pair, methyl substituted pyrazines and methyl substituted benzenes. These sets are arranged in order of increasing molecular complexity. Thus ethylene has six atoms, cyclopentadiene has eleven, methylpyrazine has thirteen and toluene has fifteen. The lifetimes of the excited molecules increase in this order. When excited with the same 193 nm photon, ethylene was calculated to have a lifetime of about 3 ps, cyclopentadiene about 150 ps, and methylpyrazine and toluene have measured lifetimes of 50 and 300 ns respectively. Virtually all the rate data in Table I confirm this relation of lifetime to the complexity of the molecules. Thus, adding a methyl group invariably slows down the rate of decomposition by almost an order of magnitude.(Fig.2) The slight structural variation from toluene to pyrazine in which two C-H bonds are replaced by two nitrogen atoms substantially decreases the rate of decomposition of an attached methyl group. Indene which is benzocyclopentadiene has a much longer lifetime than cycclopentadiene.

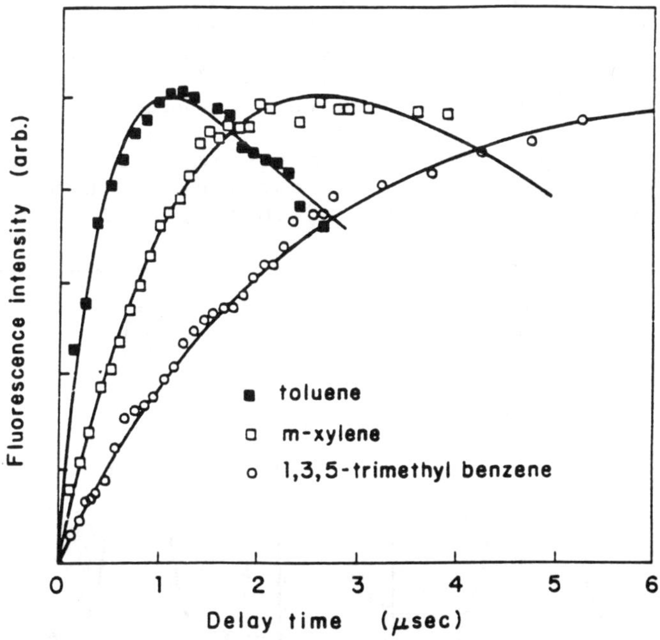

Figure 2. Growth curves of H-atom signals from toluene, m-xylene and 1,3,5-trimethylbenzene irradiated at 193.3 nm.

The vibrational temperatures listed in Table 1 are all calculated by substituting the known or estimated vibrational frequencies into Eq.(2). For the same total energy the vibrational temperature decreases as the number of vibrational modes increases. The same statement is also true about the translational temperature or equivalently, the average kinetic energy. This can be seen readily in Fig.3 in which the fluorescence excitation curve of the H atoms from cyclopentadiene is clearly broader than that from indene. Even more remarkable is the fact that these two temperatures, the vibrational temperature of the molecule before dissociation and the translational temperature of the product H atom are usually within 10% of each other. However the experimental second moments may well have this much error in their determination. As a first approximation then we can say that the two temperatures are approximately equal. This equality is found for a large number of molecules. It is not true for ethylene whose vibrational temperature of 7410 K is substantially more than the translational temperature of 3650 K. This is presumably because in this small molecule dissociation, predicted to require 3 ps. on average is faster than complete vibrational redistribution. In this one case the hot molecule may not live long enough to be described by a temperature. However the equality $T_V = T_T$ is valid for all the other molecules. What is the meaning of this equality? Is it a new important discovery or is it obvious? This is an extraordinary result which relates to the general concept of mode specific chemistry.

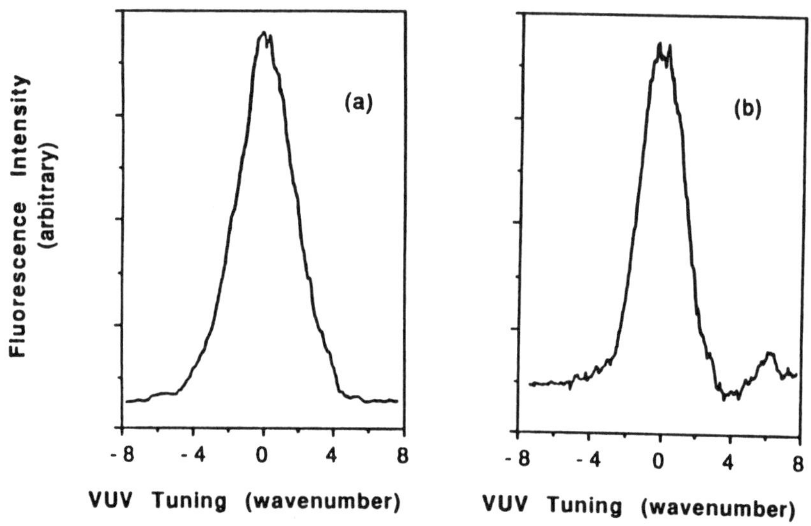

Figure 3. Fluorescence excitation spectrum of H atoms from a) cyclopentadiene and b) indene irradiated at 193.3 nm.

First, consider a classical ensemble of extremely large hot molecules at a temperature T_V. Let p,q stand for the collection of momentum and position coordinates respectively describing the internal motion of the molecules. By choosing the right set of coordinates the Hamiltonian can be written as T(p) + V(q). The distribution of the molecules in phase space is given by the Boltzmann function $\exp\{-[T(p) + V(q)]/kT_v\}$. In classical mechanics T(p) and V(q) commute and thus the distributions over momentum and coordinates are factorable. Even if the coordinates are such that the molecule is about to cross the transition state, the distribution of kinetic energy ofrelative motion of the two fragments is the same as it was when all internal motions were periodic. From this perspective it seems obvious that the translational temperatures of the released H atoms should be the same as the vibrational temperature of the energized molecule.

A similar example might be that of a molecule desorbing from the surface of a large solid.[4] It is often found that the temperature of the desorbing molecule is the same as that of the surface. A corollary is that microscopic reversibility requires that all collisions of the molecules with the surface result in capture. On the other hand when there is a barrier to release of the adsorbed molecule the desorbing molecules are much colder than the surface. It also follows that slow molecules with insufficient energy to surmount the barrier would not be adsorbed on collision.

The situation of the medium sized hot molecules which we treat is different from the canonical ensemble or from adsorbed molecules on a warm macroscopic surface. In the latter cases there is a reservoir of energy such that the total energy of the system is far greater than the energy taken away by the departing species. Our set of hot molecules is more properly described by a microcanonical enesemble in which every member of the ensemble has a fixed energy E. Toluene, for example, has thirty nine vibrational modes and therefore at the energies in question (6-8 eV) has a very dense manifold of states. Internal vibrational equilibrium seems assured with a Boltzmann distribution over states just as for a molecule of unlimited size. However, the molecule is not infinite in the sense that the energy required for dissociation of a hydrogen atom, 88 kcal/mol is not negligible compared to the total energy of the molecule, typically 150 kcal/mol. The photoproduct, in this case, benzyl has a vibrational temperature much less than that of the original excited molecule.

One key hypothesis of the Rice-Rampsberger-Kassel-

Marcus (RRKM) theory of unimolecular dissociation is that an activated molecule is in internal thermal equilibrium up to the transition state.[5,6] While the RRKM theory makes no predictions about final state distributions, it follows from the hypothesis that at the transition state the H atom relative motion should have cooled relative to the initial vibrational temperature.

In principle, the H atoms could speed up again if there were a barrier in the exit channel. There are however several arguments against the existence of such a barrier. The reverse process, radical-H atom recombination appears to have no observable barrier and hence a substantial barrier in the forward direction would be ruled out. If there were such a barrier, the same for all molecules, it would add aconstant amount of kinetic energy to each H atom as it slid down the repulsive surface. The fluorescence excitation curve would no longer be a Gaussian but would have a broad flat top. However a Gaussian shape is always observed. Finally one sees a steady decrease of translational temperature with increasing molecular size even though mechanically the same dissociation process is occurring. For example, the CH_2 group in cyclopentadiene and indene have almost identical local environments but the H atoms from indene have significantly lower energy. (Fig.3). Similarly, in the series toluene, 1,3-dimethylbenzene and 1,3,5-trimethylbenzene or 2-methylpyrazine, 2,6-dimethylpyrazine and 2,3,5-trimethylpyrazine the average kinetic energies decrease even though the environment of each methyl group is almost identical. We therefore assume that no barrier exists in the exit channel though of course the necessary bond dissociation energy must be present in the reaction coordinate.

As it passes through the transition state the H atom preserves the memory of the original vibrational temperature of the molecule when it was far from the transition state. During the tens of thousands if not millions of vibration periods which precede its dissociation the hot molecule spends most of the time in regions remote from the transition state and there achieves thermal equilibrium. What the experiments tell us is that the fluctuations which lead to dissociation are on such a short time scale that the light H atom does not have time to equilibrate, that is, to reach a new temperature. In other words the energy required for bond breaking is obtained from all the <u>other</u> modes of the molecule and not from the three involving the leaving H atom. We speculate that this would not be true if a heavier atom were dissociated. (A few experiments with D atoms show that they have the same translational temperatures as H atoms dissociated from the same molecule.) There is an implication here that there are two time scales for internal vibrational redistribution, one for the heavy atoms and a significantly longer one for the light hydrogen atoms.

The primitive RRK theory in which all frequencies are assumed to be have the same value ν leads to a rate of unimolecular dissociation, $k(E) = \nu(1-E/E_0)^s$ where E_0 is the bond energy and s is the number of vibrational modes. In order to achieve agreement with experiment it is customary to adopt a value for s rather smaller than the actual number of vibrational modes. The experimental results described here are a step toward understanding why the higher frequencies involving the H atoms can be neglected. At the crucial moment they are apparently not strongly enough coupled to the other modes.

Acknowledgments

This work was supported by the U.S.Department of Energy and the Petroleum Research Fund. Discussions with I.Oref and M.Shapiro were very helpful.

References

1. Park, J., Bersohn, R. and Oref,I. (1990) 'Unimolecular dissociation of methylsubstituted benzenes into benzyl radicals and hydrogen atoms", J.Chem.Phys. 93,5700-5708
2. Satyapal, S., Johnston, G. W., Bersohn, R. and Oref,I. (1990) 'The hydrogen atom channel in the photodissociation of ethylene', J.Chem.Phys. 93,6398-6404
3. Yi, W., Chattopadhyay,A. and Bersohn,R. (1991) 'The photodissociation of cyclopentadiene and indene at 193 nm', J.Chem.Phys.94,5994-5999
4. Cosma, G. and David, R. (1985) 'Dynamical parameters of desprbing molecules', Surface Sci.Repts. 5,145-172
5. Marcus, R.A. (1952) 'Unimolecular dissociations and radical recombination reactions' J.Chem.Phys. 20,359-370
6. Marcus, R.A. (1965) 'Dissociation and isomerization of vibrationally excited species III.' J.Chem.Phys. 43,2658-2670

THEORY OF EXCITATION, BOND BREAKING, AND RAMAN SCATTERING WITH SHORT LASER PULSES

MOSHE SHAPIRO
Chemical Physics Department
Weizmann Institute of Science
Rehovot, Israel

ABSTRACT. Theory of photodissociation and continuum-Raman with coherent laser pulses is developed with special emphasis to the nature of the prepared state under various modes of excitation.

Applications to the femtosecond transition state spectra of NaI are presented and it is shown that the pulse characteristics cannot be disentangled from the observed curve-crossing decay rates.

The formulation is extended to treating the "continuum Raman" experiments. Spontaneous emission from a system capable of dissociating, while excited (on- or off-resonance) by a pulse, is shown to be composed of two transients - a "resonance-fluorescence" term which decays via intramolecular (wavepacket) dynamics and natural radiative processes, and a Raman term which follows the pulse. The Kramers-Heisenberg formula, is shown to be the long-time limit of these two terms.

Introduction

With the advent of chemistry with short laser pulses,[1] it is often stated that "it is now possible to probe photodissociation processes while the products fall apart". The feeling is that pulses of 50-100 femtoseconds duration allow a slower reaction or dissociation process to be followed in real time. However, in quantum mechanics the term "real time" is difficult to define because it is not possible to disentangle the preparation from the dissociation step. It is widely agreed that there is no unique bond-breaking "time" when the laser pulse is as fast or slower than that "time". It has perhaps not been appreciated that the "real-time" concept is not very useful even if the laser is much faster than the process studied.

The purpose of this paper is to give a precise treatment of photodissociation with pulsed coherent sources and to examine in detail the nature of the state created by such lasers. As an example of the type of questions addressed here is the extent to which an excitation process can be considered "instantaneous". This topic is associated with viewing photodissociation as creating at $t=0$, $\Psi(0) = \mu\psi_B$, where μ is transition-dipole

J. Jortner et al. (eds.), Mode Selective Chemistry, 283–305.

operator and ψ_B is the initial (bound) state of the system prior to the laser excitation. After the pulse, the wavepacket thus created is assumed to propagate under the action of a field-free Hamiltonian, $\Psi(t) = \exp(-iH_0t/\hbar)\mu\psi_B$. Often, the implicit assumption is that with 50-100 femtosecond pulses one is indeed in a situation where the above is an appropriate description of the dynamics. We shall show that more often than not this is not the case.

In what follows we give explicit expressions to the form of the wavepacket created by a laser pulse *during* and *after* the excitation process. We study in detail the effect of the detuning of the pulse center with respect to the level positions. We show that, while the pulse is on, the coefficient of preparation of a given level displays a non trivial dependence on time, dependent on the detuning. Such transients, whose temporal behavior has nothing to do with wavepacket dynamics, are observable via the introduction of a second (probe) photon (such as in the pump-dump or spontaneous or stimulated Raman experiments).

We supplement this study by developing a consistent time-dependent theory for spontaneous Raman scattering of a pulse of light, with special emphasis to the case of a continuum of intermediate states (continuum Raman). We show that the Raman signal is composed of two transients, one following the pulse, which may be termed *(resonance or off-resonance) Raman*, and the other, decaying via the combination of the natural radiative decay and wavepacket dynamics, representing *resonance fluorescence*. Our derivation, which yields both processes as a natural outcome, constitutes extension of the well known two-levels CW-excitation problem[2] to the multi-level pulsed-excitation case.

We illustrate the use of the above formulation by examining the predissociation of NaI with short laser pulses. We show that, in principle, for this case the CW measurements give *superior* information to the "femto-second transition state" measurements, for which the disentanglement of the laser characteristics from the decay dynamics is virtually impossible

I. Preliminaries of Pulse-Molecule Interaction

We consider a molecule interacting with a pulse of coherent light. Classically, a general pulse can be written as an integral over all the modes,

$$\epsilon(r,t) = \int_0^\infty d\omega\epsilon(\omega)\left\{\exp[i\omega z/c - i\omega t] + c.c.\right\} \equiv \epsilon_+(r,t) + \epsilon_-(r,t). \qquad (I.1)$$

where z is the direction of propagation of the light, ω - the mode frequency and $\mathbf{k}$ - the mode's *wave-vector* ($k = \omega/c$). In the above we have used the fact that,

$$\epsilon(\omega) \equiv |\epsilon(\omega)|\exp[i\phi(\omega)] = \epsilon^*(-\omega) \qquad (I.2).$$

$\phi(\omega)$ is a constant (time and space independent) real (for real ω) phase which incorporates our (arbitrary) definition of the coordinate-origin and zero-time.

We denote the molecular Hamiltonian as H_0 and its (discrete or continuum) set of energy eigenvalues and eigenfunctions as E_n^0 and ψ_n^0,

$$H_0\psi_n^0 = E_n^0\psi_n^0. \qquad (I.3)$$

In the presence of the field, the temporal evolution is governed by the *full* Hamiltonian H, defined as,

$$H(t) = H_0 + V(t) = H_0 - \mu\cdot\epsilon(r,t), \qquad (I.4)$$

where μ is the electric dipole operator. The time-dependent Schrödinger equation,

$$i\hbar\frac{d\Psi}{dt} = H(t)\Psi = [H_0 + V(t)]\Psi, \qquad (I.5)$$

can be solved in the usual manner by expanding the full time dependent wavefunction - $\Psi(t)$ in terms of ψ_n^0 -

$$\Psi(t) = \sum_n b_n(t)\psi_n^0 \exp[-i(E_n^0 - i\hbar\gamma_n/2)t/\hbar]. \qquad (I.6)$$

The addition of a small $-i\hbar\gamma_n/2$ imaginary part to the energies is done to allow for decay via spontaneous emission.

Using the orthonormality of the basis functions, and by substituting Eq. (6) in Eq. (5) we obtain a set of O.D.E. for $b_n(t)$,

$$\frac{db_m}{dt} = (1/i\hbar)\sum_n b_n(t)\exp(i\omega_{mn}t) < \psi_m^0 |V(t)| \psi_n^0 > , \qquad (I.7)$$

where $\omega_{mn} \equiv (E_m^0 - E_n^0)/\hbar - i(\gamma_m - \gamma_n)/2$. If $| < \psi_i^0 |V(t)| \psi_j^0 > /\hbar| \ll 1$, we obtain that,

$$b_m(t) = -\mu_{ms}/i\hbar\{\int d\omega\epsilon(\omega)\exp(i\omega z/c)\int_{-\infty}^{t} dt' \exp[i(\omega_{ms} - \omega)t'] +$$

$$\int d\omega\epsilon^*(\omega)\exp(-i\omega z/c)\int_{-\infty}^{t} dt' \exp[i(\omega_{ms} + \omega)t']\}. \qquad (I.8)$$

where

$$\mu_{ms} \equiv < \psi_m^0 |\mu| \psi_s^0 >, \qquad (I.9)$$

and we have assumed that the molecule is initially in state ψ_s^0,

$$b_s(t = -\infty) = 1, \text{ and } b_k(t = -\infty) = 0 \text{ for } k \neq s. \qquad (I.10)$$

As $t \to +\infty$, we can use the equality

$$\int_{-\infty}^{\infty} dt' \exp[i(\omega_{ms} \pm \omega)t'] = 2\pi\delta(\omega_{ms} \pm \omega), \qquad (I.11)$$

to obtain that

$$b_m(+\infty) = \frac{2\pi i}{\hbar}\mu_{ms}\{\bar{\varepsilon}(\omega_{ms}) + \bar{\varepsilon} * (-\omega_{ms})\}. \qquad (I.12)$$

where $\bar{\varepsilon}(\omega) \equiv \epsilon(\omega)\exp(i\omega z/c)$.

Although $\omega_{m,s}$ has a small imaginary part $(= -i\gamma_m/2)$, at least for Lorentzian and Gaussian pulses, the analytic continuation of $\epsilon(\omega)$ to the complex plane merely corresponds to a slightly changed pulse-shape. In the $\gamma \to 0$ limit Eq. (12) spells out the *resonance* (energy conservation) condition: For a given energy level E_m, the only mode absorbed (emitted) at infinite time is the one for which $\omega = \omega_{ms}$ $(-\omega_{ms})$. For real ω the argument of $\epsilon(\omega)$ must be positive, - see Eq. (1). Hence, the second term vanishes if $\omega_{ms} > 0$ (i.e., absorption), and the first term vanishes if $\omega_{ms} < 0$ (emission).

Some interesting questions arises: How long does it take the resonance condition to be established? Obviously for a pulse of finite duration, no transition can take place after the pulse is over. By energy conservation, this means that under these circumstances the resonance condition must hold true even at very short times, if the pulse is very short. Yet Eq. (11) seems to imply that many optical cycles must pass before we can safely conclude that Eq. (12) holds true and this can be very long. Does this mean that the resonance condition is established faster for a higher-frequency light? And to what extent is the resonance condition fulfilled *during* the pulse? In the next section we address some of these issues.

II. Real and Virtual States

We wish to explore the behavior of the system at finite times, and especially while the pulse is on. In order to perform the integrals over t' in Eq. (I.8), to obtain that,

$$b_m(t) = -\mu_{ms}/i\hbar\left\{\int_{-\infty}^{\infty} d\Delta\bar{\varepsilon}(\Delta)\frac{\exp(i\Delta t)}{i\Delta} - lim_{T\to\infty}\int_{-\infty}^{\infty} d\Delta\bar{\varepsilon}(\Delta)\frac{\exp(-i\Delta T)}{i\Delta}\right\}, \quad (II.1)$$

where $\Delta \equiv \omega_{ms} \mp \omega$.

It is possible to show,[3] using contour integration, that the second term of Eq. (1) is equal to zero irrespective of the pulse shape. The first term of Eq. (1) depends on the shape of $\bar{\varepsilon}(\Delta)$. We find it convenient to partition the problem into a *laser-dependent* part and a *molecular* part, by writing $b_m(t)$ as,

$$b_m(t) = (i/\hbar)\mu_{ms}\bar{\varepsilon}(\omega_{m,s})c_m(\mathbf{t}), \quad (II.2a)$$

where $\mathbf{t} \equiv t - z/c$ is the retarded time and $c_m(t)$, is defined as,

$$c_m(t) = \frac{1}{|\epsilon(\omega_{m,s})|}\int d\omega|\epsilon(\omega)|\frac{\exp[i(\omega_{m,s}-\omega)t]}{i(\omega_{m,s}-\omega)}, \quad (II.2b)$$

For a Lorentzian profile,

$$|\epsilon(\omega)| = (1/2\pi)\frac{\epsilon_a\Gamma}{(\omega-\omega_a)^2+\Gamma^2/4}, \quad (II.3a)$$

$c_m(t)$ is given, using the Residue theorem, as,

$$c_m(t) = 2\pi\left\{\theta(t) - [i(\omega_{m,s}-\omega_a) + sgn(t)\Gamma/2]\exp[-(\Gamma+\gamma)|t|/2 + i(\omega_{m,s}-\omega_a)t)/\Gamma\right\}, \quad (II.4)$$

where $sgn(t) = 1$ for $t \geq 0$, $= -1$ for $t < 0$, and $\theta(t)$ is the Heaviside function, ($=$ $[1 + sgn(t)]/2$).

The choice of a Lorentzian frequency profile is equivalent to choosing

$$\epsilon(r,t) = \epsilon_0(r)\exp(i\omega_a t - \Gamma|t|/2), \quad (II.3b)$$

which unfortunately is unrealistic because of the cusp at $t = 0$. A better choice for the shape of the pulse is a Gaussian function,

$$\epsilon(r,t) = \epsilon_a(r)\exp\{i\omega_a t - 4\ln 2[(t-t_a)/\tau]^2\}, \quad (II.5a)$$

(where $4\ln 2$ has been introduced to ensure that τ is exactly the FWHM of the pulse). Choosing $t_a = 0$, we have that,

$$|\epsilon(\omega)| = (\pi^{\frac{1}{2}}\epsilon_a/\alpha)\exp\{-[\alpha(\omega-\omega_a)]^2\}. \qquad (II.5b)$$

where $\alpha \equiv \tau/4(\ln 2)^{1/2}$.

In this case $c_m(t)$ is related to the $W[z]$ - *the complex error function*, (see, Ref. 4, Eqs. (7.1.3.) and (7.1.8.)),

$$c_m(t) = 2\pi\left\{\theta(t) - (1/2)\exp[\beta^2]W[sgn(t)\beta]\right\}, \qquad (II.6a)$$

where

$$\beta \equiv \alpha(\omega_{m,s}-\omega_a) + i\frac{t}{2\alpha}. \qquad (II.6b)$$

Similar formulae were derived for the Gaussian case by Rhodes[5], and Taylor and Brumer.[6]

It follows from Eq. (6) that

$$c_m(t) \to 0, \text{ for } t \ll -\tau/2(\ln 2)^{1/2}$$

and

$$c_m(t) \to 2\pi \text{ for } t \gg \tau/2(\ln 2)^{1/2}. \qquad (II.7)$$

This gives us the desired criterion as to how long it takes the resonance condition (Eq. (I.12)) to be established. Contrary to the CW domain, in which the goodness of Eq. (I.12) depends only on the passage of a large number of optical cycles, it follows from Eq. (7) that in the pulsed case the relevant parameter is the pulse duration, τ. This quantity can, in principle, be *shorter* than a single optical cycle.

In order to gain physical insight from this derivation we have computed

$$c'_m(t) \equiv c_m(t)|\epsilon(\omega_{m,s})/\epsilon(\omega_a)|, \qquad (II.8)$$

using Eqs. (6a,b), at a number of $\Delta_a \equiv \omega_{m,s} - \omega_a$ values, for a Gaussian pulse whose *intensity* band-width - ($\Delta_\omega \equiv 2(\ln 2)^{1/2}/\alpha$) is 120 cm^{-1}. The results for $|c'_m(t)|$, $R_e c'_m(t)$ and $I_m c'_m(t)$ are presented in Fig. 1(a-c).

It is immediately evident from Fig. 1a that while at the end of the pulse the amplitude for populating a state with a given transition frequency - ω_{ms} - is proportional to $|\epsilon(\omega_{ms})|$, the path leading to this value is remarkably different for different values of Δ_a. For ω_{ms} near the line center, $c'_m(t)$ rises smoothly to its asymptotic value. However at off-center energies $c'_m(t)$ does not rise so monotonically: At early times all the c'_m's respond to the field in almost the same manner since the system has insufficient "information" to determine the true spectral composition of the pulse. It therefore "thinks" that it is exposed to a much broader band of frequencies, hence a more slowly varying $\epsilon(\Delta)$. Only at later times does the system "realize" its "mistake" and corrects for it by depleting the off-center $c'_m(t)$'s.

In the extreme case of detuning i.e., when $\epsilon(\omega_{m,s}) \approx 0$, although the probability of observing the level after the pulse is essentially zero, it can be easily verified from Eq. (6) and from Fig. 1 that $c'_m(t)$ during the pulse is not necessarily zero. This means that the

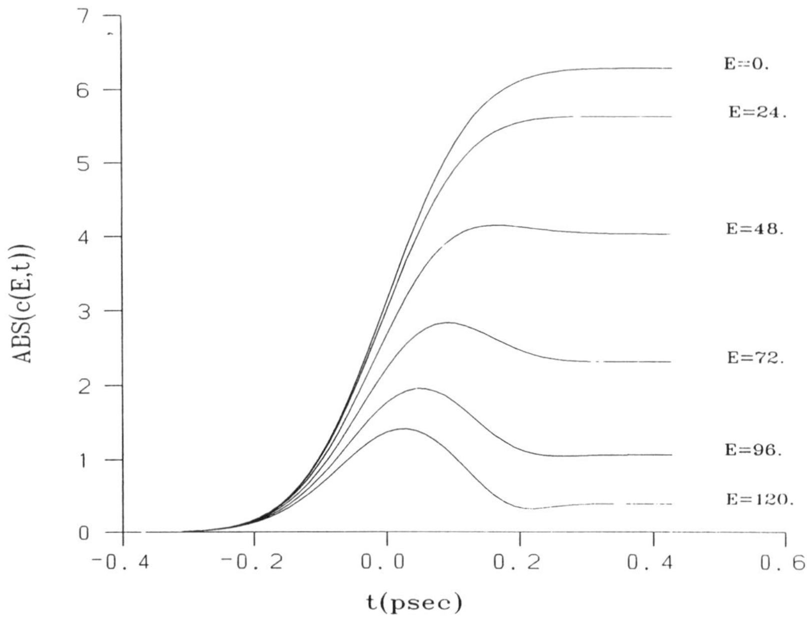

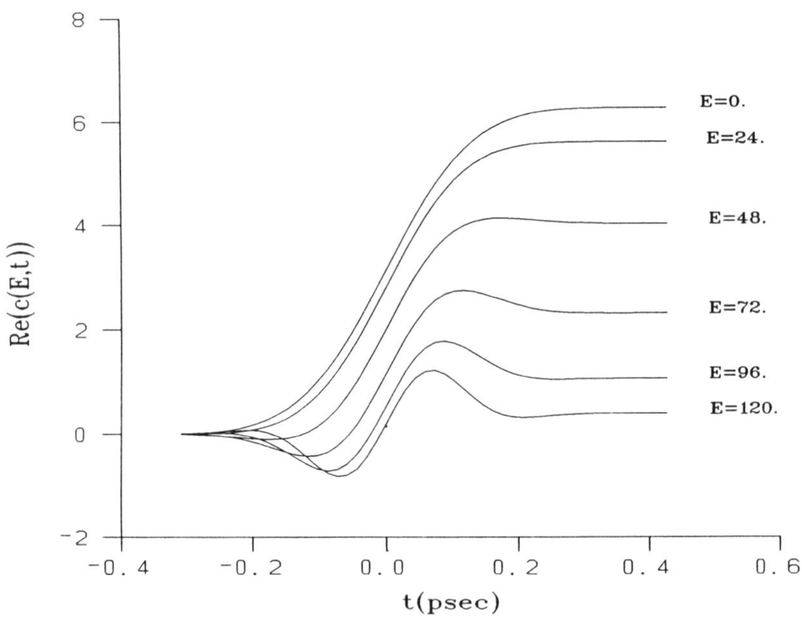

1. Time evolution of the $c'_m(t)$ coefficients at different detunings from the center of the pulse for a Gaussian pulse with FWHM of 120 cm^{-1}.

a) $|c'_m(\mathbf{t})|$, b) $R_e c'_m(\mathbf{t})$, c) $I_m c'_m(\mathbf{t})$.

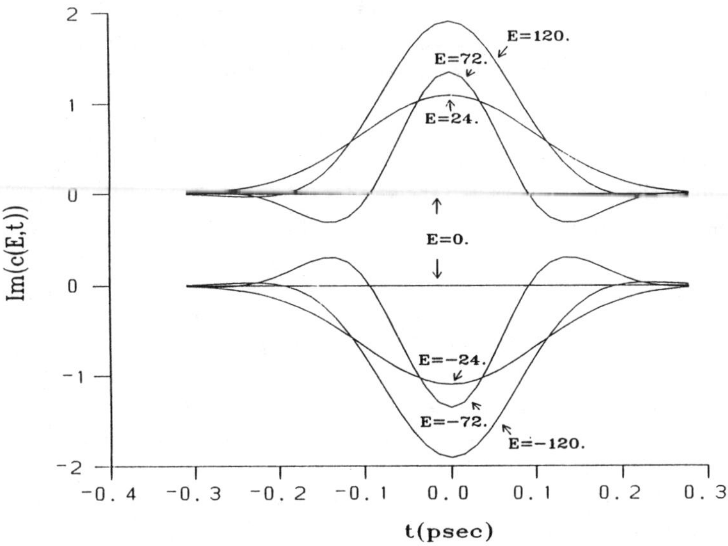

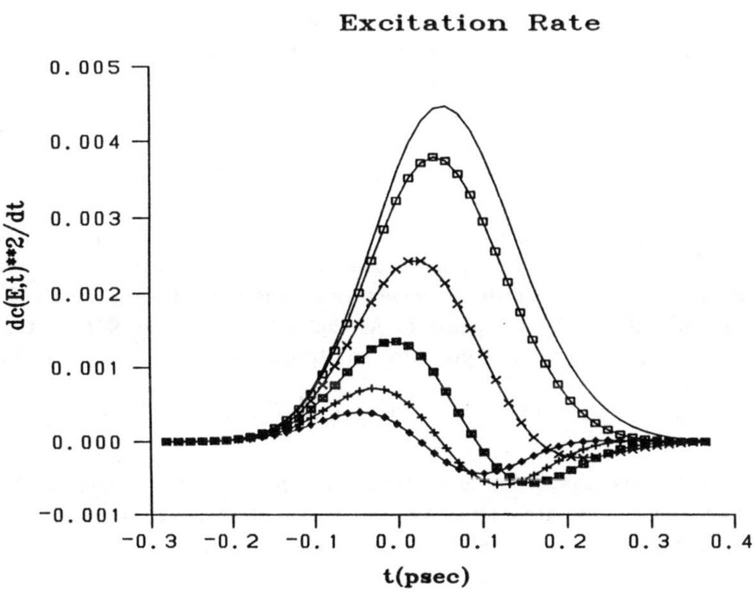

2. Excitation rate ($d|c'_m(t)|^2/dt$) for a 120 cm^{-1}-wide pulse at different detunings from the pulse center.

____ $\Delta_a = 0$, ◻ $\Delta_a = 24\text{cm}^{-1}$, × $\Delta_a = 48\text{cm}^{-1}$,
☒ $\Delta_a = 72\text{cm}^{-1}$, + $\Delta_a = 96\text{cm}^{-1}$, ⟨⟩ $\Delta_a = 120\text{cm}^{-1}$.

level in question gets populated and de-populated during the pulse. In the usual jargon what we have described here is a *virtual* state. The above description gives a physically meaningful content to the concept of a virtual state, which is often treated as a pure mathematical construct.

The role of the phase is also interesting: The phase of $b_m(t)$ at the end of the pulse is guaranteed by Eq. (I.12) to be that of $\bar{\epsilon}(\omega_{m,s})$, which means, using Eq. (2b) that $c_m(t)$ is real at the end of the pulse. As Figs. (1b) and (1c) show, $c'_m(t)$ is real *at all times* for zero detuning ($\Delta_a = 0$), whereas at finite detunings, $c'_m(t)$ is complex during the pulse. In the extreme wings of the pulse, the imaginary part of $c'_m(t)$ essentially follows the pulse shape, reaching a maximum at $t = 0$, while the real part exhibits a temporal "dispersion-like" curve. At intermediate detuning from the pulse center the behavior is more complicated: The imaginary part changes sign several times whereas the real part starts varying more monotonically.

It is also of interest to look at the *rate* of populating a given level, defined as $d|c'_m(t)|^2/dt$. If the pulse is incoherent it is possible to derive the existence of a constant rate at long times by first calculating the rate for a single mode and then averaging over some frequency spread. Since this procedure presupposes that there are no multi-mode coherent effects, it obviously cannot be adopted in the present case. With coherent pulses we must, as done here, first calculate the action of the whole pulse and *then* compute the rate.

The results of the rates derived in this way are given in Fig. 2. We see that for a Gaussian pulse (contrary to a square pulse), the rate reaches a steady state over a very short duration only. Interestingly, the steady-state points depend on the detuning. For finite detunings the steady state points are *not* at the peak of the pulse. The off-center behavior is substantially different than the near-center behavior: Whereas at near-center frequencies the rate is always positive, at off-center frequencies the rate actually becomes negative towards the later part of the pulse. This corresponds to an actual *depletion* of the excited levels, which, for extreme detunings approach the status of virtual state.

The behavior noted above for $c'_m(t)$ and $d|c'_m|^2/dt$ is common to all systems, because $c'_m(t)$ was defined to contain no molecular attributes (save for its implicit dependence on ω_{ms}). For a particular system we must of course take into account the μ_{ms} matrix elements and the m summation in the $\Psi(t)$ expansion. At the end of the pulse, $\Psi(t)$, obtained by substituting Eq. (I.12) in Eq. (I.6), is given for an absorption process ($\omega_{ms} > 0$) by,

$$\Psi_s(t \gg \tau) = \psi_s^0 \exp(-iE_st/\hbar) + (2\pi i/\hbar)\sum_m \bar{\epsilon}(\omega_{ms})\mu_{ms}\psi_m^0 \exp(-iE_m^0 t/\hbar - \gamma_m t/2). \quad (II.9)$$

Thus, after the pulse, the absorption of a photon has created a wavepacket in which the coefficients of preparation are proportional to the field amplitude at the ω_{ms} frequency.

During the excitation pulse, the above picture must be corrected via the use of Eqs. (2a,b). We obtain that,

$$\Psi_s(t) = \psi_s^0 \exp(-iE_s^0 t/\hbar) + (i/\hbar)\sum_m c_m(t - z/c)\bar{\epsilon}(\omega_{ms})\mu_{ms}\psi_m^0 \exp(-iE_m^0 t/\hbar - \gamma_m t/2). \quad (II.10)$$

In the next section we extend these results to the case of excitation in the continuous spectrum.

III. Photodissociation States

Photodissociation arises when the states accessed after absorption of a photon are in the continuous spectrum. Equation (I.6) can be easily extended to that regime by using *scattering* states as our basis functions . Allowing for possible degeneracy of the scattering states, we denote these states by their energy E and a set of extra quantum numbers $\mathbf{n}$. Degeneracy always occurs in a (polyatomic) molecule because it can break apart to fragments with different internal states. Using the energetically accessible internal states of the fragments (also called *open channels*) to distinguish between the degenerate continuum states, we denote the continuum eigenstates of H_0 of interest here as $|\,\psi_{\mathbf{n}}^{-}(E)>$,

$$lim_{\eta\to 0}(E - i\eta - H_0)|\,\psi_{\mathbf{n}}^{-}(E)>= 0. \qquad (III.1)$$

Each $|\,\psi_{\mathbf{n}}^{-}(E)>$ state correlates in the long-time limit to a well defined internal state $|\,\mathbf{n}>$ and translational state $|\,\mathbf{k_n}>$ (where $\mathbf{k_n}$ is the asymptotic wave-vector of the freely moving fragments),

$$lim_{t\to\infty}|\,\psi_{\mathbf{n}}^{-}(E)>=|\,\mathbf{k_n}>|\,\mathbf{n}>. \qquad (III.2)$$

We can now modify Eq. (II.10) to include continuum states by replacing the m summation by integration over the energy, plus summation over the open channels. We obtain that,

$$\Psi_s(t) = \psi_s^0 \exp(-iE_s^0 t/\hbar)+$$

$$(i/\hbar)\sum_{\mathbf{n}}\int dE c_E(\mathbf{t})\bar{\varepsilon}(\omega_{Es}) < \psi_{\mathbf{n}}^{-}(E)|\mu|\,\psi_s^0 > \psi_{\mathbf{n}}^{-}(E)\exp(-iEt/\hbar - \gamma_E t/2). \qquad (III.3a)$$

where $\omega_{Es} = (E - E_s^0)/\hbar - i\gamma_E/2$. In analogy to Eq. (II.9) we can write $\Psi_s(t)$ after the pulse as,

$$\Psi_s(t) = \psi_s^0 \exp(-iE_s^0 t/\hbar)+$$

$$(2\pi i/\hbar)\sum_{\mathbf{n}}\int dE\bar{\varepsilon}(\omega_{Es}) < \psi_{\mathbf{n}}^{-}(E)|\mu|\,\psi_s^0 > \psi_{\mathbf{n}}^{-}(E)\exp(-iEt/\hbar - \gamma_E t/2). \qquad (III.3b)$$

The boundary conditions implied by $\psi_{\mathbf{n}}^{-}(E)$, allow for an immediate evaluation of $P_{\mathbf{m}}(E')$ - the photodissociation probability. Defining the photodissociation *probability-amplitude* - $A_{\mathbf{m}}(E')$, as the amplitude of observing a $|\,\mathbf{k}'_{\mathbf{m}}>|\,\mathbf{m}>$ free state in the long-time limit,

$$A_{\mathbf{m}}(E') = lim_{t\to\infty}\exp[i(E' - i\gamma_{E'}\hbar/2)t/\hbar] < \mathbf{k}'_{\mathbf{m}}\mathbf{m}|\Psi_s(t)>, \qquad (III.4)$$

we have that,

$$P_{\mathbf{m}}(E') = |A_{\mathbf{m}}(E')|^2. \qquad (III.5)$$

Letting $t\to\infty$, we obtain from Eqs. (2) and (3b) that the excited part of the wavepacket is,

$$\Psi'_s(t\to\infty) = \frac{2\pi i}{\hbar}\sum_{\mathbf{n}}\int dE\bar{\varepsilon}(\omega_{Es}) < \psi_{\mathbf{n}}^{-}(E)|\mu|\,\psi_s^0 > |\,\mathbf{k_n}>|\,\mathbf{n}> \exp[-iE - i\gamma_E\hbar/2)t/\hbar]. \qquad (III.6)$$

Hence from Eq. (4)

$$A_{\mathbf{m}}(E) = \frac{2\pi i}{\hbar}\bar{\varepsilon}(\omega_{E_s}) < \psi^-_{\mathbf{m}}(E)\,|\mu|\,\psi^0_s >, \qquad (III.7)$$

where we have used the orthonormality of the $|\,\mathbf{k_n} > |\,\mathbf{n} >$ functions.

$\sigma_{\mathbf{n}}(E)$ - the photodissociation cross-section is defined as *the photon energy absorbed due to a transition to a final fragment state, divided by the incident intensity of light per unit energy.* The energy absorbed is $P_{\mathbf{n}}(E)\hbar\omega_{E_s}$, and the incident intensity $I(E)$ is given as $|\bar{\varepsilon}(\omega)|^2 c/(2\pi\hbar)$. We thus have from Eqs. (4) and (5), for negligible γ_E, that,

$$\sigma_{\mathbf{n}}(E) = P_{\mathbf{n}}(E)\hbar\omega_{E_s}/I(E) = \frac{8\pi^3\omega_{E_s}}{c}| < \psi^-_{\mathbf{n}}(E)\,|\mu|\,\psi^0_s > |^2. \qquad (III.8)$$

This formula forms the basis for many of the computations of detailed photodissociation cross-sections and the angular distribution of photofragments reported in the literature.[7,9]

We next turn our attention to investigating the actual states created under various excitation conditions. We first look at an $ultra-short$ laser pulse. We see from Eq. (I.1) that assuming an ultra-short laser pulse, i.e., that

$$\epsilon(t) \approx \epsilon_0\delta(t) \qquad (III.9)$$

is equivalent to choosing a completely white pulse,

$$\epsilon(\omega) = \epsilon_0/2\pi. \qquad (III.10)$$

Neglecting γ_E, we can write the excited portion of the wavepacket after the pulse as,

$$\Psi'_s(t) = \frac{i\epsilon_0}{\hbar}\left\{\sum_{\mathbf{n}}\int dE\exp(-iEt/\hbar)|\,\psi^-_{\mathbf{n}}(E) >< \psi^-_{\mathbf{n}}(E)\,|\right\}\mu|\,\psi^0_s > . \qquad (III.11)$$

We recognize that the term in the curly brackets above is just the (field-free) time evolution operator expanded in the continuum states,

$$\exp(-iH_0t/\hbar) = \sum_{\mathbf{n}}\int dE\exp(-iEt/\hbar)|\,\psi^-_{\mathbf{n}}(E) >< \psi^-_{\mathbf{n}}(E)\,|, \qquad (III.12)$$

and rewrite Eq. (11) as,

$$\Psi'_s(t) = \frac{i\epsilon_0}{\hbar}\exp(-iH_0t/\hbar)\mu|\,\psi^0_s > . \qquad (III.13)$$

Equation (11) is the basis for a popular picture of photodissociation[10] according to which the light creates, at time $t = 0$, a wavepacket given by $\mu|\,\psi^0_s >$ which subsequently evolves under the action of $\exp(-iH_0t/\hbar)$. It should however be noted that in order for Eq. (11) to hold, $\epsilon(\omega)$ must vary more slowly with energy than any other variable in Eq. (3b). In particular, it should vary more slowly than the energy (or frequency) dependence of the photodissociation amplitudes - $< \psi^-_{\mathbf{n}}(E)\,|\mu|\,\psi^0_s >$.

An easy way of deciding whether this is the case, for a given laser pulse, is to compare its frequency spread with that of the absorption spectrum of the molecule. If, (as is typically the case for direct dissociations), the absorption spectrum extends over 4000 cm^{-1}, $\epsilon(t)$, in order to have a broader band than this, must, by Eq. (I.1), be as short as $\approx$ 2 femtoseconds. Since most pulses used in real photodissociation experiments are much longer than that, Eq. (13) is not a true description of direct photodissociation experiments.

Fourier transforming of Eq. (13) is however, as shown by Heller,[10] a viable computational route for calculating $\sigma_{\mathbf{n}}(E)$. Thus, strangely enough, wavepacket propagation with ultra-short pulses according to Eq. (13) find their greatest utility in calculating *frequency resolved* quantities.

The above comments do not apply to time-dependent methods which explicitly include the external field in the Hamiltonian (cf. Imre et al,[11] Metiu et al,[12], Manz et al,[13] and Rabitz et al,[14]). Such computations are however much more expensive, because they involve both the ground and excited electronic states and must follow faithfully the oscillation of the field (at optical - 10^{15} sec^{-1} - frequencies). If the field is not too strong, (i.e., perturbation theory can be used) the route described by Eq. (3), in which one first calculates the frequency-resolved amplitudes by some time-independent method[15,7,8,9] and then folds them with the field profile, is a much faster one.

In most direct photodissociations the opposite limit to Eq. (13) is realized: The laser's bandwidth is usually much narrower than that of the absorption spectrum. Under these circumstances we can replace in Eq. (3b) the narrow range of energies accessed by the laser by a *single* continuum energy level E_0, and write $\Psi_s'(t)$ as,

$$\Psi_s'(t) \approx \frac{2\pi i}{\hbar} \sum_{\mathbf{n}} | \psi_{\mathbf{n}}^-(E_0) >< \psi_{\mathbf{n}}^-(E_0) |\mu| \psi_s^0 > \left\{ \int dE \bar{\epsilon}(\omega_{Es}) \exp(-iEt/\hbar) \right\}. \quad (III.14)$$

We recognize the term in the curly bracket as $\epsilon_+(r,t)$ - the rotating wave part of the laser pulse, (see Eq. I.1)). Hence we see that under these circumstances the wavepacket created by the laser simply follows the time dependence of the pulse. Classically, this corresponds to a situation in which the dissociation is "faster" than the laser: Every photon absorbed leads to instantaneous dissociation, and the only time evolution that can be observed is due to the laser pulse. The case of NaI, studied in the next section, is intermediate between these two limits discussed above.

It is of interest to look at the wavepacket created by the laser while the pulse is *on*. The complete answer is given in Eq. (3a), but it contains the spatial dependence of $\Psi_s'(t)$, which is not always desirable. We can integrate out the spatial dependence by considering the *autocorrelation* function, defined as

$$F_s(t, t_0) \equiv < \Psi_s'(t_0) | \Psi_s'(t) > . \quad (III.15)$$

It follows from Eq. (3a) that $F_s(t, t_0)$ can be calculated as,

$$F_s(t, t_0) = (i/\hbar)^2 \sum_{\mathbf{n}} \int dE c_E(\mathbf{t}) c_E^*(\mathbf{t_0}) |\epsilon(\omega_{Es})|^2 | < \psi_{\mathbf{n}}^-(E) |\mu| \psi_s^0 > |^2 \cdot$$

$$\exp[-iE(t - t_0)/\hbar - \gamma_E(t + t_0)/2]. \quad (III.16)$$

Notice that this form contains the dependence on the position of the pulse because $c_E(\mathbf{t})$ contain $\mathbf{t}(=t - z/c)$.

If both $\mathbf{t}$ and $\mathbf{t_0}$ occur after the pulse (i.e., $\mathbf{t}, \mathbf{t_0} \gg \tau$) we obtain, using Eq. (3b), that

$$F_s(t,t_0) = (2\pi i/\hbar)^2 \sum_{\mathbf{n}} \int dE |\epsilon(\omega_{E_s}) < \psi_{\mathbf{n}}^-(E)|\mu|\psi_s^0 > |^2 \exp[-iE(t-t_0)/\hbar - \gamma_E(t+t_0)/2]. \tag{III.17}$$

We see that after the pulse the autocorrelation function is essentially a Fourier transform of the pulse-modulated absorption spectrum,

$$I_\epsilon(E) \equiv \sum_{\mathbf{n}} |\epsilon(\omega_{E_s}) < \psi_{\mathbf{n}}^-(E)|\mu|\psi_s^0 > |^2, \tag{III.18}$$

therefore it "remembers" the shape of the pulse that created it, at all times, as illustrated below for the NaI case.

IV. The Photodissociation of NaI

To illustrate the nature of the photodissociation autocorrelation functions we present here a study of the photo-predissociation of NaI. This system has been extensively studied experimentally, using a "pulse-probe" technique (called FTS - Femtosecond Transition-State Spectroscopy) with short ($\approx$100 fsec) pulses, by Zewail et. al.[16] In addition, a number of theoretical studies[12,17] aimed at explaining the NaI FTS experiments, were recently presented.

In this paper we do not attempt to exactly reproduce Zewail's experiments, which involve an additional "probe" laser pulse used to promote the system to a higher state correlating with the fluorescent Na(^{2}P) atomic state. Computations pertaining to the entire pump+probe experiment are reported in a forthcoming publication[18]. Our purpose here is to use the NaI system as an example of how attributes of a short pulse affect the long-time dynamics. We shall therefore concentrate on the study of $F_s(t,t_0)$.

The potential surfaces we have used for the NaI system, given in the *diabatic* representation, are illustrated in Fig. 3. In the FTS experiment the NaI was excited at $\approx 32000 - 33300$ cm^{-1}. Therefore, initially, the system is deposited almost exclusively on the repulsive branch of both the excited $\Omega = 0^+$ and $\Omega = 1$ states[19]. Following Engel and Metiu[12] we neglect the rapidly dissociating $\Omega = 1$ state and concentrate on the 0^+ and the ground states.

In the Franck-Condon region, the 0^+ state is dominantly covalent and repulsive. At larger internuclear separations its potential is affected by the attractive branch of the ionic structure. As a result, the 0^+ Born-Oppenheimer state is binding, with a short-range covalent repulsion and a long-range ionic attraction. Due to non-adiabatic coupling, and the proximity of the ground state, a slow leakage (resulting in dissociation) of molecules from the 0^+ state to the ground state occurs.

In Fig. 3 we show the original ionic and covalent (diabatic) potentials, as well as the next excited state, correlating to ^{2}P Na atoms. The exact form of the potentials which are based on available spectroscopic information (following Berry[20] and Grice and Hershbach,[21] and Rittner[22]) is reported elsewhere [3]

In Fig. 4 we present exact quantum computations, using the artificial channel method,[15,7] of the photodissociation cross-section - ($\sigma_{\mathbf{n}}(E)$ - of Eq. (III.8)) for the above set of potentials. The range of frequencies chosen is that spanned by some of the pulses used in

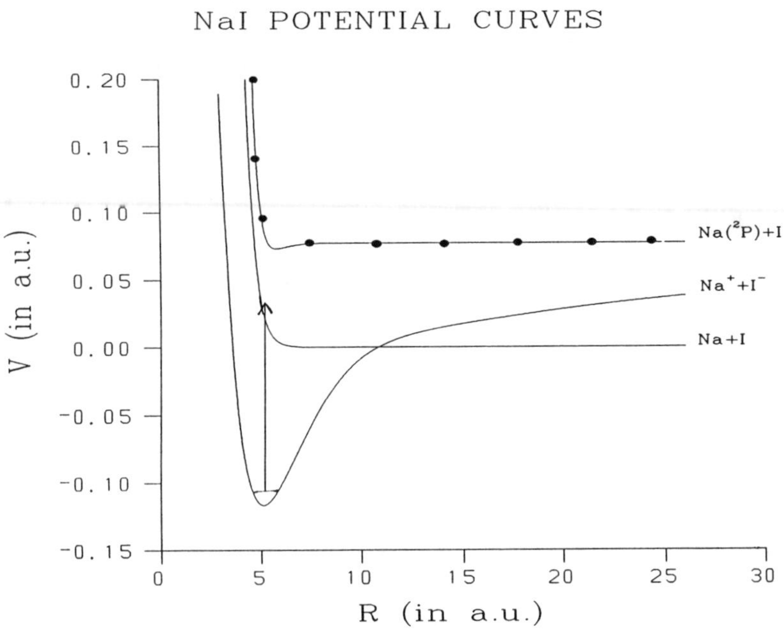

3. NaI diabatic curves.

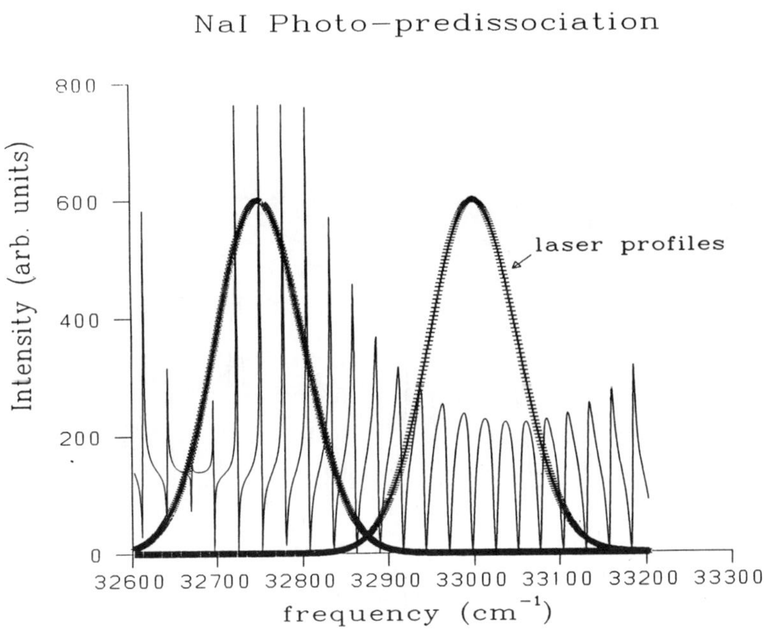

4. Photodissociation cross-section of NaI in the 32600 cm^{-1} - 33200 cm^{-1} range.

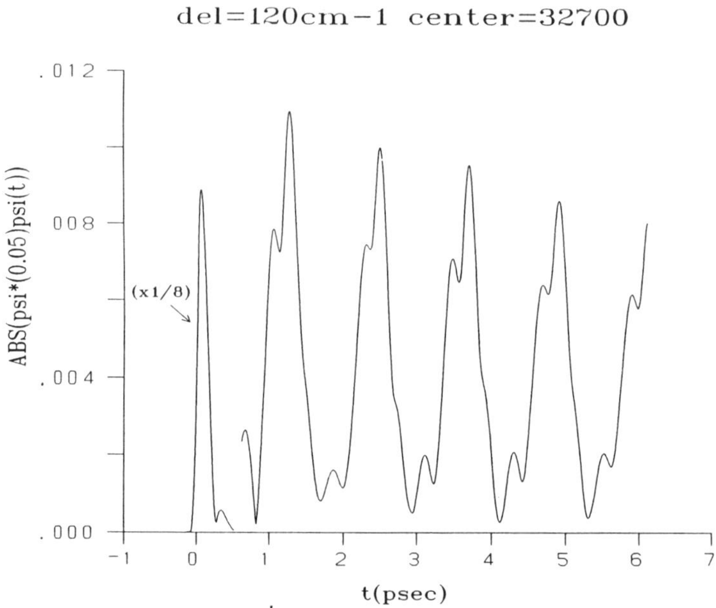

5. $|F_s(t, t_0 = 0.05\text{psec})|$ NaI autocorrelation function for a 120 cm^{-1} wide Gaussian pulse centered at 32700 cm^{-1}.

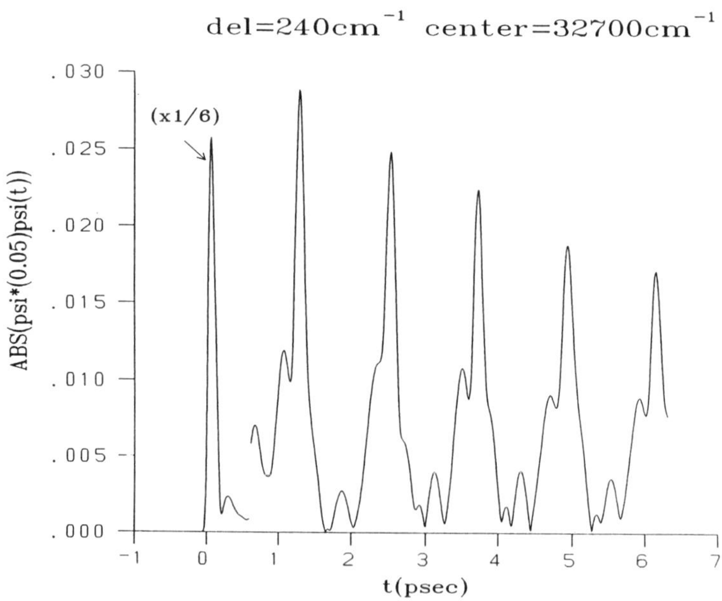

6. The same as in Fig. 5 for a $240cm^{-1}$-wide pulse.

the FTS experiments[16]. The displayed high-resolution spectrum reveals a very interesting sequence of narrow *Fano-type* lines, slowly changing to a series of broader features. The complexity of the spectrum is due to interferences induced by the curve crossing processes.

In order to obtain the actual wavepacket created by the laser (Eq. (III.3)) or its autocorrelation function (Eqs.III.15-17)) we need to "fold" the laser amplitude $\epsilon(\omega_{E_s})$ with the above spectrum. Two laser frequency-profiles whose FWHM is 120 cm^{-1} (i.e. pulses of $\approx$ 80 femtosecond duration) , one centered at 32750 cm^{-1}, the other at 33000 cm^{-1} are shown superimposed on the spectrum. This illustrates the portion of the spectrum actually "seen" by the pulse. Thus, one pulse "sees" narrower resonances (hence much slower decays) than the other.

In Fig. 5 we study the behavior of $F_s(t, 0.05\text{psec})$ with a 120cm^{-1}-wide excitation pulse, centered at 32700 cm^{-1}, for t up to 6 psec. The initial rise and fall of $F_s(t, 0.05\text{psec})$, is followed by a sequence of structured recurrences, spaced $\approx$ 1.3 psec apart. The train of recurrences decays, due to the dissociation, with half-life ($\tau_{1/2}$) of $\approx$ 8 psec. The spacings and decay rate are in reasonable agreement with the FTS experiments.[16]

The width and structure of each recurrence peak is determined by a combination of the laser pulse-shape and the shapes of the individual resonances lines (shown in Fig. 4) accessed by the laser pulse. In contrast, the temporal spacings between successive recurrences are thought to be a measure of "pure" molecular attributes: Zewail et. al.[16] interpreted them to be due to reflections from the walls and their spacings to be given by the classical travel time between the two turning points of the adiabatic 0^+ potential.

It is interesting to see whether the decay of the recurrences-train is determined solely by the molecular attributes. To this end, we have computed $F_s(t, 0.05\text{psec})$ for a *shorter* pulse, whose width is 240cm^{-1}, centered at the same (32700cm^{-1}) frequency. The results are shown in Fig. 6. While the fact that the shape of each recurrence has changed and its duration has shrunk is in accordance with the above expectations, the shortening of the decay-time of the whole recurrence *pattern* to $\approx$ 6 psec, is surprising. How is this possible? A glance at Fig. 4 reveals that a 240cm^{-1}-wide pulse centered at 32700cm^{-1} encompasses more of the broad resonances than the 120cm^{-1}-wide pulse. This results in a faster decay by some of the resonances. We may conclude that, for NaI, sub-picosecond laser pulses only give us information about an average decay time of many individual resonances. The shorter the pulse, the more averaged is this information.

We therefore turn back to Fig. 4, as the one closer to the detailed dynamics, and make the immediate prediction that a shift of the pulse to higher frequencies, while maintaining its width at 120cm^{-1}, would further decrease the recurrence-train decay-time. In Fig. 7 we display $|F_s(t, 0.05\text{psec})|$ obtained with a 120cm^{-1}-wide pulse centered at 33000cm^{-1}. This is the center of the wide resonances of Fig. 4 and indeed, the recurrences decay time has shrunk to $\approx$ 1.5 psec.

Some of the frequencies studied by Zewail et. al. (e.g., 32900cm^{-1}) are in an in-between region. In this region we expect to see an effect due to *two* essentially different life-times, those of the narrow 32700cm^{-1} resonances and those of the broad 33000cm^{-1} resonances. This is in fact the case, as can be seen from Fig. 8 where we plot $|F_s(t, 0.05\text{psec})|$ at 32900cm^{-1} . The decay time is intermediate ($\tau_{1/2} \approx$ 2.5 psec). The FTS spectra of Fig. 8a does not reveal however that the decay pattern is influenced by two types of resonances. The general shape of its envelope looks very similar to the 32700cm^{-1} or 33000cm^{-1} cases.

V. Raman Scattering via Virtual and Dissociative States

The "continuum-Raman" experiments[23,24,25] as performed by Imre et al..[23] involve excitation of a molecule to a dissociative electronic state and monitoring the emission emanating from the molecule as it falls apart. The theoretical framework used by most researchers in the field,[10,27] is based on the Kramers-Heisenberg expression[26,2] which is correct, in the CW case, to second order perturbation theory, according to which, the observed emission rate is proportional to

$$(\omega_a - \omega_{i,s})^3 \left| \int dE \sum_{\mathbf{n}} < \psi_i^0 | \mu_{\hat{\epsilon}} | \psi_{\mathbf{n}}^-(E) >< \psi_{\mathbf{n}}^-(E) | \mu_{\hat{\mathbf{k}}} | \psi_s^0 > / (\omega_{E,s} - \omega_a) \right|^2 ,$$

where as in sections II and III, ω_a is the laser frequency, $\omega_{i,s}$ is the transition frequency between the final and initial states (the Stokes shift), ψ_s^0 and ψ_i^0 are the initial and final states, and $\mu_{\hat{\mathbf{k}}}$ is the component of the transition dipole in the direction of the field of the scattered light.

This formula however ignores the fact that the excitation is usually done with pulses whose duration is often shorter than that of the radiative lifetime of the continuum states. Thus, transient effects may contribute to the observed signal. In what follows we show that that the continuum Raman signal is composed of three terms, the (resonance+off-resonance) Kramers Heisenberg term, (giving the probability as $t \to \infty$), a *resonant-fluorescence* transient and a *Raman* transient which follows the pulse.

In order to include the possibility of spontaneous emission from the dissociating wavepacket we generalize the expansion of the wavefunction to include both zero-photon and one-photon states. The field cannot be treated in a completely classical manner as we have done above, and we find it useful to define two types of radiation states:

1) $|\bar{\alpha}> \equiv |\alpha_1, \alpha_2, ..., \alpha_N, 0, ..., 0>$ composed of $|\alpha_1>|\alpha_2>...|\alpha_N>$ - a multimode coherent state representing the incoming pulse, and zero photons in all other modes.

2) $|\bar{\alpha} + \bar{1}_{\mathbf{k}}> \equiv |\alpha_1, \alpha_2, ..., \alpha_N, 0, ..., 0, 1_{\mathbf{k}}, 0, ..., 0>$, representing the pulse plus one (spontaneously emitted) photon in the $\mathbf{k}$-th mode.

Using these radiation states we expand the system's wavefunction - $\Psi(t)$ as

$$\Psi(t) = |\bar{\alpha}> \left\{ b_s(t)\psi_s^0 \exp(-iE_s^0 t/\hbar) + \sum_{\mathbf{n}} \int dE b_{\mathbf{n}E}(t)\psi_{\mathbf{n}}^-(E) \exp[-i(E - i\gamma_E\hbar/2)t/\hbar] \right\}$$
$$+ \sum_{\mathbf{k},i} |\bar{\alpha} + \bar{1}_{\mathbf{k}}> b_i^{\mathbf{k}}(t)\psi_i \exp(-iE_i t/\hbar). \qquad (V.1)$$

This expansion implicitly assumes that the absorption of one pulse photon does not affect the pulse, which is justified for $|\alpha_l| \gg 1$.

We now add to $V_1(t)$ - the interaction of the molecule with the pulse, (Eq. (I.4)) which, written in the quantum form, is,

$$V_1(t) = -|\bar{\alpha}> \mu \cdot \epsilon(r,t) < \bar{\alpha}|, \qquad (V.2)$$

the *spontaneous emission* term,

$$V_2(t) = \sum_{\mathbf{k}} |\bar{\alpha} + \bar{1}_{\mathbf{k}}> i \left(\frac{2\hbar\omega_k}{\epsilon_0 V} \right)^{\frac{1}{2}} \mu_{\hat{\mathbf{k}}} \{\exp[i\mathbf{k}\cdot\mathbf{r} - i\omega_k t] + c.c.\} < \bar{\alpha}|. \qquad (V.3)$$

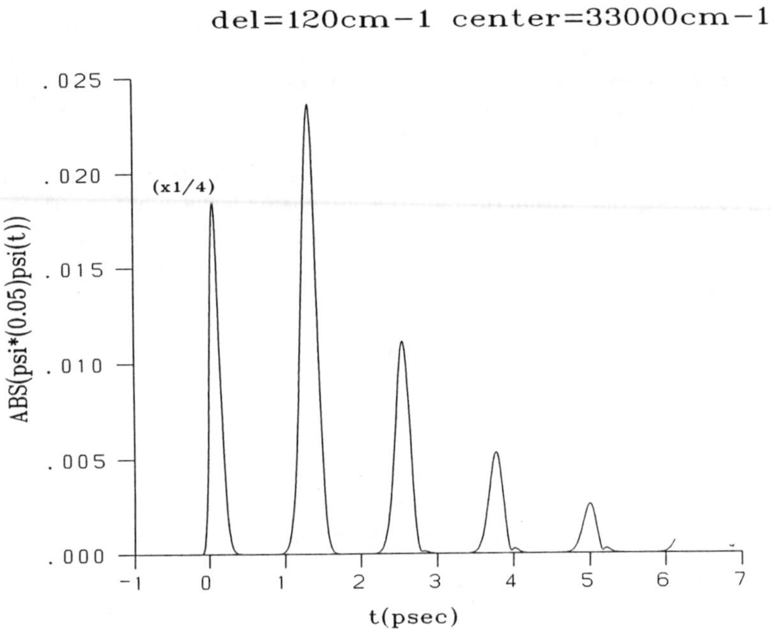

7. The same as in Fig. 5 for a pulse centered at 33000cm^{-1}.

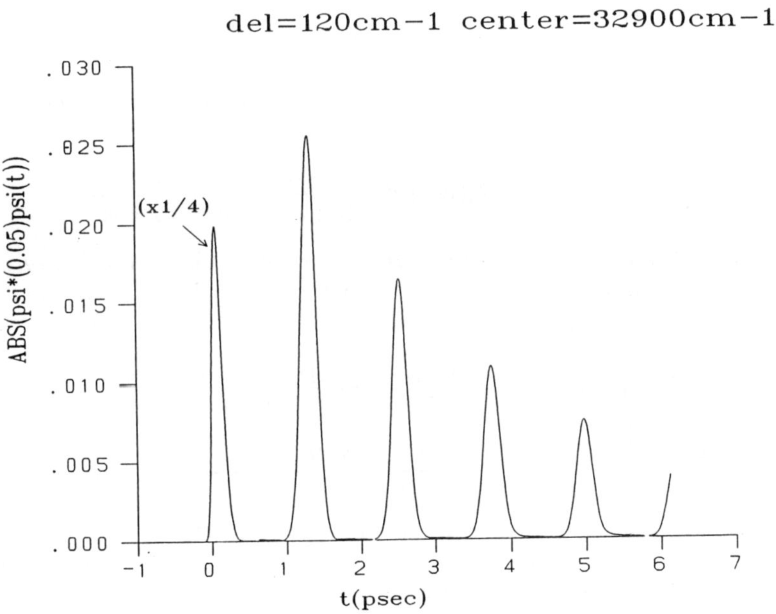

8. The same as in Fig. 5 for a pulse centered at 32900cm^{-1}.

In the above, v is the cavity volume and ϵ_0 is the permittivity of free space. $\mu_{\hat{\mathbf{k}}}$ is the component of the electric dipole operator in the direction of the scattered light field. With the use of coherent states, the expectation value of the pulse's electric-field, - $\epsilon(r,t)$ of Eq. (2) - assumes the classical form of Eqs. (I.1).

In the continuum-Raman experiment,[23] only a small fraction of the excited molecules manage to fluoresce. Basically, the wavepacket of continuum states created by the pulse dissociates so quickly that within the radiative lifetime typically only 1 in 10^{6-7} molecules emits. Under these conditions, we can use the solution of section III to solve the augmented time-dependent Schrödinger equation, in which $V(t) = V_1(t) + V_2(t)$.

Substituting Eq. (II.2a) in Eq. (1) we obtain,

$$\Psi(t) = |\,\bar{\alpha} > \{\psi_s^0 \exp(-iE_s^0 t/\hbar) + (i/\hbar)\sum_{\mathbf{n}} \int dE c_E(\mathbf{t})\epsilon(\omega_{E,s}) < \psi_{\mathbf{n}}^-(E)\,|\mu_{\hat{\epsilon}}|\,\psi_s^0 > \cdot$$

$$\psi_{\mathbf{n}}^-(E)\exp(-iEt/\hbar - \gamma_E t/2)\} + \sum_{\mathbf{k},i} |\,\bar{\alpha} + \bar{1}_{\mathbf{k}} > b_i^{\mathbf{k}}(t)\psi_i \exp(-iE_i t/\hbar). \qquad (V.4)$$

Equation (4) is valid when only a small fraction of the molecules manage to emit.

In order to determine the $b_i^{\mathbf{k}}(t)$ coefficients we denote

$$a_{i,s}^{\mathbf{k}}(E) \equiv -\left(\frac{2\omega_k}{\hbar\epsilon_0 \mathrm{v}}\right)^{\frac{1}{2}} \exp(i\mathbf{k}\cdot\mathbf{r})\sum_{\mathbf{n}} < \psi_i\,|\mu_{\hat{\mathbf{k}}}|\,\psi_{\mathbf{n}}^-(E) >< \psi_{\mathbf{n}}^-(E)\,|\mu_{\hat{\epsilon}}|\,\psi_s^0 >, \qquad (V.5)$$

and substitute Eq. (4) into Eq. (I.5) to obtain,

$$i\hbar\frac{db_i^{\mathbf{k}}}{dt} = \int dE a_{i,s}^{\mathbf{k}}(E)\epsilon(\omega_{E,s})c_E(\mathbf{t})\exp[-i(\omega_{E,i} - \omega_k)t]\ , \qquad (V.6)$$

where $\omega_{E,i} \equiv (E - E_i)/\hbar - \gamma_E/2$. Using the definition of $c_E(\mathbf{t})$ (Eq. (II.2b)) and performing the time-integration we have that,

$$b_i^{\mathbf{k}}(t) = -i/\hbar \int dE a_{i,s}^{\mathbf{k}}(E)\int d\omega \frac{\bar{\epsilon}(\omega)\exp[-i(\omega - \omega_{i,s} - \omega_k)t]}{(\omega_{E,s} - \omega)(\omega - \omega_{i,s} - \omega_k)}, \qquad (V.7)$$

Assuming a Lorentzian pulse, (Eq. (II.3a)), we obtain by contour integration (for $t > 0$ the contour of integration must be closed in the lower half ω-plane and must include the $\omega = \omega_{E,s}$, $\omega = \omega_{i,s} + \omega_k$ and $\omega = \omega_a - i\Gamma/2$ poles), that,

$$b_i^{\mathbf{k}}(t) = -2\pi/\hbar \int dE a_{i,s}^{\mathbf{k}}(E)\{\frac{\bar{\epsilon}(\omega_{E,s})\exp[-i(\omega_{E,i} - \omega_k)t]}{-\omega_{E,i} + \omega_k} +$$

$$\frac{\bar{\epsilon}(\omega_k + \omega_{i,s})}{\omega_{E,i} - \omega_k} + \frac{\bar{\epsilon}_a \exp[i(\omega_{i,s} + \omega_k - \omega_a)t - \Gamma t/2]}{2\pi(\omega_{E,s} - \omega_a + i\Gamma/2)i(\omega_{i,s} + \omega_k - \omega_a + i\Gamma/2)}\}\ ,\ t > 0\ , \qquad (V.8a)$$

where $\omega_{i,s} \equiv (E_i - E_s)/\hbar$. For $t < 0$ the contour of integration must be closed in the upper half-plane, thus omitting the $\omega = \omega_{E,s}$ and the $\omega = \omega_{i,s} + \omega_k$ poles while including only the $\omega = \omega_a + i\Gamma/2$ pole. We obtain that,

$$b_i^{\mathbf{k}}(t) = 1/\hbar \int dE a_{i,s}^{\mathbf{k}}(E)\frac{\bar{\epsilon}_a \exp[i(\omega_{i,s} + \omega_k - \omega_a)t + \Gamma t/2]}{(\omega_{E,s} - \omega_a - i\Gamma/2)i(\omega_{i,s} + \omega_k - \omega_a - i\Gamma/2)}\}\ ,\ t < 0\ . \qquad (V.8b)$$

The first two terms in Eq. (8a) are written in a form that suggests their independence on our choice of the pulse shape. We can do the same thing for the third term by noting that for a Lorentzian frequency-profile the numerator of the third term is simply $\epsilon_+(r,t)\exp[i(\omega_{i,s}+\omega_k)t]$ where $\epsilon_+(r,t)$ is the + portion of the pulse, as defined in Eq. I.1).

The first term of Eq. (8a) is easily recognizable as representing *resonance fluorescence*: It exists only if $\epsilon(\omega)$ is non-negligible in the range, spanned by the continuum, of the $\omega_{E,s}$ resonance-frequencies. For CW excitation mode, the resonance fluorescence term decays exponentially with γ (the radiative decay rate). In the pulsed mode, for $\Gamma \gg \gamma$, the resonance fluorescence term is shown below to decay at a much faster rate.

The third term of Eqs. (8a,b) is the Raman transient, which is seen to follow the pulse. The second term corresponds to the (resonance or off-resonance) Kramers-Heisenberg expression. In order to see this we calculate explicitly below $|b_i^{\mathbf{k}}(t\to\infty)|^2$ - summed over all the field modes. At this stage it is already clear that as $t\to\infty$ only the second term remains. It is therefore related to the *total* number of photons emitted.

If our photo-detector detects the (averaged or non-averaged) *rate* of emission, it will also be influenced by the transient terms. For example, if the pulse varies much more slowly with E than $a_{i,s}^{\mathbf{k}}(E)$ (i.e. a broad band pulse or a predissociating state) then the first and second terms which are opposite in sign essentially cancel at very short times and

$$d|b_i^{\mathbf{k}}(t)|^2/dt = -\left(\frac{2\epsilon_a}{\hbar}\right)^2 \exp(-\Gamma t)\left|\int dE \frac{a_{i,s}^{\mathbf{k}}(E)}{[\omega_{E,s}-\omega_a+i\Gamma/2]}\right|^2 . \qquad (V.9)$$

This term is significantly different than the total number of photons emitted.

We now consider explicitly the two most interesting limiting cases.

a) The $t\to\infty$ limit with pulses of finite duration

In the $t\to\infty$ limit, the third term of Eq. (8a) obviously vanishes. The first term also vanishes but often at a rate much faster than that dictated by the spontaneous decay. Performing the contour integration, we have that,

$$lim_{t\to\infty}\exp[i(\omega_k-\omega_{E,i})t]/(\omega_k-\omega_{E,i}) = 0. \qquad (V.10)$$

The reason this term vanishes in the long time limit is that the contour of integration can be closed in the upper half plane where (because $t\to\infty$) the integrand is always zero, irrespective of the shape of the contour. Thus, the segment of the contour in the upper half-plane can be constructed to avoid the poles of $a_{i,s}^{\mathbf{k}}(E)$ and $\bar{\epsilon}(\omega_{E,s})$ (the $\omega_k=\omega_{E,i}$ pole is in the *lower* half-plane). As a result, the real-axis segment must also be zero.

With Eq. (10) we obtain that,

$$b_i^{\mathbf{k}}(\infty) = -2\pi\epsilon(\omega_k+\omega_{i,s})/\hbar\int dE\frac{a_{i,s}^{\mathbf{k}}(E)}{\omega_{E,i}-\omega_k}. \qquad (V.11)$$

$P_{i,s}$ - the *total* number of spontaneously emitted photons due to an $s\to E\to i$ cycle is obtained by integrating $|b_i^{\mathbf{k}}(\infty)|^2$ over all the possible radiation modes,

$$P_{i,s} = (\mathrm{v}/\pi^2c^3)\int d\omega_k\omega_k^2|b_i^k(\infty)|^2. \qquad (V.12)$$

The $\mathrm{v}\omega_k^2/\pi^2c^3$ factor comes from the $\hat{\mathbf{k}}$ integration over all the cavity modes. Using Eqs. (12), (11) and (5) we obtain that,

$$P_{i,s} = \frac{8}{c^3\hbar^3\epsilon_0}\int d\omega\omega^3 \left|\epsilon(\omega+\omega_{i,s})\int dE \frac{A_{i,s}(E)}{\omega_{E,i}-\omega}\right|^2 , \qquad (V.13)$$

where

$$A_{i,s}(E) \equiv \sum_{\mathbf{n}} < \psi_i\,|\mu|\,\psi_{\mathbf{n}}^-(E) >< \psi_{\mathbf{n}}^-(E)\,|\mu_{\hat{\epsilon}}|\,\psi_s^0 > . \qquad (V.14)$$

Contrary to the CW case, the ω integration completely washes away the $\omega_{E,i}=\omega$ singularity in the integral over E.

b) The $\Gamma \to 0$ (CW) limit

For a CW excitation we can set (see Eq. (II.3a)),

$$|\epsilon(\omega)| = \epsilon_a\delta(\omega-\omega_a). \qquad (V.15)$$

Substituting Eq. (15) in Eq. (7) and remembering that when $\epsilon(\omega)$ has a singularity the second term in Eq. (II.1) is not zero, we obtain that,

$$b_i^{\mathbf{k}}(t) = (i\epsilon_a/\hbar)\int dE \frac{a_{i,s}^{\mathbf{k}}(E)}{\omega_{E,s}-\omega_a}\cdot$$

$$lim_{T\to\infty}\frac{\exp[i(\omega_{i,s}+\omega_k-\omega_a)t]-\exp[-i(\omega_{i,s}+\omega_k-\omega_a)T]}{\omega_{i,s}+\omega_k-\omega_a}. \qquad (V.16)$$

The rate of emission, given as,

$$\frac{dP_{i,s}(t)}{dt} = (\mathrm{v}/\pi^2c^3)\int d\omega_k\omega_k^2\frac{d|b_i^k(t)|^2}{dt}, \qquad (V.17)$$

is obtained, using Eq. (16) as

$$\frac{dP_{i,s}(t)}{dt} = (\mathrm{v}/\pi^2c^3)\int d\omega_k\omega_k^2(\epsilon_a/\hbar)^2\left|\int dE\frac{a_{i,s}^{\mathbf{k}}(E)}{\omega_{E,s}-\omega_a}\right|^2\cdot$$

$$lim_{T\to\infty}\frac{4\sin[(\omega_{i,s}+\omega_k-\omega_a)(t+T)]}{\omega_{i,s}+\omega_k-\omega_a}. \qquad (V.18)$$

Using the identity,

$$lim_{T\to\infty}\frac{\sin[(\omega_{i,s}+\omega_k-\omega_a)(t+T)]}{\omega_{i,s}+\omega_k-\omega_a} = \pi\delta(\omega_{i,s}+\omega_k-\omega_a), \qquad (V.19)$$

we finally obtain for the CW case that,

$$\frac{dP_{i,s}(t)}{dt} = \frac{8\epsilon_a^2(\omega_a-\omega_{is})^3}{\epsilon_0\pi\hbar^3c^3}\left|\int dE\frac{A_{i,s}(E)}{\omega_{E,s}-\omega_a}\right|^2 . \qquad (V.20)$$

This is the Kramers-Heisenberg expression with continuum intermediate states, in which (as explained above) we have for simplicity ignored the counter rotating term.

VI. Conclusion

We have presented a theoretical analysis of the type of states created in a photodissociation process. We have shown how the coefficient of preparation evolve under the action of the laser pulse and what form they take after the pulse is over. We have examined the effect of the preparation process on the wavepacket after the pulse, and how the subsequent dynamics is influenced by the "memory" of the act of excitation. The formulation has been applied to Zewail's FTS experiments on the predissociation of NaI. It has been shown that even for this relatively "slow" process the measured transients in the FTS experiments depend on the pulse characteristics in a very strong way. Finally we have shown that the usual theoretical framework for understanding Raman scattering must be augmented when pulses are employed.

Our present treatment of transients has yielded a uniform theory of virtual and "real" states. We have shown that these states give rise to both resonance and off-resonance Raman, as well as resonance fluorescence, phenomena.

References

1. A. H. Zewail, Science **242**, 1645 (1988); A. H. Zewail and R. B. Bernstein, Chem. Eng. News No. 45, 24 (1988).

   For early work on transition state spectroscopy, see, e.g., P. Arrowsmith, P.E. Bartoszek, S.H.P. Bly, T. Carrington, P.E. Charters and J.C. Polanyi, J. Chem. Phys. **73**, 5895 (1980); P. Arrowsmith, S.H.P. Bly, P.E. Charters and J.C. Polanyi, J. Chem. Phys. **79**, 283 (1983); R.P. Brooks, R.F. Curl and T. Maguire, Ber. Bunsenges. Phys. Chem. **86**, 401 (1982).

2. R. Loudon, "The Quantum Theory of Light" 2nd Edition, (Clarendon Press, Oxford 1983).

3. M. Shapiro, J. Phys. Chem. , (), submitted.

4. M. Abramowitz and I. A. Stegun, *Handbook of Mathematical Functions,* (Dover, New York, 1965).

5. W. Rhodes, in "Radiationless Transitions", S.H. Lin Editor, (Academic Press, N.Y. 1980).

6. R. D. Taylor and P. Brumer, Faraday Disc. Chem. Soc. **75**, 117 (1983).

7. M. Shapiro and R. Bersohn, Ann. Rev. Phys. Chem. **33**, 409 (1982);

8. G.G. Balint-Kurti and M. Shapiro Chem. Phys. **61**, 137 (1981).
   G.G. Balint-Kurti and M. Shapiro Adv. Chem. Phys. **60**, 403 (1985).

9. V. Engel, R. Schinke and V. Staemmler, Chem. Phys. Lett. **130**, 413 (1986);
   P. Andresen and R. Schinke, in *Molecular Photodissociation Dynamics* Edited by M.N.R. Ashfold and J.E. Baggott (The Royal Society of Chemistry, London, 1987) p. 61.

10. E.J. Heller, in *Potential Energy Surfaces and Dynamics Calculations*, edited by D. Truhlar (Plenum, New York, 1981);
    E.J. Heller, Acc. Chem. Res. **14**,368 (1981).

11. J. Zhang and D.G. Imre Chem. Phys. Lett. **149**, 233 (1988).
    D.G. Imre and J. Zhang, J. Chem. Phys. **89**, 139 (1989).

12. a) V. Engel, H. Metiu, R. Almeida, R. Marcus and A. H. Zewail Chem. Phys. Lett. **152**, 1 (1988).

    b) V. Engel and H. Metiu, J. Chem. Phys. **90**, 6116 (1989).

13. W. Jakubetz, J. Manz and H.-J. Schreier, Chem. Phys. Lett. **165**, 100 (1990).

14. S. Shi, A. Woody and H. Rabitz J. Chem. Phys. **88**, 6870 (1988).
    S. Shi and H. Rabitz J. Chem. Phys. **92**, 2927 (1990).

15. M. Shapiro , J. Chem. Phys. **56**, 2582 (1972).

16. M. J. Rosker, T. S. Rose and A. H. Zewail, Chem. Phys. Lett. **146**, 175 (1988); T. S. Rose, M. J. Rosker and A. H. Zewail, J. Chem. Phys. **88**, 6672 (1988); T. S. Rose, M. J. Rosker and A. H. Zewail, J. Chem. Phys. **91**, 7415 (1989). A. H. Zewail, Faraday Disc. Chem. Soc. **91** (1991)

17. a) S.E. Choi and J.C. Light, J. Chem. Phys. **90**, 2593 (1989).
b) R.A. Marcus, Chem. Phys. Lett. **152**, 8 (1988).
c) S.H. Lin and B. Fain Chem. Phys. Lett. **155**, 216 (1989).
d) S.Y. Lee, W.T. Pollard and R.A. Mathies, J. Chem. Phys. **90**, 6146 (1989).

18. M. Shapiro, to be published.

19. N.J.A. van Veen, M.S. de Vries, J.D. Sokol, T. Baller and A. E. de Vries, Chem. Phys. **56**, 81 (1981).

20. R.S. Berry in *Alkali Halide Vapors*, edited by P. Davidovitz and D.L. McFadden (Academic, New York, 1979)

21. R. Grice and D. R. Hershbach, Mol. Phys. **27**, 159 (1974).

22. E.S. Rittner, J. Chem. Phys. **19**, 1030 (1951).

23. D. Imre, J. L. Kinsey, A. Sinha and J. Krenos, J. Phys. Chem. **88**, 3956 (1984).

24. R.J. Sension, R.J. Brudzynski and B.S. Hudson, Phys. Rev. Lett. **61**, 694 (1988).

25. K.Q. Lao, M.D. Person, P. Xayariboun and L.J. Butler, J. Chem. Phys. **92**, 823 (1990).

26. H. A. Kramers and W. Heisenberg, Zs. f. Phys. **31**, 681 (1925).

27. R.L. Sundberg, D. Imre, M.O. Hale, J.L. Kinsey and R.D. Coalson, J. Phys. Chem. **90**, 5001 (1986).

28. M. Shapiro, J. Phys. Chem. **90**, 3644 (1986).

29. I. Levy, Ph.D. thesis - submitted to the Feinberg graduate school of the Weizmann Institute of Science, (1991).
I. Levy A. Yogev and M. Shapiro "The Wavelength Dependence of the Resonance Raman Scattering of Dissociating IBr: Theory and Experiment", to be published.

COHERENT EXCITATION EFFECTS IN PHOTODISSOCIATION: STRONG ORIENTATION OF MOLECULAR PRODUCTS

J.A. BESWICK
LURE, [a] Université de Paris Sud, 91405 Orsay, France

M. GLASS-MAUJEAN
Laboratoire de Spectroscopie Hertzienne, [b]
Université Pierre et Marie Curie, Paris, France

O. RONCERO
Instituto de Física Fundamental, CSIC, Serrano Madrid, Spain

ABSTRACT. The theory of the degree of alignment and orientation of photofragments angular momenta in the dissociation of polyatomic molecules is presented. It is shown that strong orientation of molecular products can be obtained by simultaneous excitation of electronic states with different helicities and this is the reason for the anomalously high degree of orientation observed in recent experiments on ICN.

---

[a] Laboratoire du CNRS, CEA et MEN
[b] Laboratoire de l'ENS, associé au CNRS

J. Jortner et al. (eds.), Mode Selective Chemistry, 307–322.

1. INTRODUCTION

Molecular fragmentation induced by the absorption of light is one of the most basic processes in photochemistry. A detailed understanding of the half-collision dynamics undergone by the fragments after the photon absorption is necessary for the development of mode selective photochemistry and laser control of half-reactions (see refs. 1-12). The experimental observables can be classified as scalar or vectorial. Among the former we have:

1. Total photodissociation and photoabsorption cross sections;
2. Branching ratios into different final arrangement channels;
3. Final electronic states distributions;
4. Final vibrational and rotational states distributions;
5. Fluorescence quantum yields;
6. Lifetimes;

On the other hand, some of the photoproduct vectorial observables are:

1. Angular distributions of photoproducts;
2. Alignment and orientation;
3. Angular momenta correlations;
4. Polarization of the fluorescence.

From the theoretical point of view there has been growing interest on the study of vector properties as a mean to obtain more detailed information on the dynamics of photodissociation (see refs. 13-15 for general theoretical reviews). Consider for instance the angular distribution which is associated to the spatial direction of the relative velocity $\boldsymbol{v}$ of the fragments. Even for initially randomly oriented molecules the distribution of $\boldsymbol{v}$ is anisotropic because the incident light (through its polarization or its propagation direction) introduces a privileged axis in space. Since there is a $|\boldsymbol{d} \cdot \boldsymbol{e}|^2$ dependence of the photon absorption probability ($\boldsymbol{e}$ being the polarization vector and $\boldsymbol{d}$ the transition dipole moment), light will preferentially excite molecules with $\boldsymbol{d}$ parallel to $\boldsymbol{e}$. Now $\boldsymbol{d}$ has a well defined orientation with respect to the internuclear axis $\boldsymbol{R}$ of the molecule, and in rapid dissociation $\boldsymbol{v}$ will be parallel to $\boldsymbol{R}$. Therefore there will be in general a strong correlation between the $\boldsymbol{e}$-$\boldsymbol{d}$-$\boldsymbol{R}$-$\boldsymbol{v}$ vectors and the angular distribution of the photofragments will be nearly always anisotropic. Determination of this anisotropy provides information on the symmetry of the dissociative molecular states and dissociation lifetimes.

Another consequence of the $|\boldsymbol{d} \cdot \boldsymbol{e}|^2$ dependence of the absorption probability is that the electronic and rotational angular momentum vectors will also be anisotropically distributed in space. This anisotropy corresponds to the so called alignment and orientation and it is due to the $\boldsymbol{e}$-$\boldsymbol{d}$-$\boldsymbol{j}$ correlation. Actually, since $\boldsymbol{j}$ and $\boldsymbol{v}$ are both correlated to $\boldsymbol{d}$, $\boldsymbol{j}$ and $\boldsymbol{v}$ are also strongly correlated with each other independently of the polarization $\boldsymbol{e}$ of the incident light. This has an important effect on the Doppler profile spectra of individual fragment rotational transitions.

The alignment and orientation of fragments by molecular photodissociation also provides an alternative technique to produce polarized atoms and radicals which can be used to study stereochemical effects in collisions with other species. Alignment refers to even spatial moments of the angular momentum vector $\boldsymbol{j}$ distribution, *i.e.* a non-statistical distribution of the populations of the different magnetic sublevels m_j but with positive and negative m_j sublevels having equal populations. Orientation on the other hand, refers to odd moments and hence to distributions where the population of positive m_j sublevels is not equal to that of the corresponding negative m_j. Because of symmetry, orientation can only be obtained by the use of circularly polarized light.

Although the production of spin oriented (spin polarized) electrons using circularly polarized light has been known for many years now (ref. 16), it is only very recently that experiments have been conducted in molecules to produce oriented atoms or radicals. The first evidence for orientation in atomic photofragments has been given by Vasyutinskii (ref. 17). The oriented ground state cesium atoms produced in the photodissociation of CsI molecules by circularly polarized UV radiation were monitored by optical dichroism. More recently, oriented ground state Tl atoms were produced by dissociating TlBr and in this case it has been shown that non adiabatic effects in the excited electronic states have a profound influence on the degree of orientation (ref. 18).

Another consequence of the alignment and orientation of the photofragments is that they will preferentially absorb light of an specific polarization. This has to be taken into account when performing laser-induced fluorescence measurements. Also, when produced in an excited state which fluoresces, the light emitted by the fragments will be polarized as predicted many years ago by van Brunt and Zare (ref. 19).

Recently, the importance in this area of coherence effects due to simultaneous excitation of several continua has been demonstrated (refs. 20-26). One particular striking example is the anomalous polarization of the $Ca^*(^1P)$ fluorescence observed in the photodissociation of Ca_2 (refs. 20-21). This effect is due to the interference between the emission from the coherently populated $\Lambda = \pm 1$ states in the dissociation through a Π electronic manifold. Consider a diatomic molecule AB excited by photon absorption to a dissociative $^1\Pi$ state leading to $A^*(^1P) + B(^1S)$ fragments. Since $\Lambda = \pm 1$ for a Π state, the excited $A^*(^1P)$ fragment will be populated exclusively in the magnetic sublevels $m = \pm 1$ (in the molecular frame) if the dissociation proceeds adiabatically. Thus $A^*(^1P)$ will be fully aligned in the molecular frame. If the photon absorption probability would have been equally probable in all directions in space then, for initially randomly oriented molecules, the alignment would average to zero in the space fixed frame. But actually the absorption probability has a $|\boldsymbol{d} \cdot \boldsymbol{e}|^2$ dependence and $\boldsymbol{d}$ is strongly correlated to the internuclear axis $\boldsymbol{R}$

(for a diatomic molecule $\boldsymbol{d}$ is parallel to $\boldsymbol{R}$ for a $\Sigma \rightarrow \Sigma$ or $\Pi \rightarrow \Pi$ transition, and perpendicular to $\boldsymbol{R}$ for a $\Sigma \rightarrow \Pi$ transition). Hence $A^*(^1P)$ is aligned in the space fixed frame. In addition the simultaneous excitation of the $\Lambda = \pm 1$ states creates a coherent superposition of the $m = \pm 1$ magnetic sublevels and the fluorescence is dramatically affected by the interference between the emissions from these two levels. In the case of Ca_2 for instance, the predicted degree of polarization is 0.78 (in good agreement with the experimental value) but it drops to a merely .14 if the interference terms are not taken into account (ref. 21).

Other coherent excitation effects can also be of importance. Consider the case discussed above namely, a molecule dissociating into $A^*(^1P) + B(^1S)$ fragments. Two molecular singlet states ($^1\Pi$ and $^1\Sigma$) are correlated to this limit. They can both be excited by optical absorption from a $^1\Sigma$ ground state and they will be coherently populated. Since the $^1\Pi$ state produces adiabatically $A^*(^1P, m = \pm 1)$ fragments while the $^1\Sigma$ state produces $A^*(^1P, m = 0)$ fragments, there will be an additional coherence between the $m = \pm 1$ and the $m = 0$ sublevels. As opposed to the case discussed before, this coherence depends on the ratio of the photoabsorption amplitudes to the $^1\Sigma$ and $^1\Pi$ states as well as on the relative phases of their corresponding vibrational continuum wave functions. Recently, a quantum mechanical treatment of H_2 photodissociation in the region of the $C(^1\Pi)$, $B(^1\Sigma)$ and $B'(^1\Sigma)$ states, has shown that these coherence effects can produce pronounced oscillations of the Ly_α fluorescence as a function of the excitation photon energy (ref. 25).

All the examples discussed up to now concern diatomic molecules and therefore the alignment and orientation of the electronic angular momenta only. Recently, the first observation of orientation of a rotational angular momentum of a molecular photofragment has been performed (ref. 27). By photolyzing ICN with circularly polarized light, Zare and coworkers have detected by laser-induced fluorescence the orientation of the CN fragments. For a rotational angular momentum, orientation is equivalent to the existence of a preferred sense of rotation (either clockwise or counterclockwise around the laboratory Z-axis). In the ICN experiment, not only the CN fragment rotational angular momentum has been found to be oriented, but in addition the degree and sign of the orientation changed as a function of the rotational state N being considered. Low N states for instance, have positive orientation but as N increases the the orientation vanishes and then it becomes negative.

It has also been noted (ref. 28) that the degree of orientation found in these experiments for large N correspond to an average value of $|\langle N_z \rangle| \simeq 7$ which is much larger than the unit angular momentum transferred to the molecule by the photon. A model called vibronic angular momentum "amplification" has been proposed (ref. 28) to explain this effect. An electronic dissociative excited state corresponding to a

Σ configuration in the colinear geometry, is coupled by a bending-angle dependent interaction (see also ref. 29) to a Π state. This allows the excitation of a coherent superposition of $v = 0$ and $v = 1$ bending levels which can then produce oriented fragments even for large rotational angular momentum.

Very recently Yabushita and Morokuma (ref. 30) have performed *ab initio* calculations for the ICN molecule in the ground and in several excited electronic states. Based in these calculations they have also proposed that two of their calculated excited electronic states (the $^3\Pi_0^+$ and the $^1\Pi_1$), both of them optically active from the ground state are responsible for all the experimental findings so far. However, the angular dependence of the potential energy surfaces and the nature of the coupling are very different from those proposed before. The two excited surfaces are bent and they undergo a conical intersection in the exit valley outside the Franck-Condon region.

It is clear that a general theoretical treatment taking into account both the coherent effects due to the simultaneous excitation of several excited states and the non adiabatic transitions during the dissociation, is highly desirable (ref. 31). It is the purpose of this paper to present such a theory.

2. GENERAL THEORY

In the framework of the electric dipole approximation the photon excitation dynamics of a molecular system from an initially discret state $|\Psi^{J_iM_i}\rangle$ can be computed from the excitation density operator

$$\rho = \boldsymbol{d}\cdot\boldsymbol{e}^*|\Psi^{J_iM_i}\rangle\langle\Psi^{J_iM_i}|\boldsymbol{d}\cdot\boldsymbol{e} \tag{1}$$

where $\boldsymbol{d}$ is the electronic dipole moment of the system and $\boldsymbol{e}$ is the polarization of the incident light. In order to compute the relevant matrix elements of $\boldsymbol{\rho}$ one needs to specify the basis sets used to describe the final wave functions.

For the case of dissociation in two fragments A and B we shall have

$$\Psi^{JM}_{\alpha j\Omega} = \sum_{\alpha' j'\Omega'} \varphi^{J}_{\alpha j\Omega,\alpha' j'\Omega'}(R)\, W^{JM}_{\alpha' j'\Omega'} \tag{2}$$

where

$$W^{JM}_{\alpha j\Omega} = \sqrt{\frac{2J+1}{4\pi}}\, D^{J*}_{M\Omega}(\phi_R,\theta_R,0)\,\phi_{\alpha j\Omega} \tag{3}$$

with θ_R, ϕ_R being the polar angles of the molecule-fixed $\boldsymbol{R}$ vector (which is the vector joining the center of mass of the two fragments), and $\phi_{\alpha j\Omega}$ a wavefunction of all the degrees of freedom of the fragments. In these equations j denotes the total fragments angular momentum ($\boldsymbol{j} = \boldsymbol{j}_A + \boldsymbol{j}_B$ for the dissociation into two fragments A and B), $\boldsymbol{J}$ the total angular momentum of the system, and M, Ω its projections on the laboratory and the molecular z-axis, respectively. We have $\boldsymbol{J} = \boldsymbol{j} + \boldsymbol{\ell}$, $\boldsymbol{\ell}$ being the orbital angular momentum associated to the relative motion of the fragments (*i.e.* to $\boldsymbol{R}$). We note that J and M are good quantum numbers, while j and Ω are good quantum numbers only in the limit $R \to \infty$. Finally α is a collection of quantum numbers associated to the remaining electronic and vibrational degrees of freedom of the fragments.

The initial state $\Psi^{J_iM_i}$ can be similarly expanded as

$$\Psi^{J_iM_i} = \sum_{\alpha_i j_i\Omega_i} \varphi^{J_i}_{\alpha_i j_i\Omega_i}(R)\, W^{J_iM_i}_{\alpha_i j_i\Omega_i} \tag{4}$$

Using Eqs. (2) and (4), the matrix elements of the electric dipole operator will be given by

$$\langle\Psi^{J_iM_i}|\boldsymbol{d}.\boldsymbol{e}|\Psi^{JM}_{\alpha j\Omega}\rangle = \sum_p (-)^{J_i-M_i+p}(2J_i+1)^{\frac{1}{2}}(\boldsymbol{e})_{-p} \times \begin{pmatrix} J & 1 & J_i \\ M & p & -M_i \end{pmatrix} \langle J_i\|\boldsymbol{d}\|\alpha j\Omega J\rangle \tag{5}$$

where we have defined the reduced matrix elements

$$\langle J_i \| \boldsymbol{d} \| \alpha j \Omega J \rangle = \sum_{\alpha' j' \Omega'} \sum_{\alpha_i j_i \Omega_i} (-1)^{J_i - \Omega_i} (2J+1)^{\frac{1}{2}} \begin{pmatrix} J & 1 & J_i \\ \Omega' & q & -\Omega_i \end{pmatrix} \times \int dR\, \varphi^{J_i *}_{\alpha_i j_i \Omega_i}(R) \, \langle \phi_{\alpha_i j_i \Omega_i} | (\boldsymbol{d})_q | \phi_{\alpha' j' \Omega'} \rangle \, \varphi^{J}_{\alpha j \Omega, \alpha' j' \Omega'}(R) \tag{6}$$

In Eq. (5),

$$(\boldsymbol{e})_0 = (\boldsymbol{e})_Z, \qquad (\boldsymbol{e})_{\pm 1} = \mp \frac{1}{\sqrt{2}} \left[(\boldsymbol{e})_X \pm i (\boldsymbol{e})_Y \right] \tag{7}$$

are the tensorial components of the polarization vector in the space-fixed frame. Thus $p = 0$ for linear polarization and $p = \pm 1$ for circular polarization. In Eq. (6) on the other hand,

$$(\boldsymbol{d})_0 = (\boldsymbol{d})_z, \qquad (\boldsymbol{d})_{\pm 1} = \mp \frac{1}{\sqrt{2}} \left[(\boldsymbol{d})_x \pm i (\boldsymbol{d})_y \right] \tag{8}$$

are the tensorial components of the electric dipole operator in the molecule-fixed frame.

Using Eq. (5) in Eq. (1) and after averaging over M_i assuming the molecule is initially randomly oriented, one obtains

$$\rho_{\alpha j \Omega J M, \alpha' j' \Omega' J' M'} = \sum_K (-)^{K - Q - J_i - M'} (2K+1)\, F_{KQ}(\boldsymbol{e}) \times \begin{Bmatrix} 1 & 1 & K \\ J' & J & J_i \end{Bmatrix} \begin{pmatrix} J & J' & K \\ M & -M' & -Q \end{pmatrix} \langle J_i \| \boldsymbol{d} \| \alpha j \Omega J \rangle^* \langle J_i \| \boldsymbol{d} \| \alpha' j' \Omega' J' \rangle \tag{9}$$

where K can only take the values 0,1 and 2, and

$$F_{KQ}(\boldsymbol{e}) = \sum_{pp'} (-)^p (\boldsymbol{e})^*_{-p} (\boldsymbol{e})_{-p'} \begin{pmatrix} 1 & 1 & K \\ -p' & p & Q \end{pmatrix} = \sum_{pp'} (\boldsymbol{e})_p (\boldsymbol{e}^*)_{p'} \begin{pmatrix} 1 & 1 & K \\ p & p' & Q \end{pmatrix} \tag{10}$$

is a function which depends only on the polarization vector $\boldsymbol{e}$. For linearly polarized light along Z for instance, $p = p' = 0$ and F_{KQ} is different from zero only for $K = 0$ and $K = 2$.

The next step is to transform the excitation density matrix to the space-fixed frame where the observables are measured. In the space-fixed reference system the angular momenta basis set is defined as

$$W^{\ell m_\ell}_{\alpha j m_j} = Y_{\ell m_\ell}(\theta_R, \phi_R) \phi_{\alpha j m_j} \tag{11}$$

where the $Y_{\ell m_\ell}(\theta_R, \phi_R)$ functions are spherical harmonics. In Eq. (11), the $\phi_{\alpha j m_j}$ wave functions refer now to the space-fixed system. Between the molecule-fixed $W^{JM}_{\alpha j \Omega}$ and the space fixed $W^{\ell m_\ell}_{\alpha j m_j}$ basis functions there is the transformation

$$W_{\alpha j m_j}^{\ell m_\ell} = \sum_{\Omega JM} \mathcal{T}_{\Omega JM, j m_j \ell m_\ell} W_{\alpha j \Omega}^{JM} \tag{12}$$

with

$$\mathcal{T}_{\Omega JM, j m_j \ell m_\ell} = (-)^{M-\Omega}(2\ell+1)^{\frac{1}{2}}(2J+1)^{\frac{1}{2}} \begin{pmatrix} j & \ell & J \\ \Omega & 0 & -\Omega \end{pmatrix} \begin{pmatrix} j & \ell & J \\ m_j & m_\ell & -M \end{pmatrix} \tag{13}$$

The reduced excitation density matrix to produce a particular j level in the space-fixed frame can now be obtained using the relationship

$$\rho_{m_j,m_j'}^{(\alpha j)} = \sum_{\Omega JM} \sum_{\Omega' J' M'} \mathcal{T}_{\Omega JM, j m_j \ell m_\ell}\ \rho_{\alpha j \Omega JM, \alpha j \Omega' J' M'}\ \mathcal{T}_{\Omega' J' M', j m_j' \ell m_\ell} \tag{14}$$

Using Eqs. (13) and (9) into (14) and performing the sum over ℓ and m_ℓ, one obtains

$$\rho_{m_j,m_j'}^{(\alpha j)} = \sum_K (-)^{j+m_j}(2K+1) \begin{pmatrix} j & j & K \\ m_j' & -m_j & Q \end{pmatrix} P_K^{(\alpha j)} F_{KQ}(\boldsymbol{e}) \tag{15}$$

with

$$P_K^{(\alpha j)} = \sum_{\Omega J} \sum_{\Omega' J'} (-)^{j+J_i} (2J+1)^{\frac{1}{2}} (2J'+1)^{\frac{1}{2}} \begin{pmatrix} J & J' & K \\ \Omega & -\Omega' & Q' \end{pmatrix} \begin{pmatrix} j & j & K \\ -\Omega' & \Omega & Q' \end{pmatrix}$$
$$\times \begin{Bmatrix} 1 & 1 & K \\ J' & J & J_i \end{Bmatrix} \langle J_i \| \boldsymbol{d} \| \alpha j \Omega J \rangle^* \langle J_i \| \boldsymbol{d} \| \alpha j \Omega' J' \rangle \tag{16}$$

The excitation density matrix Eq. (15) provide all the elements to calculate the experimental observables we are interested in. The dissociation probability into fragments in state (αj) for instance, is clearly related to the total population of channel (αj)

$$\mathcal{W}^{(\alpha j)} = \sum_m \rho_{m,m}^{(\alpha j)} \tag{17}$$

from which the partial photodissociation cross section into channel (αj) can be computed using $\sigma^{(\alpha j)} = (4\pi^2\omega/c)\mathcal{W}^{(\alpha j)}$ with ω being the frequency of the incident photon. Similarly, the orientation $\mathcal{O}$ and alignment $\mathcal{A}$ (which in the literature are also denoted $A_0^{\{1\}}$ and $A_0^{\{2\}}$, respectively) can be defined as

$$\mathcal{O}^{(\alpha j)} \equiv \langle (j_z/|\boldsymbol{j}|) \rangle = \sum_m \frac{m}{\sqrt{j(j+1)}} \rho_{m,m}^{(\alpha j)} / \mathcal{W}^{(\alpha j)} \tag{18a}$$

$$\mathcal{A}^{(\alpha j)} \equiv \langle (3j_z^2/j^2) - 1 \rangle = \sum_m \left(\frac{3m^2}{j(j+1)} - 1 \right) \rho_{m,m}^{(\alpha j)} / \mathcal{W}^{(\alpha j)} \tag{18b}$$

Using Eq. (15) into Eqs. (17) and (18), one obtains

$$\mathcal{W}^{(\alpha j)} = -\sqrt{(2j+1)/3}\, P_0^{(\alpha j)} \tag{19a}$$

$$\mathcal{O}^{(\alpha j)} = p \frac{P_1^{(\alpha j)}}{\sqrt{2}\, P_0^{(\alpha j)}} \tag{19b}$$

$$\mathcal{A}^{(\alpha j)} = (3p^2 - 2)\left[\frac{(2j+3)(2j-1)}{10j(j+1)}\right]^{\frac{1}{2}} \frac{P_2^{(\alpha j)}}{P_0^{(\alpha j)}} \tag{19c}$$

where $p = 0$ for linear polarization and $p = \pm 1$ for circular polarization. From an inspection of Eqs. (19) we note that $P_K^{(\alpha j)}$ is directly related to the population ($K = 0$), orientation ($K = 1$), and alignment ($K = 2$) of the fragments. It is also clear that orientation cannot be obtained with linearly polarized light as expected from the axial symmetry of the photon field. On the other hand, orientation can be obtained by the use of circularly polarized light except for a final state with $\Omega = \Omega' = 0$ since one of the 3-j coefficients in Eq. (16) vanishes in that case.

3. APPLICATION TO A TRIATOMIC CASE

Let consider a triatomic molecule ABC dissociating into A + BC fragments. The total angular momentum of the fragments $\boldsymbol{j}$ can be written as $\boldsymbol{j} = \boldsymbol{j}_A + \boldsymbol{j}_{BC}$. For simplicity, we shall assume here that the BC fragment is in a Σ state and that its rotational angular momentum without spin $\boldsymbol{N}$ is sufficiently large that the identification $j \equiv N$ is a good approximation. Since the BC fragments are assumed to be in a Σ state, the ϕ fragments wavefunctions defined in Eq. (3) can be written as the product $Y_{N\Omega_N}(\theta,\phi)|\Omega_e\rangle$ where the (θ,ϕ) are the polar angles of the BC internuclear axis in the molecule-fixed frame, and $|\Omega_e\rangle$ is the electronic (plus BC vibrational) fragments wavefunction. Thus $\alpha \equiv \Omega_e$ in our equations above. The final dissociative eigenstates Eq. (2) can then be rewritten as

$$\Psi^{JM}_{\Omega_e N\Omega} = \sum_{\Omega'_e N'\Omega'} \varphi^{J}_{\Omega_e N\Omega,\Omega'_e N'\Omega'}(R)\, W^{JM}_{\Omega'_e N'\Omega'} \tag{20}$$

with

$$W^{JM}_{\Omega_e N\Omega} = \sqrt{\frac{2J+1}{4\pi}} D^{J*}_{M\Omega}(\phi_R,\theta_R,0)\, Y_{N(\Omega-\Omega_e)}(\theta,\phi)\, |\Omega_e\rangle \tag{21}$$

The $\varphi^{J}_{\Omega_e N\Omega,\Omega'_e N'\Omega'}(R)$ functions in Eq. (20) are obtained by solving the system of coupled equations

$$\begin{aligned}&\left[-\frac{\hbar^2}{2m}\frac{\partial^2}{\partial R^2} - \frac{\hbar^2\left[J(J+1)+N(N+1)-2\Omega^2\right]}{2mR^2} + V^{J}_{\Omega_e N\Omega} - E\right]\varphi^{J}_{\Omega_e N\Omega,\Omega_e N\Omega} = \\ &-\sum_{\Omega'_e N'} V^{J}_{\Omega_e N,\Omega'_e N'}\,\varphi^{J}_{\Omega_e N\Omega,\Omega'_e N'\Omega} - \sum_{\Omega'\neq\Omega} \langle W^{JM}_{\Omega_e N\Omega}|\frac{\boldsymbol{\ell}^2}{2mR^2}|W^{JM}_{\Omega_e N\Omega'}\rangle \varphi^{J}_{\Omega_e N\Omega,\Omega_e N\Omega'}\end{aligned} \tag{22}$$

where we have denoted by $V^{J}_{\Omega_e N\Omega}(R)$ and $V^{J}_{\Omega_e N,\Omega'_e N'}(R)$ the potential energy curves and couplings, respectively. In Eq. (22) m is the reduced mass for the relative motion of the fragments and $E = E_i + \hbar\omega$ the final total energy. The $\varphi^{J}_{\Omega_e N\Omega,\Omega'_e N'\Omega'}(R)$ functions are specified by imposing the asymptotic boundary condition

$$\begin{aligned}\varphi^{J}_{\Omega_e N\Omega,\Omega'_e N'\Omega'}(R\to\infty) \simeq \left(\frac{m}{2\pi k\hbar^2}\right)^{\frac{1}{2}} &\left[\frac{e^{ikR}}{R}\delta_{\Omega_e\Omega'_e}\delta_{NN'}\delta_{\Omega\Omega'}\right. \\ &\left. + \left(\frac{k}{k'}\right)^{\frac{1}{2}} S^{J*}_{\Omega_e N\Omega,\Omega'_e N'\Omega'}\frac{e^{-ik'R}}{R}\right]\end{aligned} \tag{23}$$

where $\boldsymbol{S}$ is the scattering matrix, and

$$k = \sqrt{2m(E - E_{\Omega_e N\Omega})}/\hbar \tag{24}$$

with $E_{\Omega_e N\Omega}$ being the asymptotic threshold energy. Actually it is useful in order to reduce the number of coupled equations to be solved, to define parity adapted wavefunctions of the form

$$W^{JM\epsilon}_{\Omega_e N\Omega} = \frac{\sqrt{2-\delta_{0\Omega}\delta_{0\Omega_e}}}{2}\left\{W^{JM}_{\Omega_e N\Omega} + \epsilon(-1)^{J+\epsilon'} W^{JM}_{-\Omega_e N-\Omega}\right\} \tag{25}$$

where ϵ is the total parity and ϵ' the intrinsic electronic parity. However, for the clarity of the general presentation we shall keep the notation with positive and negative values of Ω and Ω_e.

The initial state $|\Psi^{J_i M_i}\rangle$ can be similarly written as

$$\Psi^{J_i M_i} = \sum_{N_i\Omega_i} \varphi^{J_i}_{N_i\Omega_i}(R)\, W^{J_i M_i}_{0N_i\Omega_i} \tag{26}$$

where we have assumed that the ground electronic state is described by only one electronic state (denoted below by $|X\rangle$) with $\Omega_e = 0$.

Using this notation we shall have from Eqs. (19)

$$\mathcal{W}^{(\Omega_e N)} = -\sqrt{(2N+1)/3}\, P_0^{(\Omega_e N)} \tag{27a}$$

$$\mathcal{O}^{(\Omega_e N)} = p\frac{P_1^{(\Omega_e N)}}{\sqrt{2}\, P_0^{(\Omega_e N)}} \tag{27b}$$

$$\mathcal{A}^{(\Omega_e N)} = (3p^2-2)\left[\frac{(2N+3)(2N-1)}{10N(N+1)}\right]^{\frac{1}{2}} \frac{P_2^{(\Omega_e N)}}{P_0^{(\Omega_e N)}} \tag{27c}$$

The multipole coefficients $P_K^{(\Omega_e N)}$ defined by Eq. (16) are then given by

$$P_K^{(\Omega_e N)} = \sum_{\Omega J}\sum_{\Omega' J'}(-)^{N+J_i}(2J+1)^{\frac{1}{2}}(2J'+1)^{\frac{1}{2}}\begin{pmatrix} J & J' & K \\ \Omega & -\Omega' & Q' \end{pmatrix}\begin{pmatrix} N & N & K \\ -\Omega' & \Omega & Q' \end{pmatrix}$$
$$\begin{Bmatrix} 1 & 1 & K \\ J' & J & J_i \end{Bmatrix}\langle J_i\|\boldsymbol{d}\|\Omega_e N\Omega J\rangle^*\langle J_i\|\boldsymbol{d}\|\Omega_e N\Omega' J'\rangle \tag{28}$$

with

$$\langle J_i\|\boldsymbol{d}\|\Omega_e N\Omega J\rangle = \sum_{\Omega'_e N'\Omega'}(2J+1)^{\frac{1}{2}}\begin{pmatrix} J & 1 & J_i \\ \Omega' & -\Omega'_e & \Omega'_e-\Omega' \end{pmatrix}\mathcal{D}_{\Omega'_e}$$
$$\times\, \delta_{N_i N'}\,\delta_{\Omega_i(\Omega'-\Omega'_e)}\int dR\, \varphi^{J_i *}_{N_i\Omega_i}(R)\,\varphi^{J}_{\Omega_e N\Omega,\Omega'_e N'\Omega'}(R) \tag{29}$$

where we have assumed the Franck-Condon approximation and defined by

$$\mathcal{D}_{\Omega_e} = \langle X|(\boldsymbol{d})_{-\Omega_e}|\Omega_e\rangle \tag{30}$$

the electronic transition dipole moment between the ground state $|X\rangle$ and the excited dissociative states $|\Omega_e\rangle$.

We shall now consider a particularly simple case corresponding to an initial state with $J_i = 0$. Since the transition to $J = 0$ is forbidden in this case, we shall have $J = 1$ in the excited manifold with two possible values of Ω (0 and 1). Usually, for fast dissociation it is a very good approximation to neglect the Coriolis coupling between the $\Omega = 0$ and $\Omega = 1$ states induced by the ℓ^2 term in Eq. (22). It should be noted that in the framework of the Coriolis decoupling approximation, the sum over Ω' in Eq. (29) reduces to only one term with $\Omega' = \Omega$. Since in addition $\Omega_i = 0$, we have $\Omega_e' = \Omega$ and this implies that the $\Omega = 0$ calculation corresponds to the dissociation dynamics associated to the parallel optical transition $\mathcal{D}_0$ while the $\Omega = 1$ calculation deals with the perpendicular transition $\mathcal{D}_1$ only. This does not necessarily implies that the final Ω_e is unique for each one of these calculations. Since there could be electronic transitions between the two surfaces, an initial parallel transition may produce both $\Omega_e = 0$ as well as $\Omega_e = 1$ channels

From Eqs. (27), (28) and (29) with $J_i = 0$ and the Coriolis decoupling approximation, we get

$$\mathcal{W}^{(\Omega_e N)} = \frac{1}{3}\left\{\left|\mathcal{M}_0^{(\Omega_e N)}\right|^2 + 2\left|\mathcal{M}_1^{(\Omega_e N)}\right|^2\right\} \tag{31}$$

$$\mathcal{O}^{(\Omega_e N)} = p\frac{\left\{\left|\mathcal{M}_1^{(\Omega_e N)}\right|^2/\sqrt{N(N+1)} - \left(\mathcal{M}_0^{(\Omega_e N)*}\,\mathcal{M}_1^{(\Omega_e N)} + c.c.\right)/\sqrt{2}\right\}}{\left\{\left|\mathcal{M}_0^{(\Omega_e N)}\right|^2 + 2\left|\mathcal{M}_1^{(\Omega_e N)}\right|^2\right\}} \tag{32}$$

and

$$\mathcal{A}^{(\Omega_e N)} = \frac{(3p^2-2)}{5}\left\{\left|\mathcal{M}_0^{(\Omega_e N)}\right|^2 + \left(2 + \frac{3}{N(N+1)}\right)\left|\mathcal{M}_1^{(\Omega_e N)}\right|^2 - \frac{3}{\sqrt{2N(N+1)}}\left(\mathcal{M}_0^{(\Omega_e N)*}\mathcal{M}_1^{(\Omega_e N)} + c.c.\right)\right\} \Big/ \left\{\left|\mathcal{M}_0^{(\Omega_e N)}\right|^2 + 2\left|\mathcal{M}_1^{(\Omega_e N)}\right|^2\right\} \tag{33}$$

with

$$\mathcal{M}_\Omega^{(\Omega_e N)} = \delta_{\Omega_e',\Omega}\,\mathcal{D}_{\Omega_e'}\sum_{N'}\int dR\;\varphi^{(J_i=0)*}_{(N_i=N')(\Omega_i=0)}(R)\;\varphi^{J}_{\Omega_e N\Omega,\Omega_e' N'\Omega}(R) \tag{34}$$

and where we have assumed that the transition dipole moment lies on the plane of the molecule (*i.e.*, that $\mathcal{M}_\Omega^{\Omega_e N} = \mathcal{M}_{-\Omega}^{-\Omega_e N}$).

Several comments are now in order concerning Eq. (32). When only the $\Omega = 1$ state is excited

$$\mathcal{O}^{(N)} = p\frac{1}{2\sqrt{N(N+1)}} \tag{35}$$

and the orientation is maximum for $N = 1$ with $\mathcal{O} = 0.5/\sqrt{2}$. For large N the orientation diminishes as $1/N$ in this case. On the other hand when both $\Omega = 0$ and $\Omega = 1$ states are coherently excited, the degree of orientation for large N will be given by

$$\mathcal{O}^{(N)} = \frac{r + r^*}{\sqrt{2}(2 + |r|^2)} \qquad \text{with}\, r = \mathcal{M}_1/\mathcal{M}_0 \tag{36}$$

and we note that in this case the orientation can be as large as $\mathcal{O} = 0.5$, for $r = \sqrt{2}$. This corresponds to a large angular momentum transfer in the z-direction. From the definition of the orientation given in Eq. (18), we have

$$\langle N_z \rangle = \mathcal{O}^{(N)} \sqrt{N(N+1)} \tag{37}$$

and therefore, if $\mathcal{O} = 0.5$ we get $\langle N_z \rangle$ much larger than one if N is large. From the above considerations we conclude that the large degree of orientation of the CN fragments in the photodissociation of ICN (ref. 27) is the result of coherent excitation of the $^3\Pi_0^+$ and $1\Pi_1$ dissociative states calculated by Yabushita and Morokuma (ref. 30).

4. CONCLUSIONS

We have presented the general theory of the dynamics of photodissociation. Using the excitation density matrix formalism, we have obtained the final expressions for both scalar as well as vector properties such as the alignment and orientation of photofragments. This formalism can be applied to diatomics as well as to polyatomic molecules dissociating into two fragments. We have shown that coherent excitation of several continua can produce significant effects on the degree of alignment and orientation of the fragments. In particular, they provide a simple explanation of the large degree of orientation of CN fragments observed in the experiments on ICN (ref. 27).

REFERENCES

1 A.H. Zewail, in Proceeedings of Faraday Symposium 24, J. Chem. Soc. Far. Trans. II, Molecular and Chemical Physics, 85 (1989) 1221-1242.
2 R.B. Bernstein, S.E. Choi, and S. Stolte, in ref. 1 pp. 1097-1113.
3 T.J. Butenhoff, K.L. Carleton, M-C. Chuang, C.B. Moore, in ref. 1 pp. 1155-1169.
4 M.P. Docker, A. Ticktin, U. Bruhlmann, and J.R. Huber, in ref. 1 pp. 1169-1184.
5 R. Ogorzalek Loo, C.E. Strauss, H-P. Haerri, G.E. Hall, P.L. Houston, I. Burak, and J.W. Hepburn, in ref. 1 pp. 1185-1206.
6 M. Brouard, M.T. Martinez, J. O'Mahony, and J.P. Simons, in ref. 1 pp. 1207-1220.
7 J.F. Black, J.R. Waldeck, E. Hasselbrink, and R.N. Zare, in ref. 1 pp. 1044-1047.
8 I. Nenner and J.A. Beswick, in G.V. Marr (Editor), Handbook of Synchrotron Radiation, Vol. 2, North Holland, 1987, p. 355.
9 P. Houston, J. Phys. Chem., 91 (1987) 5388-5397.
10 P. Andresen and R. Schinke, and several other contributions in M.N.R. Ashfold and J.E. Baggott (Editors), Molecular Photodissociation Dynamics, Advances in Gas-Phase Photochemistry and Kinetics, Royal Soc. of Chem., London, 1987.
11 W.M. Jackson and H. Okabe, in D.H. Volman, K. Gollnick, and G.S. Hammond (Editors), Advances in Photochemistry, Vol. 13, Wiley, New York, 1986.
12 S. Buclow, M. Noble, G. Radharderishnan, H. Reisler, C. Wittig, and G. Hancock, J. Phys. Chem., 90 (1986) 1015.
13 C.H. Greene and R.N. Zare, Annu. Rev. Phys. Chem., 33 (1982) 119.
14 S.J. Singer, K.F. Freed, and Y.B. Band, Adv. Chem. Phys., 61 (1985) 1.
15 M. Shapiro and G.G. Balint-Kurti, in K.P. Lawley (Editor), Photodissociation and Photoionization, Wiley, New York, 1985.
16 N.A. Cherepkov, Adv. Atom. Mol. Phys., 19 (1983) 395.
17 O.S. Vasyutinskii, Sov. Phys. JETP Lett., 31 (1980) 428.
18 O.S. Vasyutinskii, Z. Phys. D, 15 (1990), 105-115.
19 R.J. Van Brunt and R.N. Zare, J. Chem. Phys., 48 (1968) 4304.
20 J. Vigué, P. Grangier, G. Roger, and A. Aspect, J. Phys. Lett., 42 (1981) L531.
21 J. Vigué, J.A. Beswick, and M. Broyer, J. Phys. (Paris), 44 (1983) 1225.
22 S.J. Singer, K.F. Freed, and Y.B. Band, J. Chem. Phys., 72 (1983) 6060.
23 S.J. Singer, K.F. Freed, and Y.B. Band, J. Chem. Phys., 84 (1986) 3762.
24 M. Glass-Maujean and J.A. Beswick, Phys. Rev., A36 (1987) 1160.
25 M. Glass-Maujean and J.A. Beswick, Phys. Rev., A38 (1988) 5660.
26 M. Glass-Maujean and J.A. Beswick, in ref. 1 pp. 983-1002.
27 E. Hasselbrink, J.R. Waldeck, and R.N. Zare, Chem. Phys., 126 (1988) 191.
28 J. Vigué, B. Girard, G. Gouédard, and N. Billy, Phys. Rev. Lett., 62 (1989) 1358.

29 R.N. Dixon, Mol. Phys., 68 (1989) 263.
30 S. Yabushita and K. Morokuma, Chem. Phys. Lett., 175 (1990) 518.
31 J.A. Beswick, M. Glass-Maujean, and O. Roncero, to be published.
32 R.N. Zare, Angular Momentum: Understanding Spatial Aspects in Chemistry and Physics, Addison Wesley, 1988.

QUANTUM INTERFERENCE AND THE CONTROL OF MOLECULAR PROCESSES

PAUL BRUMER
Chemical Physics Theory Group
Department of Chemistry
University of Toronto
Toronto, Ontario M5S 1A1, Canada

and

MOSHE SHAPIRO
Chemical Physics Department
Weizmann Institute of Science
Rehovot, Israel

ABSTRACT. We describe the current status of coherent radiative control, a quantum-interference based approach to controlling molecular processes by the use of coherent radiation. Emphasis is on providing an overview of proposed laboratory scenarios, ongoing experimental studies and recent theoretical developments.

1. Introduction

Control over the yield of chemical reactions is the ***raison d'etre*** of practical chemistry and the ability to use lasers to achieve this goal is one primary thrust of modern Chemical Physics. The theory of coherent radiative control of chemical reactions, which we have developed over the last five years [1-16], affords a direct method for controlling reaction dynamics using coherence properties of weak lasers, with a large range of yield control expected in laboratory scenarios. In addition, the theory of coherent control provides deep insights into the essential features of reaction dynamics, and of quantum interference, which are necessary to achieve control over elementary chemical processes. Below we provide a schematic overview of coherent radiative control including an example which emphasizes the general principles (Section II), a summary of several experimental scenarios which implement the principles of coherent control (Section III) and a summary of the current status of theory and experiment. Readers seeking a qualitative introduction to coherent control should consult reference [9]. An alternative approach, based upon the use of shaped light pulses has been advocated by Tannor and Rice [17] and Rabitz[18].

J. Jortner et al. (eds.), Mode Selective Chemistry, 323–331.

II. Interference and Control: An Example

Consider a chemical reaction which, at total energy E produces a number of distinct products. The total Hamiltonian is denoted $H = H_q^0 + V_q$, where H_q^0 is the Hamiltonian of the separated products in the arrangement channel labeled by $q, (q = 1, 2, ...)$ and V_q is the interaction between products in arrangement q. We denote eigenvalues of H_0^q by $|E, n, q^0\rangle$, where n denotes all quantum numbers other than E; eigenfunctions of H, which correlate with $|E, n, q^0\rangle$ at large product separation, are labeled $|E, n, q^-\rangle$. By definition of the minus states, a state prepared experimentally as as superposition $|\Psi(t = 0)\rangle = \sum_{n,q} c_{n,q}|E, n, q^-\rangle$ has probability $|c_{n,q}|^2$ of forming product in channel q, with quantum numbers n. As a consequence, control over the probability of forming a product in any asymptotic state is equal to the probability of initially forming the appropriate minus state which correlates with the desired product. The essence of control lies, therefore, in forming the desired linear combination at the time of preparation. The essential *modus operandi* of coherent radiative control is to utilize phase and intensity properties of laser excitation to alter the character of the prepared state so as to enhance production of the desired product.

Many of the proposed coherent control scenarios rely upon a simple way of achieving active control over the prepared state and product. Specifically, active control over products is achieved by driving an initially pure molecular state through two or more independent coherent optical excitation routes. The resultant product probability displays interference terms between these two routes, whose magnitude and sign depend upon laboratory parameters. As a consequence, product probabilities can be manipulated directly in the laboratory.

The scheme outlined above has, as a well-known analogy, the interference between paths as a beam of either particles or of light passes through a double slit. In that instance source coherence leads to either constructive or destructive interference, manifest as patterns of enhanced or reduced probabilitites on an observation screen. In the case of coherent control the overall coherence of a pure state plus laser source allows for the constructive or destructive manipulation of probabilities in product channels.

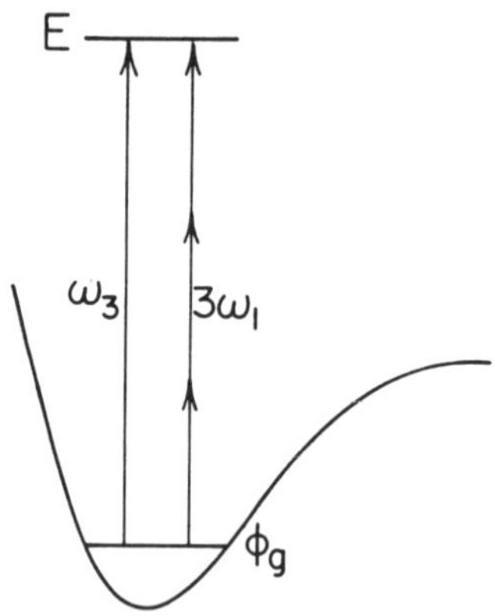

Figure 1: One-photon plus three-photon coherent control scenario.

We note, as an aside relevant to the present conference, that practitioners of "mode-selective chemistry" often advocate the use of lasers to excite specific bonds or modes in the reactant molecule in order to drive a reaction in a desired direction. The brief remarks above, elaborated upon elsewhere[10], make clear that the proper modes to excite in order to produce product in arrangement q are the eigenfunctions $|E,n,q^-\rangle$, the system's "natural modes", if you will. Thus the extent to which excitation of some zeroth order state $|\chi\rangle$ is successful in promoting reaction to the desired product q depends entirely on the extent to which $|\chi\rangle$ overlaps $|E,n,q^-\rangle$.

Consider, as an example of coherent control, a specific scenario for unimolecular photoexcitation[5,13] (Figure 1) where a system, initially in pure state ϕ_g (or $|E_i\rangle$ below), is excited to energy E, by simultaneous application of two CW frequencies ω_3 and $3\omega_1$ ($\omega_3 = 3\omega_1$), providing two independent optically driven routes from ϕ_g to $|E,n,q^-\rangle$.

Straightforward perturbation theory, valid for the weak fields under consideration, gives the probability $P(E,q;E_i)$ of forming product at energy E in arrangement q as:

$$P(E,q;E_i) = P_3(E,q;E_i) + P_{13}(E,q;E_i) + P_1(E,q;E_i)\,. \tag{1}$$

with terms defined as follows:

$$P_3(E,q;E_i) = (\frac{\pi}{\hbar})^2 \epsilon_3^2 \sum_n |\langle E,n,q^- \,|\, (\hat{\underset{\sim}{\epsilon}}_3 \cdot \underset{\sim}{\mu})_{e,g} \,|\, E_i \rangle|^2\,. \tag{2}$$

Here $\underset{\sim}{\mu}$ is the electric dipole operator, and

$$(\hat{\underset{\sim}{\epsilon}}_3 \cdot \underset{\sim}{\mu})_{e,g} = \langle e \,|\, \hat{\underset{\sim}{\epsilon}}_3 \cdot \underset{\sim}{\mu} \,|\, g \rangle\,, \tag{3}$$

where $|\,g\,\rangle$ and $|\,e\,\rangle$ are the ground and excited electronic state wavefunctions, respectively. The second and third terms in Eq. (1) are

$$P_1(E,q;E_i) = (\frac{\pi}{\hbar})^2 \epsilon_1^6 \sum_n |\langle E,n,q^- \,|\, T \,|\, E_i \rangle|^2\,, \tag{4}$$

with

$$T = (\hat{\underset{\sim}{\epsilon}}_1 \cdot \underset{\sim}{\mu})_{e,g}(E_i - H_g + 2\hbar\omega_1)^{-1}(\hat{\underset{\sim}{\epsilon}}_1 \cdot \underset{\sim}{\mu})_{g,e}(E_i - H_e + \hbar\omega_1)^{-1}(\hat{\underset{\sim}{\epsilon}}_1 \cdot \underset{\sim}{\mu})_{e,g}\,. \tag{5}$$

and

$$P_{13}(E,q;E_i) = -2(\frac{\pi}{\hbar})^2 \epsilon_3 \epsilon_1^3 \cos(\theta_3 - 3\theta_1 + \delta_{13}^{(q)})|F_{13}^{(q)}| \tag{6}$$

with the amplitude $|F_{13}^{(q)}|$ and phase $\delta_{13}^{(q)}$ defined by

$$|F_{13}^{(q)}|\exp(i\delta_{13}^{(q)}) = \sum_n \langle E_i \,|\, T \,|\, E,n,q^- \rangle\langle E,n,q^- \,|\, (\hat{\underset{\sim}{\epsilon}}_3 \cdot \underset{\sim}{\mu})_{e,g} \,|\, E_i \rangle\,. \tag{7}$$

The branching ratio $R_{qq'}$ for channels q and q', can then be written as

$$R_{qq'} = \frac{P(E,q;E_i)}{P(E,q';E_i)} = \frac{\epsilon_3^2 F_3^{(q)} - 2\epsilon_3\epsilon_1^3\cos(\theta_3 - 3\theta_1 + \delta_{13}^{(q)})|F_{13}^{(q)}| + \epsilon_1^6 F_1^{(q)}}{\epsilon_3^2 F_3^{(q')} - 2\epsilon_3\epsilon_1^3\cos(\theta_3 - 3\theta_1 + \delta_{13}^{(q')})|F_{13}^{(q')}| + \epsilon_1^6 F_1^{(q')}}\,, \tag{8}$$

where

$$\begin{aligned} F_3^{(q)} &= (\tfrac{\hbar}{\pi})^2 \frac{P_3(E,q;E_i)}{\epsilon_3^2} , \\ F_1^{(q)} &= (\tfrac{\hbar}{\pi})^2 \frac{P_1(E,q;E_i)}{\epsilon_1^6} , \end{aligned} \tag{9}$$

with $F_3^{(q')}$ and $F_1^{(q')}$ defined similarly.

The numerator and denominator of Eq. (8) each display what we regard as the canonical form for coherent control. That is, independent contributions from two routes, modulated by an interference term. Since the interference term is controllable through variation of laboratory parameters (here the relative intensity and relative phase of the two lasers), so too is the product ratio R.

This 3-photon + 1-photon scenario has now been experimentally implemented[19] in studies of HCl ionization through a resonant bound Rydberg state. Specifically, HCl is excited to a selected rotational state in the $^3\Sigma^-(\Omega^+)$ manifold using $\omega_1 = 336$ nm; ω_3 is obtained by third harmonic generation in a Krypton gas cell. The relative phase of the light fields was then varied by passing the beams through a second Krypton cell and varying the cell gas pressure. The population of the resultant Rydberg state was interrogated by ionizing to HCl^+ with an additional photon. This REMPI type experiment showed that the HCl^+ ion probability depended upon both the relative phase and intensity of the two exciting lasers, in accord with the theory described above. Note that in this case, control is achieved by simultaneous excitation to a nondegenerate bound state. As a consequence there is no sum over n in Eqs. (2), (4) and (7) and hence $|F_3^{(q)} F_1^{(q)}| = |F_{13}^{(q)}|^2$. Satisfaction of this Schwartz equality suffices to ensure that the probability of forming the HCl Rydberg state can be varied over the full range of zero to unity[1]. Specifically, setting $\theta_3 - 3\theta_1 = -\delta_{13}^{(q)}$ and $\epsilon_3^2/\epsilon_1^6 = F_1^{(q)}/F_3^{(q)}$ gives zero probability in channel q.

A similar phase control experiment has been performed on atoms[20], in the simultaneous 3-photon + 5-photon ionization of Hg. Although the effect of the relative laser intensity was not studied, the Hg^+ ionization probability was shown to be a function of relative phase of the two lasers. Since in this case excitation is nonresonant, this experiment shows the feasibility of control over continuum channels via interfering optical excitation routes.

These experiments clearly show that coherent control of simple molecular processes through quantum interference of multiple optical excitation routes is both feasible and experimentally observable. Further experimental studies designed to show control over processes with more than one product channel are in progress by a number of experimental groups. Further, our recent computational results [13] on photodissociation of IBr, to produce I + Br and I + Br* suggest that such experiments should show huge control over the product ratio. For example, 1-photon plus 3-photon photodissociation of the ground vibrotational state in the ground electronic state of IBr showed variations in the Br*/Br ratio of 45% to 95% with variations of the relative laser intensity and phase. Similar results were obtained for higher J, including $J = 42$ where extensive M_J averaging was required.

3. The Essential Principle and Assorted Coherent Control Scenarios

The 3-photon plus 1-photon case is but one example of a scenario which embodies the essential principle of coherent control, i.e. that ***coherently driving a pure state through multiple optical excitation routes to the same final state allows for the possibility of control.*** Given this general principle, numerous scenarios may be proposed to obtain control in the laboratory. Such proposals must, however, properly account for a number of factors which reduce or eliminate control. Amongst these are the need to (a) adhere to selection rule requirements, (b) minimize extraneous and parasitic uncontrolled satellites, (c) insure properly treated laser spatial dependence and phase jitter and (d) allow maintainance of coherence over time scale associated with any true relaxation processes, e.g. collisions. Briefly:

(a) control requires that the interference term (e.g. $F_{13}^{(q)}$ above) arising between optical routes is nonzero. In general, this is the case only if the various optical excitation routes satisfy specific rules regarding conserved integrals of motion. For example[21], consider the 1-photon + 3-photon case discussed above. Here excitation from, e.g. $J = 0$, where J is the rotational angular momentum, yields $J = 1$ for the 1-photon route and $J = 1$ and $J = 3$ for the three photon route. In this case selection rules are such that only final states with the same J, here $J = 1$, interfere;

(b) the $J = 3$ state created by three photon absorption in this scenario is an example of a satellite state, i.e. an uncontrolled (and hence undesired) product state arising in the course of the photoexcitation. Effective scenarios must insure that such contributions are small compared to the controlled component;

(c) since the ability to accurately manipulate the laser relative phase is crucial to coherent control, one must account for all laboratory features which affect the phase. For example, scenarios must be designed so as to eliminate effects due to the $\mathbf{k}\cdot\ \mathbf{R}$ spatial dependence of the laser phase. Not doing so would results in the dimunition of control resulting from the variation of this term over the molecule beam-laser beam intersection volume.

(d) proposed scenarios must maintain phase coherence in the face of possibly dephasing effects (e.g. collisions or partial laser coherence). Studies indicate[7,22] that control can survive moderate levels of such phase destructive processes.

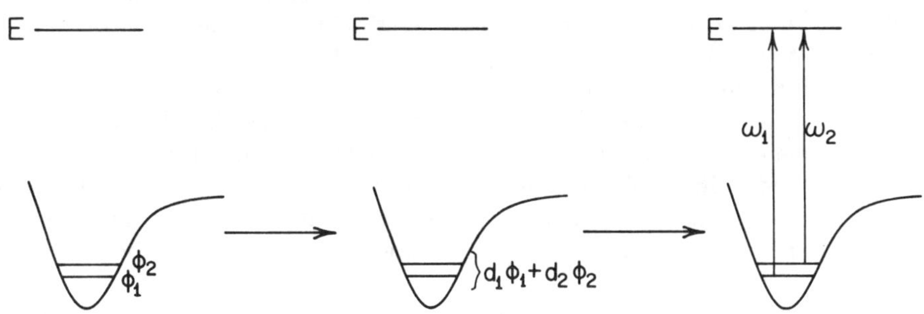

Figure 2: Basic control scenario.

We have, thus far, proposed a number of different scenarios for coherent control.

Below we comment on some of them, with the intent of providing only the briefest of "roadmaps" for the proposed schemes. Detailed discussions are provided in the literature, along with extensive computations. The original coherent control scenario is shown in Figure 2 wherein a superposition of two bound states is subjected to two cw frequencies which lift the system to energy E. An analysis of this very basic scenario shows that the control parameters are the relative intensity and relative phase of the two indicated electromagnetic fields. This basic mechanism has been examined in the gas phase, both in the presence and in the absence of collisions and this scenario has been adapted to control currents in semiconductors. In all cases control was extensive.

Developments in pulsed laser technology may be used to good advantage in a straightforward modification of the above scenario. Specifically, the superposition state preparation and subsequent excitation with frequencies ω_1, ω_2 may be both carred out using pulsed lasers in accord with Figure 3:

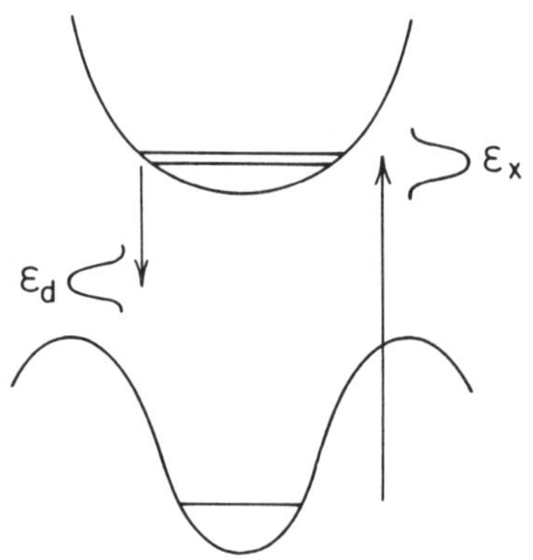

Figure 3: Pump-Pump scenario

In this instance the frequencies required to excite to energy E are contained within the second excitation pulse, which also creates product, and associated interference contributions over a wider energy range. Multiple excitation paths are contained within this overall pump-pump excitation scheme and serve to introduce the necessary interfering coherent paths. A straightforward analysis[8] in this case shows that the laboratory control parameters which affect the relative yield are the time delay between the pulses and the detuning of the initial excitation pulse which, in turn, determines the relative population and phase of the intermediate superposition state. We have applied this scenario to both model DH_2[8] and to IBr photodissociation [1]. In both cases relatively small changes in the detuning or in the time delay, which is in the convenient picosecond domain, resulted in extremely large variations of the yield (e.g. variations in the Br yield in IBr dissociation case over a range of 4% to 96%).

We have also applied this pump-dump type scheme to control bimolecular reactions [14]. In this case the initial excitation is from the continuum ground electronic state to a bound excited electronic state, followed by a dump back down to the ground electronic state. Significant control results only in the energy domain which is essentially below reaction threshold. Above threshold the interference term does not effectively compete with the energetically allowed reaction.

It is important to note that the scenarios involving two bound levels described above are not "restricted-two-state problems". Rather, we recognize that two levels are sufficient to carry the necessary phase information to allow control over the relative yield. That is, lasers with narrow frequency profiles (i.e. long pulses in time) are sufficient. Indeed as the molecules increase in size, and the density of states increases, longer and longer pulses could be used, and two intermediate levels retained. There appears little advantage to the inclusion of more than two levels, a situation which would result if the pulses were considerably shorter in time.

4. The Current State of Affairs

We have provided a capsule summary of the current status of coherent control of chemical reactions and molecular processes. Three essential points should be emphasized:

a) theoretical studies show that control over molecular processes, via simple quantum interference, is possible;

b) computational results show that such control can be extensive, with achievable variations in yield close to the full possible range of 0% to 100%;

c) experimental studies have shown that the proposed coherent control scenarios are feasible in the laboratory.

In summary, the age of mode selective experiments, based upon the principle of coherent control, is upon us.

Acknowledgements

We acknowledge support from the U.S. Office of Naval Research under contract numbers N00014-87-J-1204 and N00014-90-J-1014. Work carried out in conjunction with R.J. Gordon was supported by NSF PHY-8908161.

References

1. P. Brumer and M. Shapiro (1986), "Control of Unimolecular Reactions Using Coherent Light", Chem. Phys. Lett. **126**, 541-545.

2. P. Brumer and M. Shapiro (1987), "Coherent Radiative Control of Unimolecular Reactions: Three Dimensional Results", Faraday Disc. Chem. Soc. **82**, 177-185.

3. M. Shapiro and P. Brumer (1986), "Laser Control of Product Quantum State Populations in Unimolecular Reactions", J. Chem. Phys. **84**, 4103-4104.

4. C. Asaro, P. Brumer and M. Shapiro (1988), "Polarization Control of Branching Ratios in Photodissociation", Phys. Rev. Lett. **60**, 1634-1637.

5. M. Shapiro, J.W. Hepburn and P. Brumer (1988), "Simplified Laser Control of Unimolecular Reactions: Simultaneous (ω_1, ω_3) Excitation", Chem. Phys. Lett. **149**, 451-454.

6. G. Kurizki, M. Shapiro and P. Brumer (1989), "Phase Coherent Control of Photocurrent Directionality in Semiconductors", Phys. Rev. B**39**, 3435-3437.

7. M. Shapiro and P. Brumer (1989), "Laser Control of Unimolecular Decay Yields in the Presence of Collisions", J. Chem. Phys. **90**, 6179-6186.

8. T. Seideman, M. Shapiro and P.Brumer (1989), "Coherent Radiative Control of Unimolecular Reactions: Selective Bond Breaking with Picosecond Pulses", J. Chem. Phys. **90**, 7132-7136

9. For an introductory discussion see P. Brumer and M. Shapiro (1989), "Coherence Chemistry: Controlling Chemical Reactions With Lasers", Accounts Chem. Res. **22**, 407-413.

10. P. Brumer and M. Shapiro (1989), "One Photon Mode Selective Control of Reactions by Rapid or Shaped Laser Pulses: An Emperor Without Clothes?", Chem. Phys., **139**, 221-228.

11. I. Levy, M. Shapiro and P. Brumer (1990), "Two-Pulse Coherent Control of Electronic States in the Photodissociation of IBr: Theory and Proposed Experiment", J. Chem. Phys. **93**, 2493-2498.

12. M. Shapiro and P. Brumer (to be published).

13. C.K. Chan, P. Brumer and M. Shapiro (1991), "Coherent Radiative Control of IBr Photodissociation via Simultaneous (ω_1, ω_3) Excitation", J. Chem. Phys. **94**, 2688-2696.

14. J. Krause, M. Shapiro and P. Brumer (1990), "Coherent Control of Bimolecular Reactions", J. Chem. Phys. **92**, 1126-1131.

15. M. Shapiro and P. Brumer (to be published).

16. P. Brumer, X-P. Jiang and M. Shapiro (to be published).

17. See, e.g., S.H. Tersegni, P. Gaspard and S.A. Rice (1990), " On Using Shaped Laser Pulses to Control the Selectivity of Product Formation in a Chemical Reaction", J. Chem. Phys. **93**, 1670-1680 and references therein.

18. See, e.g., S. Shi and H. Rabitz (1990), "Optimal Control of Selective Vibrational Excitation of Harmonic Molecules: Analytic Solution and Restricted Forms for the Optimal Field", J. Chem. Phys. **92**, 2927-2937 and references therein.

19. S.M. Park, S-P. Lu and R.J. Gordon, "Coherent Laser Control of the Resonant Enhanced Multiphoton Ionization of HCl", J. Chem. Phys. (in press)

20. C. Chen, Y-Y. Yin and D.S. Elliott (1990), "Interference Betweeen Optical Transitions", Phys. Rev. Lett. **64**, 507.

21. As another interesting case consider the use of two components of laser polarization to excite a bound state $|E_i\rangle$ to the continuum[4]. These two components will interfere constructively and destructively and allow control over product ratios in the differential cross section. However, the interference contribution to the *total* cross section vanishes since these two routes reach different values of M_J.

22. X-P. Jiang, P. Brumer, and M. Shapiro (to be published)

Design of Femtosecond Pulse Sequences to Control Photochemical Products

*David J. Tannor*and Yijian Jin*
Department of Chemistry and Biochemistry
University of Notre Dame
Notre Dame, IN 46556

Abstract The methodology of Optimal Control Theory (OCT) is applied to the design of optical femtosecond pulse sequences. Two studies are reported, involving different but related objectives. The first study concerns the design of pulses or pulse sequences to optimize electronic population inversion in a system consisting of two Born-Oppenheimer states. The second study involves the design of pulse sequences to control chemical arrangement channels in photodissociation. This latter study is an extension of earlier work by Kosloff et al. [4], but the techniques used here are refined and the yields vastly improved. The second study can be considered, in some sense, as the doubled version of the first study: the first study maximizes the pump pulse while the second study maximizes the pump-dump sequence to obtain a particular chemical objective.

1 Introduction

Is it possible to take advantage of the unique properties of lasers – monochromaticity, high power, and phase coherence – to bring about selective and energetically efficient photochemical reactions? Many intuitive approaches to laser selective chemistry have been tried in the past twenty years. Most of these approaches have focussed on depositing energy in a sustained manner, using monochromatic radiation, into a particular state or mode of the molecule. Virtually all such schemes have failed, due to rapid intramolecular energy redistribution. In the last six years a new approach has emerged in which laser pulse shapes and pulse sequences, designed to achieve a desired chemical objective, are calculated systematically. In 1985, Tannor and Rice [1](see also [2] and [3]) formulated the problem of the search for optimal optical pulses, subject to suitable constraints, as a problem in the calculus of variations. Their formulation was based on a perturbation

*Alfred P. Sloan Foundation Fellow 1991-93

J. Jortner et al. (eds.), Mode Selective Chemistry, 333–345.

theory expression for the time dependent wavepacket amplitude, and led to a complicated integral equation expression for the optimal pulse(s). A formal and computational breakthrough was achieved in 1988 by Rabitz and coworkers, [5],[6],[7],[8] who recognized the utility of optimal control techniques, used widely in engineering, for the calculation of optimal pulses. Optimal control theory (OCT) is nothing more and nothing less than the extension of the classical calculus of variations to problems with differential equations constraints. In the context of molecular control, the time dependent Schrödinger equation (TDSE) is introduced as an additional constraint into the variational equations. The OCT methodology, in addition to providing a simplified algorithm for obtaining optimal fields numerically, has no difficulty in handling fields of arbitrary strength. This results in a significant generalization of the perturbative formalism, not only because greater chemical yields can be obtained with strong fields, but also because many new mechanisms are available to strong fields that are not accessible in the weak field regime. From the mathematical standpoint, the formalism may actually be simpler in the strong field regime in that it may be possible to remove the somewhat artificial constraint or penalty on the energy in the field adopted by both Tannor and Rice [1] and Rabitz et al. [5]. In the weak field regime the stronger the field the higher the yield and some penalty function is required to keep the value of the field finite; in the strong field regime, due to nonlinear effects, increasing field strength may decrease yield and no penalty on energy may be necessary. Many of the components of the TDSE/OCT formalism developed by Rabitz et al. were discovered independently by Kosloff et al. [4], who developed an intuitively based iteration procedure to optimize pulses.

The first part of this paper focusses on electronic population inversion in a two electronic state model system. Starting from an off resonance guess for the initial pulse, which gives negligible excited state population, a final pulse is obtained which achieves nearly 100 % excited state population. The final pulse has moved to resonance, and has intriguing periodic structure which is related to the period of motion on the excited Born-Oppenheimer potential.

This second part of this paper focusses on a model triatomic AAB (masses H,H,D), which has been been previously studied by Tannor and Rice [1], Tannor Kosloff and Rice [2], and Kosloff et al. [4]. The optimal control formalism is applied to breaking the triatomic into H+HD vs. HH+D. By developing an efficient optimization algorithm (and by removing a long standing bug from the program), the initial branching ratio of 1.2 %-1.0% can be changed to 56.4%-1.8% with one pulse, and 4.5%-45.6% with a second pulse. These pulses are not been completely optimized: the iterative procedure to produce optimal pulses was simply truncated after significant amounts of selectivity into each of the desired channel was demonstrated. This truncated iteration strategy was adopted for three reasons: 1) continued refinement of the pulses in general makes the pulses less accessible experimentally, 2) the mechanism by which the pulse shapes or pulse sequences manipulate molecular motion are expected to be evident in any good pulse, not just the optimal pulse, and 3) the extra yield provided by an optimal pulse is not of immediate practical interest, so long as the molecular parameters are not perfectly known and the technology to make the precise pulses is not immediately available. Where a *model* system is being studied, as in this paper, the practical value of knowing the precise optimal pulses is even more questionable. On the other hand, the unique mathematical properties of the optimal pulses is a rich topic for further study. Such a study really needs to be set in the context of an analysis of the global solution space (i.e. knowledge of the full set of local optima and their basins of attractions), and is beyond the scope of this paper. The emphasis in this paper then, is on the physical characteristics of *good* but not optimal pulses, and a preliminary attempt to understand the mechanism by which they control molecular motion.

Two alternative pulse sequences are obtained, depending on the definition of the objective: one favoring A+AB and one favoring AA+B. Surprisingly, the two sets of pulse sequences display a great deal of temporal overlap, despite the fact that they direct amplitude into opposite chemical arrangement channels. This is in contrast with the early Tannor and Rice work [1], which suggested the use of disjoint time windows for the pulses to achieve selectivity. To highlight the differences between the pulse sequences, the optimal pulses are analyzed in time-frequency space using the Husimi transform. Although some differences between the pulses are apparent in the Husimi transforms, the time-frequency windows remain unexpectedly similar. This increases our perplexity: one would expect pulses leading to disjoint chemical objectives to occupy disjoint regions of time-frequency space. The resolution to this paradox is speculated to be as follows: that orthogonality between the pulses can obtain even in cases where the windows in time (and frequency) are not disjoint, if the phases of the amplitude in the different regions conspire to cancel. This orthogonality concept represents a generalization of the disjoint time window ideas of Tannor and Rice: indeed, pulses with disjoint time windows obey a rudimentary form of orthogonality. It is likely that orthogonality, if it holds in the usual sense, will hold only for optimal pulses. It is also conceivable that some weighting factor, or some functional of the optimal fields, will have to be suitably defined in order that orthogonality be completely satisfied. Nevertheless, it is evident that some significant degree of orthogonality between the pulses must obtain in order for them to achieve disjoint chemical objectives.

2 Optimal Control Methodology

The time dependent Schrödinger equation for a two electronic state system with transition dipole moment μ can be written as:

$$i\hbar\frac{\partial}{\partial t}\begin{pmatrix}\psi_a(t)\\ \psi_b(t)\end{pmatrix}=\begin{pmatrix}H_a & \mu\varepsilon(t)\\ \mu\varepsilon^*(t) & H_b\end{pmatrix}\begin{pmatrix}\psi_a(t)\\ \psi_b(t)\end{pmatrix} \tag{1}$$

The subscripts a and b refer to the ground and excited electronic state indices, respectively. $H_{a/b}$ refers to the Born-Oppenheimer Hamiltonian for the ground/ excited electronic state, respectively. The two electronic states are coupled by the transition dipole moment, μ, which interacts with the electric field $\varepsilon(t)$ associated with the laser pulse. Complex values of the field are considered admissible, in keeping with the spirit of the rotating wave approximation in which the complex conjugate of the active part of the field is neglected on the grounds that it is off resonant. $\psi_{a/b}$ is the wavefunction (wavepacket) associated with the ground/excited Born-Oppenheimer potential energy surface.

One seeks to maximize

$$J \equiv \lim_{T\to\infty} < \psi(T)|P_\alpha|\psi(T) > \tag{2}$$

where P_α is a projection operator for the excited electronic state in the population inversion study, or for chemical channel α in the bond breaking study (here, α takes on two values, referring to arrangement channels A+AB and AA+B; in general, in a triatomic molecule ABC, α takes on three values, 1,2,3, referring to arrangement channels A+BC, AB+C and AC+B). The quantity J is a *functional* of $\varepsilon(t)$, and the problem of maximizing J with respect to $\varepsilon(t)$ falls into the class of problems belonging to the calculus of variations. It has been the practice, starting with the original work in the perturbative regime [1], to introduce a constraint on the energy of the field:

$$||E|| \equiv \int_0^T dt|\varepsilon(t)|^2 = E \tag{3}$$

where E is a constant. This leads to a maximization problem for the modified objective functional,

$$\bar{J} \equiv \lim_{T\to\infty} < \psi(T)|P_\alpha|\psi(T) > -\lambda\int_0^T dt|\varepsilon(t)|^2 \tag{4}$$

where λ is a Lagrange multiplier. Alternatively, it is possible to let the norm of the field vary, but put a penalty on the size of the norm. In this case, one treats λ as fixed, the higher the value of λ the higher the penalty. As discussed above, it is plausible that the problem can be formulated with neither constraint nor penalty on the energy in the field; this paper, however, adopts the penalty factor formulation.

In the OCT formulation, the time dependent Schrödinger equation written as a 2x2 matrix in an Born-Oppenheimer basis set, is introduced into the objective functional with a Lagrange multiplier, $\chi(x,t)$. The modified objective functional may now be written as:

$$\bar{J} \equiv \lim_{T\to\infty} < \psi(T)|P_\alpha|\psi(T) > +2Re\int_0^T dt < \chi(t)| - i\frac{\partial}{\partial t} - \frac{iH}{\hbar}|\psi(t) > -\lambda\int_0^T dt|\varepsilon(t)|^2 \tag{5}$$

where

$$H \equiv \begin{pmatrix} H_a & \mu\varepsilon(t) \\ \mu\varepsilon^*(t) & H_b \end{pmatrix} \tag{6}$$

It is clear that as long as ψ satisfies the time dependent Schrödinger equation the new term in $\bar{J}$ will vanish for any $\chi(x,t)$. The particular form of the new term is chosen so that the variations of $\bar{J}$ with respect to ε and with respect to ψ are independent, to first order in $\delta\varepsilon$.

The requirement that $\frac{\delta \bar{J}}{\delta\psi} = 0$ leads to the equations:

$$i\hbar\frac{\partial\chi}{\partial t} = H\chi \tag{7}$$

$$\chi(x,T) = P_\alpha\psi(x,T) \tag{8}$$

i.e., the Lagrange multiplier must obey the time dependent Schrödinger equation, subject to the boundary condition at the *final* time T that χ be equal to the projection operator operating on the Schrödinger wavefunction. These conditions 'conspire', so that a change in ε, which would ordinarily change $\bar{J}$ through the dependence of $\psi(T)$ on ε, does not do so to first order in the field.

For a physically meaningful solution it is *required* that

$$i\hbar\frac{\partial\psi}{\partial t} = H\psi \tag{9}$$

$$\psi(x,0) = \psi_0(x) \tag{10}$$

Some workers prefer to derive this latter equation from the condition that $\frac{\delta \bar{J}}{\delta\chi} = 0$ The physical interpretation is that if one removes the constraint that ψ must satisfy the time dependent Schrödinger equation, one can only attain a maximum value of J higher than or equal to the physical maximum. If one then proceeds to minimize this 'fictitious' $\bar{J}$ with respect to χ one can hope to recover the maximum value of $\bar{J}$ corresponding to the physical problem. This forms the basis for a powerful, but unconventional approach to functional maximization.

Equations 7, 8, 9 and 10 form the basis for a double ended boundary value problem. ψ is known at t=0, while χ is known at t=T. One can propagate ψ forward in time to obtain $\psi(t)$, and propagate χ backwards in time to obtain $\chi(t)$, once one has a form for $\varepsilon(t)$. The optimal $\varepsilon(t)$ is given by the condition that $\frac{\delta J}{\delta \varepsilon} = 0$ which leads to the equation:

$$\varepsilon(t) = \frac{i}{\lambda}[< \psi_b|\mu|\chi_a > - < \chi_b|\mu|\psi_a >] \tag{11}$$

Note that knowledge of the optimal $\varepsilon(t)$ involves knowledge of $\psi(t)$ and $\chi(t)$, while knowledge of $\psi(t)$ and $\chi(t)$ requires knowledge of $\varepsilon(t)$. In general the equations are solved iteratively. For a deeper physical understanding of the set of equations (7)-(11) the reader is strongly encouraged to see Ref. [4].

The gradient of the objective with respect to the field is given by (gradient is used here and below to refer to the direction of maximum *positive* slope):

$$\frac{\delta \bar{J}}{\delta \varepsilon} = \varepsilon(t) - \frac{i}{\lambda}[< \psi_b|\mu|\chi_a > - < \chi_b|\mu|\psi_a >] \tag{12}$$

Therefore, given the field at the k iteration, the field at the k+1 iteration may be written as: $\varepsilon^{k+1}(t) = \varepsilon^k + \alpha\frac{\delta J}{\delta \varepsilon}$. If α is small (α a positive real parameter) the objective functional must increase or stay the same at each new iteration. However, the second order dependence of the objective on the field is not completely known a priori (particularly because of the nonlinear dependence of $\psi(T)$ on the field). This implies that there is no a priori way of knowing the value of alpha which will maximize the objective in the direction of the gradient. One approach to the selection of α is to perform a line search along the direction of the gradient, i.e., the best value of α is found by empirical comparison of many different values. Once the best value of α is found the new gradient is calculated and a new search performed. This is the approach adopted here, although only a very crude line search has been performed in each direction. The reasoning behind performing a crude, rather than a fine search, is that any residual component of the gradient which is neglected in one iteration will automatically be incorporated into the definition of the gradient at the next iteration. It was confirmed empiricially that 1) small positive values of α always led to an increase in the objective and 2) the dependence of the objective on α for small values of α was strictly linear. At each iteration the range of α was empirically adjusted so that the objective functional had a maximum between the first and last value.

Some workers have reported rapid convergence to the optimal pulse using a conjugate gradient (CG) method [6] [11] [12]. This was not tested here, but was not expected to provide any savings over the method adopted for several reasons: 1) the CG method is really intended for use near the extremum, while the initial pulse here is presumably far from the extremum, 2) the CG method requires an expression for the second derivative matrix (or second variation function) in order to compute the conjugate direction; as discussed above such an expression is not available here, and 3) the CG method never goes back to directions it has already explored; i.e. the new direction is 'Q-orthogonal' to all previous directions. This implies the method should properly be used in conjunction with a careful line search at each iteration, since any missing component of the CG direction cannot be recovered at subsequent iterations. Hence, the CG method is apparently incompatible with the crude line search method described above.

3 Results

Figures 1 and 2 show the effect of pulse design on population inversion in a simple system, which consists of two displaced harmonic oscillators in one dimension. Starting with an initial guess for the pulse which is far off resonance, the program iteratively improves the pulse until the pulse in Fig. 1a is obtained. The pulse achieves orders of magnitude more excited state population than the initial guess. The periodic form of the pulse is striking, and reveals an interesting mechanism at work. The period of the pulse is equal to the excited state vibrational period. Every period the pulse transfers additional wavepacket amplitude from the ground to the excited electronic state, during the time interval that the excited state wavepacket is returning to the Franck-Condon region. The phase of the pulse is precisely determined by the condition that the new amplitude interfere constructively with the returning wavepacket. That this is the mechanism is

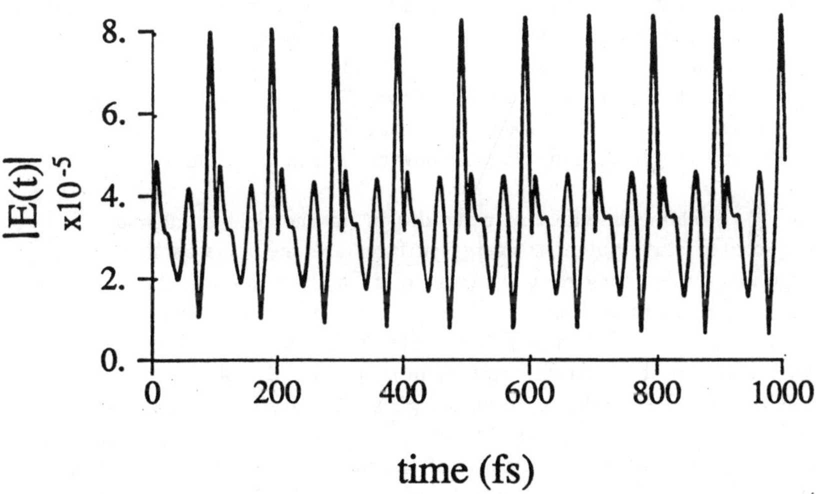

Fig. 1. a) Final pulse designed to achieve maximum population inversion in a system consisting of two displaced harmonic oscillators in one dimension.

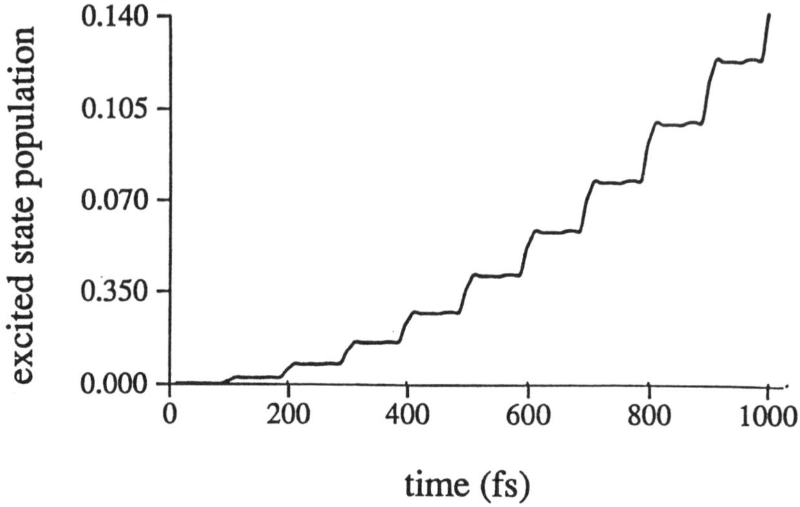

b) Excited state population vs. time using the pulse from a).

confirmed by inspection of Fig. 1b, which shows the norm on the excited state surface as a function of time. The population jumps sharply every vibrational period, when the new amplitude is transferred to the excited electronic state and adds constructively with the existing amplitude. The height of successive 'steps' increases quadratically; this is because the amplitude adds linearly every period, leading to a quadratic growth in excited state *population*. This pulse sequence provides a nice example of how a simple mechanism often underlies the result of a TDSE/OCT calculation. The mechanism may bear no resemblance to that at work in the initial guess. As a result, one can use the TDSE/OCT machinery to actually *discover* new mechanisms to achieve desired chemical objectives.

Figures (2)-(4) show the effect of pulse design on chemical branching. The model ground and excited state potential energy surfaces are the same as those in Ref. [4]. Note the two inequivalent exit channels (visible in Figs. 3c and 4c) which differ only by an isotopic substitution (the channel to the right corresponds to D+H_2 while the channel to the left corresponds to H+HD). Figure 2 shows the initial guess for the pulse (two gaussians), together with its Husimi transform (gaussians in both time and frequency). Figure 3a shows the pulse designed to yield high probability in channel 1, together with its Husimi transform. Figure 3b shows the norm on the ground and excited state surfaces as a function of time, under the influence of the pulse in Fig. 4. Figure 3c shows the wavepacket on the ground state potential surface after 1500 a.u. (37.5 fs), under the influence of the pulse in Fig. 3a. Note the high degree of selectivity for exit out of channel 1 (D+H_2). Figure 4a shows the pulse designed to yield high probability in channel 2, together with its Husimi transform. Figure 4b shows the norm on the two surfaces as a function of time, under the influence of the pulse in Fig. 4a. Figure 4c shows the wavepacket on the ground state potential surface after 1500 a.u. (37.5 fs), under the influence of the pulse in Fig. 4a.

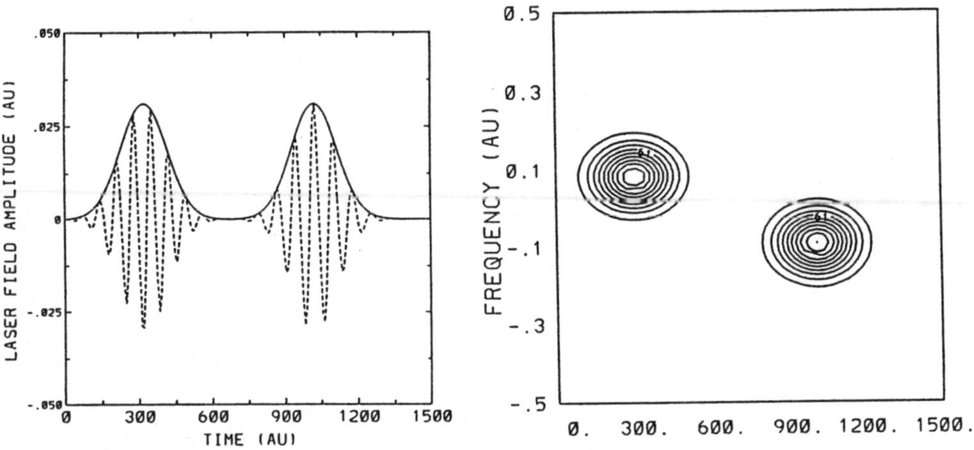

Fig. 2. Initial guess for the pulse sequence to control product branching, consisting of two gaussions. The Husimi distribution corresponding to this pulse sequence is shown on the right.

It is clear that these pulses are remarkably effective in directing wavepacket amplitude out one channel or the other. Equally intriguing is the superficial similarity between the two pulses, and between both pulses and the initial guess. The pulses in Figs. 3a and 4a occupy identical windows in time, in contrast with the simple pump-dump control scheme of Tannor and Rice [1] which makes use of different windows in time to achieve different products. The Husimi plots show that the two pulses are even similar in their joint time-frequency distributions. There are significant differences, however. The pulse in Fig. 3a is still primarily a two pulse, pump-dump sequence. The pulse in Fig. 4a, however, has substantial amplitude between the pump and dump, and is much closer to a

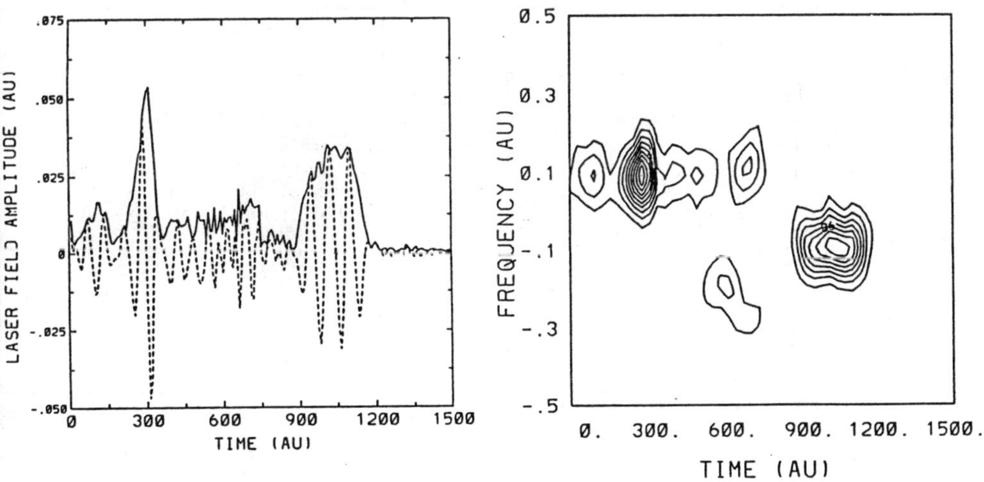

Fig. 3. a) Final pulse designed to direct amplitude out channel 1 (D+H_2). The Husimi distribution corresponding to this pulse is shown on the right.

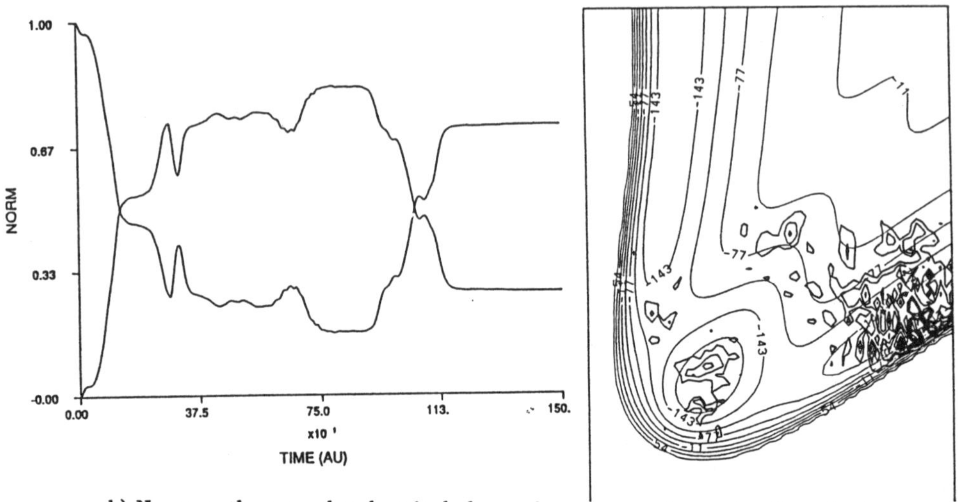

b) Norm on the ground and excited electronic states vs. time using the pulse in a). c) Absolute value of the ground state wavefunction at 1500 a.u. (37.5 fs) using the pulse of a), superimposed on equipotential contours of the ground state surface. The exit out channel 1 is extremely selective.

being a continuous excitation/deexcitation. This is confirmed by inspection of the norm in Figs. 3b and 4b. The norm in Fig. 3b changes once from the pump pulse and once again from the dump pulse. The norm in Fig. 4b, corresponding to the pulse in Fig. 4a, undergoes three full cycles. It should be noted that there may be significant phase differences between the two pulses n Figs. 3a and 4a which are not evident in the Husimi plots, the latter showing only the absolute value of the distribution.

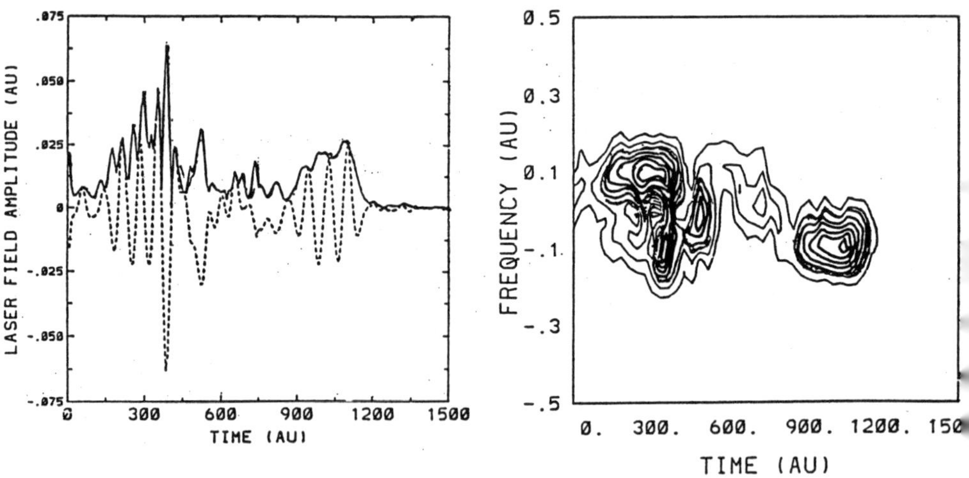

Fig. 4. a) Final pulse designed to direct amplitude out channel 2 (H+HD). The Husimi distribution corresponding to this pulse is shown on the right.

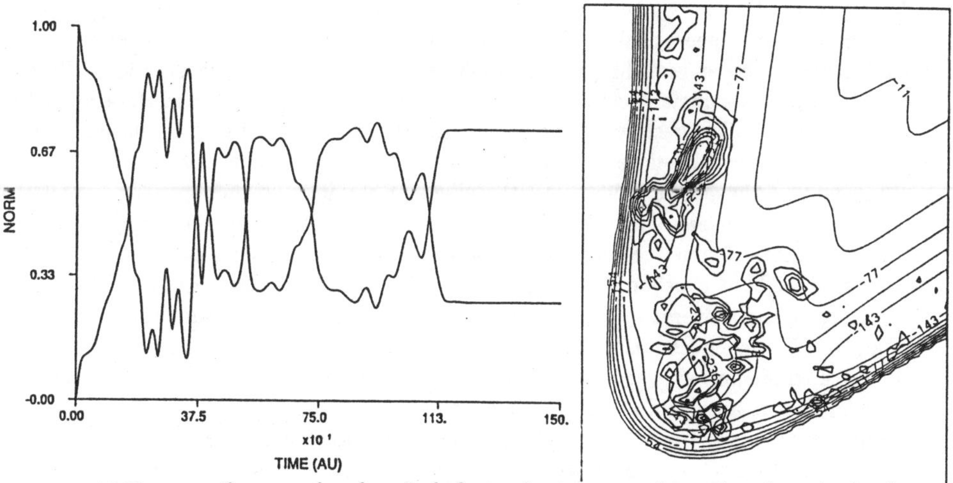

b) Norm on the ground and excited electronic states vs. time using the pulse in a). Note the three full Rabi cycles. c) Absolute value of the ground state wavefunction at 1500 a.u. (37.5 fs) using the pulse of a), superimposed on equipotential contours of the ground state surface. The exit out channel 2 is extremely selective.

The pulses in Figs. 3a and 4a provide additional examples of how simple mechanisms may underlie the pulses calculated using the TDSE/OCT machinery. The pulse sequence in Fig. 3a is qualitatively of a simple pump-dump form, i.e. the time delay between the two pulses is such that the wavepacket on the excited state surface has the correct position and momentum to exit out of channel 1 on the ground state surface [1], [2]. The pulse sequence in Fig. 4a reflects a different mechanism, closer to a cyclically repeating pump and dump. The cyclical pump-dump pulse sequence provides a natural mechanism for preparing highly excited symmetric stretch vibration, since this is the coordinate with greatest difference in forces on the two potential surfaces and thus this motion is amplified on each cycle. This can be seen in pictures of the wavepacket motion (not shown). If amplified sufficiently this motion will lead to selective three body fragmentation (A+A+C). In the present study, however, with the objective of A+AB fragmentation, the symmetric stretch motion is amplified only in the first stage of the pulse sequence. In the second stage of the pulse sequence the energy in the symmetric stretch is channeled into the A+AB (H+HD) channel.

4 Conclusions

The power of the TDSE/OCT methodology to achieve population inversion and control photochemical products is evident from the results presented here. As explained in the Introduction, the motivation for the calculation of the pulses here is twofold. 1) First and foremost, the hope is that these pulses attract the interest of experimentalists: that something akin to them ultimately be produced in the laboratory, and that the laboratory pulses produce qualitatively the same effects as the calculated pulses. 2)

That the mechanism by which the OCT designed pulses produce their objective will be amenable to theoretical analysis and understanding, either illustrating the use of known principles or revealing novel principles of laser-molecule interaction. To study these issues it is sufficient, and perhaps desirable, to use the truncated iteration scheme adopted here. The mechanism by which the pulses control the molecular motion can already be discerned before the iterative optimization process is converged. Continued refinement of the pulses could have the effect of making the pulses more complicated, and consequently less attractive to experimentalists. The very interesting question of the mathematical structure of the solution space of optimal pulses can not, of course, be addressed without continuing the optimization procedure to convergence.

The pulses discovered in the course of the population inversion study reveals the utility of an interference mechanism in achieving population transfer. By using a periodic pulse sequence whose period is equal to the period of the excited state oscillator, it is possible to ensure that the amplitude that is transferred at later times adds constructively with the amplitude at earlier times. In particular, during the time interval that the wavepacket on the excited state returns to the Franck-Condon region additional ground state amplitude is transferred up, with a phase that matches that of the returning wavepacket.

The pulses obtained in the study of control of chemical branching are equally intriguing. Husimi representations of the pulses allow one to compare, both in time and frequency simultaneously, the pulses that lead to different fragments. The pulses are remarkably similar,although the differences are certainly significant. Both pulses retain a good deal of the character of the initial guess, i.e. a two pulse, pump-dump appearance. However, the pulse leading to exit out of channel 2 has almost continuous amplitude between the pump and the final dump, and generates multiple Rabi cycles between the two electronic states. Thus, the optimal pulses again reveal a simple underlying mechanism that was not present in the initial guess.

It is speculated that the optimal pulses leading out the different channels are orthogonal, in either a straightforward or in some generalized sense. Although the Husimi plots suggest that there is in fact significant overlap between the pulses, note that the Husimi plots show only the absolute value of the time-frequency distributions, while the phase can be extremely important in determining orthogonality. The possibility of orthogonality between the pulses, in either a straightforward or in some generalized sense, will be explored analytically and checked numerically in the near future.

DJT wishes to acknowledge the many significant contributions of his collaborators, Professors Ronnie Kosloff, Pierre Gaspard, Stuart Rice and Andras Lorincz, to this research and the hospitality of the Hebrew University in Jerusalem where this manuscript was completed. This work was supported by a grant from the US Office of Naval Research.

References

[1] D.J.Tannor and S.A.Rice, J.Chem.Phys. **83**, 5013 (1985).

[2] D.J.Tannor, R.Kosloff and S.A.Rice, J.Chem.Phys. **85**, 5805 (1986).

[3] D.J.Tannor and S.A.Rice, Adv.Chem.Phys. **70**, 441 (1988).

[4] R.Kosloff, S.A.Rice, P.Gaspard, S.Tersigni and D.J.Tannor, Chem.Phys. **139**, 201 (1989).

[5] A.P.Pierce, M.A.Dahleh and H.Rabitz, Phys.Rev.A. **37**, 4950 (1988).

[6] S.Shi, A.Woody and H.Rabitz, J.Chem.Phys. **88**, 6870 (1988).

[7] S.Shi and H.Rabitz, J.Chem.Phys. **92**, 364 (1990).

[8] S.Shi and H.Rabitz, Comp.Phys.Comm. **63**, 71 (1991).

[9] P.Brumer and M.Shapiro, Chem.Phys.Lett. **126**, 54 (1986).

[10] M.Shapiro and P.Brumer, J.Chem.Phys. **84**, 4103 (1986).

[11] W.Jakubetz, B.Just, J.Manz and H.-J.Schreier, J.Phys.Chem. **94** 2294 (1990).

[12] J.E.Combariza, B.Just, J.Manz and G.K.Paramonov (preprint).

Impulsive Iterative Inversion.

Ronnie Kosloff and Roi Baer

Department of Physical Chemistry and The Fritz Haber Research Center
for Molecular Dynamics, The Hebrew University, Jerusalem 91904, Israel.

Abstract

An inversion procedure which can obtain molecular potentials from impulsive experiments is presented. It is based on the time reversal symmetry of the mechanical equations of motion which allow propagation backwards to the original state. Because the experimental data does not supply the full picture necessary for backward propagation and is therefore inconclusive, an iterative procedure has been developed which is a combination of a forward and backward propagation. This allows the inclusion of data distributed in time from the beginning to the end of the experiment. The full power of time dependent quantum mechanical propagation methods is employed. The inversion procedure is applied to the coloumb explosion experiment. The collinear coloumb explosion of DHD^+ is used to demonstrate the procedure step by step form the experimental velocity distribution to the molecular potential. A full quantum mechanical inversion procedure for impulsive ultra short laser pulse experiments is developed. A demonstration for the ICN molecule is supplied.

J. Jortner et al. (eds.), Mode Selective Chemistry, 347–357.

I. Introduction

Inverting experimental data can be looked upon as similar to the task of the historian who views a complex situation and tries to infer its reasons and causes. This process is complicated because usually the current picture is blurred which means that the inference is based only on a partial knowledge of the situation. Moreover the inference is not unique, meaning that it may not be the only explanation. The physical chemist who faces these problems has the advantage he can use time reversal symmetry. If formally the sign of time is reversed in the equations of motion, the current picture should return to its original form. It is this tool of time reversal combined with high quality time dependent quantum propagation techniques, that allow an effective inversion procedure. The present work focusses on inferring the molecular potential from experiments which have an impulsive character i.e. in which large changes happen in very short time.

The general problem of inversion has been addressed by many authors [1-4]. Mathematically the inversion problem is ill posed since more than one solution can exist for a set of experimental data. A typical example is the inversion of spectra by the one dimensional RKR semiclassical method. Already in this example, the inversion is not unique; a transformation of the potential that will preserve the action between two turning points will produce the same spectra. Most successful inversion methods are one dimensional, the reason being the large degree of over-determination in higher dimensional problems. Mathematical approaches based on formal scattering theory have been proposed [1], but it is rare that the experiment can supply the required data. An exception is a multiple dimensional approach based on an iterative self consistent field reduction to coupled one dimensional problems. It has been applied successfully to the inversion of spectral data [2].

In physical chemistry The usual situation of insufficient experimental data has lead to heuristic approach in which the experiment is modeled by first assuming a known potential. The computed results from the model are compared to the experiment. At this point the potential is modified for another try. The lack in this procedure is that there is no guaranty of convergence. Moreover good simulation of multidimensional problems are numerically expensive. The advantage, on the other hand, of the heuristic approach is that there are no restrictions on the quality or completeness of the experimental data.

The development of new experimental tools, in particular, impulsive methods and new theoretical techniques have supplied the motivation for reconsidering the inversion problem. The best known example of an impulsive experiment is the coloumb explosion experiment [5]. A molecule is accelerated and then passed through a very thin film which on a timescale of 10^{-17} seconds is ionized. In this process the nuclei exhibit a sudden change from the bound molecular potential to a repulsive coloumb potential. Experimentally the time of flight of all nuclei is monitored. From this data the asymptotic relative momentum distribution of all participating nuclei is obtained. Another example of an impulsive experiment is photodissociation of a molecule using light pulses in the femtosecond regime. For a very short light pulse the nuclei are moved instantaneously from the ground potential surface to an excited one. Experimental data can be obtained either by time of flight spectra or by monitoring the dynamics using a second short pulse.

Inversion schemes for impulsive experiments exclusively based on classical mechanics have been applied to these problems [6]. A critical evaluation of this procedure becomes concerned with the applicability of classical mechanics to these dynamical molecular processes. Another concern is the validity of the impulsive assumption. At least for existing femtosecond experiments the pulses are not short enough in relation to molecular motion to be considered impulsive.

In this presentation, a full quantum scheme for inversion of incomplete experimental data is presented. The insight for this study has come from the development of laser selective chemistry and in particular the use of optimal control theory [7][8]. Similar conceptual problems arise in trying to infer the optimal light pulse which induces a particular chemical outcome. The rest of this paper is devoted to illustrate these ideas.

II. Coloumb explosion inversion

The coloumb explosion technique serves as an ideal example for an impulsive experiment [5]. The process can be described by the following steps: First, the molecule is prepared in a well defined initial state. This is usually done by a jet cooled source and a lazer preparation method. For the analysis it is assumed that the molecule is either in the ground state or in a well defined eigenstate. The second step is to accelerate the molecule. It is assumed during this step that the state of the molecule does not change. The third step is the stripping stage where the molecule is passed through a thin foil in which the field is strong enough to ionize the molecule in a very short time so that the nuclei do not have time to change their positions. Once the molecule is ionized, it moves under the influence of a repulsive coloumb potential. The final stage is a time of flight measurement of all nuclei constituting the molecule. The experiment is then repeated many times to obtain a velocity distribution. The relative momentum distribution of the nuclei can be constructed by subtracting the center of mass velocity. Two points should be emphasized: 1) the impulsive nature of the change from the unknown potential to the known coloumb potential. 2) The complete momentum distribution measurement of all participating particles.

The main purpose of the experiment is to invert the experimental data and obtain the molecular potential. The inversion proceeds as follows: First, from the asymptotic momentum distribution the initial wavefunction $\psi(R,t=0)$ is inferred. Once the initial wavefunction is known, the molecular potential can be obtained. Examining the first stage with the aid of microscopic reversibility, it would be expected that the asymptotic data could be propagated backward, but a typical problem of inversion emerges. The experimental data which is supplied as the asymptotic relative momentum distribution, is insufficient to invert the data. In quantum mechanics the momentum observable is equivalent to the absolute value squared of the asymptotic wavefunction in momentum space at $t=\infty$: $P(k)=|\psi(k,t=\infty)|^2$. The crucial part that is missing in order to reconstruct the wavefunction is the phase factor $e^{i\phi(k)}$. On the other hand if the wavefunction at $t=0$ is an eigenstate, the initial phase is known to be constant but the absolute value of the wavefunction is not known and is the object of inversion. At this point one is faced with the problem that the known facts about the system are split between the times $t=0$ and $t=\infty$. The iterative inversion procedure is designed to let the ends meet to obtain a wavefunction consistent with all known data.

The inversion procedure is as follows:

A. Construct the best possible guess for the molecular potential. One can use previous potential fits from spectroscopic data or ab-initio potential calculations.
B. Use the relaxation procedure [9] to obtain the ground state wavefunction of the trial potential of step A. This wavefunction is the input for the iterative procedure and therefore becomes $\psi_0(R,t=0)$.
C. Propagate the initial wavefunction of the n th iteration $\psi_n(R,t=0)$ from $t=0$ to $t=\infty$ using the coloumb potential obtaining $\psi_n(R,t=\infty)$.
D. Transfer the wavefunction into momentum space: $\psi_n(k,t=\infty)$.
E. Combine the phase of the wavefunction of D and the amplitude in momentum space obtained from the experimental data, to produce a new final wavefunction:

$$\psi_n^f(K,t=\infty) = |\psi_E(k)| \frac{\psi_n(k,t=\infty)}{|\psi_n(k,t=\infty)|} \tag{2.1}$$

where: $|\psi_E(k)|$ is the square root of the experimental momentum distribution. Transform the wavefunction back to configuration space: $\psi_n^f(R,t=\infty)$.
F. Propagate backwards from $t=\infty$ to $t=0$, the wavefunction of E, to obtain: $\psi_n^f(R,t=0)$
G. Adjust the phase of the wavefunction of F to be constant $\psi_{n+1}(R,t=0) = |\psi_n^f(R,t=0)|$. This wavefunction is the input for the $n+1$ iteration.
H. Repeat steps C to G until convergence is obtained.

I. From the iterative final wavefunction $\psi_f^l(R,t=0)$ of the l th iteration obtain the potential by rewriting the stationary Schrödinger equation:

$$V(R) = \left[\frac{1}{2m} \nabla^2 \psi_f^l(R,t=0) \right] \frac{1}{\psi_f^l(R,t=0)} + E \tag{2.2}$$

Comments: 1) Although the algorithm described is general, there is a computational advantage for using the Fourier method representation [10]. This is because of the ease of transformation from coordinate to momentum space. 2) The final time $t=\infty$ can be chosen to be finite, provided the asymptotic momentum distribution has been reached. It can be tested by calculating the residual coloumb potential energy expectation: $<\psi | V_{col} | \psi>$. 3) High quality time propagation methods in which the phase error is minimal, have to be used [11]. In particular the last inversion step (step I) is sensitive to the absolute accuracy of the propagation method. 4) The use of interaction representation [12] can be advantageous because momentum is constantly shifted to higher values.

A two dimensional study of the coloumb explosion of collinear DHD^+ will be used to illustrate the procedure. Other one and two dimensional studies have been carried out with the conclusion that the following illustration is typical. Figure 1 displays the final relative momentum distribution function.

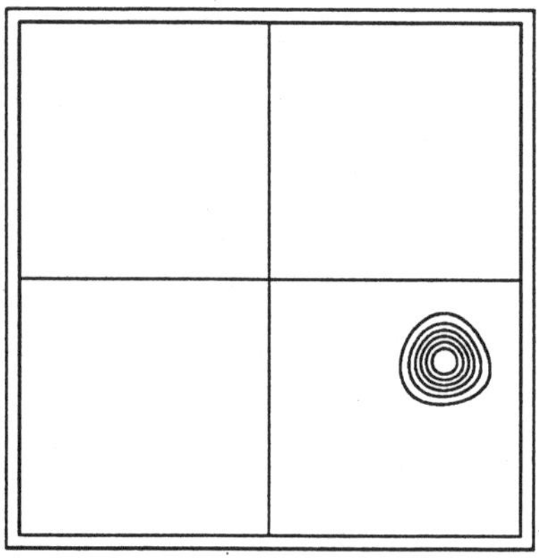

Figure 1 : countour plot of the final momentum distribution, the cross hairs point to the origin (0,0) .

This distribution represents the coloumb explosion experimental data. A grid of 128x128 points was used in coordinate R space or in momentum k space. Jackobi coordinates were used and a Fourier representation scheme. The time propagation was carried out using the Chebychev method [10]. (all units used are atomic units).

The algorithm is now followed step by step.

Step A: To emphasize the fast convergence, a very poor initial guess of the molecular potential was chosen. Figure 2 displays this potential. In comparison to the real molecular potential the potential minimum is shifted by 0.4 a.u. and the vibrational frequencies are only half the value of the original potential.

Step B: The initial wavefunction $\psi_0(R,t=0)$ is produced on the potential of A by the relaxation method. It is shown in figure 2 superimposed on the potential of step A. To further emphasize the convergence the kinetic energy operator in the Hamiltonian was calculated with a 1/3 of the reduced mass. Figure 3 shows the wavefunction superimposed on the true molecular potential. The energy of this wavefunction in comparison to the real ground state, is shifted by 0.15 a.u. (4. ev).

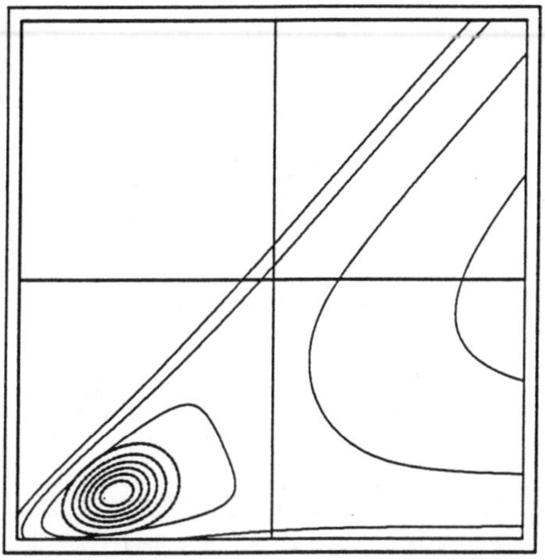

Figure 2: contour plot of trial potential (solid line) and the ground state of this potential (dashed line)

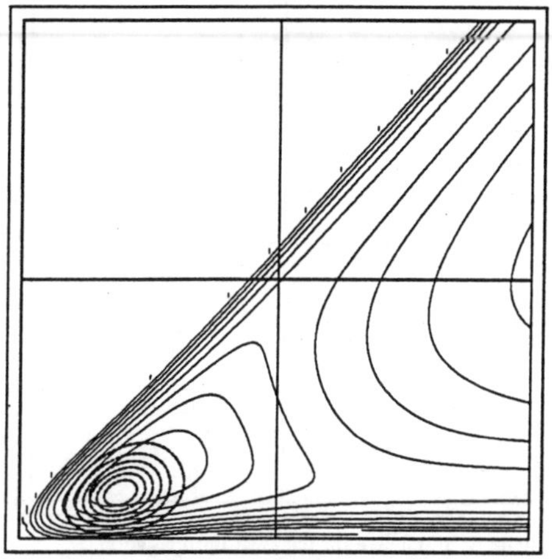

Figure 3: contour plot of the true potential (solid line) and the ground state of trial potential (dashed line)

Step C: The wavefunction is propagated to the final asymptotic time chosen to be 120 a.u.. The wavefunction $\psi_0(R,t=120)$ is displayed in figure 4.

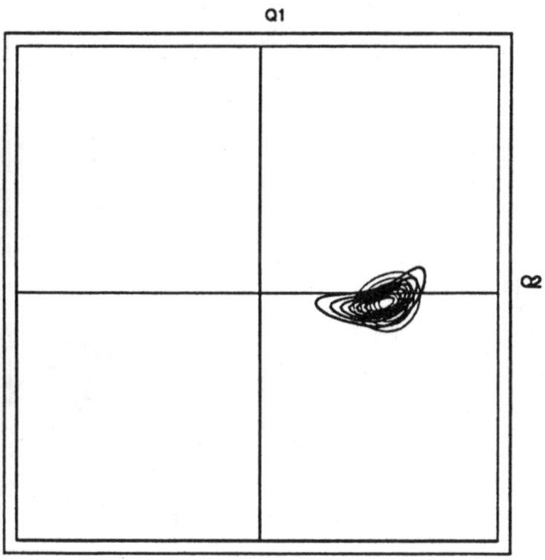

Figure 4: asymptotic wavefunction in coordinate space.

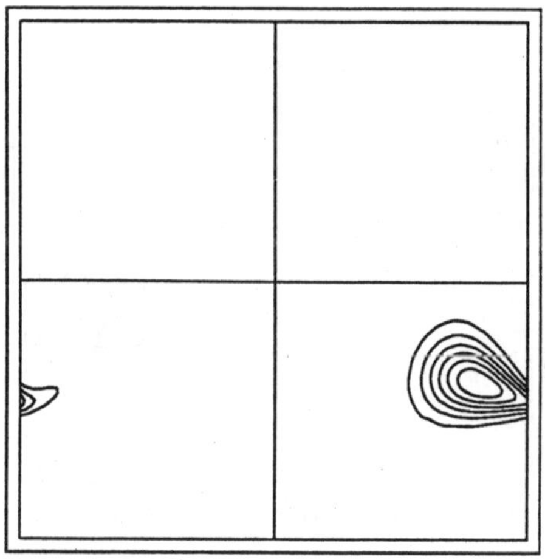

Figure 5: asymptotic wavefunction in momentum space.

Step D: Figure 5 shows the wavefunction in momentum space $\psi_0(k,t=120)$. The momentum distribution is clearly broader in comparison to the experimental data of figure 1.

Step E: The combined wavefunction using the phase of D and the amplitude of figure 1 is produced: $\psi_0^f(k,t=120)$. This is the focal point of the inversion in which the forward propagated wavefunction which carries the phase is matched with the asymptotic data which carries the amplitude. This wavefunction in coordinate space is displayed in figure 6.

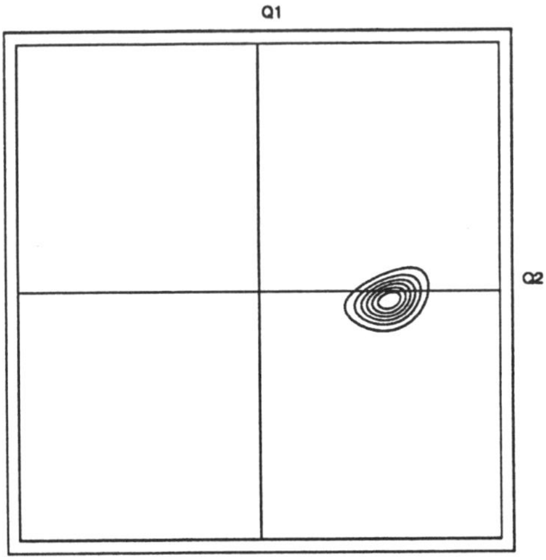

Figure 6: initial sate for backward propagation.

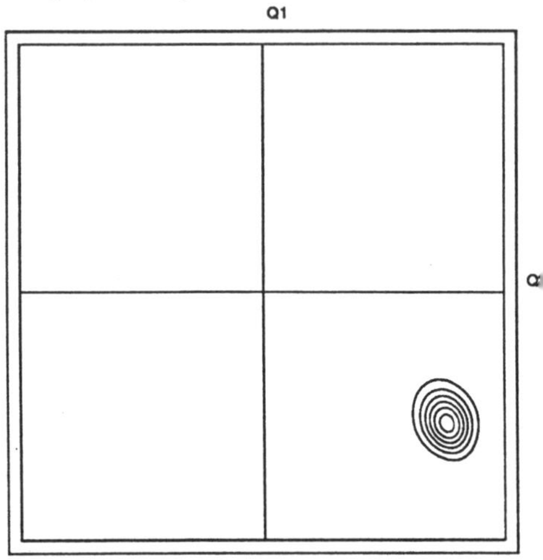

Figure 7: backward propagated wavefunction first iteration

Step F: Propagation backward in time to obtain $\psi_0^f(R,t=0)$ shown in figure 7.
Step G: Adjustment of phase to obtain the input of the next iteration.
Step H: Figure 8 shows the convergence of $\psi_n(k,t=120)$ while figure 9 shows $\psi_n^f(R,t=0)$.

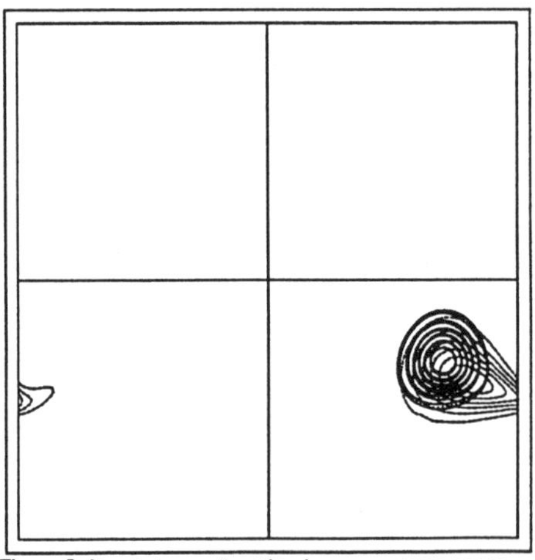

Figure 8: iterated wavefunction in momentume space

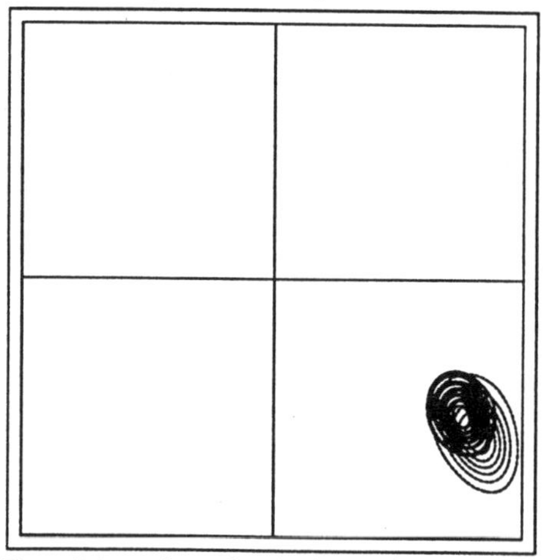

Figure 8: iterated wavefunction in coordinate space

Five iterations were enough to converge the results.

Step I: Figure 10 compares the potential obtained by the inversion to the true potential used to produce the "experimental" data of figure 1.

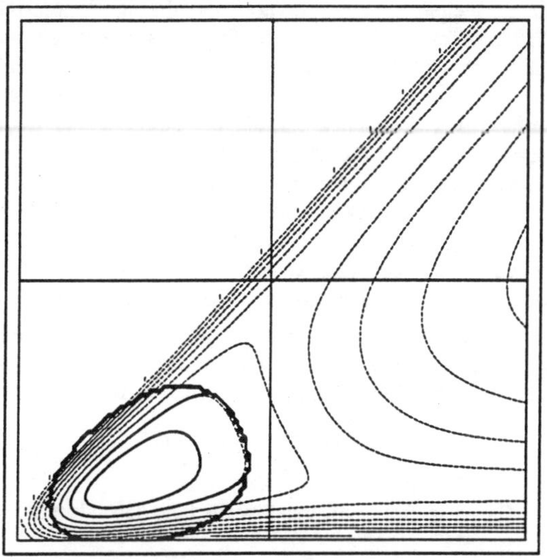

Figure 10: The inverted potential (solid) superimposed on the tru molecular potential (dashed).

The figure displays clearly the window in coordinate space where the wavefunction has significant amplitude in which meaningful results can be obtained. In order to shift this window, an excited initial state should be employed.

On examining the evolution of the wavepacket both in coordinate and momentum space (figures 2-10), it is evident that the wavefunction is compact. This fact makes it possible to use dynamical grid approaches in which the grid follows the wavefunction in coordinate as well as momentum space. This shift can be carried out either at every time step such as in the work of Feldman [12], or continuously by the use of the interaction representation.

The amazing fact about the inversion procedure is its robustness and fast convergence. Four iterations were enough to obtain converged results of the wavefunction and of the potential in regions of space of up to 10% of the maximum wavefunction. After ten iterations, the potential could be obtained in regions which are up to 2% of the maximum wavefunction. The frequent transformations from coordinate to momentum space creates an effective filter able to eliminate experimental noise.

At this point the procedure should be compared to the classical inversion procedure. The main difference is that the classical inversion procedure trajectories are sent backwards one by one. The difficulty is to identify the position in which to stop the integration of each trajectory. In multidimensional systems such a stopping point is not unique. The global quantum iterative procedure avoids the difficulty. Another problem arizes from the omission of the zero point motion in the classical procedure. This in turn leads to wavefunction in coordinate space which is too broad and which manifests itself with a molecule which appears to be floppy. This effect is particularly important in molecules containing hydrogen.

Inversion of short pulse experimental data.

The impulsive experimental process can be carried out by using light as the driving force. Based on the Frank Condon principle, the nuclei do not move during the absorption of light. The introduction of ultra fast pulsed light sources enables the use of light in an impulsive inversion procedure. A classical procedure based on these ideas has been suggested by Bernstien and Zewail [6]. The ICN molecule is used to illustrate the procedure. The experiment starts with the molecule in the ground electronic surface assumed to be known (surface g). Then impulsively, the molecule is transferred to the first excited surface which is repulsive (surface v). The forces on this potential photodissociate the molecule to I + CN. After a preset time delay another impulsive pulse transfers the system to the second excited potential which is assumed to be known (surface u). The reason is that this excited state is associated with the CN chromophore and therefore is almost independent of the I-CN distance. The purpose of the inversion procedure is to obtain the shape of the first excited potential surface using a series of two pulse experiments where the time delay between pulses and central frequency of the second pulse is varied.

At this point, one should summarize the known facts about the experiment. As the ground potential surface is known, the initial wave function of the experiment $\psi_g(t{=}0)$ can be calculated. The final measurement supplies the final probability on the second excited surface:

$$|<\psi_{u_l}(t{=}\infty)\,|\,\psi_{u_l}(t{=}\infty)>|^2 \tag{3.1}$$

for each pulse sequence l. The dynamics is governed by the time dependent Schrödinger equation for three surfaces:

$$i\hbar\frac{\partial}{\partial t}\begin{pmatrix}\psi_u\\ \psi_v\\ \psi_g\end{pmatrix} = \begin{bmatrix}\hat{\mathbf{H}}_u & \hat{\mathbf{V}}_{vu} & \hat{\mathbf{V}}_{gu}\\ \hat{\mathbf{V}}_{uv} & \hat{\mathbf{H}}_v & \hat{\mathbf{V}}_{gv}\\ \hat{\mathbf{V}}_{ug} & \hat{\mathbf{V}}_{vg} & \hat{\mathbf{H}}_g\end{bmatrix}\cdot\begin{pmatrix}\psi_u\\ \psi_v\\ \psi_g\end{pmatrix} \tag{3.2}$$

where $\hat{\mathbf{H}}_i = \dfrac{\hat{\mathbf{P}}^2}{2m} + V_i$ is the diagonal part of the Hamiltonian, $\mathbf{V}_{ij} = \mu^{ij}\varepsilon(t)$ where μ^{ij} is the dipole moment between surface i and j and $\varepsilon(t)$ represents the amplitude of the electromagnetic field. The pulse envelope is assumed to be Gaussian:

$$\varepsilon(t) = Ae^{\frac{(t-t_l)^2}{2\sigma}}\cos(\omega_l t) \tag{3.3}$$

where the maximum amplitude is obtained at $t = t_l$, and the central frequency of the pulse is ω_l. The first excitation pulse has the same form but a constant frequency.

The inversion procedure seeks the potential of $\hat{\mathbf{H}}_v$ ie : $\hat{\mathbf{V}}_v$, the first excited state potential. Once again the known facts about the system are distributed through time. At $t = 0$ the system is known to be in the ground state $\psi_g(t{=}0)$ which can be calculated knowing the ground state potential $\hat{\mathbf{V}}_g$ and the relaxation method [9]. At $t{=}0$ the amplitude of the other states is zero. For each experiment the time and central frequency of the two pulses is known. At the final time $t = \infty$ the probability to be on the second excited state is measured. From this data obtained from a set of L independent experiments, the excited state potential is inferred. Since the amount of experimental data is limited, a potential is sought which has minimum deviation from a known potential form. Mathematically this is equivalent to seeking the minimum of the functional J_1:

$$J_1 = \int_0^\infty\left[\hat{\mathbf{V}}_v^o(R) - \hat{\mathbf{V}}_v^n(R)\right]^2 dR \tag{3.4}$$

Where $\hat{\mathbf{V}}_v^o$ is the reference potential which can be obtained from ab-initio calculations or some other method. $\hat{\mathbf{V}}_v^n$ is the inferred new potential. Once a potential is set the dynamics can be solved for all sequences of pulses. The results are compared to the experiment. The objective is then to minimize the difference between the experimental and calculated data:

$$J_2 = \sum_{l=1}^{L} \left[<\psi_{u_l} | \psi_{u_l}>_l - A_l \right]^2 \tag{3.5}$$

where the sum is on all experiments performed and the wavefunction ψ_{u_l} is the result of the calculation of the lth experiment and A_l is the measured data of the l th experiment. The two functionals J_1 and J_2 which are to be minimized are subject to the constraint that the dynamics is governed by the Schrödinger equation. This in turn leads to the minimalization of the functional J_3:

$$J_3 = \alpha J_1 + J_2 + \sum_{l=1}^{L} \int_0^{\infty} \left[<\chi_l | (i\frac{\partial}{\partial t} - \hat{\mathbf{H}}) | \psi_l> + c.c \right] dt \tag{3.6}$$

Where χ_l is the Lagrange multiplier arising because the motion is constrained by the Schrödinger equation. α is a weight which emphasizes the minimization of J_1 or J_2. To minimize equation (3.6) both the wavefunction ψ_l and the potential $\hat{\mathbf{V}}_v$ are varied independently. The result which is obtained by integrating by parts and taking the variations $\delta\psi_l$ and $\delta\hat{\mathbf{V}}_v$ to zero is a set of Euler Lagrange equations. From the variation of ψ one obtains:

$$i\frac{\partial \psi_l}{\partial t} = \hat{\mathbf{H}}_l \psi_l \tag{3.7}$$

where $\hat{\mathbf{H}}_l$ is the hamiltonian of the lth pulse sequence. Equation (3.7) is subject to the initial conditions:

$$\psi_l(0) = \begin{bmatrix} 0 \\ 0 \\ \psi_g(0) \end{bmatrix} \tag{3.8}$$

and the backward equation of motion for χ:

$$-i\frac{\partial \chi_l}{\partial t} = \hat{\mathbf{H}}_l \chi_l \tag{3.9}$$

subject to the final conditions:

$$\chi_l(t{=}\infty) = \begin{bmatrix} A_l \dfrac{\psi_{u_l}(t{=}\infty)}{<\psi_{u_l} | \psi_{u_l}>} \\ 0 \\ 0 \end{bmatrix} \tag{3.10}$$

From the variation of the potential one obtains:

$$\Delta\hat{\mathbf{V}}_v(R) = \frac{1}{L}\sum_{l=1}^{L}\int_{t=0}^{\infty}\left[\ \psi_{v_l}(R,t)\chi^*_{v_l}(R,t) + c.c.\right]\ dt \tag{3.11}$$

where $\Delta\mathbf{V}_v$ is the difference between the reference potential and the new potential.

Examining equation (3.11) the final potential difference is obtained by the overlap of the forward moving wavefunction ψ which originated on the ground state and the backward moving wavefunction χ which originated on the second excited state u. The meeting ground is the first excited state potential which is determined by this process. An iterative procedure to solve these equation for the ICN molecule is under way [14].

The inversion procedure just described can be complicated if one adds the lack of knowledge of the dipole functions $\mu^{ij}(R)$. Also the molecule has two more degrees of freedom which have not been considered. Nevertheless this procedure can be the first step in a more elaborate assault on the full complexity of the molecular potentials and dipole functions.

Conclusions

Time reversal symmetry which allows propagation backwards in time has enabled a new inversion procedure. Because the experimental data is inconclusive a full picture which can be used as an initial state for backward propagation cannot be constructed. The problem is overcome by a combination of forward and backward propagation which enables additions to the experimental data from sources of knowledge about the molecule. The two examples presented both rely on the same principle. The coloumb explosion data is more complete than the ICN impulsive pulse experiment. The result is faster convergence and less input from guessed potentials. The illustration shows that multidimensional inversion converges rapidly. The iterative process serves both as a filter for noisy experimental data and as a way to incorporate knowledge distributed through the sequence of events. The extension of these ideas to more degrees of freedom is closely related to the development of multidimensional time dependent propagation technique. A method using bound coordinates for triatomic molecules is under development [12]. Other methods such as the use of hyperspherical coordinates are also under investigation.

The main obstacle in the development of inversion procedures has been the inconclusiveness of experimental data. The mathematical inversion procedures use strong analytic properties to fill in the gap. This in turn leads to very strong demands on the quality of the experimental data. The iterative procedure developed here succeeds because the inversion goal is more modest. Instead of relying completely on the experimental findings the method incorporates intelligent guesses on the potential. Convergence of the iterative procedure, unlike the mathematical inversion, only means that a local minimum has been reached. This weaker result on the other hand has the advantage that the requirements on the experimental data are relaxed guaranteeing that a molecular potential is obtained. A similar approach has been proposed by Shi and Rabitz [15] for inversion of scattering data. The method is based on the stationary solution of the scattering problem and therefore does not utilise directly the time reversal symmetry.

To conclude, the work presented here only begins to reveal the wealth of phenomena which be tackeled. With the appearence of new experimental possibilities and the development of new three-dimensional time dependent propagation methods many molecules can be investigated. These methods will lead eventually to new insight on the structure of many moleculs.

Acknowledgements

We wish to thank the help and many stimulating discussions with David Tannor, Stuart Rice, Audrey Hammerich, Ron Naaman, Zeev Veger and Haim Feldman. This research was partially supported by a grant from the Binational United States - Israel Foundation. The Fritz Haber Research Center for Molecular Dynamics is supported by the Minerva Gesellschaft für die Forschung, GmbH München, Federal Republic of Germany.

References

[1] R. G. Newton, *"Invers Schrödinger Scattering in Three Dimensions"* (Springer-Verlag 1989).

[2] R. B. Gerber, *Comm. Atom. Mol. Phys.* **17**,65 (1985).

[3] J. H. Schulte *" The Theory of Molecular Spectroscopy "* (North Holland 1976).

[4] M. S. Child, *"Molecular Collision Theory"* (Academic Press 1974).

[5] Z. Veger, R. Naaman, E. P. Kantor *Science* **244**, 436 (1989).

[6] R. B. Bernstein and A. H. Zewail, *J. Chem. Phys.* **87**, 2397 (1987).

[7] R. Kosloff, S. A. Rice, P. Gaspard, S. Tersgini and D. Tannor, *Chem. Phys.* **139** 201 (1989).

[8] S. Shi, A. Woody, and H. Rabitz, *J. Chem Phys.* **88**, 6870 (1988).

[9] R. Kosloff and H. Tal-Ezer, *Chem. Phys. Lett.* **127**, 223 (1986).

[10] R. Kosloff, *J. Phys. Chem.* **92**, 2087 (1988).

[11] C. LeForestier, R. Bisseling, C. Cerjan, M. Feit, R. Friesner, A. Guldberg, A. D. Hammerich, G. Julicard, W. Karrlein, H. D. Meyer, N. Lipkin, O. Roncero and R. Kosloff, *J. Comp. Phys.* **93**, (1991).

[12] C. A. Williams J. Qian and D. Tannor *J. Chem. Phys.* (1991).

[13] H. Feldman, Msc Thesis Weizman Institute (1991).

[14] Roi Baer and R. Kosloff, to be pulished (1991).

[15] S. Shi and H. Rabitz, *Comp. Phys. Rep.* **10**, 1 (1989).

TRIPLET STATES DYNAMICS IN ISOLATED MOLECULES

Ofer Sneh and Ori Cheshnovsky
School of Chemistry, Reymond and Beverly Sackler Faculty of Exact Sciences
Tel Aviv University, Tel Aviv 69978, Israel.

ABSTRACT. We have experimentally studied the dynamics of triplet states in isolated molecules. Central in this project is the development of a sensitive detector for the triplet states. This detector is based on the Surface Penning-Ionization of low work-function surfaces by the triplet excited states. Mounting the detector on a translational stage enabled us to measure the decay times of excited triplet states in the range of 5 μsec - 3.5 msec. The combination of this detector with phosphorescence and fluorescence measurements has allowed us to determine various dynamic quantities, such as the relative pure radiative rates and the triplet formation (by intersystem crossing) quantum yield.

The absolute values of the intersystem crossing and internal conversion processes in isoquinoline excited to its S_2 vibronic levels were obtained. The triplet decay rates of pyrazine and methylpyrazine in the unexplored range between the T_1 and the S_1 origins were determined. We have attributed the very strong increase in the intersystem decay rate and the large methyl substitution effect to the proximity effect of the $n\pi^*$ and $\pi\pi^*$ triplet states. The decay times and the relative pure radiative lifetimes in isolated benzaldehyde over the T_1 and S_1 manifolds were obtained. The fine details of the singlet-triplet coupling were measured.

1. Introduction

The phenomena of energy redistribution in isolated excited molecules has been central in chemical kinetics over many decades [1]. A large revival of the subject followed the introduction of supersonic beam techniques which enabled the study of well isolated ultra-cold molecules in beams [2,3].

In optically excited molecule isolated in supersonic-beams, the idea of *temperature* as a gauge for the molecular internal energy is replaced by the definite quantity of *excess energy*. The supercooled molecules are gas-phase isolated and thus the experiments are not subjected to molecule-solvent interactions [4,5] or to vibrational relaxation [6,7] which characterize bulb and cold matrices experiments. The nonradiative dynamics in isolated molecules is isoenergetic. There is no dissipation of energy associated with the dynamics.

The triplet state serves as an important intermediate in nonradiative processes of molecules. The dynamics of the triplet-state-generation, and its decay, plays an important role in photochemical processes. In spite of its importance, the investigation of the triplet states of cold isolated molecules in supersonic expansions has been limited [8], due to the low oscillator strength and the low emission rates of the triplet state. Other experimental complications, which origin from the long lifetimes of triplet states dynamics, are the finite travel of the molecules during the long time-period of the experiment, intra-beam collisions and the effects of beam divergence.

J. Jortner et al. (eds.), Mode Selective Chemistry, 359–370.

Recently we have introduced an experimental technique for the interrogation of isolated molecules excited by lasers to electronic metastable states [8,9]. The method is based on the phenomena of electron ejection from low work-function surfaces induced by the excited molecules. The basic idea has been developed by Klemperer, Freund and their coworkers [10,11] for the detection of metastables which were generated by electron impact in effusive beams. The information extracted from their experiments was crude due to the high internal temperature of the molecules in the effusive beam and due to the undefined excitation energy.

By combining the techniques of supersonic beams and laser excitation we have developed the method of Surface Ejection of Electrons by Laser Excited Metastables (SEELEM) as a versatile spectroscopic technique in the study of the triplet state. Mounting the detector on a translational stage enabled us to vary the excitation-detection distance and to measure the decay times of excited triplet states in the range of 5 μsec - 3.5 msec. Fig. 1 displays the time of flight dependence of the SEELEM signals.

Our SEELEM method can sensitively detect triplets regardless of their radiative rate (detection of the triplet population). By combining SEELEM with Laser Induced Fluorescence (LIF), Laser Induced Phosphorescence (LIP) or with absorption in molecular beams we have performed a wide range of spectroscopic and dynamics measurements: the relative triplet formation quantum yields [9,12,13], the spectroscopy and the absolute total decay rates of directly excited triplet states [12-16], the relative pure radiative lifetimes of directly excited triplet states [16]. These experimental methods have enabled us to study several scientific subjects: the branching ratio of internal conversion (IC) and ISC [13], the dynamics of ISC decay of triplets to the ground state [14,15], and singlet-triplet electronic states mixing [16].

This work summarizes our experimental studies of triplet-states dynamics in supersonic-beams-isolated-molecules. The experiments were performed in a differentially pumped collimated supersonic beam apparatus. The experimental system is described in detail elsewhere [8,16]. More details on the SEELEM technique are given in appendix A. The following scientific projects were studied by the above techniques:

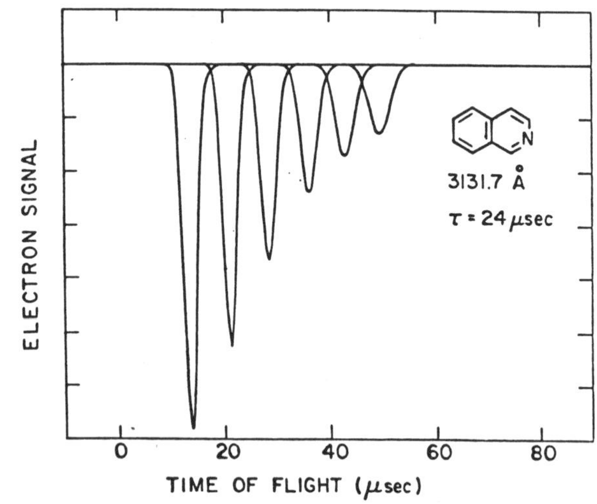

Fig. 1: Time of flight traces of SEELEM, taken at several excitation-detector distances, resulting from exciting isoquinoline to S_2 0-0. The dependence of the integrated amplitudes on the flight time fits well a single exponential decay of 24 μsec.

2. The Branching of Nonradiative Processes in Isoquinoline

The distinction between internal conversion (IC) and intersystem crossing (ISC) is central to the understanding of the photophysics of large aromatic molecules. Some isolated molecules show a dramatic increase in the rate of of nonradiative processes at some excess energy [17]. This phenomena was attributed to the strong energy dependence of internal conversion which takes over the weak dependence of intersystem crossing at some value of the excess vibrational energy. The validity of this model was never directly proved.

We have measured the absolute rates of internal conversion and of intersystem crossing in isoquinoline excited to its S_2 vibronic states (3131-3062 Å, Fig. 2) in supersonic beams. The combination of direct detection of the triplets, absorption and fluorescence-excitation spectra together with published lifetimes [18] allowed us to extract the IC and ISC rates [13]. We have found that the intersystem crossing rate barely depends on the excess vibrational energy in the S_2 manifold, being about $2x10^9$ sec^{-1}. The internal conversion rates grow by more than two orders of magnitude over the range of 0-800 cm^{-1} of excess vibrational energy in the S_2 state, up to $5x10^{10}$ sec^{-1}. At the S_2 electronic origin, intersystem crossing is the dominant nonradiative channel, while at vibrational energy exceeding 400 cm^{-1} internal conversion becomes dominant. These results confirm our general view on nonradiative processes in molecules with large electronic energy gap.

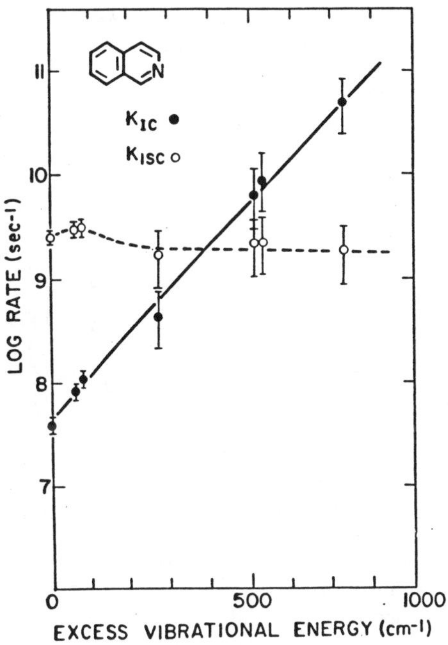

Fig. 2: A semi-logarithmic plot of the calculated nonradiative rates as a function of excess vibrational energy above the S_2 origin. On the logarithmic scale k_{ISC} is practically constant while k_{IC} fits well an exponential increase. The exponential coefficient is $4.3x10^{-3}$ cm and is 20 times larger than in naphthalene.

ISC, which is governed by a relatively small singlet-triplet energy gap, is Franck-Condon allowed and consequently has a weak vibrational energy dependence. IC, on the other hand, is electronically allowed but due to the larger singlet-singlet energy gap has small Franck-Condon coefficients at the origin which grow fast with the excess vibrational energy. The consequences are IC rates which grow steeply with excess vibrational energy.

Lim and his coworkers attributed this phenomena to a "channel three"-like behavior [19]. Our experimental rates are consistent with their physical model which assumed that k_{IC} is much smaller than k_{ISC} at the S_2 origin. However behind their model lies the assumption that the IC channel opens sharply because of IVR onset [19]. We, however, fitted a smooth exponential rise in the IC rate which competes with the relatively constant ISC. Whatever the exact mechanism is, we succeeded to show that the sharp change in the excess energy dependence of the nonradiative rates is indeed the onset of the dominance of IC. The question whether this channel branching is stimulated by IVR, as was proposed for benzene [20], is still open.

3. The decay of Isolated Triplet Pyrazine and Methylpyrazine

The dynamics of the $T_1 \rightarrow S_0$ intersystem crossing (ISC) processes are particularly interesting because they provide a physical system which should fit a simple theoretical model, with only one decay channel to consider. Recent experimental results have indicated that the nonradiative decay rates of triplet states in many heteroaromatic large molecules grow surprisingly fast with the excess vibrational energy above the triplet electronic origin [21]. These findings were in clear contradiction to the expected exponential behavior, predicted by the conventional theory of nonradiative transitions.

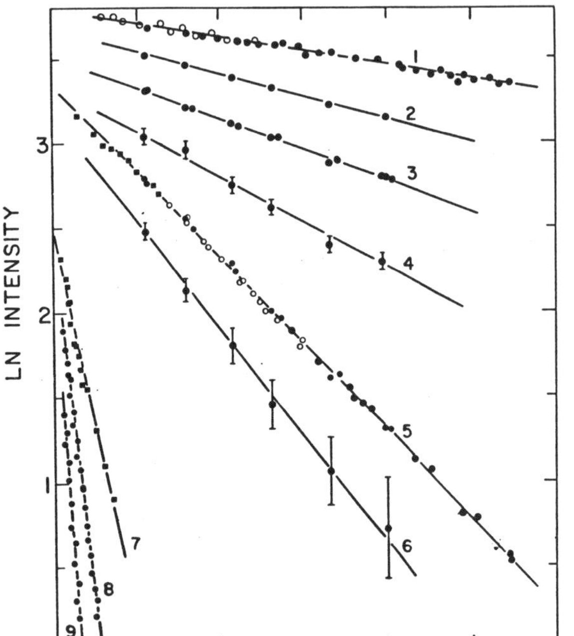

Fig. 3: Decay plots of several excited states of pyrazine T_1.
(1) 3727 Å, τ=1850 μsec ;
(2) 3643 Å, τ=1000 μsec ;
(3) 3596 Å, τ=620 μsec ;
(4) 3588 Å, τ=430 μsec ;
(5) 3574 Å, τ=200 μsec ;
(6) 3564 Å, τ=165 μsec ;
(7) 3535 Å, τ=40 μsec;
(8) 3522 Å, τ=24 μsec ;
(9) 3497 Å, τ=10 μsec.

Until recently, all studies on the decay rates of triplets in isolated molecules have been limited to energies above the origin of the S_1 state. The development of a satisfying theory for the observed dynamics has been hindered by the lack of measurements on most of the energy range between the T_1 origin and the S_1 origin. This energy range, which varies from 2000 to 9000 cm^{-1} in different molecules, is of utmost importance because it is expected to contain a steep rise in the decay rates. We have performed decay-rate measurements of optically excited triplet states of pyrazine [14] and methylpyrazine [15] in this "unvisited" energy range, and confronted them with existing theoretical models.

The decay rates of optically excited triplet states of pyrazine in supersonic expansion were measured. Fig. 3 presents the decay curves of vibronic levels of the triplet manifold in the energy range between the T_1 and the S_1 electronic origins. The excess energy dependence of the radiationless rate constants is displayed in Fig. 4. Decay rates between $5x10^2$-10^5 sec^{-1} were found in the 1800 cm^{-1} range of excess vibrational energy from the origin of the T_1 state. The decay rates are free of mode specificity and rotational effects. The pure radiative lifetime in the measured energy interval is vibronic independent.

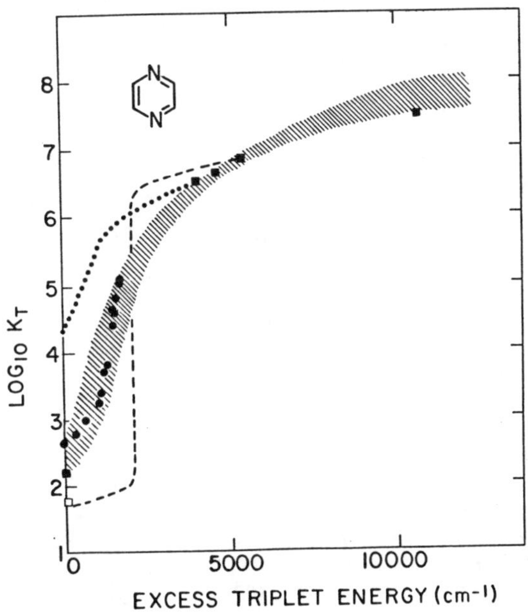

Fig. 4: The decay-rates of triplet pyrazine as a function of excess vibrational energy above the T_1 origin, superimposed with the predictions of some theoretical models. The doted line displays the predictions of the conventional models adapted from the calculations of Morse et al. [24]. The dashed line presents the analysis of Terazima et al. [25]. The hatched area represents our interpolation of the results of Sobolewski [22]. We interpolated his model calculations for the parameters of pyrazine using the values of 2.3-2.6 for the frequency ratios of the ν_{10a} accepting modes, and ΔE_{ST} = 26830 cm^{-1} for the singlet-triplet energy gap. All the models were made to fit the data taken at high excess vibrational energy. Note the much better correspondence of Sobolewski simulations.

The measured nonradiative rates support a model which suggests that certain vibrational modes, those which undergo large frequency changes in the excited state, control the strong vibrational energy dependence of the $T_1 \rightarrow S_0$ intersystem crossing of pyrazine [22]. This situation probably results from the proximity of the $^3n\pi^*$ $^3\pi\pi^*$ excited states in this molecule [23]. The results of this model along with two other accounts of the results (which have been rejected by us [24,25]) are displayed in Fig. 4.

We have studied the effect of methyl substitution on the $T \rightarrow S_0$ ISC dynamics of pyrazine. The decay rates of optically excited triplet states of methylpyrazine in supersonic beam were measured. The excess energy dependence of the radiationless rates in the energy range between the T_1 and the S_1 electronic origins of the isolated molecules is displayed in Fig. 5. The decay rates increase from $8x10^2$ to $3.3x10^4$ in the first 600 cm^{-1} excess energy range above the T_1 origin. Again, the decay rates are free of mode specificity and rotational effects. The large scale $T_1 \rightarrow S_0$ radiationless dynamics of pyrazine, is substantially enhanced by the methyl substitution. The results support the mechanism which rely on the large frequency changes of the vibrational modes in controlling the strong vibronic energy dependence of the $T_1 \rightarrow S_0$ intersystem crossing.

The enhancement of the excess energy dependence of the decay rates by the methyl substitution is rationalized by the physical picture of the "proximity effect" between nearby $^3n\pi^*$-$^3\pi\pi^*$ states. In the case of pyrazine this mechanism can cause a large reduction in the frequency of the $^3n\pi^*$-$^3\pi\pi^*$ vibronic coupling modes, which become the prominent T-S_0 accepting modes at the first few thousand cm^{-1} of excess triplet vibrational energy. Further reduction in the frequency of the coupling modes, induced by methyl substitution, is responsible for the increase in the decay rates over those of pyrazine.

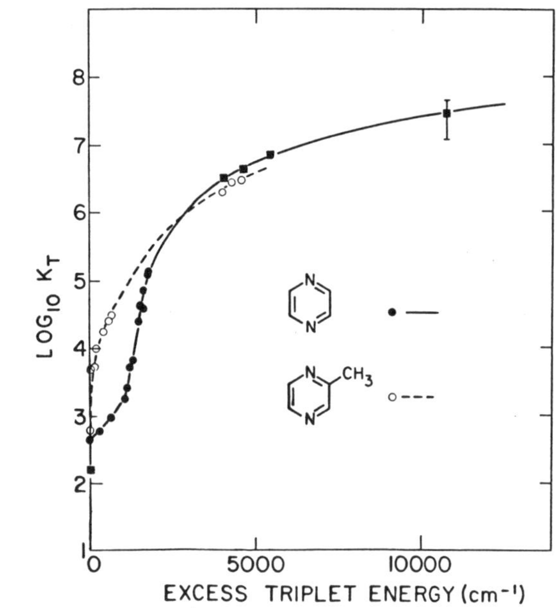

Fig. 5: The triplet decay-rates of pyrazine and methylpyrazine as a function of excess vibrational energy above the T_1 origin. The full circles represent our measurements for pyrazine. The open circles represent our measured decay rates of methylpyrazine. The full square at the origin is the cold matrices phosphorescence lifetime of pyrazine. The full squares at more than 4000 cm^{-1} are taken from ref. [21]. The smooth lines in the range 1450-4050 cm^{-1} (pyrazine) and 600-4020 (methylpyrazine) were interpolated in accordance with the simulations of ref. [22].

4. Triplet-Triplet and Singlet-Triplet Coupling and Dynamics in Benzaldehyde

Radiative processes in triplet benzaldehyde, the simplest aromatic carbonyl molecule, were extensively examined in condensed and gaseous phases. The phosphorescence emission from benzaldehyde and its derivatives is the most studied case of *dual phosphorescence* phenomena, where the phosphorescence is composed of two different (in both spectra and lifetime) components [26]. The ratio and temperature dependence of the two dual phosphorescence components is very sensitive to the solvent and to substitutions. It was attributed to the existence of two nearly degenerate $^3n\pi^*$ and $^3\pi\pi^*$ triplet states.

The spectroscopy of directly excited triplet states in supersonic beams [27-29] did not reveal pure $^3\pi\pi^*$ spectral features, probably due to their much lower oscillator strength.

The radiative and nonradiative dynamics of benzaldehyde in the 0-3750 cm^{-1} energy interval above the T_1 origin were explored by us [16]. Figs. 6 and 7 present the simultaneous LIP and SEELEM spectra of benzaldehyde in the triplet and singlet regions respectively. The relative radiative rates were extracted by combining two different simultaneous measurements: Laser Induced Phosphorescence (LIP) and SEELEM. The LIP signal is proportional to the product of the triplet population and the pure radiative rate, while the SEELEM signal is proportional to the triplet population only. Consequently, the ratio of the LIP signal and the SEELEM signal is proportional to the pure radiative rate.

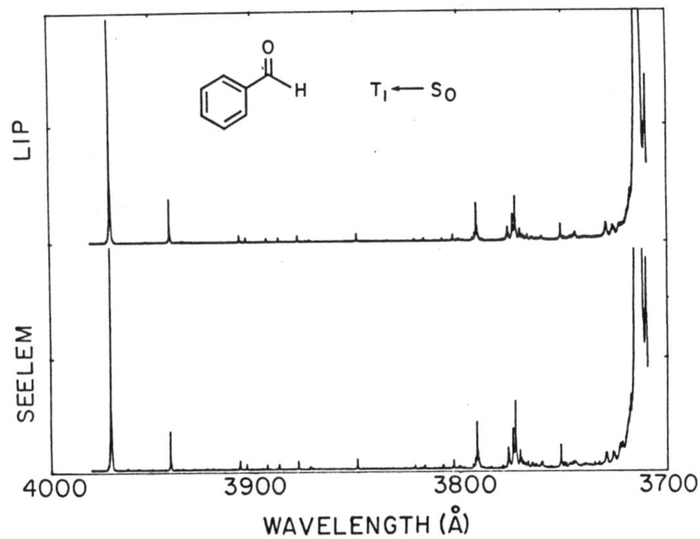

Fig. 6: Simultaneous LIP (top) and SEELEM (bottom) spectra of benzaldehyde excited at the first 1800 cm^{-1} above the T_1 origin. 1 torr of benzaldehyde was expanded in 0.5 b. of He. The flight times of the LIP and the SEELEM signals were 108 and 133 μsec respectively

Fig. 8 presents the relative pure radiative rate of benzaldehyde spanning the energy range of the T_1 and S_1 excitations. The details of the dynamics were rationalized as a development in the coupling of three electronic states. The $T_1(^3n\pi^*)$-$T_2(^3\pi\pi^*)$ vibronic coupling increases regularly in the region between 400 and ≃2000 cm^{-1} of excess vibrational energy above the T_1 origin. Starting from the excess energy of the S_1 origin (1730 cm^{-1}), the S_1 excitations are coupled to the sparse T_1-T_2 background, resulting in the increase of the pure radiative lifetime. The S_1 content in the coupled states is gradually diluted in the 1730-2500 cm^{-1} region due to the congestion of the triplet states. Finally, the T_2 character is eroded probably due to the larger density of T_1 states and the decrease in the T_1-T_2 isoenergetic Franck-Condon overlap.

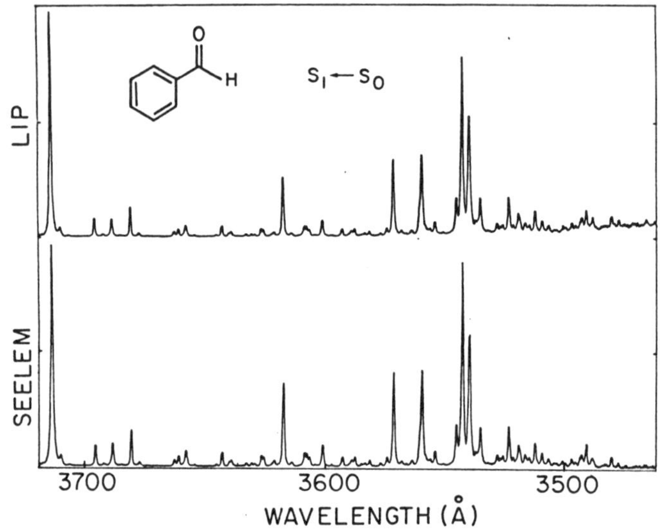

Fig. 7: Simultaneous LIP (top) and SEELEM (bottom) spectra of benzaldehyde excited at the first 2000 cm^{-1} above the S_1 origin. 1 torr of benzaldehyde was expanded in 0.5 b. of He. The flight times of the LIP and the SEELEM signals were 108 and 133 μsec respectively

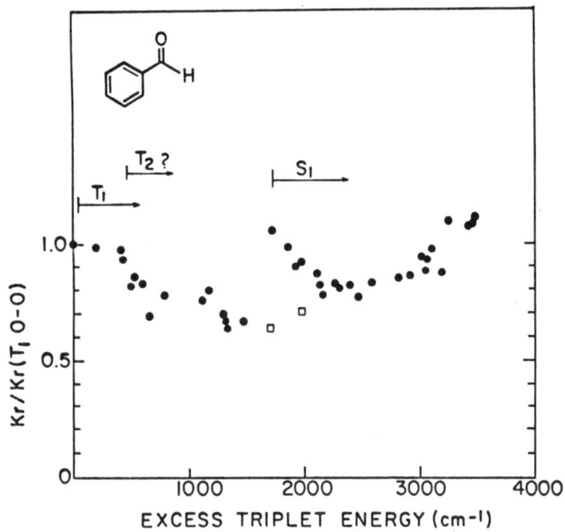

Fig. 8: The excess energy dependence of the radiative rates (normalized to the rate k_r at the T_1 0-0). The open squares represent the estimated pure radiative decay rates of the triplet manifold [16].

5. Quantitative Study of Singlet-Triplet Coupling in Benzaldehyde.

We have shown that the line profile and dynamics of the S_1 manifold in the 1730-2500 cm^{-1} can be described in terms of the interaction of an optically active state with a background of almost dark states, in accordance with the theory of radiationless transitions [30].

Monitoring the radiative rate profiles over the S_1 vibronic excitations in benzaldehyde enabled us to quantitatively study singlet-triplet coupling. In this molecule we have the rare opportunity to monitor at the same experiment the relative radiative rate of the diluted singlet level (doorway state) and the background triplet. Fig. 9 presents the lorentzian dilution of the doorway state (the S_1 origin) by the dark triplet manifold . We could evaluate for the first time the parameters of radiationless transitions theory: the density of states and the coupling strength. We have estimated the density of coupled states ρ, to be about 200 states/cm^{-1} and the coupling strength V, to be 0.05 cm^{-1}.

Actually, the applicability of the theory of RT is not obvious for benzaldehyde. In this "intermediate level structure case" ($\rho \cdot V \simeq 10$) the coupling strength varies randomly. Thus, the simple model of equally spaced equally coupled states of Bixon and Jortner may be inadequate. This fact was demonstrated in the high resolution domain, in the case of pyrazine S_1 excitations [31]. It is gratifying that a 0.3 cm^{-1} (our laser) excitation resolution averages the individuality of the states dynamics completely (only 45-90 states averaging). This means, that the applicability of the simple model in predicting the general coupling and dynamics is beyond the limits of its basic assumptions.

We have observed the further dilution of the doorway state with the excess S_1 energy caused by the growth of the triplet density of states. Consequently the relative contribution of the singlet state to the oscillator strength is completely washed out.

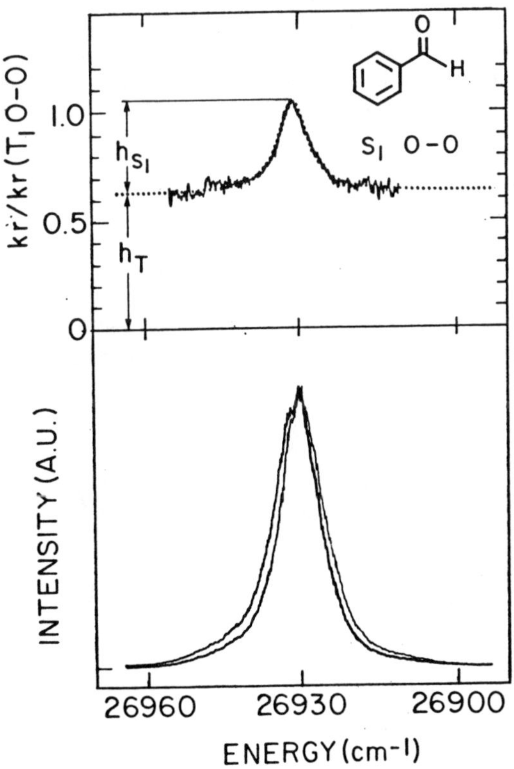

Fig. 9: Simultaneous rovibronic LIP and SEELEM spectra (bottom) at the vicinity of the S_1 0-0. The LIP spectrum is narrower than the SEELEM spectrum. The LIP spectrum was divided by the SEELEM one to provide the pure radiative rate line profile (top). The k_r line profile was fitted to a constant plus Lorentzian shape [16].

Appendix A: The SEELEM detector.

The SEELEM detector assembly is displayed in Fig. 10. Its central element is a low work-function surface, which emits electrons with high efficiency when it is hit by the excited molecules. This surface, which serves as the detector cathode, is inclined in three directions: towards the molecular beam, the alkali oven and an electron-multiplier. A stable low work-function surface is maintained by a careful continuous deposition of alkali vapors on the metallic substrate (0.01-10 monolayers per second). The alkali metal (usually potassium or potassium/sodium alloys) is evaporated from a small oven and is baffled by a double radiation shield. The cathode structure is closed by a fine mesh and is $\simeq$ -400 V biased relative to the electron multiplier, in order to maximize the detection efficiency. A two multichannel plates stack, with a 10^5-10^6 gain, amplifies the signal.

The sensitivity of the detector is stable within 3 % over six hours of continuous work. The detection efficiency depends on the metastable molecule and varies between 10^{-3} to 0.05 for molecules in which the triplet-electronic energy exceeds the surface work-function. A base pressure of better than $5x10^{-7}$ torr is maintained in the detector chamber for a reliable operation. The whole assembly is very compact, with nominal dimensions of 90x90x30 mm.

The mechanism of SEELEM has not been studied by us, yet. Following Klemperer Freund and coworkers [10,11], we believe that the mechanism is surface Penning ionization. Basically, it is a resonant energy-transfer process between the metastable molecule and the surface, which occurs via a short-range exchange interaction. The process ends with an electron from the metal occupying the ground state of the molecule and the ejection of the excited metastable electron.

The detection efficiency was found to be sensitive only to the excess electronic energy. We have evaluated the relative quantum yields of several molecules, optically excited to their S_1 vibronic states, from the simultaneous spectroscopy of SEELEM and LIF [9,12,13].

Fig. 10: A sketch of the SEELEM detector assembly. (1) emitting surface; (2) electron multiplier assembly ; (3) anode ; (4) radiation shield ; (5) the alkali oven ; (6) electrical connectors ; (7) translation stage ; (8) the supersonic beam direction.

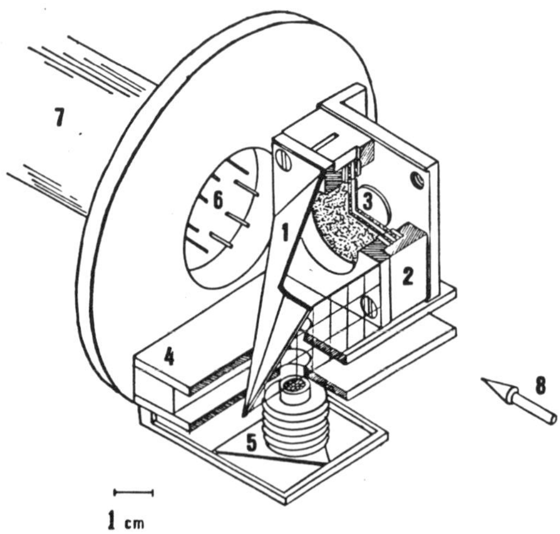

The fluorescence quantum yield values of the S_1 excitations, taken with the SEELEM/LIF method for several molecules where ISC is the dominant nonradiative process, are similar to those taken in supersonic beam by absorption/fluorescence comparison. Thus, the SEELEM sensitivity is constant over all the vibronic manifold.

Additional support to this mechanism is the relative sensitivity ratio of the three anthracene derivatives: anthracene, 9-cianoanthracene and 9,10-dichloroanthracene which is 10.8 : 4.4 : 1 respectively[53]. The excess energy of the triplet in these molecules is about 2.4 eV which is close to the work-function of the potassium surface. The S_1 origins of 9-cianoanthracene and of 9,10-dichloroanthracene are red shifted from the anthracene S_1 origin by 1520 and 1750 cm^{-1} respectively. The origins of the respective T_1 states are expected to have similar shifts. Because of the proximity of the triplet origins to the SEELEM threshold, the sensitivity should decline monotonically with the red shifts.

The threshold condition for electron ejection requires that the electronic excitation exceeds the surface work-function. This is consistent with a fast electronic-energy transfer mechanism such as Penning ionization. In some molecules, which possess triplet-electronic energy which is less than the work-function, the SEELEM sensitivity is low (10^{-5} for tetracene) and depends strongly on the kinetic energy. In cases where the triplet energy exceeds the work-function, the sensitivity is independent of the kinetic energy.

The strongest support to the assumption that surface-Penning-ionization is the mechanism for SEELEM, lies in the specific sensitivity of the electron ejection process to the electronic part of the internal energy. The decays of the triplet states that we have measured are nonradiative. Namely, the molecules remain isoenergetic while the energy is exchanged from electronic to vibrational form. The fact that the SEELEM signal decays to zero proves that the sensitivity of the surface to vibrational energy is orders of magnitude lower than its sensitivity to electronic energy.

The applicability of the SEELEM technique depends on its inherent selectivity to excess electronic energy. In singlet excitation the assumption is, of course, that after several microseconds of flight time, none of the excited molecules which reach the detector remain in the singlet state. Thus, the detector exclusively detects excited triplet states [12-16].

Acknowledgements: We wish to thank Joshua Jortner, Aviv Amirav and Uzi Even for stimulating discussions throughout this project. Aviv Amirav has cooperated with us in the isoquinoline study. Dana Dünn-Kittenplon has participated in the methylpyrazine project. This research was supported by the Fund for Basic Research, administered by the Israel Academy of Sciences and Humanities and the James Franck German-Israeli Binational Program in Laser-Matter Interaction.

References

1. Rice S. R., (1977), "Some Comments on the Dynamics of Primary Photochemical Processes". in *Excited States*, ed. E. C. Lim, Academic, New-York, Vol. III, p. 111.
2. Levy D. H., Wharton L., and Smalley R. E., (1977), "Molecular Optical Spectroscopy with Supersonic Beams and Jets", *Acc. Chem. Res.*, **10**, 134.
3. Scoles G., ed., (1988), "*Atomic and Molecular Beam Methods*", Oxford U. press, New York, Oxford.
4. Madej S. L., Gillispie G. D., and Lim E. C., (1978), "Proximity Effects in the $T_1 \rightarrow S_0$ Radiationless transitions of Pyrazine and Its Methyl Derivatives", *Chem. Phys.*, **32**, 1.
5. Richards J. L., and Rice S. A., (1971), "Study of Impurity-Host Coupling in Shpolskii Matrices", *J. Chem. Phys.*, **54**, 2014.

6. Lim E. C., Laposa J. D., and Yu J. M. H., (1966), "Temperature Dependence of Intersystem Crossing in Substituted Anthracenes", *J. Mol. Spect.*, **19**, 412.
7. Lin S. H., (1966), "Rate of Interconvension of Electronic States and Vibrational Energy", *J. Chem. Phys.*, **44**, 3759.
8. Sneh O., Cheshnovsky O., (1990), "Triplet State Interrogation in Supersonic Beams by Surface Electron Ejection", *J. Israel Chem. Soc.*, **30**, 13.
9. Sneh O., and Cheshnovsky O., (1986), "Surface Ejection of Electrons by Laser Excited Metastables of Anthracene Molecules", *Chem. Phys. Lett.*, **130**, 53.
10. Freund R. S., and Klemperer W., (1967), "Molecular Beam Time of Flight Measurements for the Study of Metastable and Repulsive Electronic States", *J. Chem. Phys.*, **47**, 2897.
11. Hemminger J. C., Wicke B. G., and Klemperer W., (1976), Delocalization of Electronic States Energy in The Lowest Triplet States of Molecules", *J. Chem. Phys.*, **65**, 2798.
12. Sneh O., and Cheshnovsky O., (1988), "Triplet Selectivity in Surface Ejection of Electrons by Laser Excited Metastables", *Chem. Phys. Lett.*, **146**, 216.
13. Sneh O., Amirav A., and Cheshnovsky O., (1989), "The Branching of Nonradiative Processes in Isoquinoline", *J. Chem. Phys.*, **91**, 3532.
14. Sneh O., Dünn-Kittenplon D., and Cheshnovsky O., (1989), "The Decay of Triplet Pyrazine in Supersonic Jets", *J. Chem. Phys.*, **91**, 7331.
15. Sneh O., and Cheshnovsky O., (1991), "The Decay Rates of Triplet Pyrazine and Methylpyrazine - Substitution Effects", Submitted to *J. Chem. Phys.*.
16. Sneh O., and Cheshnovsky O., (1991), "The Dynamics of Triplet States in Beam Isolated Benzaldehyde", *J. Phys. Chem.*, in press, to be published in September 1991.
17. Hsieh J. C., Huang C. S., and Lim E. C., (1974), "Radiationless Singlet Deactivation in Isolated Large Molecules. I - Naphthalene, Naphthol, and Naphthylamine", *J. Chem. Phys.*, **60**, 4345.
18. Felker P. M., and Zewail A. H., (1983), "Jet Spectroscopy of Isoquinoline", *Chem. Phys. Lett.*, **94**, 448.
19. Forch B. E., Okajima S., and Lim E. C., (1984), "Channel Three Like Behaviour of Photoexcited Isoquinoline Vapors: A Model for Electronic and Vibrational Relaxation", *Chem. Phys. Lett.*, **108**, 311.
20. Riedle E., Weber Th., Schubert U., Neusser H. J., and Schlag E. W., (1990), "Back to the Roots of "Channel Three": Rotationally Resolved Spectra of the $6^1_0 1^1_0$ Band of C_6H_6", *J. Chem. Phys.*, **93**, 967.
21. Smalley R. E., (1982), "Vibrational Randomization Measurements with Supersonic Beams", *J. Phys. Chem.*, **86**, 3504.
22. Sobolewski A. L., (1987), "On the Excess Energy Dependence of Radiationless Decay Rate Constant", *Chem. Phys.*, **115**, 469.
23. Lim E. C., (1977), "Vibronic Interactions and Luminescence in Aromatic Molecules with Nonbonding Electrons". in *Excited States*, ed. E. C. Lim, Academic, New-York, Vol. III, p. 305.
24. Morse M. D., Puiu A. C., and Smalley R. E., (1983), "Intramolecular Vibrational Relaxation: Effects on Electronic Nonradiative Relaxation Rates", *J. Chem. Phys.*, **78**, 3435.
25. Terazima M., Yamauchi S., and Hirota N., (1986), "Temperature Dependence of the Triplet Lifetime of Pyrazine", *J. Phys. Chem.*, **90**, 4294.
26. Duben A. J., Goodmann L., and Koyanagi M., (1974), "Interstate Interaction in Aromatic Aldehydes and Ketones", In "Excited States", Ed. E. C. Lim, Vol. I p. 295, Academic Press, New York and London.
27. Ohmori N., Susuki T., and Ito M., (1988), "Why Does Intersystem Crossing Occur in Isolated Molecules of Benzaldehyde, Acetophenone, and Benzophenone", *J. Phys. Chem.*, **92**, 1086.

BOND-SELECTIVE MEMORY EFFECTS FOR 1s-LEVEL EXCITATION OF MOLECULES: ZERO KINETIC ENERGY (ZEKE) ELECTRON-PHOTOION COINCIDENCE MEASUREMENTS

Klaus Müller-Dethlefs* and Edward W. Schlag

Institut für Physikalische und Theoretische Chemie
Technische Universität München, Lichtenbergstrasse 4
W-8046 Garching, Germany

Abstract

Bond-selective fragmentation of the molecules N-N-O, $CF_2=CH_2$ and CF_3-CH_3 is observed for site-specific 1s excitation. Excitation at either the terminal or central N-atom in N_2O, or at either the fluorinated or hydrogenated C-atom in $CF_2=CH_2$ and CF_3-CH_3 shows that the molecule memorizes the initial site of energy deposition, hence a chemical "memory effect".

To follow the site-specific fragmentation channels, we employ a new "Zero Kinetic Energy (ZEKE)" photoelectron - photoion coincidence technique using monochromatized X-ray synchrotron radiation from BESSY. The detection of ZEKE electrons ejected from a 1s-orbital serves to select the site-specific excitation of an atom within the molecule.
The fragmentation channels could be assigned with high mass resolution by using a reflectron time-of-flight mass spectrometer.

Introduction

Bond selective chemistry, or rather the search for it, has attracted considerable attention during the last two decades. For core-level excitation of molecules, site-specific fragmentation effects have been observed that indicate some bond selective chemistry /1, 2/. Compared to outer-shell (e.g. valence) excitation, the dynamics of inner-shell excitation, ionization and relaxation processes is far more complex.

From the experimental point of view, the study of core-level

* Author to whom correspondence should be addressed

J. Jortner et al. (eds.), Mode Selective Chemistry, 371–392.

excitation and photoionization processes in free molecules in low density molecular beams relies on the availabilty of high intensity tunable monochromatic X-ray radiation in the soft X-ray region from synchrotron radiation storage rings.

For the search for bond selective chemistry, the excitation of inner-shell orbitals in molecules by soft X-ray synchrotron radiation offers a particular advantage since core orbitals like C (1s), N (1s) etc. are generally strongly localized at a specific atom and do virtually not participate in the chemical bonding. For the same kind of atom a **site-specific** excitation (i.e. the formation of a $1s^{-1}$ core hole localized at a specific atom) is made possible due to the "chemical shift" /3/. This shift in energy of the 1s inner-shell orbitals for different ligands can be interpreted as an electron correlation effect between the valence electronic structure and the core level which depends on ligand properties like their electronegativity. For example, the binding energy of the C_F (1s) orbital of fluorinated hydrocarbons is some eV higher compared to the C_H (1s) orbital. To understand bond selective chemistry and core level excitation it is of great interest to determine whether fragmentation channels depend on the **site of excitation** or randomization of energy distribution destroys the molecule's **memory** of the excitation process.

Site specific fragmentation processes in acetone following C 1s $\rightarrow$ Rydberg orbital and 1s $\rightarrow$ π^* excitation were first reported by Eberhardt et al. /4,5/ in a mass spectrometric study using monochromatized tunable synchrotron radiation. In this example, the 1s $\rightarrow$ π^* transition leads to a localized C_O ($1s^{-1}$) core hole (i.e. site selectivity), since the total cross section in the below threshold region is dominated by that resonance. The fragmentation patterns observed in the π^*-resonance were quite different to those observed in the threshold region. Other studies on site-selective fragmentation processes were carried out by Nenner et al. /6/, Murakami et al. /7/ and Murphy et al. /8/. Many aspects of chemistry induced by core orbital excitation were recently reviewed by Hanson /9/. The trend of such investigations is to use electron - ion coincidence techniques /10, 11, 12, 13/ and to resolve specific channels e.g. by use of Auger electron-photoion coincidences /7, 8, 9, 14, 15 / or by photoion-photoion coincidences /16, 17/.

In recent papers /18, 19, 20/ we have shown that "Zero Kinetic Energy" (ZEKE) spectroscopy /21, 22, 23, 24/ and ZEKE photoelectron-photoion-coincidence (ZEKE-PEPICO) measurements are powerful tools to investigate the inner-shell excitation and relaxation processes in atoms and molecules. With the ZEKE - PEPICO technique, site-selectivity is not only obtained in resonances, when the total cross-section is dominated by the transition into a bound orbital, but also at the ionization thresholds. This allows to distinguish atoms (of the same kind) in different parts of the molecule by their different 1s ionization energies and hence serves to selectively produce a localized $1s^{-1}$ core hole at that atom. In particular, ZEKE-PEPICO studies are very well suited since

i) the theoretical detectivity for ZEKE photoelectrons is one and in practice only limited by the efficiency of the detector
ii) the ionization thresholds can easily be determined and hence the emission of ZEKE electrons from a specific **excitation site** can be used to distinguish selected atoms in the molecule
iii) the relaxation and fragmentation dynamics of such **localized** core-hole states, as defined by the ZEKE electron emission, can be studied by measuring the fragments in coincidence with the ZEKE electron.

To study the dependence of the molecular fragmentation processes on the site of the excitation we chose small hydrocarbons with two carbon atoms in a different chemical environment. Fluorinated hydrocarbons offer a model system for such investigations as shown in earlier coincidence experiments /1, 2/. The fluorine atom has a very strong electronegativity and hence produces a high chemical shift of the C (1s) orbital. In the earlier atom-specific fragmentation studies on fluorinated hydrocarbons a static field geometrical electron analyzer, a linear time-of-flight mass-spectrometer and tunable synchrotron radiation as excitation source were used. For several molecules the observed fragmentation patterns show a strong dependence on the excitation site and the molecule seems to memorize the intial localized energy deposition, hence a chemical "memory effect" /1, 2/.

In contrast to ionization from valence orbitals, the formation of ZEKE electrons from the excitation of core orbitals (here we consider only 1s orbitals), can occur in the **photoexcitation** and in the **relaxation** channel. In the 1s excitation channel, ZEKE electrons are emitted by direct ionization, here denoted 1s → cont, when the photon energy matches the binding energy of the 1s electron. The simultaneous excitation of a valence electron, the so called "shake-up" process, leads to satellite peaks in the ZEKE spectra at energies above the 1s ionization threshold, as described in refs. 18, 19. For the experiments reported here, shake-up satellites are not considered. The $1s^{-1}$ core hole state formed by direct 1s ionization relaxes by emission of a high energy Auger electron, leading to the formation of a **doubly charged**, highly excited ion.

The formation of ZEKE electrons in the relaxation channel is important for excitation into bound resonances, here denoted 1s → res or 1s → π^* excitation. The most probable mechanisms leading to ZEKE electrons from resonance excitation is the Auger "shake-off" and, very important, the "participator valence Auger" decay, which follows the relaxation of the $1s^{-1}$ core hole by the primary core Auger process. Hence also for excitation into bound resonances, the emission of an energetic Auger electron (from the core Auger relaxation) and of a ZEKE electron from the valence Auger process leads to a **doubly** charged ion. The final states of the doubly charged ion will be different for ZEKE electron formation in the **relaxation** channel (i.e. for resonance excitation) compared to ZEKE electron formation in the **excitation** channel, e.g. by direct ionization. Fig. 1 exemplifies the mechanisms leading to ZEKE electrons for N_2O.

This model is suggested from ZEKE photoelectron-photoion coincidence measurements of nitrous oxide, reported recently /19/, photoelectron spectra /25/ and the theoretical calculation of Auger relaxation rates from F. Larkins /26/.

These results suggest that the localized Auger process is preferred for the main relaxation mechanism of the $1s^{-1}$ core hole state. This can be rationalized in terms of the coupling strength between the 1s shell and the different orbitals in the valence band. In first order the coupling strength is proportional to the electron density at the site of excitation. This implies that the probability for core Auger transitions depends on the electron density of the valence orbitals near the excitation site (the $1s^{-1}$ core hole state) and hence a core Auger process is favored which is preferentially localized **at the site of excitation**. From this model, depending on the electronic structure of the doubly charged ion formed, one would expect more or less bond-specific fragmentation of the molecule. The model of localized core Auger relaxation has also implicitly been invoked in other work /7, 8/.

In this paper, we report the ZEKE-photoelectron spectra of CF_2=CH_2 and CF_3-CH_3 at the carbon K-edge and fragmentation mass-spectra from ZEKE-PE-photoion coincidence measurements for site-specific excitation.

Fig.1 (following page): Formation of ZEKE electrons: a) for resonant excitation by Auger "shake-off" or "participator valence Auger"; b) by direct ionization

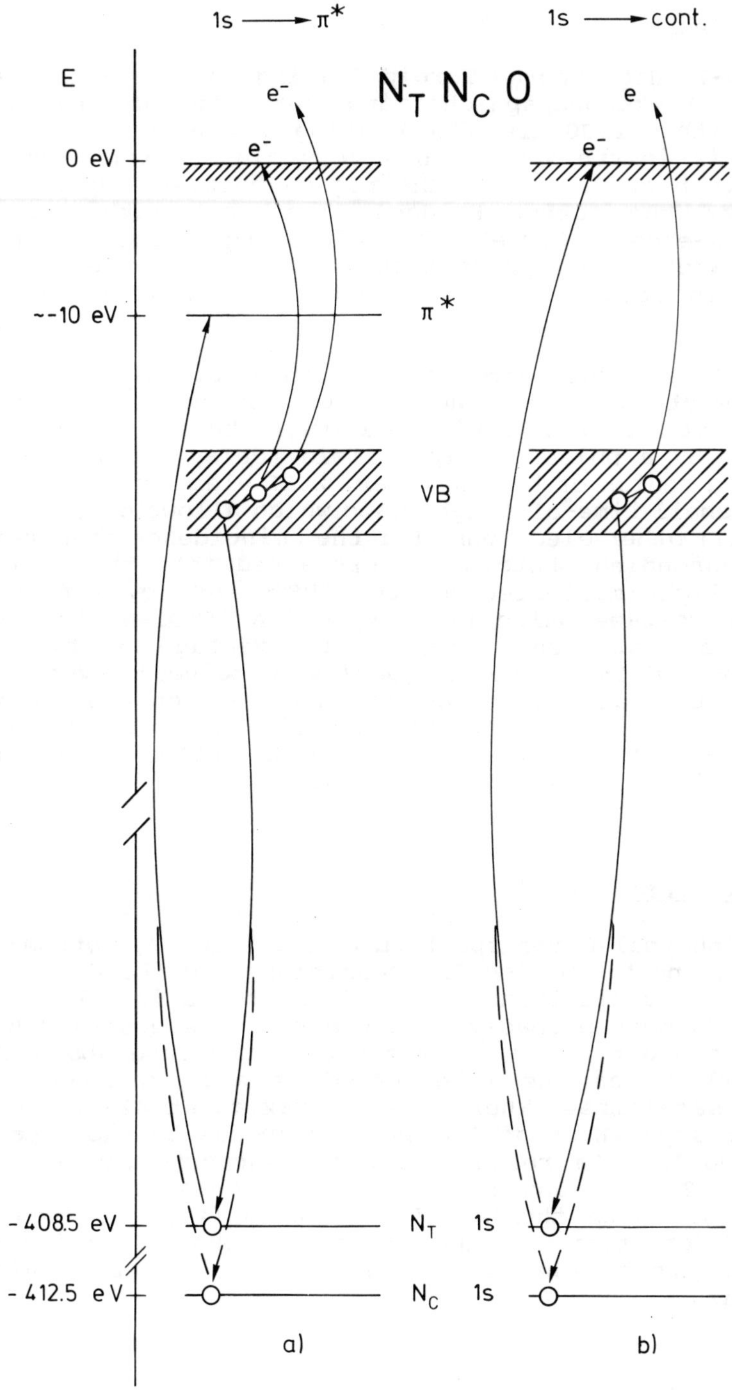
1s ⟶ π*
1s ⟶ cont.
E
N_T N_C O
e⁻
e
0 eV
~-10 eV
π*
VB
- 408.5 eV
- 412.5 eV
N_T 1s
N_C 1s
a)
b)

Experimental

The HETGM-2, High Energy Toroid Grating Monochromator, at the "Berliner Elektronenspeicherring für Synchrotronstrahlung" (BESSY) with a 1000 lines/mm grating was used in the experiments. This instrument gives a very high photon flux within a band-width of 0.3 eV in the region between 270 eV and 310 eV (around the carbon K-edge). The band-width around the nitrogen K-edge is 0.9 eV. The beam emerging from the HETGM-2 has a diameter of approximately 1 mm in the experimental chamber. In brief, the experimental set-up is based on the following:

ZEKE-electrons (the energy resolution of 0.1 eV is adapted to the bandwidth of the monochromator of appr. 0.3 eV) are extracted at a rate of 5 MHz (the single-bunch repetition rate at BESSY) by a pulsed delayed field (delay 1-3 ns) from the ionization region of the analyser. ZEKE electrons are detected within a small time-of-flight gate which allows discrimination against all other electrons. For the coincidence measurements, the corresponding photoion is extracted into the reflectron time-of-flight mass-spectrometer (RETOF-MS) by a fast rise-time high-voltage pulse (1.1 kV/cm). A RETOF-MS is employed since this instrument offers the advantage of high mass-resolution ($m/\Delta m$ = 1000 is easily obtained) even if the fragments that are to be observed have a high kinetic energy of some eV /19, 27, 28/. The experimental set up incorporating the ZEKE electron detector and the RETOF-MS has already been described in refs. 18, 19, 20.

Results

Nitrous oxide

The ZEKE-photoelectron spectrum of N_2O (fig. 2, bottom) shows the N_T (a) and N_C (b) 1s→ 3π transitions and the N_T (c) and N_C (d) 1s → threshold transitions. Transitions into Rydberg states lead most probably to the structures observed between peaks b, c and peaks c, d. The broad structure above the N_C-I.P. (peak d) can be interpreted as a σ^* resonance or as shake-up satellites. There are two maxima at 423.5 and 426.9 eV. Their separation of 3.4 eV corresponds to the separation of the two 1s → 3π resonances which supports the assignment as a N_T (423.5 eV) and N_C (426.9 eV) resonance.
A comparison with the N_2 ZEKE spectrum (fig. 2, top) shows that ZEKE detection provides an unambiguous definition of the **localized excitation**, in particular at the **1s ionization thresholds.**

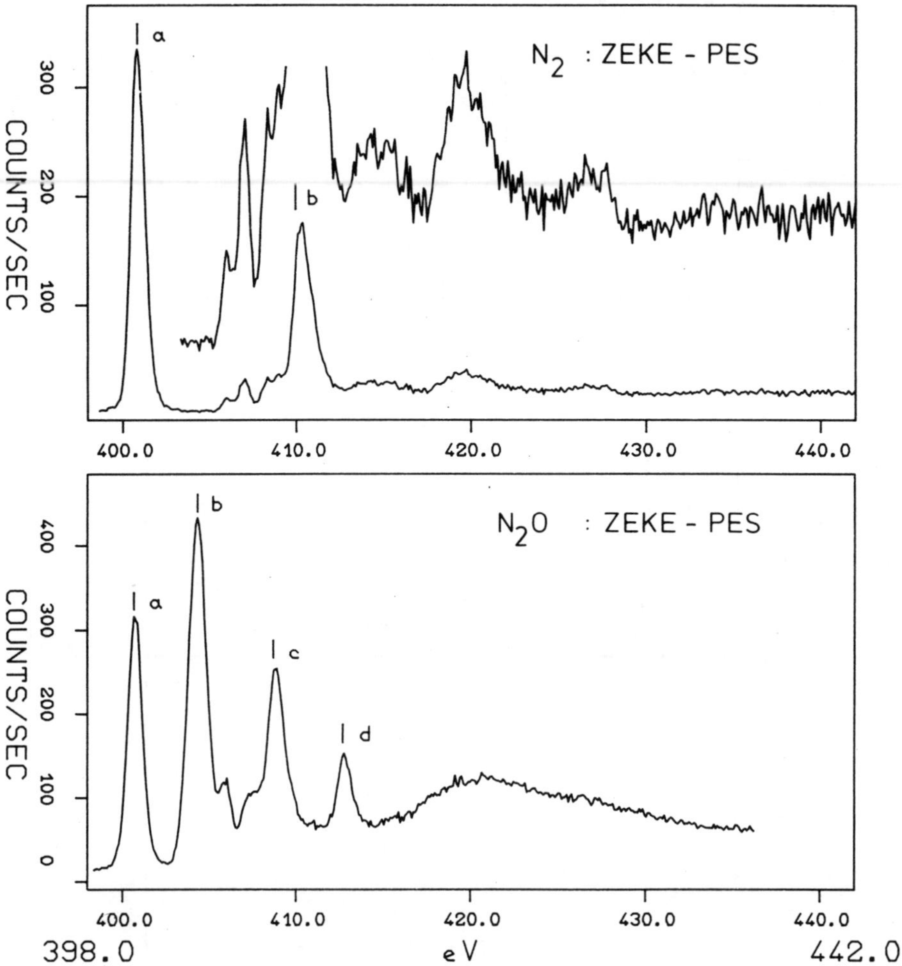

Fig. 2: Zero kinetic energy electron spectrum of N_2 (top) and N_2O (bottom). Peak a: N_T 1s → 3π; b: N_c 1s → 3π; c: N_T 1s → threshold and d: N_c 1s → threshold transition /18, 19/

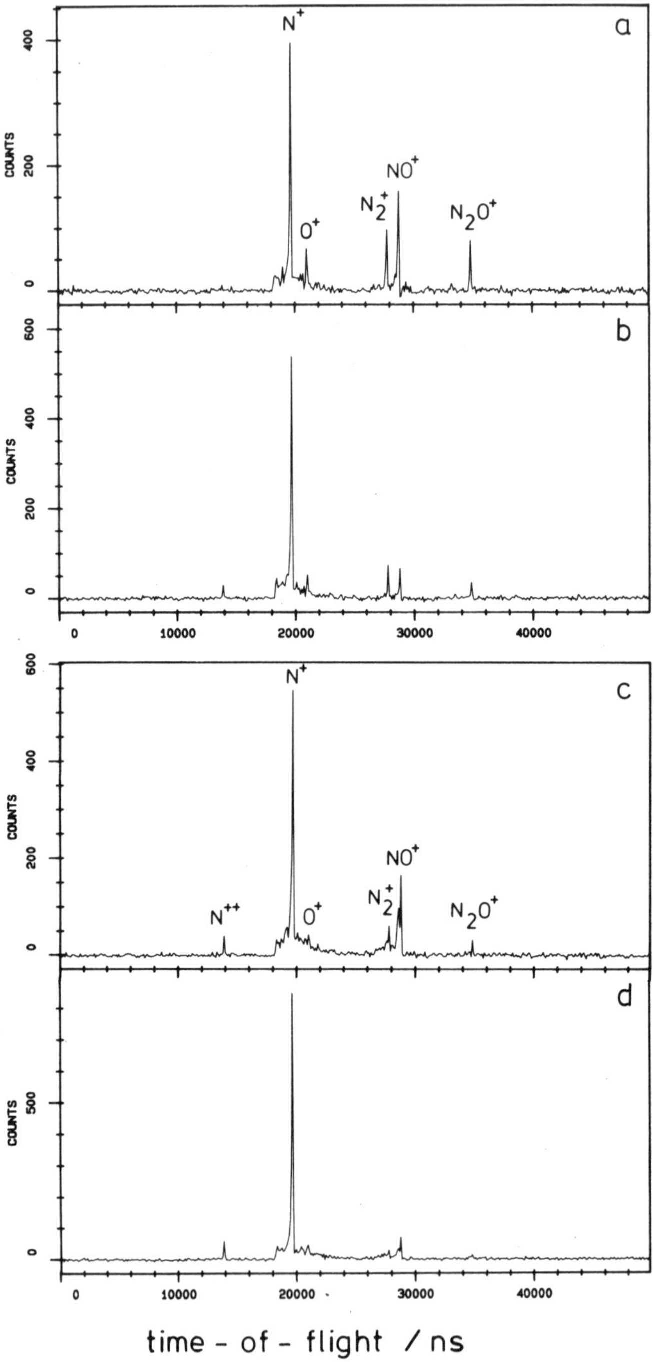

a
b
c
d
N+
O+
N2+
NO+
N2O+
N++
COUNTS
0
200
400
600
500
10000
20000
30000
40000
time - of - flight / ns

Fig. 3 (preceding page):Photoion mass-spectra in coincidence with ZEKE-photoelectrons from a) N_T 1s → π^*; b) N_C 1s → π^*; c) N_T 1s → threshold and d) N_C 1s → threshold transition /19/

ZEKE-PEPICO measurements were obtained for the N_T 1s → 3π (fig. 3a), N_C 1s → 3π (fig. 3b), N_T 1s → threshold (fig. 3c) and N_C 1s → threshold (fig. d) transitions. A strong dependence of the fragmentation pattern on excitation site and the kind of excitation is observed. For NO^+, we observe the strongest dependence on excitation. At threshold, we observe sixfold more NO^+ for terminal excitation (fig. 3c) than for central (fig. 3d) excitation. In the π^* resonance, N_T (fig. 3a) yields 2.5 times more NO^+ than N_C (fig. 3b) excitation.

Excitation at the terminal N-atom (N_T) enhances the fragmentation channel NO^+ + N^+. The N_2^+ + O^+ channel shows a reversed behavior. Hence, the electron in the π^* orbital plays a role in the Auger relaxation process.

Excitation of the central N-atom (N_C) leads to much more complex mass-spectra. We find only a marginal difference between the N_2^++O^+ and NO^++N^+ channels. The most important channel is fragmentation into N_2^{2+}+O and, contrary to N_T excitation, the contribution of this channel does not change with π^* or threshold excitation. For N_C π^* excitation, we conclude that the electron in the π^* orbital acts as a "spectator" and is not significantly involved in the Auger relaxation.

1,1-Difluoroethene , CF_2=CH_2

The chemical shift due to the two F-atoms in the unsaturated hydrocarbon 1,1-difluoroethene causes a lowering in energy of the C_F (1s) orbital of 7 eV compared to the C_H (1s) orbital. This can be seen from the ZEKE spectrum in fig. 4. The main features of the spectrum are peaks attributed to the 1s ionization thresholds and transitions into the π^* resonance. We determine a 1s ionization energy for the C_H atom of 292.7 eV and for C_F of 297.0 eV. The structure below the C_F 1s threshold can be interpreted as transitions to Rydberg series. Above the C_F 1s threshold there are two additional peaks that could be explained by "shake-up" satellites. The observed ZEKE electron transitions are broadened towards higher energies, which can be attributed to unresolved vibrational transitions /29/. From the two ionization thresholds (here denoted 1s → cont) a chemical shift of the C 1s orbitals of approx. 2.2 eV per F-atom is determined. An inspection of fig. 4 shows that the intensity for C_F 1s → cont is higher than for C_H 1s → cont. This is plausible from the different electron density which is a measure for the absorption cross-section. Since the chemical environment of the C-atoms is different, the electron density at the C-atoms is influenced as well. For ligands like strongly electronegative F-atom, the electron density higher at the C_F-atom compared to the C_H-atom.

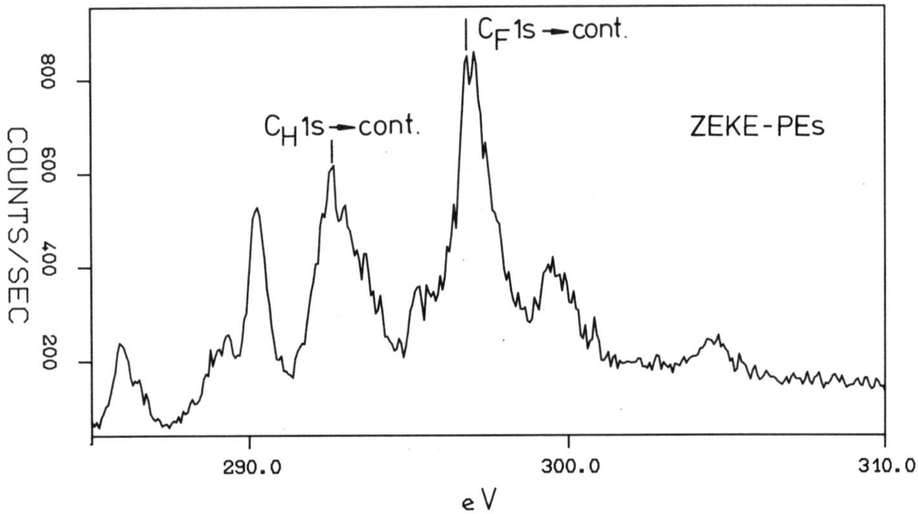

Fig. 4: $CF_2{=}CH_2$ ZEKE spectrum: C_H 1s → π^* 285.7 eV, C_F 1s → π^* 290.3 eV, C_H 1s → cont. 292.7 eV, C_F 1s → cont. 297.0 eV /20/

Coincidence measurements were carried out in both π^*-resonances (C_H 1s → π^* or C_F 1s → π^*) (fig. 5) and at both thresholds (C_H 1s → cont or C_F 1s → cont) (fig. 6), using the RETOF-MS. A comparison of the mass-spectra shows characteristic differences depending on the **site** of the excitation (C_F or C_H, marked with arrows in the figs.) and the **kind** of excitation (1s → π^* or 1s → cont).

Coincidence rates for H^+ are small. This is probably caused by the small detection probability of H^+ compared to heavier fragments of approx. 1 %.

The mass-spectra can be divided into six main groups, with

C^+, C_2^+, CF^+, C_2F^+, CF_2^+ and $C_2F_2^+$

as basis, within which the fragments differ by only one mass-unit and with up to two H atoms bound to a fragment.

For the coincidence rates in each of these mass-groups one observes a dependence on excitation **site** (C_H or C_F) and the **kind** of excitation (π^*-excitation or threshold ionization).

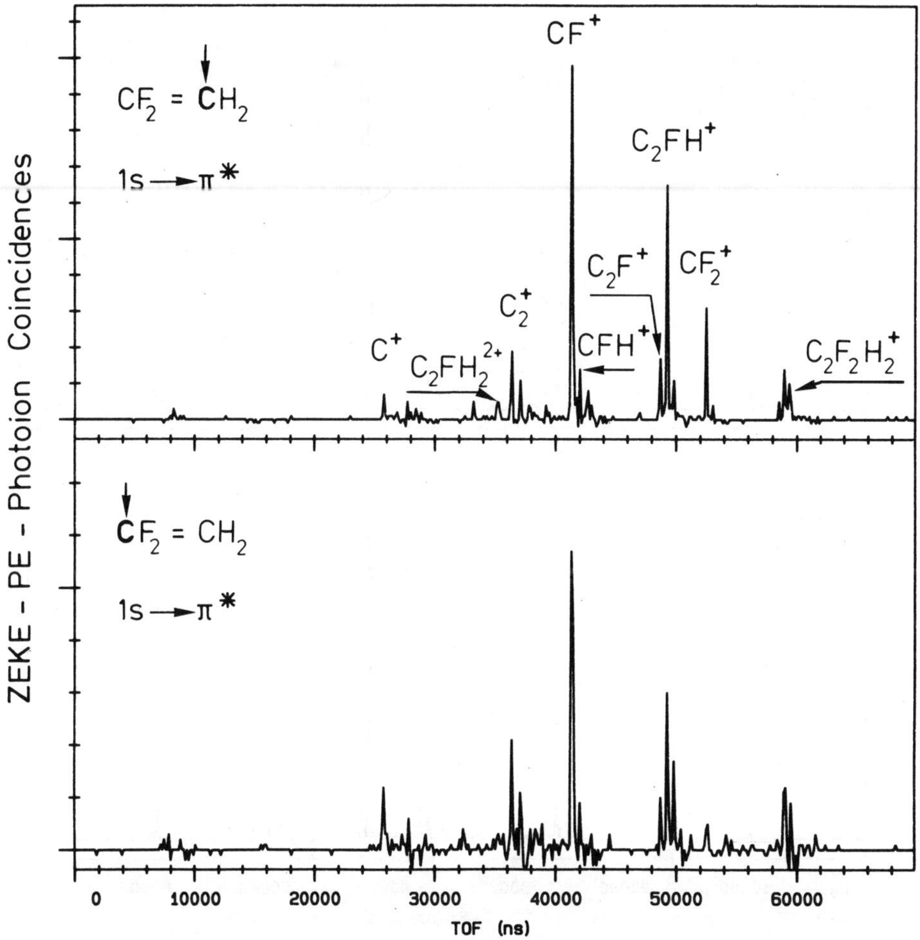

Fig. 5: CF_2=CH_2: coincidence mass spectra taken at C_H 1s → π^* at 285.7 eV, C_F 1s → π^* at 290.3 eV /20/

For the C^+ group one observes a strong site-specific dependence independent of the kind of excitation. For C_F 1s → π^* or 1s → cont one finds twice as many coincidences within this mass-group as for C_H 1s → π^* or 1s → cont excitation. The formation of CH^+ and CH_2^+ is very small and independent of excitation. For C^+ (see table 1), the formation rate is strongly site-specific - a twofold higher C^+ yield is observed for C_F compared to C_H 1s → cont. excitation.

For the C_2^+ group, the excitation site has a similar effect.

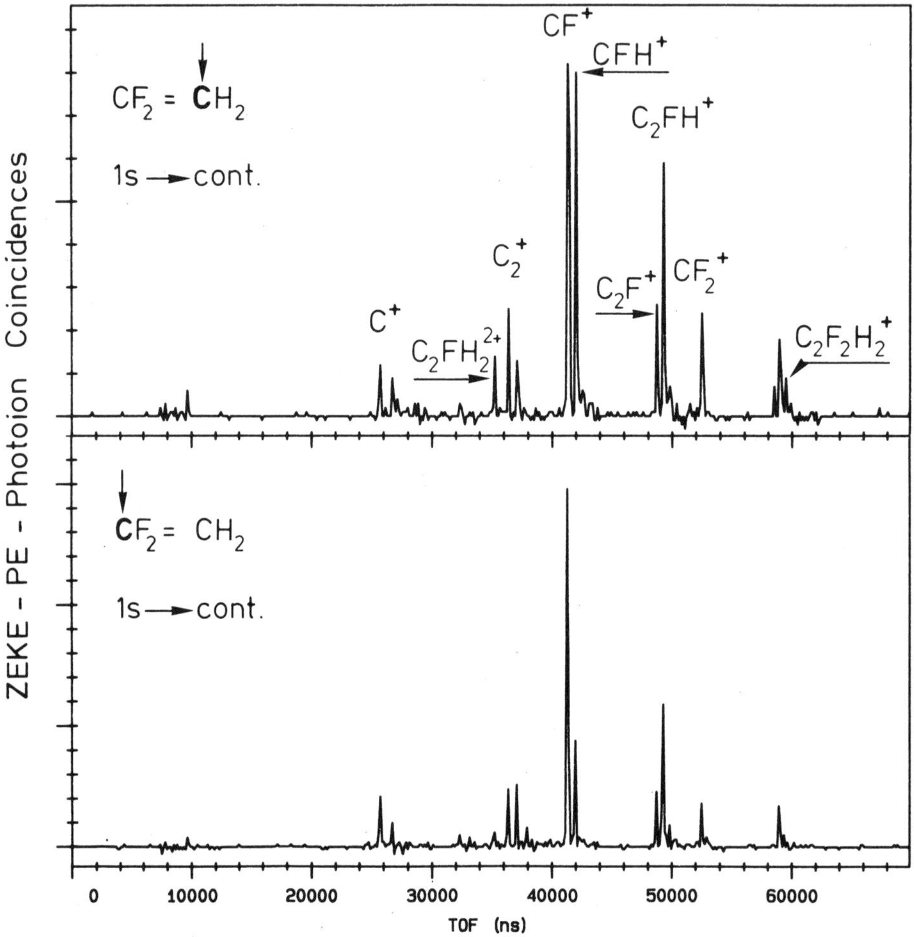

Fig. 6: CF_2=CH_2 coincidence mass spectra taken at C_H 1s → cont. at 292.7 eV, C_F 1s → cont. at 297.0 eV /20/

C_F 1s → π^* produces about 50% more fragments within this group compared to C_H 1s → π^* excitation. Within the C_2^+ mass-group, the fragment C_2H_2 is the least abundant. For C_2^+ and C_2H^+, one observes twofold higher coincidence rates for C_F 1s → π^* compared to C_H 1s → π^* excitation but only a small increase for C_F 1s → cont compared to C_H 1s → cont excitation.

The CF^+ group shows a different behavior. The sum of all coincidences in this group show no dependence on excitation site but a dependence on the **kind** of excitation (π^*-excitation or threshold ionization). For the sum of all coincidences within the CF^+-group we find 38% (see table 1) for C_H 1s → π^* compared to 46% for C_H 1s → cont. With C_F-excitation we find a rate for the CF-group of 38% for 1s → π^* compared to 48% for C_F 1s → cont. For the fragments within this group, the CFH_2^+ is the least abundant. The fragments CF^+ and CFH^+ show, for 1s → π^*, no dependence on excitation site. However, for 1s → cont, one observes an increase of CF^+ for C_F compared to C_H excitation. The opposite effect takes place for CFH^+, which is much more prominent for C_H 1s → cont than for C_F 1s → cont. A very important observation is that CFH^+ coincidence rates are much lower for 1s → π^* than for 1s → cont. The occurance of substantial coincidence rates for CFH^+, which must originate from a molecular rearrangement, and its dependence on excitation site for threshold excitation is one of the important features of the fragmentation observed.

The behavior of the C_2F^+ group is again different. For the sum of all coincidences within this group one finds 26% for C_H 1s → π^* compared to 21% for C_H 1s → cont and 24% for C_F 1s → π^* compared to 20% for C_F 1s → cont. The total counting rate for the group C_2F^+ is independent of the site of the excitation, similarly as observed for the CF^+ group, but 1s → cont leads to lower coincidence rates than 1s → π^*. For $C_2FH_2^+$ one observes a nearly twofold increase for C_F compared to C_H 1s → π^*, whereas the opposite occurs for 1s → cont excitation (but with lower absolute count rates). The effect of excitation site on the formation of C_2FH also seems to become inverted from 1s → π^* (more C_2FH^+ for C_H than for C_F) to 1s → cont (more C_2FH^+ for C_F than for C_H).

For the CF_2H^+ group, only CF_2^+ is an important fragment, showing a strong dependence on excitation site. For both 1s → π^* and 1s → cont the CF_2^+ coincidence rates are twice for C_H compared to C_F excitation.

Within the $C_2F_2^+$ group, the presence of the fragment C_2F_2H is twofold higher for C_F 1s → π^* compared to C_H 1s → π^*, whereas for 1s → cont one observes the opposite, more C_2F_2H for C_H than for C_F excitation.

The effect of site and kind of excitation on the fragment distribution within the mass groups is revealed in much more detail due to the use of the RETOF mass-spectrometer. This shows clearly the necessity for sufficient mass resolution when observing the fragmentation patterns.

1,1,1-Trifluoroethane , CF_3-CH_3

Fig. 7 shows the ZEKE spectrum of CF_3-CH_3 (fig. 7a), the "total" electron spectrum (fig. 7b) and the ion-yield spectrum (fig. 7c) which corresponds to the total ionization cross-section. In a similar way as for $CF_2=CH_2$ the observed peaks in the ZEKE spectrum can be assigned. The two peaks at 289.7 eV and at 296.7 eV are transitions from C_H (1s) resp. C_F (1s) into Rydberg- or σ^*-resonances. The transitions into either σ^*- or Rydberg-states observed have a lower cross-section compared to transitions into π^* orbitals. The 1s → cont transitions are found at 293.4 eV for C_H and at 299.9 eV for C_F. This yields a difference between the two ionization energies of 6.4 eV which again is in agreement with the chemical shift of 2.2 eV per F-atom as observed for $CF_2=CH_2$. Another resonance is observed at 298.2 eV as a shoulder in the C_F 1s → cont peak in the ZEKE electron spectrum.

For the four excitation energies, corresponding to the resonance transitions which are denoted C_H 1s → res and C_F 1s → res, and the threshold transitions denoted C_H 1s → cont and C_F 1s → cont, mass spectra were measured in coincidence with the ZEKE electrons. Similarly as already described for $CF_2=CH_2$ we observe a strong dependence of the fragmentation patterns on the excitation site.

The mass-spectra of CF_3-CH_3 can be separated into 7 main groups with the basis

$$C^+,\ C_2^+,\ CF^+,\ C_2F^+,\ CF_2^+,\ C_2F_2^+ \text{ or } CF_3^+.$$

Within these groups there can be up to three H-atoms bound to the fragments.

Within the C^+-group, one finds a very strong dependence on excitation site. For the C^+ group, similarly as for the observed mass-spectra of $CF_2=CH_2$, we find more C^+-ions for C_F-excitation than for C_H-excitation. This applies for excitation to the 1s ionization threshold as well as for resonant excitation. For excitation to the ionization threshold (1s → cont) the effect is biggest for the C^+ mass. Compared to that, for 1s → res excitation, the biggest site-specific effect is observed for the mass CH_2^+, which occurs in much higher rates for C_F compared to C_H 1s → res excitation.

A similar site-specific effect of the fragmentation patterns shows the C_2^+ mass group. The coincidence rate for fragments in this group, similarly as already observed for the C^+ group, are bigger for 1s → res than for 1s → cont. For 1s → res excitation the fragments C_2^+ and C_2H^+ show double the counting rates for excitation at the C_F-atom compared to C_H. For 1s → cont the fragment ion C_2H_2 shows the biggest site-specific effect.

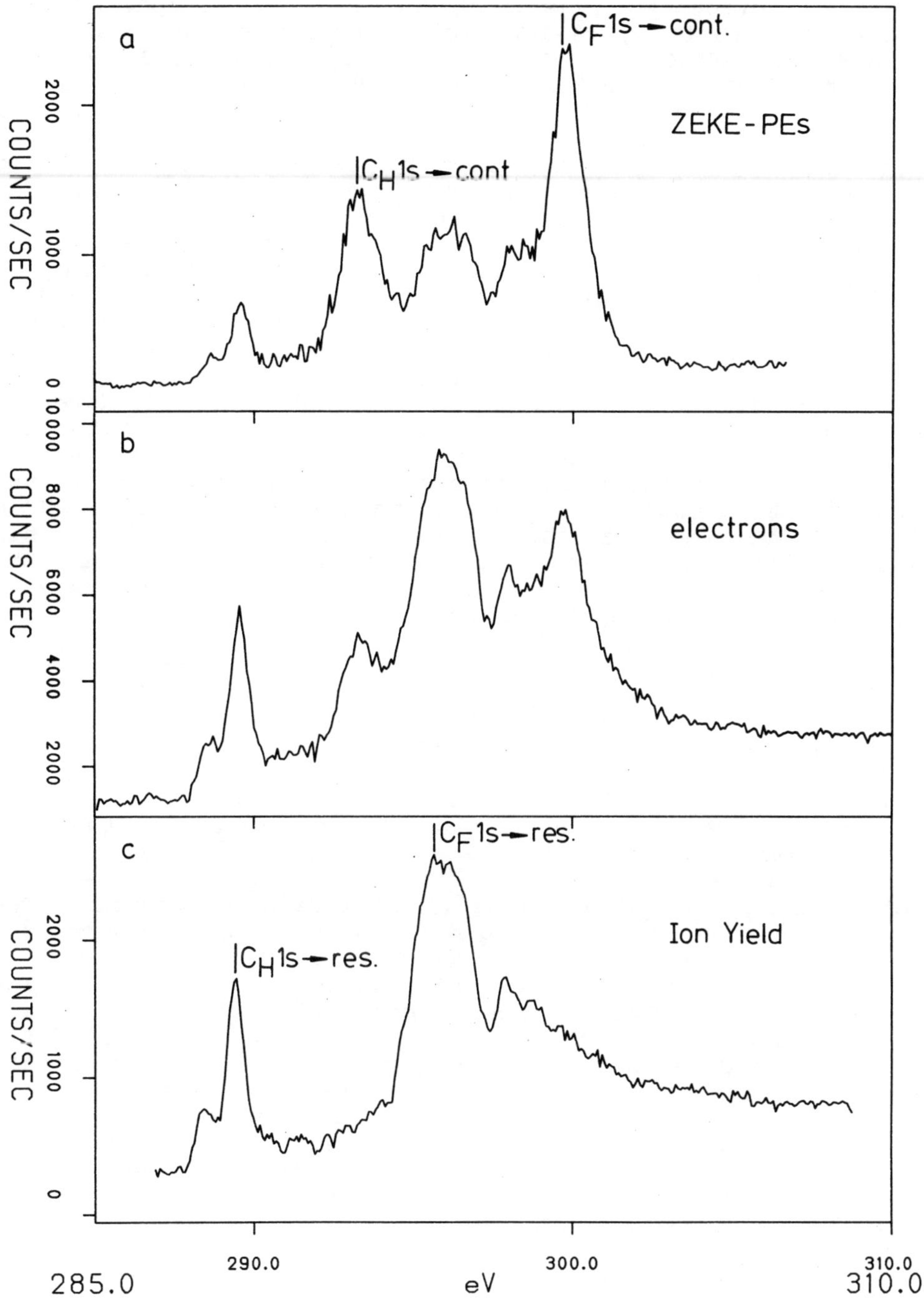
a
|C_F1s → cont.
ZEKE-PEs
|C_H1s → cont
COUNTS/SEC
0
1000
2000
b
electrons
COUNTS/SEC
2000
4000
6000
8000
10000
c
|C_F1s → res.
Ion Yield
|C_H1s → res.
COUNTS/SEC
0
1000
2000
285.0
290.0
eV
300.0
310.0
310.0

Fig. 7 (preceding page): CF_3-CH_3: a) ZEKE-PE spectrum, b) electron spectrum, c) ion yield spectrum versus energy. C_H 1s → res. 289.7 eV, C_F 1s → res. 296.7 eV, C_H 1s → cont. 293.4 eV, C_F 1s → cont. 299.9 eV /20/

Similarly as for CF_2=CH_2, the CF^+-peak is always the most intense. CF^+ is the last fragment in a chain of fragmentation processes and it is a very stable ion, being isoelectronic to CO, which explains its high abundance in the spectra. The CF^+ group shows a very strong dependence on excitation site, in contrast to the observations made for CF_2CH_2 where no dependence on excitation site was found. Within the CF^+ mass-group, for C_F-excitation (1s → res) and -ionization (1s → cont) the formation of CF^+ is significantly higher than for C_H-excitation and -ionization. The fragment CFH^+ is not observed. However, we observe the fragment CFH_2^+ /30/. The occurance of the ion CFH_2^+, which is five times stronger for C_H 1s → cont than for C_F 1s → cont is quite remarkable. The bound electron in the excited state could play an important role here, since CF_2H^+ is not observed for the C_F 1s → cont. transition. This observation, shown here experimentally, implies that the migration of atoms, i.e. a unimolecular rearrangement, depends strongly on the site of excitation.

For the C_2F^+ group, only for the ion C_2FH_2 a strong effect of excitation site is observed. The formation of C_2F^+ is very small and independent of excitation site. A much higher counting rate is observed for $CFCH^+$. The counting rate of $CFCH^+$-ions is considerably lower for excitation into the resonances compared to threshold excitation. $CFCH_3^+$ is not observed, $CFCH_2^+$ is more abundant for C_H-excitation than for C_F-excitation.

Within the group CF_2^+ one finds an interesting observation. The coincidence rate for the fragment ion CF_2^+ is much bigger, for resonant excitation, for C_H compared to C_F. In contrast to this, one observes the opposite tendency for the ion CF_2H. For C_F 1s → res one finds a four times higher coincidence rate compared to C_H. For excitation to the ionization threshold (1s → cont) one does not observe a significant effect of excitation site within this mass group. Whereas the formation of CF_2^+ as well as of C_2F^+ is unspecific concerning the site of excitation, we find CF_2H^+ more abundant for the C_F 1s → res excitation than for other excitations. This implies that H-rearrangements are favored with C_F-excitation. Further masses with CF_2 as basis are not observed.

Within the mass group $C_2F_2^+$ there is a remarkable effect for the fragment ion $C_2F_2H_2$. For excitation into the resonance, there is no site-specific effect observed. For excitation to

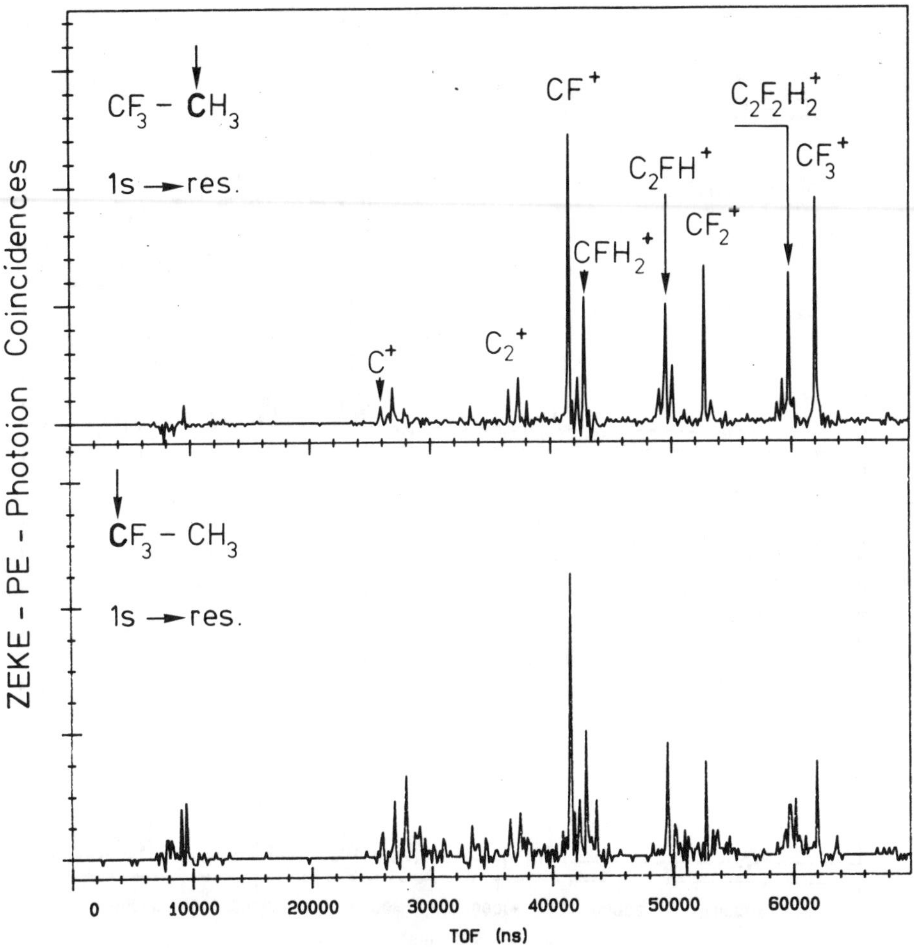

Fig. 8 CF_3-CH_3: coincidence mass spectra taken at C_H 1s → res. at 289.7 eV, C_F 1s → res. at 296.7 eV /20/

the ionization threshold, this ion does not occur at all for C_F 1s → cont, whereas one finds a coincidence rate of 7 % for C_H 1s → cont.

For the CF_3 mass group, the CF_3^+ counting rates are much higher for C_H- than for C_F-excitation. This corresponds to the analogous observation for CF_2=CH_2 where a higher abundance of CF_2^+ for C_H-excitation compared to C_F-excitation was found. For the ion CF_3, which is the only ion observed in this group, there is a big effect on excitation site for both kinds of

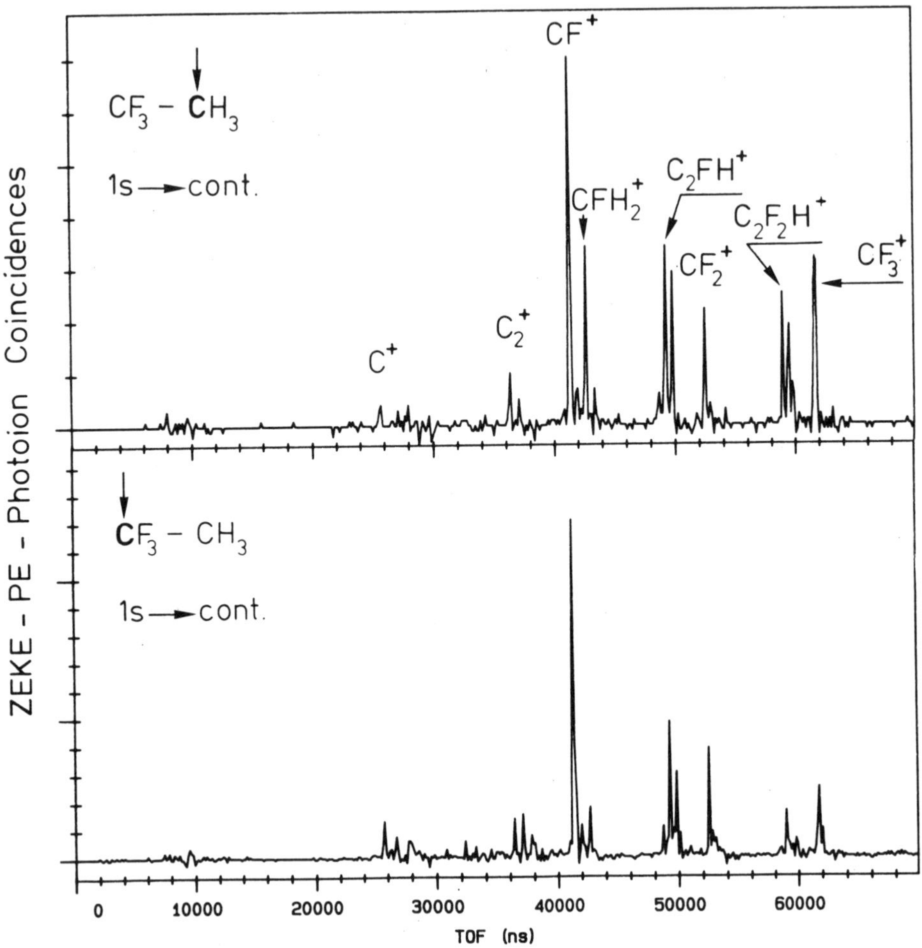

Fig. 9: CF_3-CH_3 coincidence mass spectra taken at C_H 1s → cont. at 293.4 eV, C_F 1s → cont. at 299.9 eV /20/

excitation. For 1s → cont one observes more than twice as much CF_3^+ for C_H compared to C_F. This site-specific fragmentation effect is nearly as strong for 1s → res excitation.

In a similar way as for CF_2=CH_2, it is also observed for CF_3-CH_3 that the high mass resolution offered by the reflectron time-of-flight mass spectrometer allows to obtain a clear distribution of the fragmentation channels. Within the different mass groups, one obtains a remarkable dependence of the coincidence rates for certain fragments on excitation site. The strongest effect on excitation site is observed for the fragment CF_3^+. For C_H excitation one observes three to four times as much CF_3^+ compared to C_F excitation. The fragment CFH_2^+ is significantly observed for C_H 1s → cont and not very important for C_F.

In conclusion, this molecule shows a stronger bond breaking effect on the C-C bond for C_F excitation compared to C_H excitation, similarly as observed for CF_2=CH_2. Within the model of the localized core Auger relaxation, this can be rationalized by the higher electron density at the C_F atom.

Salman et al. have made a predictive computation using a Coulomb explosion model with a site-specific constraint, which shows that the fragmentation patterns, for an initially localized excitation, are not fully statistical. Parts of the non-statistical behavior are due to the site-selectivity of the initial excitation /31/.

Conclusion

The ZEKE electron spectra of the molecules N_2O, CF_2=CH_2 and CF_3-CH_3 show peaks from direct 1s-ionization and excitation into resonances. At the 1s-ionization threshold a ZEKE electron is produced directly by the photoabsorption process and leads to a localized $1s^{-1}$ core hole which susequently relaxes by the core Auger process. For excitation into bound resonances, below the 1s-ionization threshold, ZEKE electrons are produced not in the excitation but in the relaxation channel, by the "Auger shake-off" /6/ and the "valence participator Auger" /25, 26/ relaxation process. Either for threshold or resonance excitation, the detection of the ZEKE electron serves to indicate which atom in the molecule has been selectively excited. The ZEKE method thus allows to distinguish between atoms of the same kind in a different chemical environment within the molecule.

The fragmentation patterns measured in coincidence with ZEKE electrons show a characteristic dependence on excitation site. For N_2O, N_T excitation leads to a stronger preference for the molecular fragmentation channels NO^+ +N and N_2^+ + O compared to N_C excitation which shows a preference for the atomic fragmentation channels.

For CF_2=CH_2, for the fragment ions CF_2^+, C_2FH^+ and $C_2F_2H^+$, significantly higher coincidence rates are observed for C_H compared to C_F excitation. The fragment CFH^+ is very prominent for C_H 1s → cont compared to all other excitations. For CF_3-CH_3, the fragment ions CF_3^+, C^+, CFH_2^+, $C_2F_2H_2^+$ and CF^+ show the strongest effect on excitation site. Excitation at the C_H-atom leads predominantly to CF_3^+ whereas C_F excitation has the biggest effect on C^+ and CF^+. For CFH_2^+ and $C_2F_2H_2^+$ a strong atom-selective effect is found for 1s-ionization; C_H 1s → cont leads to more than fivefold higher coincidence rates for these fragments than C_F 1s → cont. The high mass resolution available by the reflectron TOF mass spectrometer is very important to determine the specific fragmentation channels.

For the explanation of the fragmentation process and, in particular, the observed "memory effects", possible relaxation pathways of the **localized** $1s^{-1}$ core hole formed in the photoabsorption process are to be considered. The dominant relaxation of the $1s^{-1}$ core hole is the Auger process. The core Auger relaxation rate depends on the electron density of the valence orbitals near the $1s^{-1}$ core hole (the excitation site). This suggests that the core Auger relaxation process takes place **at the site of excitation** /26/. This can be rationalized from the different electron densities at the different C-atoms. The differences in fragmentation patterns observed could then be explained from the different contributions of the valence orbitals to the **localized** core Auger relaxation process. For CF_2=CH_2, the Auger relaxation involves, for C_H-excitation, the C-H, C-C σ- and the C-C π-orbital whereas for C_F-excitation the C-F orbitals and both C-C σ-, π-orbitals are important. The electron density in the C-H σ-orbital is higher at the C-atom, whereas for the C-F σ-orbital it is higher near the F-atom. The maximum of the electronic probability density in the C-C orbitals (in particular in the π-orbital) is shifted in direction of the F-atoms. These effects could lead to a correlation between the excited $1s^{-1}$ core hole and the primary bond breakage and hence explain some of the the observed effects.

For CF_3-CH_3, the observed bond selective fragmentation processes can also be interpreted with the assumption of a localized core Auger relaxation process. Similarly as for CF_2=CH_2, the C-C bond seems to be more affected by C_F excitation or ionization than by C_H excitation or ionization.

An interesting site-specific process is the formation of the fragment CFH_2^+ from a unimolecular rearrangement, as observed for C_H ionization (C_H 1s → cont), similarly to the observation of CFH^+ for CF_2=CH_2 and C_H 1s → cont.

Acknowledgement

We gratefully acknowledge funding of this research by the Bundesminister für Forschung und Technologie, Grant No. 05366 FAI.

References

1. K. Müller-Dethlefs, M. Sander, L.A. Chewter, E.W. Schlag, J. Phys. Chem. 88, (1984) 6098

2. W. Habenicht, L.A. Chewter, M. Sander, K. Müller-Dethlefs, E.W. Schlag, J. de Physique Colloque C9, (1987) 741

3. K. Siegbahn, C. Nordling, G. Johansson, J. Hedman, P.F. Hedén, K. Hamrin, U. Gelius, T. Bergmark, L.O. Werme, R. Manne and Y. Baer, "ESCA Applied to Free Molecules", North Holland Publ. Co., Amsterdam, 1969

4. W. Eberhardt, T.K. Sham, R. Carr, S. Krummacher, M. Strongin, S.L. Wergand, D. Wesner, Phys. Rev. Lett. 50 (1983) 1038

5. W. Eberhardt, T. K. Sham, Proc. of SPIE 447 (1983)

6. I. Nenner, P.Morin, P. Lablanquie, M. Simon, J. Electr. Spect. Rel. Phen. 52 (1990) 623

7. J. Murakami, M. C. Nelson, S. L. Anderson, D. M. Hanson, J. Chem. Phys. 85 (1986) 5755

8. R. Murphy, W. Eberhardt, J. Chem. Phys. 89 (1988) 4054

9. D. M. Hanson, Adv. Chem. Phys. 77 (1990)

10. B. Brehm, E. v. Puttkamer, Z. Naturforsch. A 22a (1967) 8 and Adv. Mass. Spectrom. 4 (1968) 591

11. E. v. Puttkamer, Z. Naturforsch., A 25a (1970) 1062

12. J. H. D. Eland, Int. J. Mass Spectrom. Ion Phys., 8 (1972) 143

13. C. J. Danby, J. H. D. Eland, Int. J. Mass Spectrom. Ion Phys., 8 (1972) 153

14. R. G. Hayes, W. Eberhardt, J. Chem. Phys., 94 (1991) 397

15. D. M. Hanson, C. I. Ma, K. Lee, D. Lapiano-Smith, D.Y. Kim, J. Chem. Phys. 93 (1990) 9200

16. K. Codling, L. J. Frasiņski, P. A. Hatherly M. Stankiewicz, Physica Scripta 41 (1990) 433

17. T. LeBrun, M. Lavollee, and P. Morin, in: X-Ray and Inner-Shell Processes, ed. by T. A. Carlson, M. O. Krause, and S. T. Manson, American Inst. Phys., New York, 1990, p 846

18. W. Habenicht, K. Müller-Dethlefs, E.W. Schlag, J. Electron Spectroscopy 52 (1990) 697

19. W. Habenicht, H. Baiter, K. Müller-Dethlefs, E.W. Schlag, Phys. Scr. 41 (1990) 814

20. W. Habenicht, H.Baiter, K. Müller-Dethlefs, and E.W. Schlag, J. Phys. Chem., in press (September 1991)

21. K. Müller-Dethlefs, M. Sander, E.W. Schlag, Z. Naturforsch. A39 (1984) 1089

22. L.A. Chewter, K. Müller-Dethlefs, E.W. Schlag, Chem. Phys. Lett. 135 (1987) 219

23. M. Sander, L.A. Chewter, K. Müller-Dethlefs, E.W. Schlag, Phys. Rev. A 36, (1987) 4543

24. L.A. Chewter, M. Sander, K. Müller-Dethlefs, E.W. Schlag, J. Chem. Phys. 86 (1987) 4737

25. U. Becker, R. Wehlitz, O. Hemmers, B. Langer, A. Menzel, Phys. Rev. Lett. 63, (1989) 1054

26. F. P. Larkins, J. Elect. Spectros. Relat. Phenom., 51, 115 (1990); F.P. Larkins, in: X-Ray and Inner-Shell Processes, ed. by T. A. Carlson, M. O. Krause, and S. T. Manson, American Inst. Phys., New York, 1990, p 817

27. U. Boesl, H.J. Neusser, R. Weinkauf, E.W. Schlag, J. Phys. Chem. 86 (1982) 4857

28. B. A. Mamyrin, V. I. Karataev, D. V. Shmikk, V. A. Zagulin, Sov. Phys. JETP, 37 (1973) 45

29. W. Domcke, L.S. Cederbaum, Chem. Phys. Lett. 31 (1975) 582

30. R.M. Benito, J. Santamaría, Chem. Phys. Lett. 109 (1984) 478

31. I. Salman, J. Silberstein, and R. D. Levine, J. Phys. Chem., in press (September 1991)

RANDOM WALK MODEL FOR ENERGY TRANSFER AT HIGH TEMPERATURES

Izhack OREF
Department of Chemistry, Technion – Israel Institute of Technology
Haifa 32000, Israel

and R.C. GILBERT
School of Theoretical Chemistry, University of Sydney, Sydney, Australia

ABSTRACT. An intermolecular energy transfer transition probability function $P(E, E')$ in high temperatures regim is reported. Based on trajectory calculations results, an energy diffusion model together with a Gaussian population distribution function are used to obtain $P(E, E')$ as well as analytical expressions for the average energy transferred per collision $< \Delta E >_{all}$ and the mean square energy transferred per collision $< \Delta E^2 >$. The expressions obtained obey all the conservation and boundry conditions.

1. Introduction

Since the early days of experimentation in unimolecular reactions it was realized that understanding energy transfer processes is essential for proper interpretation of unimolecular and recombination rate coefficients.. The "strength" of a collision is indicated by the values of the first and second moments of the energy transfer transition probability distribution function, $P(E, E')$. E' and E are the initial and final energy states of the reactant molecules. Common empirical models for $P(E, E')$ are exponential, stepladder, Gaussian, and Poisson but only the first one will be discussed later on. The first moments are the average energy transferred in all collisions $< \Delta E >_{all}$ and the average energy transferred in a down collision $< \Delta E >_D$. The second moment is the mean square energy transferred per collision $< \Delta E^2 >$. The moments of the distribution are defined by

$$< \Delta E^n >= \int_0^\infty (E - E')^n P(E, E') dE \tag{1}$$

and $P(E, E')$ obeys the conservation relation

$$\int_0^\infty P(E, E') dE = 1 \tag{2}$$

The empirical models for $P(E, E')$ which are listed above are used in master equation calculations such that a fit of a calculated curve to experimental results is obtained. No physical insight of the collision and energy transfer processes is obtained.

J. Jortner et al. (eds.), Mode Selective Chemistry, 393–399.

The subject was reviewed lately by Oref and Tardy [1]. Therefore only highlights of pertinent points to the development of our model will be discussed here.

The commonest of the empirical models is the exponential down model

$$r(E, E') = e^{\frac{-(E'-E)}{<\Delta E>_D}}$$

which is applicable to weak collisions, i.e. for small values of $<\Delta E>_D$. $C(E')$ is a normalization factor such that $P(E, E') = r(E, E')/C(E')$. When the population distribution function $f(E)$ is represented by an exponential expression the master equation can be solved in a closed form and an analytical expression for $<\Delta E>_{all}$ and $<\Delta E>_D$ obtained [2]-[4]. The limitations of this approach are discussed in refs. [1] and [5]. One problem relevant to the present work is the fact that an exponential representation of the population $f(E)$ is only good for low temperatures and small and moderate size molecules. For large molecules the approximation breaks at fairly moderate temperatures [1]. The model fails to give the inversion effect where up collisions dominate over down collisions and $<\Delta E>_{all}$ changes sign. Other energy transfer models [6]-[12] assume a degree of ergodicity in the collision complex between reactant and bath molecules.

The Biased Random Walk (BRW) model is another energy transfer model developed recently to account specifically for energy transfer occurring in a collision between an atom and a polyatomic molecule [13-15]. It is based on trajectory results [16] and its extension to high temperatures is the subject of the discussion given next.

2. Theory

The BRW model has been discussed in [5] and only an outline of the model will be given. The underlying justification for the model rests on the fact that trajectory calculations indicate that during the short duration of a collision the value of the energy in the reactant molecule appears to undergo random fluctuations. This indicates the possibility that the fluctuations of the energy in the molecule can be treated as a Brownian motion in energy space. It is therefore possible to describe this motion by a Fokker-Plank diffusion equation

$$\frac{\partial P}{\partial t} = \frac{\partial}{\partial E}\left[D(E)\frac{\partial P}{\partial E} + Z(E)P\right] \tag{3}$$

$D(E)$ is an energy dependent diffusion coefficient in energy space and $Z(E)$ is a quantity related to detailed balance: the "bias". P is the probability of the molecule ending with energy E at time t having started at E'. The diffusion equation is subject to the conservation relations described before and to detailed balance.

$$P(E, E')\, f(E') = P(E', E)\, f(E) \tag{4}$$

In the solution of the diffusion equation 3 it is assumed that the diffusion coefficient D is independent of E and that $Z(E)$ is given by the linear expression

$$Z(E) = a + bE \tag{5}$$

The population distribution at high temperatures can be expressed by a Gaussian form

$$f(E) = \left(2\sqrt{\pi}\,c\right)^{-1} \exp\left[- \{(E - \bar{E})/2c\}^2 \right] \tag{6}$$

$\bar{E}$ and c are constants determined from the distribution as described later. Equation 5 is substituted in equation 3 and the latter is solved by Fourier transform. Detailed balance, equation 4 and $f(E)$, equation 6, are used in the solution of the diffusion equation to give the final expression for the collisional energy transfer probability function

$$P(E, E') = \left(\frac{1}{4c^2\pi(1-q^2)}\right)^{1/2} \exp\left(\frac{-[E - \bar{E} - (E' - \bar{E})q]^2}{4c^2(1-q^2)}\right) \tag{7}$$

where $q = \exp(-bt_c)$ and t_c is the duration of the collision. The $P(E, E')$ of equation 7 obeys all the conservation relations described before.

Typical values for bt_c for the system studied here are $\sim 10^{-4}$ therefore equation 7 can be simplified for the case $bt_c << 1$ to give the final Gaussian-like expression

$$P(E, E') \simeq (2\pi D t_c)^{-1/2} \exp\left[- (E - E')^2/2Dt_c\right] \tag{8}$$

Equation 8 obeys exactly the boundry conditions i.e. for $t_c \to 0$; $P(E, E') = 0$ except when $E = E'$ and at $t_c \to \infty$, $P(E, E') = f(E)$ equation 6. This is precisely the strong collision assumption. If the duration of the collision is infinitely long it is expected that the average amount of energy transferred will be large i.e. a "strong collision".

Using the expressions for $P(E, E')$ the various moments of the distribution can be found. For the case $bt_c << 1$ discussed before the average energy transferred per collision takes an especially simple form

$$< \Delta\, E_{all} >= (\bar{E} - E')bt_c/2 \tag{9}$$

As can be seen, as E' is varied through the average energy $\bar{E}$ the sign of $< \Delta\, E >_{all}$ changes as expected.

The second moment which yields the width of the distribution is given as well by a particularly simple expression

$$< \Delta\, E^2 >= Dt_c \qquad (bt_c << 1) \tag{10}$$

Thus, within the framework of the present approximations $P(E, E')$ can be written in a simple form

$$P(E, E') \equiv P(\Delta\, E) = \left(2\pi < \Delta\, E^2 >\right)^{-1/2} \exp\left(- \Delta\, E^2 \,/\, 2 < \Delta\, E^2 >\right) \tag{11}$$

Equation 11 represents the final Gaussian expression for the energy transfer transition probability function.

The population distribution function at high temperatures is approximated by equation 6. The two constants which define the distribution $\bar{E}$, c are found by evaluating exactly the first and second moments of $f(E)$.

$$< E^n >= \int_0^\infty E^n f(E)dE \qquad (n = 1, 2) \tag{12}$$

Introducing equation 6 into equation 12, making the transformation $\varepsilon = (E - \bar{E})/2c$ and integrating between $-\bar{E}/2c$ and ∞ yields the desired relation between $\bar{E}$, c and $< E >$, $< E^2 >$. However, for large polyatomic molecules at high temperature the following expressions for $\bar{E}$ and c give excellent results

$$\bar{E} = < E > ; \quad c = \left[\left(< E^2 > - < E >^2 \right) / 2 \right]^{1/2} \tag{13}$$

The excellence of the Gaussian approximation, equation 6, to the population distribution function of large molecules at high temperatures can be seen from Figure 1 where the Boltzmann and Gaussian populations at 2000K are given over a wide range of energies. Between 70 and 180 kcal/mol the agreement is excellent. Above and below this range $f(E)$ is orders of magnitude smaller than the maximum value and it does not contribute to the dynamics.

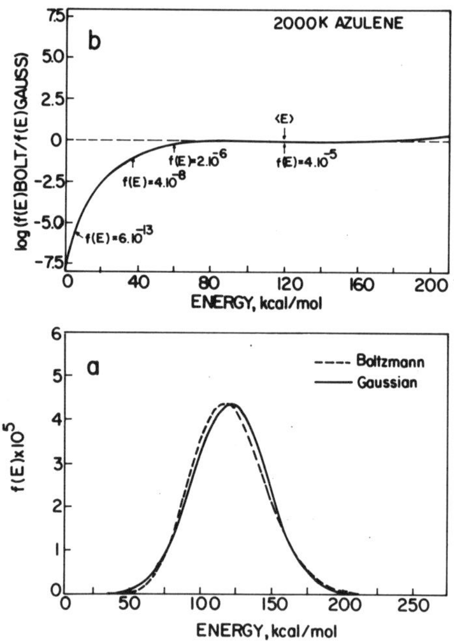

Figure 1: a) Boltzmann and Gaussian population distribution functions at 2000K; b) Log ratio of Boltzmann to Gaussian population distribution functions at 2000K.

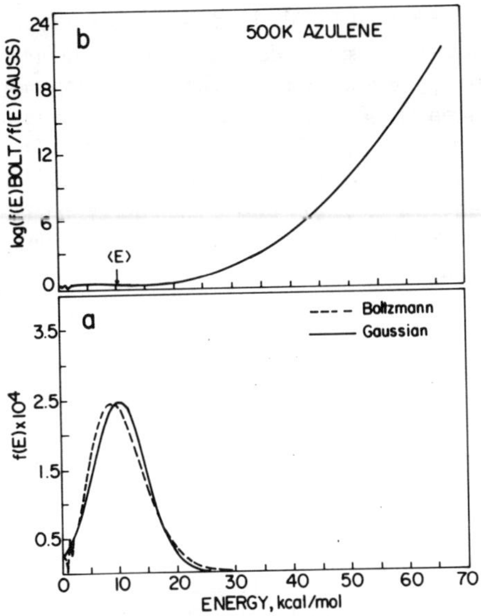

Figure 2: a) Boltzmann and Gaussian population distribution functions at 500K; b) Log ratio of Boltzmann to Gaussian population distribution functions at 500K.

At 500K (Figure 2) the agreement is only good at low energies of up to 20 kcal/mol, thus for the energies of interest around the threshold energy for reaction E_0 this approximation fails.

Application of BRW to excited azulene-noble gas atom collisions as well as trajectory calculations on the same systems indicate that the values of $< \Delta E >_{all}$ are small [19]. For excited azulene at 30 000 cm^{-1} the value of $< \Delta E >_{all}$ is around 200 cm^{-1}. This surprisingly small value for such a highly excited molecule can be explained by assuming that after rapid energy redistribution [17, 18] the azulene molecule can be looked upon as a canonical ensemble of oscillators at a temperature T_v defined by

$$< E' >= RT_v^2 \left(\frac{\partial \ell_n Q}{\partial T_v} \right) . \tag{14}$$

The internal energy E' is the sum of the photon energy (85.5 kcal/mol) and the thermal internal energy (4.3 kcal/mol at 300 K) and Q is the molecular partition function, and T_v is found to be 1581 K.

The average internal energy per mole is not divided democratically among the N normal modes of the molecule, i.e. not the classical RT. The low frequency modes have more than their share of internal energy. Figure 3 shows the energy per mode as a function of the mode energy. As can be seen the $C-H$ stretching modes have on the average 580 cm^{-1} (at $T_v = 1581K$), 170 cm^{-1} (at $T_v = 1112K$) and 0.7 cm^{-1} (at $T_v = 466K$). it is clear therefore that in spite of the high energy content of the molecule at 30 000 cm^{-1}, the average energy per high frequency mode is low and if the energy exchange is stipulated to involve only these modes it is clear that the prevailing mode of energy exchange will be up collisions i.e. $< \Delta E >_{all}$ positive.

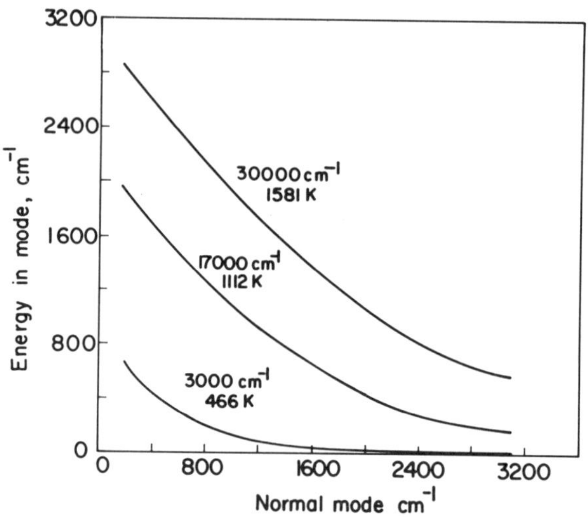

Figure 3: Energy content per mode versus normal mode for azulene at the three excitations energies.

There are two parameters, the collision duration t_c and the energy diffusion coefficient D which determine the values of the energy transfer quantities $< \Delta E >_{all}$ and $< \Delta E^2 >$. These parameters can be evaluated separately. The collision duration is the average time, the substrate molecule/atom spend in the spherically symmetric potential

$$t_c = \int_d^{r_0} dr \left(\frac{2\mu}{E_{av} - V - E_{av} \left(\frac{b_{av}}{r}\right)^2} \right)^{1/2} \tag{15}$$

where d is the distance of closest approach and r_0 is the classical turning point i.e. the point where the denominator in equation 15 is equal to zero. μ is the molecule/atom reduced mass. E_{av} is the molecule/atom average relative translational energy $E_{av} = 2RT$. b_{av} is the average impact parameter $b_{av} = 2/3d$. t_c, in this model, is independent of the initial internal energy of the molecule.

The assumption of statistical redistribution of internal energy [17, 18] is used to find D from the time integral of the autocorrelation function

$$D = \int_0^{t_c} < \dot{E}(\tau)\, \dot{E}(o) > d\tau \tag{16}$$

where $\dot{E}$ is the time derivative of the internal energy. The use of the generalized Langevin equation and assumption of exponential relaxation together with a Lennard Jones potential yield a closed form for D [18].

In summary, a collision model is presented where the internal energy is assumed to undergo random fluctuations and the energy population distribution is assumed to be Gaussian. Solving an energy diffusion equation a Gaussian energy transfer probability function and average energy transfer quantities $< \Delta E >_{all}$, $< \Delta E >_D$ and $< \Delta E^2 >$ are obtained. The model obeys exactly all the conservation relations and boundry conditions.

3. References

[1] Oref, I., Tardy, D.C., Chem. Rev., 1990, 90, 1407.
[2] Troe, J., J. Chem. Phys. 1977, 66, 4745; Gilbert, R.G., Luther, K., Troe, J., Ber. Bunsenges. Phys. Chem. 1983, 87, 169.
[3] Barker, J.R., Golden, R.E.J., Phys. Chem. 1984, 88, 1012.
[4] Tardy, D.C., Rabinovitch, B.S., (a) J. Phys. Chem., 1985, 89, 2442; (b) J. Phys. Chem., 1986, 90, 1987.
[5] Gilbert, R.C., Oref, I., J. Phys. Chem., in press.
[6] Serauskas, R.V., Schlag, E.W., J. Chem. Phys., 1966, 45, 3706 and references therein.
[7] Lin, Y.N., Rabinovitch, B.S., J. Phys. Chem., 1970, 74, 3151.
[8] Bhatacharjee, R.C., Forst, W., Chem. Phys., 1978, 30, 217.
[9] Oref, I., J. Chem. Phys., 1982, 77, 1253.
[10] Troe, J., Ber. Bunsenges. Phys. Chem., 1973, 77, 665.
[11] Nordholm, S., Freasier, B.C., Jolly, D.L. (a) Chem. Phys., 1977, 25, 433; (b) Chem. Phys., 1978, 32, 161, 169.
[12] Schranz, H.W., Nordholm, S. Int. J. Chem. Kinet., 1981, 13, 1051.
[13] Gilbert, R.G., J. Chem. Phys., 1984, 80, 5501.
[14] Lim, K.F., Gilbert, R.G., J. Chem. Phys., 1986, 84, 6124.
[15] Lim, K.F., Gilbert, R.G., J. Chem. Phys., 1990, 92, 1819.
[16] Lim, K.F., Gilbert, R.G., J. Phys. Chem., 1990, 94, 77.
[17] Oref, I., Rabinovitch, B.S., Acc. Chem. Res., 1979, 12, 166.
[18] E.g. Gilbert, R.G., Smith, S.C., Theory of Unimolecular and Recombination Reactions, Blackwell Scientific Publications, Oxford, 1990.
[19] Clark, D.L., Oref, I., Gilbert, R.C., Lim, K.F., in preparation.

ON THE LANDAU METHOD OF CALCULATING QUASICLASSICAL MATRIX ELEMENTS

E. E. NIKITIN
Institute of Chemical Physics
Academy of Sciences,
Kosygin str. 4
Moscow - 117334, USSR

ABSTRACT. We discuss the Landau-Lifshitz method of calculating quasiclassical matrix elements. It is shown that the modified quasiclassical integrand in the classically allowed region for both states can be analytically continued into the region which is classically allowed for one state but classically forbidden for the other, while the contribution from the classically forbidden region for both states can be completely ignored if the integration contour is properly chosen. This procedure allows one to completely disregard any complications connected with existence of the Stokes' lines emanating from the turning points and to displace the integration contour in order to simplify the evaluation of a matrix element.

The feasibility of mode-selective chemical reactions is based heavily on the existence of an inefficient coupling between subsystems which constitute a molecular system [1]. The slow energy exchange between different parts of the overall system can result ether from a small value of the coupling parameter (*weak coupling*) or from a considerable mismatch of frequencies which characterize the motion of uncoupled subsystems (*near-adiabatic coupling*). The latter situation is more interesting since in is more often encountered in molecular dynamics studies [1]. It is now recognized that a solution of a quantum mechanical problem can benefit very much from a study of a related classical problem by making use of the correspondence principle arguments [2,3].

Since many quantum transition in the near adiabatic limit are classically forbidden their "classical" description requires an excursion into the imaginary-valued time domain which is accomplished most easily by a manipulating the integration contour. The necessity to displace the time integration contour on the the complex time plane is often dictated not only by mathematical convenience but also by the very nature of the classical mapping of a quantum transition since the initial value classical mechanical problem becomes the boundary value problem: one wants to pass from a given initial state (a fixed value of the initial action variable and an undefined initial angle) to a given final state and not just start from fixed initial values of the action and angle variables to arrive somewhere.

J. Jortner et al. (eds.), Mode Selective Chemistry, 401–413.

The most convenient way to build a bridge between quantum mechanics and classical mechanics is based on the quasiclassical (QC) approximation (JWKB approximation) [4]. Of course, this approximation breaks down close to the turning points and it can be used only under the conditions when these non-QC regions do not contribute noticeably to the transition amplitudes. This is usually the case when a system behaves globally quasiclassically. There is yet another drawback of the QC approach which is related to the so-called Stokes' phenomenon and which shows itself in a dramatic change of a QC wave function when the coordinate crosses a certain line *even far away from the turning points*. This can be corrected by using different uniform approximations, which represent solutions of various types of comparison equations matched to appropriate JWKB asymptotics. In this context, a uniform approximation can be looked upon as a means which provides a safe crossing of a Stokes' line. Indeed, it was shown (see, e.g.[5]) that the uniform approximation coupled with the stationary phase approximation gave excellent results for a number of problems.

In this contribution we address a different question: does there exist a method of the QC level of accuracy within which one can avoid the use of the uniform approximations and still enjoy the freedom of changing integration contour? The answer is positive and goes back to the Landau paper [6]. We discuss this method described in the Landau-Lifshitz book [7] and developed further for particular applications in the theory of atomic and molecular collisions [8,9] and radiationless transitions [10,11]. For the reason of simplicity we confine ourselves to one dimensional case.

One important remark will be in order here. Though we concentrate on calculating the matrix elements which can be directly used in obtaining first order transition probabilities, the approach is quite general and may by generalized for strong coupling cases under near-adiabatic conditions. This is the principal idea of the second Landau paper [12] which relates nonadiabatic transition probabilities to the "transition" points on complex-valued adiabatic potential surfaces. As it is known now, the exponential factor here can be calculated exactly by the Landau method, and the problem of evaluating the preexponential factor boils down to the solution of a comparison equation rather than to calculating a matrix element [7,8].

Consider a matrix element V_{nm} of a function $V(x)$:

$$V_{nm} = \int V(x)\ \varkappa_n(x)\ \varkappa_m(x)\,dx \tag{1}$$

Assume for simplicity that the motion occurs in the field of a one-dimensional potential $U(x)$ with left turning points x_n and x_m for each state n and m; we assume for definiteness that $x_m > x_n$. Close to these points the QC approximation $\varkappa^{QC}$ to $\varkappa$ fails but since the singularity is integrable the integral (1) for globally quasiclassical conditions can, probably, be calculated as

$$V^{QC}_{nm} = \int V(x)\varkappa^{QC}_n(x)\varkappa^{QC}_m(x)\,dx \tag{2}$$

though some reservations were expressed about this possibility (see [7]). As shown in Fig.1, the integration domain $-\infty < x < \infty$ is partitioned in three regions: region A, $x_m < x < \infty$; region B, $x_n < x < x_m$; and region C, $-\infty < x < x_n$. Within these regions the products of the QC wave functions are (the common normalization factor is omitted here):

Region A:

$$\chi_n^{QC}\chi_m^{QC} = |p_n p_m|^{-1/2}\sin(s_n/\hbar+\pi/4)\ \sin(s_m/\hbar+\pi/4) \tag{3a}$$

Region B:

$$\chi_n^{QC}\chi_m^{QC} = |p_n p_m|^{-1/2}(1/2)\exp(-\sigma_m/\hbar)\ \sin(s_n/\hbar+\pi/4) \tag{3b}$$

Region C:

$$\chi_n^{QC}\chi_m^{QC} = |p_n p_m|^{-1/2}\ (1/4)\exp(-\sigma_n/\hbar - \sigma_m/\hbar) \tag{3c}$$

Here $s_k = s_k(x) = \int_{x_k}^{x} p_k(x)\,dx$, $\sigma_k(x) = |s_k(x)|$, and p_k is the momentum corresponding to energy E_k.

Assume now that the potential $U(x)$ is an analytic function of x across all three regions in the x axis. However, each of the functions in (3a) - (3c) are *not* analytic continuations of one another. The reason is of course well known: the passage from one piece to the other requires crossing a Stokes' line, which is not smooth at all [4]. A glance at Fig.1 confirms that "East is East, and West is West, and never the twain shall meet". This is the reason why the conventional quasiclassical formulation can not be directly used to calculate the integral (2) by analytic continuation technique.

On the other hand, it is well known [5] that matrix element V in the limit of large quantum numbers n,m and under the condition that states n and m are close to each other ($s = |n - m| \ll n,m$) becomes the Fourier-component of V corresponding to frequency $\omega_{nm} = (E_n - E_m)/\hbar$ (we ignore here the normalization constant):

$$V_{nm}^{CP} = \int dt\exp(i\omega_{nm}t)\ V[x_N(t)] \tag{4}$$

This Fourier component is calculated along a classical trajectory associated with a state N which lies somewhere close to n and m [7]. This approximation is known by different names, e.g. the Heisenberg correspondence relation, or common trajectory approximation. For our model problem the integration over dt runs from $-\infty$ to $+\infty$ and clearly corresponds to the region A of the quantum case. The representation (4) gives much freedom to manipulate the integration contour since the only singular points are now those which are associated with the interaction V, and not with the QC wave functions (Fig.2).

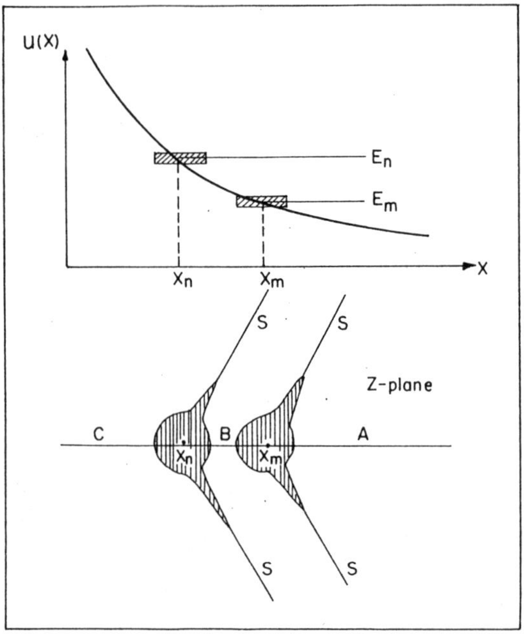

Fig.1.Regions of breakdown of the quasiclassical approximation in the U-x plane and in the complex z-plane (hatched). S are the Stokes' lines

It is also clear that calculation of the integral (4) requires knowledge of the interaction $V(x)$ within the classically accessible region of motion over trajectory N. If the integral is evaluated by the technique of analytic continuation one should make use of the analytic continuation of $V(x)$ and *not* the actual form of this potential in the classically inaccessible region.

With s increasing, this approximation becomes progressively worse since the uncertainty in the choice of the trajectory results in a large errors in V_{nm}. Various methods are described in the literature to correct the deficiency of the above approximation as applied to distant transitions (s comparable to n,m) by judicious choice of N This topic will not be discussed here. Instead, we pose the question as to whether there exists an approximation which allows one to calculate matrix elements under less restrictive conditions ($n \gg 1, m \gg 1$, $|n-m|$ comparable to n,m) and still use the analytic continuation technique.

Landau [6] and Landau and Lifshitz [7] suggested a method to evaluate quasiclassical matrix elements under the condition that the value of V_{nm} is exponentially small, with states n and m not being necessarily close to each other . The validity region of the Landau-Lifshitz (LL) method and the correspondence principle (CP) method are depicted in Fig.3.

Fig.4 illustrates typical processes to which this method has been applied in the past.

The method relies on the observation that under quasiclassical conditions the main contribution to the integral (2) (taken along the real-valued x-axis) comes from the classically allowed region A, and, possibly, from the classical-quantal region B. The contribution from the quantal region C is neglected, and the contribution from the classical region A is taken into account approximately by resolving the product of oscillating functions $\varkappa^{QC}\varkappa^{QC}$ into slowly and rapidly varying parts and by omitting the latter.

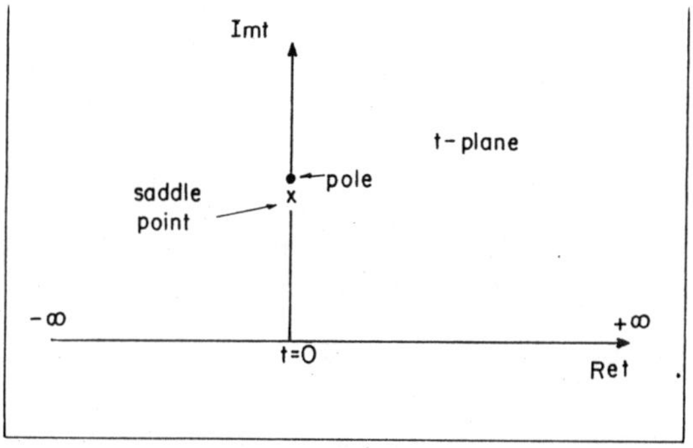

Fig.2.The correspondence principle complex t-plane. Shown are two points of interest for the Landau-Teller model: the pole of $V(t)$ and the saddle point (the stationary phase point) of the integrand in (4)

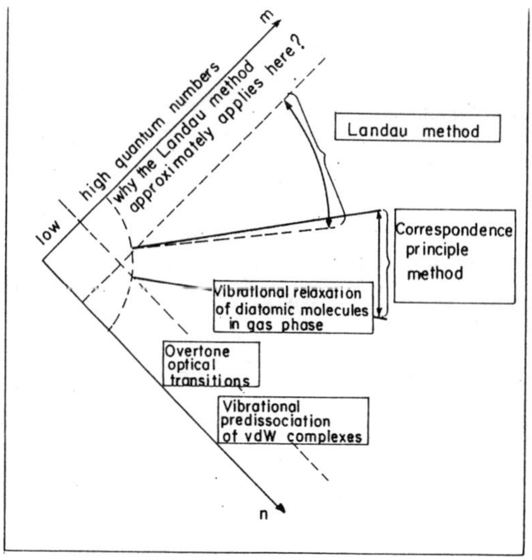

Fig.3. Correspondence principle and the Landau method.

In this way we must have the following Landau-Lifshitz counterpart of the quasiclassical approximation (3) provided the integration is carried out along the whole x axis:

Region A:

$$(x_n^{QC} x_m^{QC})^{LL} = L_{nm}(x) = |p_n p_m|^{-1/2} (1/2)\cos(s_n/\hbar - s_m/\hbar) \tag{5a}$$

Region B:

$$(x_n^{QC} x_m^{QC})^{LL} = L_{nm}(x) = |p_n p_m|^{-1/2} (1/2)\exp(-\sigma_m/\hbar) \sin(s_n/\hbar + \pi/4) \tag{5b}$$

Region C:

$$(x^{QC} x^{QC})^{LL} = 0 \tag{5c}$$

We will obtain the same result by changing the integration contour in such a way that portions (5a) and (5b) be regarded a single analytic function on this contour, and the contribution from region C be omitted irrespective of the form of the analytic continuation of this function into the region C.

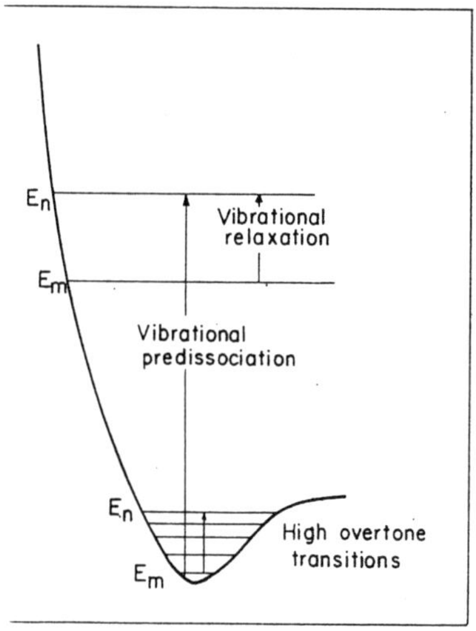

Fig.4. Types of transitions which can be described by the Landau method.

The analytic function $L_{nm}(z)$ which we are looking for is of the form

$$L_{nm}(z) = [p_n(z) p_m(z)]^{-1/2} (1/4)\exp[s_n(z)/\hbar - s_m(z)/\hbar] \tag{6}$$

with $s_k(z) = \int_{x_k}^{z} p_k(z)\,dz,\ p_k(z) = (2\mu)^{1/2}\ [E_k - U(z)]^{1/2}$

The above functions are defined on the complex z plane with branch cuts running from the turning points x_n and x_m to the right. The integration contour L circumvents this cut starting and ending at $+\infty$, see Fig. 5.

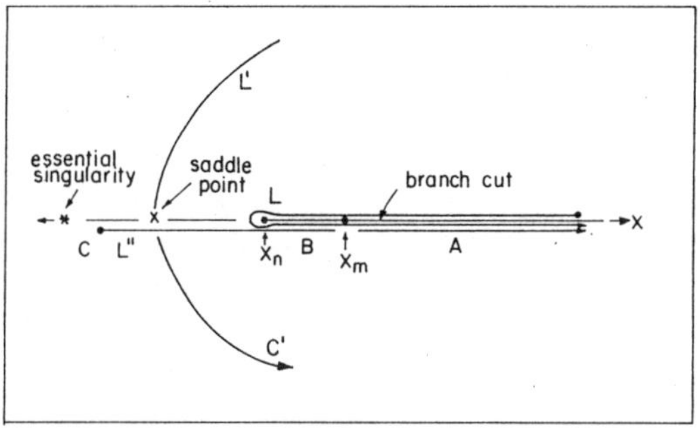

Fig.5. Integration contours for the Landau method

It is easily verified now that the sum of contributions from two parts of the contour adjacent to opposite sides of the cut will give expression (5a) in the region A, and expression (5b) in the region B provided all additional phases originating from p_k upon passing by the turning points are carefully taken into account.

As for the analytic continuation of (6) into Region C , it reads:

$$L_{nm}(x) = |p_n p_m|^{-1/2}\ (1/4)\exp(\sigma_n/\hbar - \sigma_m/\hbar\ ) \tag{7}$$

which is seen to be quite different both from (3c) and (5c). *This form of the integrand is never encountered when the integration is carried out along the original contour C shown in Fig.4. However, it will if the contour is displaced in order to evaluate the integral by the analytic continuation method (for instance, contour L')* The only restrictions for the contour displacements are now related to the branch cuts and singular points of the potential $U(x)$ and interaction $V(x)$. *There are no Stokes' lines any more*, and this is the main advantage of the Landau method.

In this way we arrive at the following expression for the Landau-Lifshitz matrix element of a function $V(x)$:

$$V_{nm}^{LL} = \int_L \mathbf{V}(z)\,L_{nm}(z)\,dz \tag{8}$$

where L is the contour shown in Fig. 5 or its equivalent. An "equivalent" contour may be quite different in shape from the original one, and only a small portion of it usually will be important since the value of the integral under near-adiabatic conditions is determined by integrand singularities which are the closest to the real axis of z. Yet another form of the integral is of interest: V^{LL} can be represented as a real part of an integral taken over the real axis from any point in the region C to infinity, $V^{LL} = V^{+} + V^{-}$, (see [7], Sect.52), e.g. contour L''.

As an example, assume, that the integral (8) can be calculated in the stationary phase approximation by a proper displacement of the contour L and that the stationary phase point $z^{\times}$ lies in the classically forbidden region of the potential , Region C, on the x axis (hence $z^{\times} = x^{\times}$); by symmetry, the displaced contour will cross the x-axis at $x^{\times}$ vertically (this situation actually occurs for the Landau-Teller model, *vide infra*). For this case we get the following exponential estimate of the matrix element

$$V^{LL}_{nm} = A^{LL}_{nm} \exp[-\sigma_m(x^{\times})/\hbar + \sigma_n(x^{\times})/\hbar] \qquad (9)$$

where the prefactor A^{LL}_{nm} depends but weakly on $x^{\times}$.

With simple expression (9) at hand, we can draw the following conclusions concerning the "nature" of the matrix element V^{LL}:

The *contribution* to the integral V_{nm} comes from the classically allowed region of motion A and from classical-quantum region B, or, equivalently, from a region near the stationary phase point $x^{\times}$ (SPP) which lies in a classically forbidden region C of the potential; this portion of the potential is constructed as the analytic continuation (AC) of the potential in the classically allowed region of the motion.

The *value* of the integral is mainly determined by the *difference* in magnitudes of (imaginary) classical actions along two branches of a (discontinuous) trajectory from their turning to SPP. Therefore, the value of the integral depends on the AC of the potential over the whole range, from the turning points to SPP.

The *exponential factor* of this integral is much larger (the absolute value of the negative exponent is much smaller) than that of a product of the exponential "tails" of two wave functions at SPP since the latter is governed by the *negative sum* of the above mentioned classical actions (see Eq.(3c)). Moreover, the integral is determined by the AC of the potential from its shape in the classically allowed region, while the exponential "tails" depend on the actual shape (AS) of the potential in the classically forbidden region. If a potential in the classically allowed region is given, one can still manipulate AS of a potential in the classically forbidden region but not its AC.

Therefore, a statement that the value of a quasiclassical matrix element depends on the behavior of the potential in the classically forbidden region is, in general, inaccurate and misleading: this value depends on AC of the potential from the classically allowed region into the classically forbidden region. If a potential is an analytic function everywhere the above statement is true but its exact meaning should be carefully interpreted.

It is clear that the correspondence principle counterpart of V^{LL} which can be derived from Eq. (4) is of the form

$$V^{CP}_{nm} = A^{CP}_{nm} \exp[-\omega_{nm}\tau^{\times}(N)] \qquad (10)$$

where $\tau^{\times}(N)$ is the imaginary time corresponding to the stationary phase point of the integrand (4) which is taken along the continuous complex classical trajectory N.

The exponent in (10) represent the first term in the expansion of the exponent in (9) in powers of the energy change, $\hbar\omega_{nm} = E_n - E_m$, and within the accuracy of this approach N can be identified either with n, or with m, or else with any state between n and m. If one wishes to go beyond the first order expansion and still keep a result simple (in the CP form) one can choose N from the condition that the second order correction to the difference in the action integrals (exponent in Eq. (9) vanishes. This is a basis for the so-called symmetrization procedure (see, e.g. [9], Eq. (9.21)) widely used now in semiclassical description of inelastic molecular collisions.

We now dwell on the question whether V^{LL} or V^{CP} integrals can be accurately calculated by the stationary phase method (of course, under the condition that they are exponentially small) since this approach is commonly used in the semiclassical collision theory. Though this is usually assumed to be the case (for instance, the asymptotic behavior of the Airy function is well described by this approximation) this approach will fail if the stationary phase point $\tau^{\times}$ (or $x^{\times}$) lies close to a singular point t_c (or z_c). An example of this situation is provided by the Landau-Teller model (matrix element of $V(x) = A \exp(-\alpha x)$ between states in the field of the potential $U(x) = B \exp(-\alpha x)$) for which t_s corresponds to a pole of second order (and z_s to essential singularity). But even in this case the stationary phase approximation gives a result which is very close to the exact one: they differ by a factor $e^2/4\sqrt{\pi} = 1.043$ [5,13]. Finally we note that the exponent in (9) may by quite insensitive to the position of the SPP if $x^{\times}$ lies in the region where $U(x^{\times})$ much exceeds both E_n and E_m. This is illustrated in Fig.6.

We now briefly discuss the question why the Landau method applies sometimes to transitions for which one of the states is not quasiclassical. For instance, the dipole matrix elements of high overtone transitions for the Morse oscillator were found [14,15] to be reasonably reproduced by the Landau method though the initial ground vibrational state is certainly not quasiclassical. The reason for this "unexpected" applicability of the Landau method is related to the fact that the stationary phase point is determined mainly by the properties of the final quasiclassical vibrational wave function. Close to this point, the deep tunneling behavior of the initial non-quasiclassical wave function is well represented by its quasiclassical tail, so that the difference between an accurate matrix element and its Landau counterpart is presumably due to the difference in the normalization factors of the ground state wave function and to the different treatment of the contribution to the integral from Region C.

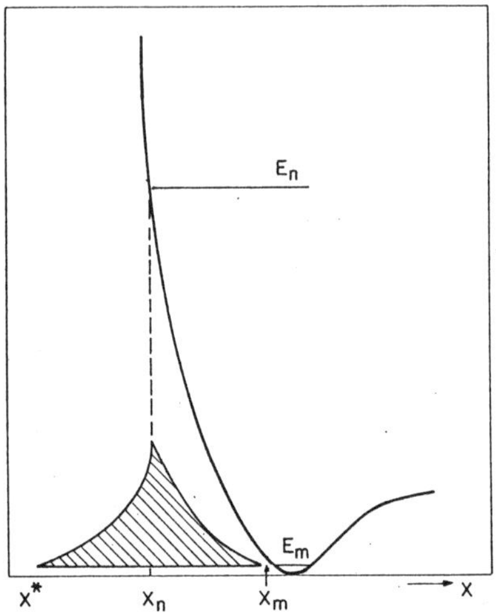

Fig.6. Where does the main contribution to the Landau exponent come from? The hatched region corresponds to the exponent integrand.

The Landau method is of course not limited to transitions between states defined for the same potential $U(x)$; it can be applied for two different potentials, $U_\mu(x)$ and $U_\nu(x)$, e.g. for electronically nonadiabatic transitions. In the latter case, the matrix element will bear four subscripts, $m\mu\ n\nu$, rather than two, mn.

The Landau-Lifshitz matrix elements can be used directly to calculate transition probabilities in the weak coupling case. What happens when V increases? Within the framework of first order perturbation theory the preexponential factor of the probability grows but then, when it becomes of the order of unity, the perturbation approach fails though the transition probability remains very small due to the Landau exponential factor. Starting from this moment, the exponent suffer a change which can be found in the framework of the adiabatic perturbation theory. Specifically, one passes adiabatically from one potential surface to the other through a complex-valued branch point at which two potential energy surfaces cross and calculates the complex-valued JWKB phase increment which determines the transition amplitudes. If there are several branch points one should take into account the interference effect; the first order perturbation result corresponds then to the case when the phase difference between trajectories passing through the neighboring branch points is small [8]. The original Landau-Zener model [7] and its generalization for transitions close to the turning points [8] provides a revealing illustration of the general approach (see Figs. 7 and 8)

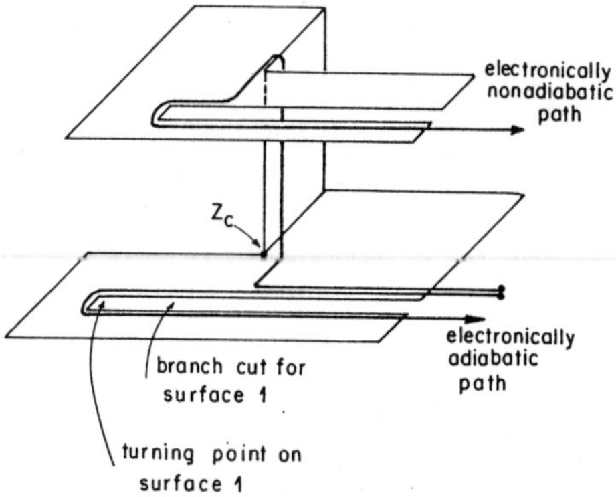

Fig.7. Two Riemann sheets for two adiabatic electronic states which cross at point z_c (the Landau-Zener model). Shown are two contours for calculating the phase shifts in the adiabatic and nonadiabatic channels in one-dimensional full collision. There exists yet another contour for the nonadiabatic channel which passes from the lower sheet to the upper sheet after circumventing the turning point on the lower sheet.

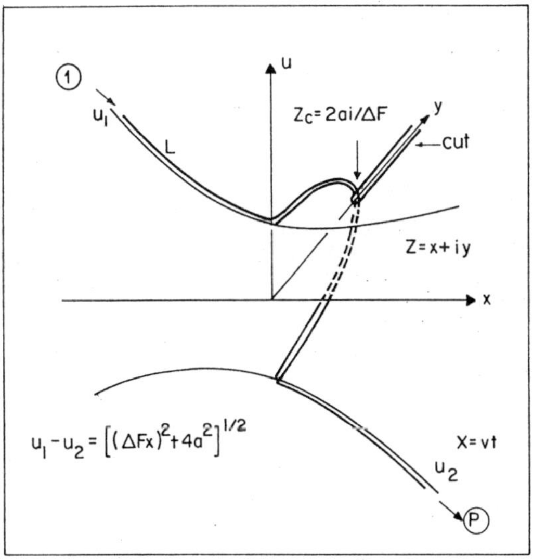

Fig.8. Two complex-valued adiabatic potential surfaces for the Landau-Zener model. Shown is the contour for calculating the Landau exponent in nonadiabatic channel of one-dimensional half collision.

In conclusion, we can outline the following hierarchy of approximations for calculation of matrix elements of $V(x)$ between states with high quantum numbers:

i. Exact calculations using accurate wave functions. In the case of one turning point for each state three regions contribute to the integral: Region A (classically allowed motion for both states), Region B (classically allowed motion for one state and classically forbidden motion for the other), and Region C (classically forbidden motion for both states)

ii. Quasiclassical approximation using JWKB wave functions with the uniform approximation near the turning points. Three regions, Regions A, B and C, contribute to the integral.

iii. Landau-Lifshitz approximation using the modified quasiclassical integrand. Two regions, Regions A and B, contribute to the integral.

iiii. Correspondence principle approximation using the Fourier transform of $V[x(t)]$ along the common classical trajectory. One region, Region A, contributes to the integral.

It is thus seen that the Landau-Lifshitz approximation constitutes an important step which bridges the gap between fully quantum mechanical and correspondence principle description of molecular systems. Needless to say that the same level of approximation can be used in solving coupled scattering equations which will be of the first order (the quantum mechanical scattering equations are of the second order) but will not rely on the common trajectory approach (the common trajectory approach does not describe effects related to the different positions of turning points on different potential curves).

Acknowledgment

Very illuminating discussions with Drs. E.A.Medvedev, V.I.Osherov, M.Ya.Ovchinnikova, D.A.Shalashilin, S.Ya.Umanskii are gratefully acknowledged.

REFERENCES:

1. A. J. Jortner, R.D.Levine, Isr.J.Chem. **30**, 207 (1990)
2. W.H.Miller, Adv.Chem.Phys. **30**, 77 (1976)
3. M.S.Child, in: *Semiclassial Methods in Molecular Scattering and Spectroscopy*, Dordrecht: Reidel, 1979, p. 127
4. N.Fröman, P.O.Fröman, *JWKB Approximation. Contributions to the Theory*, Amsterdam: North- Holland, 1965
5. T.Uzer, M.S.Child, Mol.Phys. **41**, 1177 (1980)
6. L.Landau, Phys.Z.Sow., **1**, 1 (1932)
7. L.D.Landau, E.M.Lifshitz, *Quantum Mechanics*, Oxford: Pergamon Press, Third Edition, 1977
8. E.E.Nikitin, S.Ya.Umanskii, *Theory of Slow Atomic Collisions*, Heidelberg: Springer, 1984
9. E.E.Nikitin, *Theory of Elementary Atomic and Molecular Processes in Gases*, Oxford: Clarendon Press, 1974
10 A.A.Ovchinnikov, M.Ya.Ovchinnikova, Adv.Quant.Chem. **16**, 161 (1982)
11. E.S.Medvedev, V.I.Osherov, *Theory of Radiationless Transitions in PolyatomicMolecules* (in Russian), Moscow: Nauka, 1983.
12. L.Landau, Phys.Z.Sow., **2**, 12 (1932)
13. E.E.Nikitin, M.Ya.Ovchinnikova, Zh.Eksp.Teor.Fiz. **78**,1551 (1980) English translation: Sov.Phys.JETP, **51**, 779(1980)
14. E.S.Medvedev, Chem.Phys.Lett. **120**, 173 (1985)
15. K.Lehmann, A.M.Smith, J.Chem.Phys. **93**, 6140 (1990)

MOLECULAR ELECTRONIC ENERGY TRANSFER IN A SUPERSONIC JET EXPANSION

Joel BIGMAN, Yoram KARNI and Shammai SPEISER
Department of Chemistry
Technion – Israel Institute of Technology
Haifa 32000, Israel

ABSTRACT. We present preliminary results showing electronic energy transfer (EET) between benzene and biacetyl and between naphthalene and anthracene in a supersonic jet expansion. Excitation of several vibronic levels of benzene (donor) in the presence of biacetyl (acceptor) shows quenching of benzene emission with simultaneous appearance of biacetyl fluorescence emission with similar results for the pair naphthalene (donor) and anthracene (acceptor). The quenching follows Stern-Volmer kinetics as a function of added acceptor pressure, indicative of EET in a binary donor-acceptor complex.

1. Introduction

Electronic energy transfer (EET) processes involve non-radiative transfer of electronic excitation from an excited donor molecule D* to an acceptor molecule A. The transfer may be an intermolecular process which can be described in terms of a bimolecular kinetic process

$$D^* + A \xrightarrow{k_Q} D + A^* \tag{1}$$

where the bimolecular quenching process is related to an intermolecular energy transfer rate by

$$k_{ET}^{inter} = k_Q[A] \tag{2}$$

Theoretically k_{ET} is attributed to two possible contributions. The long range coulombic contribution formulated by Förster [1]–[3] in terms of dipole-dipole interaction is well documented [4]. It is particularly suitable for describing electronic energy transfer in solution whenever conditions for favourable spectroscopic overlap conditions between the emission of D* and the absorption of A are met.

The second contribution to EET can be realized whenever these conditions are not fulfilled. A short range exchange interaction, as formulated by Dexter, can then facilitate EET [5]. It was shown to be the correct mechanism to describe intramolecular EET in

J. Jortner et al. (eds.), Mode Selective Chemistry, 415–427.

bichromophoric molecules of the type D–A [6]–[10] in a unimolecular process of the type

$$D^* - A \xrightarrow{k_{ET}^{intra}} D - A^* \tag{3}$$

These studies were performed in solutions where solvent effects cannot be ruled out. Short range EET processes in the gas phase involve collision which will lead to a collision complex formation for a period long enough for the process described by eq. (3) to take place. There is evidence to support the validity of a modified Dexter formulation for the description of these gas phase EET processes [11]–[16]. In some cases there is evidence for a formation of a collision complex as part of the EET process [16].

The unique conditions in a supersonic jet expansion suggest studying these processes in more detail. In particular the dependence of the transfer on a particular vibronic excitation of the donor moiety in a bichromophoric cluster of the type D–A can be studied. These can be van der Waals complex or a more loosely bound bichromophoric complex. One question that may be addressed is whether or not an EET process described by eq. (2) can be observed resulting as in solution by dual fluorescence spectrum [6]–[10] typical of a bichromophoric D–A molecule or cluster.

Our choices for a D–A system were the benzene (D) – biacetyl (A) pair and the naphthalene (D) anthracene (A) pair. Studies of intermolecular EET [17]–[19] and intramolecular EET [7]–[9] in solution as well as gas phase EET studies of the benzene–biacetyl pair [11]–[16] indicate that the basic EET mechanism operative in this case is the Dexter exchange interaction [5], whereas for the naphthalene-anthracene pair solution studies suggests intramolecular EET via dipole-dipole interaction [20].

We present in this paper preliminary results that demonstrate a short range EET in supersonic jet expansion. Previous studies of EET under similar conditions were performed by Poeltl and McVey [21] on dimers of benzoic acid, by Tomioka and coworkers [22] on mixed dimer of benzoic acid and p-toluic acid and by Lahamani and coworkers for the p-xylene p-difluorobenzene pair [23]. In all these studies the extent of spectral overlap rules out exchange interaction as the sole mechanism thus complicating mechanistic interpretation. The choice of benzene-biacetyl pair eliminates this complication.

2. Experimental

Samples for the benzene-biacetyl experiment were prepared in a 2 liter stainless steel cylinder. Liquid benzene and biacetyl were held at room temperature in separate containers. Specific pressures of the organic vapors were let into the cylinder, and measured by a mechanical manometer. Finally, helium was added to bring the total pressure up to 5 Atm.

The gas mixture was expanded through a General Valve series 9 pulsed valve with a 0.5 mm orifice into a stainless steel vacuum chamber. The chamber was pumped by an untrapped CVC Goldline 6 inch diffusion pump backed by a Welch model 1397 rotary pump. Typical background pressure was about 5×10^{-5} torr.

For the naphthalene – anthracene experiment mixtures were prepared by passing He at 5 atm over naphthalene held at room temperature in a stainless steel container, into a high-temperature solenoid valve. The valve, built according to Li and Lubman [24] design, serves as the hot oven and contains the added anthracene. The expansion conditions are the same as those used in the benzene-biacetyl experiment.

The samples were excited by a Quanta Ray PDL-1 Dye laser, pumped by the third harmonic of a DCR Nd:YAG laser, operated at 10 Hz. Coumarin 420 was used to excite biacetyl, while benzene excitation was provided by frequency doubling the output of the dye laser, operating with coumarin 500, using a WEX1 wavelength extender. The laser light is passed through a series of baffles into a vacuum chamber, and intersects the molecular beam about 35 mm from the valve orifice.

Excitation spectra were recorded with a Hamamatsu R1104 photomultiplier tube, using appropriate filters to exclude scattered laser light. A Bausch and Lomb 0.25 meter monochromator was used to record dispersed emission spectra. Signals were averaged over a number of laser shots with a PAR model 162 boxcar, or with a Textronix digital oscilloscope.

3. Results

3.1. BENZENE – BIACETYL PAIR

The first basic requirement of this study was to obtain the excitation and emission spectra of both benzene and biacetyl in order to examine the extent of spectral overlap under the conditions of supersonic jet expansion.

Figure 1 shows the excitation spectrum of 3 torr benzene jet expansion with 5 atm of helium in the 260 nm wavelength region. This is basically the spectrum observed by Levy and coworkers [25] and is indicative of insufficient cooling. Thus some of the lines are due to hot transitions. A low resolution dispersed emission spectrum resulting from exciting the 6_0^1 transition of benzene at 38610 cm^{-1} is shown in Fig. 2. This emission overlaps poorly with the $S_0 \rightarrow S_1$ absorption spectrum of biacetyl, as can be seen from its jet expansion excitation spectrum shown in Fig. 3. Biacetyl does not show any absorption in the range of the benzene 260 nm bands of Fig. 1 [26], which allows a specific excitation of benzene in benzene biacetyl jet expanded mixture. The dispersed emission of this mixture is shown in Fig. 4. The 350 nm emission band is the same as that of pure benzene (Fig. 2), whereas the band centered around 480 nm can be attributed to biacetyl. The dispersed fluorescence emission from pure biacetyl excited at its 444.5 nm (22497 cm^{-1}) band (Fig. 3) is shown in Fig. 5, matching the 480 nm emission band of the mixture's dual emission.

This dual fluorescence seems to indicate enlectronic energy transfer from benzene to biacetyl similar to that observed intermolecularly in solution [17]–[19], or in the gas phase [11]–[15]. This is further supported by studying quenching of the 6_0^1 excitation peak of benzene upon adding biacetyl (Fig. 6). This quenching seems to follow a Stern-Volmer dependence, shown in Fig. 7 for different excitation lines of benzene. In addition, evidence for EET through complex formation is given in Figure 8 where the excitation spectrum of the benzene is monitored by looking at laser induced fluorescence of the biacetyl band. This spectrum is different from that of pure benzene (Fig. 1).

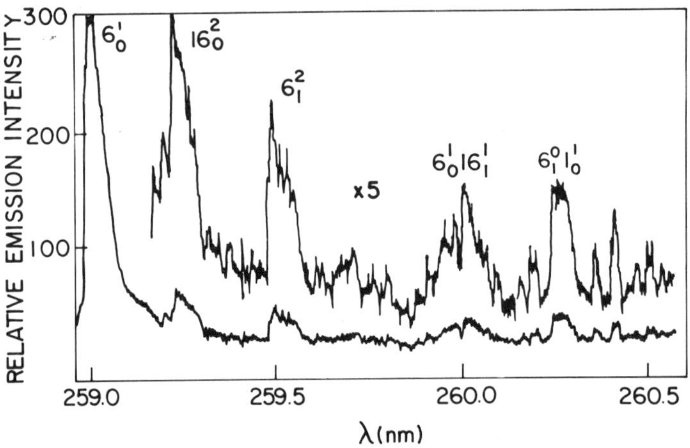

Figure 1: Benzene excitation spectrum, for 3 torr benzene in 5 atm helium. Upper trace taken at 5 times higher sensitivity.

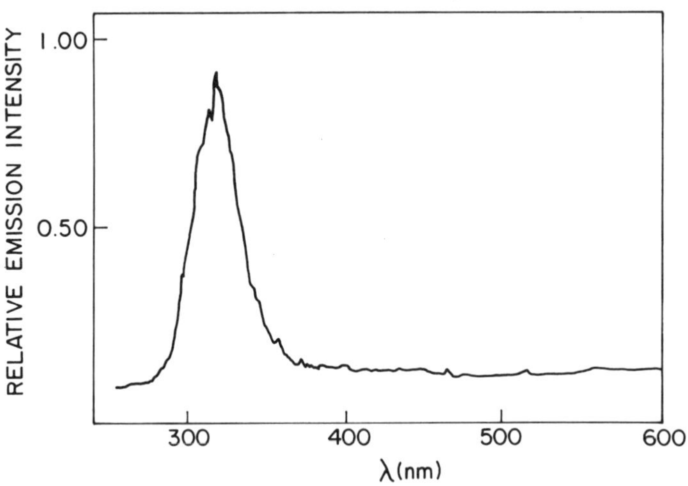

Figure 2: Low resolution (5 nm) benzene emission spectrum (10 torr benzene in 5 atm helium) excited at the 6^1_0 transition (38610 cm^{-1}).

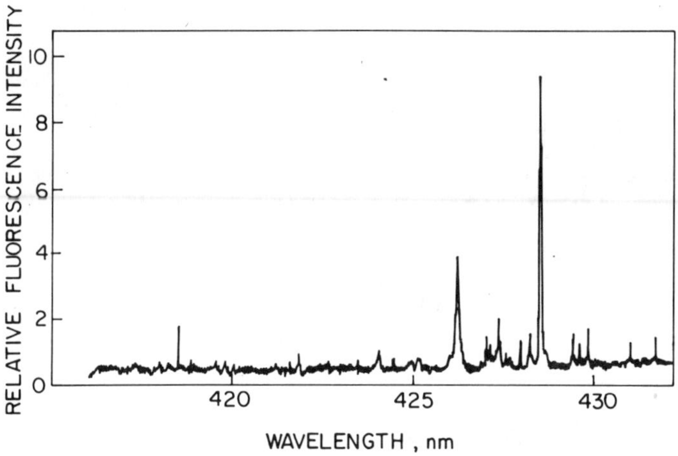

Figure 3: Biacetyl excitation spectrum, for 10 torr biacetyl in 5 atm helium.

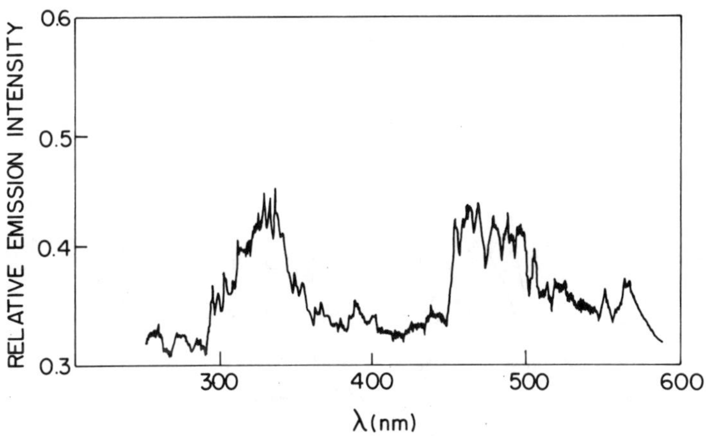

Figure 4: Low resolution (10 nm) emission spectrum of jet expanded benzene – biacetyl mixture (10 torr benzene, 16 torr biacetyl in 5 atm helium), excited at the 6_0^1 benzene transition.

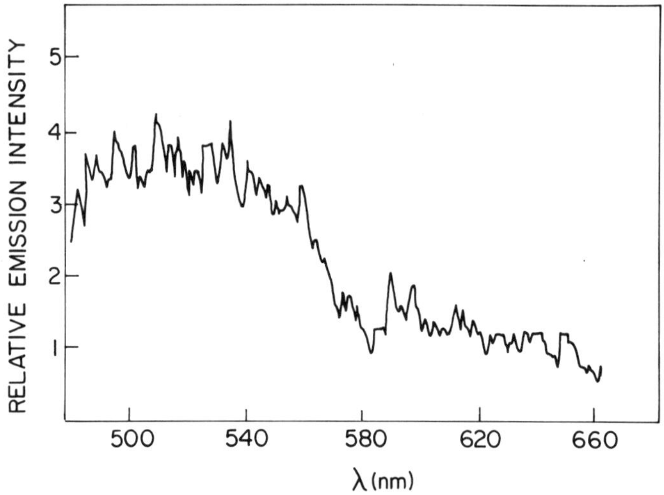

Figure 5: Low resolution (10 nm) emission spectrum of jet expanded biacetyl, excited at 444.5 nm.

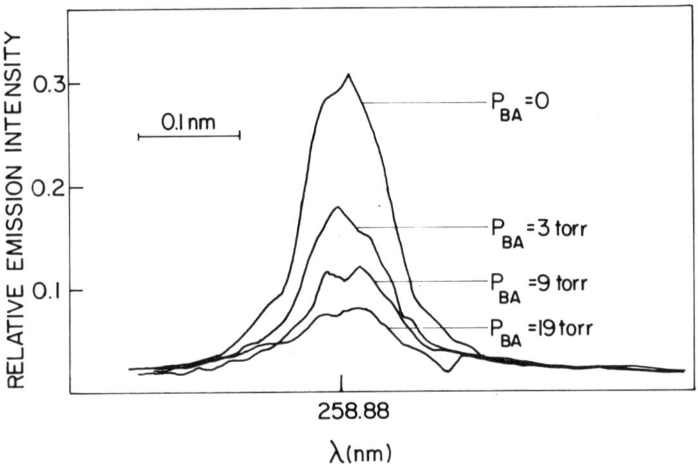

Figure 6: Progressive quenching of the 6_0^1 peak of benzene by added pressure, P_{BA}, of biacetyl (3 torr benzene in 5 atm helium).

3.2. NAPHTHALENE - ANTHRACENE PAIR

The excitation spectrum for jet expanded naphthalene and anthracene is shown in Figures 8. The results are similar to those reported previously [27], [28]. The extent of spectral overlap between naphthalene fluorescence [29] and anthracene absorption [30] may determine the mechanism promoting the transfer. Figure 10 shows quenching of the $8(b_{1g})^1_0$. Transition of naphthalene by added anthracene pressure. Again this quenching follows a Stern-Volmer dependence Fig. 11. Examination of the excitation spectrum of the jet expanded naphthalene-anthracene mixtures by looking at anthracene emission at the 380-400 nm range shows (Fig. 12) additional spectral features to those observed for pure naphthalene indicative of complex formation between these molecules in which electronic excitation of the naphthalene moiety is being transferred to the attached anthracene moiety.

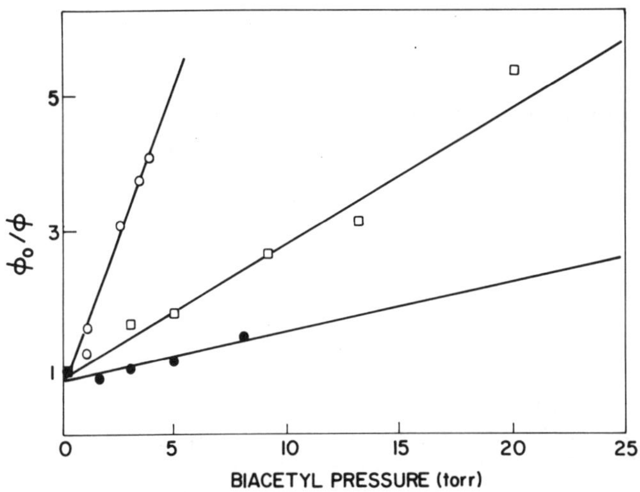

Figure 7: Stern-Volmer plot of quenching benzene emission by added biacetyl in jet expanded mixture, for three excitation wavelengths corresponding to the 16^2_0 transition •, 6^1_0 transition □ and the $6^1_0\ 16^1_1$ transition o.

4. Discussion

Our results provide evidence for intramolecular EET is a supersonic jet. In the gas phase this type of behavior is indicative of collisionally induced energy transfer, that might involve an excited D*–A complex formation prior to energy transfer [16]. In the jet complex formation precedes the excitation process and can be described by

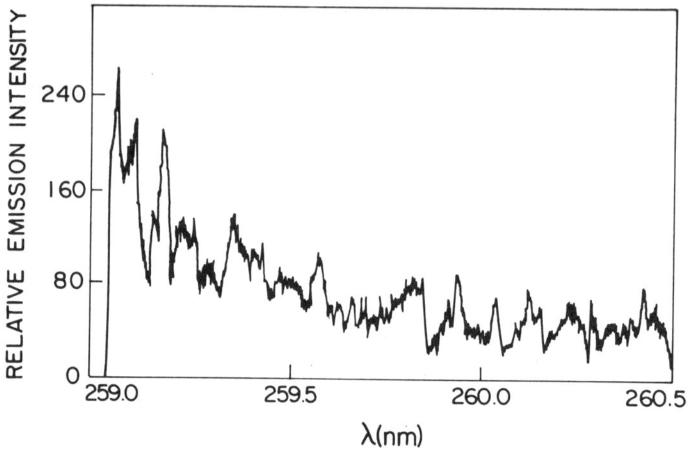

Figure 8: Excitation spectrum of the benzene-biacetyl complex, viewed through a filter that blocks the benzene emission. Sample: 3 torr benzene and 25 torr biacetyl in 5 atm helium.

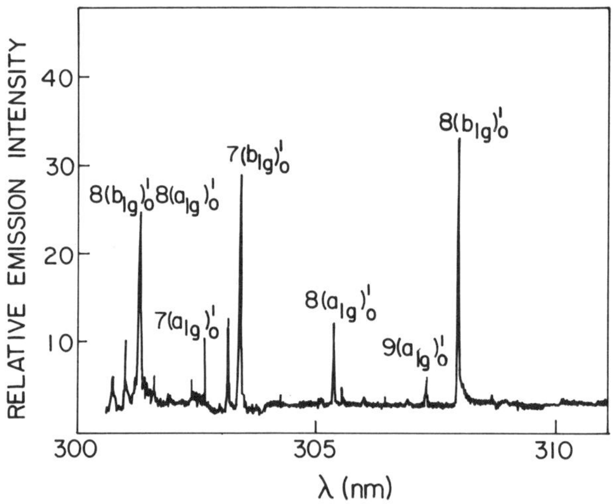

Figure 9: Excitation spectrum of 0.3 torr naphthalene in 5 atm He.

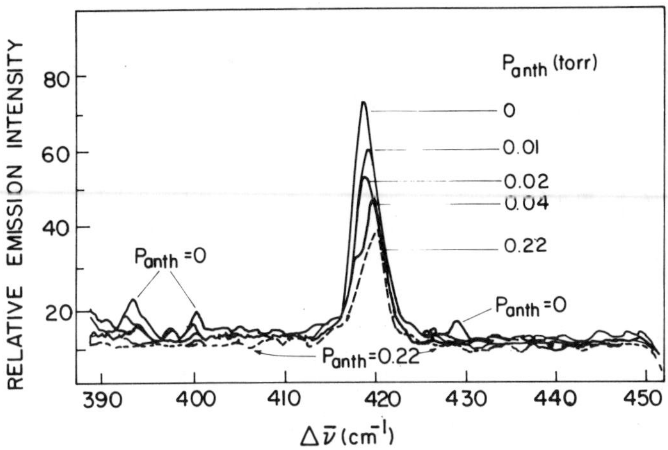

Figure 10: Progressive quenching of the $8(b_{1g})_0^1$ peak of 0.03 torr naphthalene in 5 atm He by added anthracene pressure, P_{anth}.

$$D + A \overset{K}{\rightleftharpoons} D - A \tag{4}$$

where K is the association equilibrium constant. Further assumption made here is that in the excited D* – A complex EET process is faster than dissociation and that D – A is a weakly bound complex where D and A maintain most of their spectral identity.

The EET and the excitation processes may be summarized in the following kinetic scheme:

D	$\xrightarrow{\sigma_D I}$	D*	donor excitation	(5)
D–A	$\xrightarrow{\sigma_{DA} I}$	D*–A	complex excitation	(6)
D*	$\xrightarrow{k_f^D}$	D	donor fluorescence	(7)
D*–A	$\xrightarrow{k_f^{D^*A}}$	D–A	complex donor's moiety fluorescence	(8)
D*–A	$\xrightarrow{k_{ET}^{intra}}$	D–A*	intramolecular EET	(9)
D–A*	$\xrightarrow{k_f^{DA^*}}$	D–A	complex acceptor's moiety fluorescence	(10)

where σ_D and σ_{DA} are the crossections for donor absorption for the free donor and for its complexed form respectively, I is the excitation intensity and the k_f's are the fluorescence rate constants.

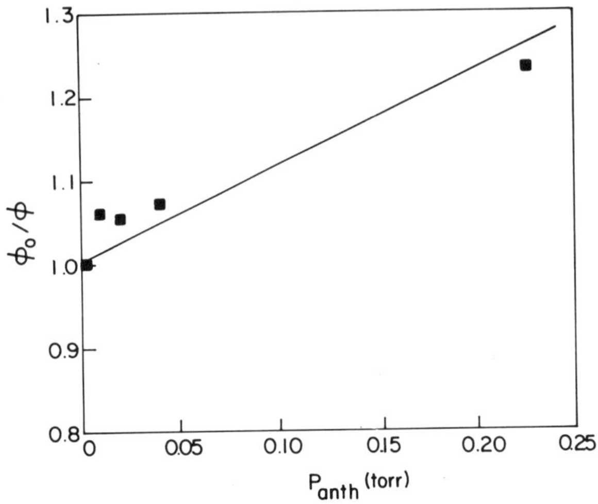

Figure 11: Stern-Volmer plot of quenching naphthalene emission by added anthracene in jet expanded mixture of 0.03 torr naphthalene at 5 atm He.

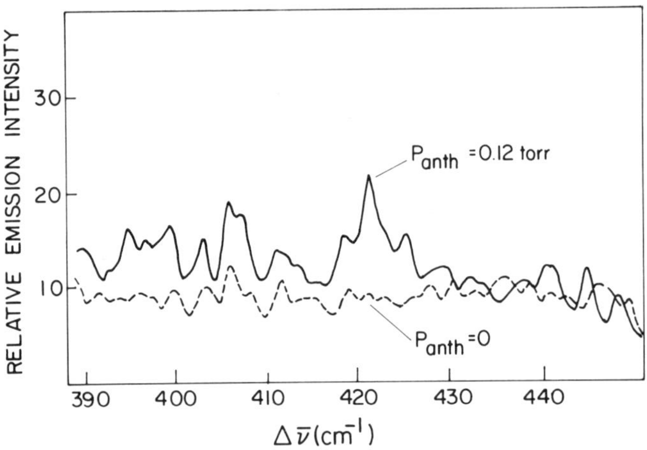

Figure 12: Excitation spectrum of naphthalene observed by monitoring the 380-400 nm spectral range of anthracene fluorescence, indicative of complex formation in 0.06 torr naphthalene with 0.12 torr anthracene in 5 atm jet expanded He. Lower trace shows results with no anthracene present in the jet.

A kinetic analysis of this scheme leads to equation (11) which has the form of a traditional Stern-Volmer relation

$$\phi^o/\phi = 1 + \tau_f^{D^*A} k_Q[A] \tag{11}$$

where $\tau_f^{D^*A} = 1/k_f^{D^*A}$ and k_Q is the apparent bimolecular quenching rate constant related to the intramolecular EET rate by

$$k_Q = (\sigma_{DA}/\sigma_D) k_{ET}^{intra} K \tag{12}$$

Table I summarizes k_Q values for the different excited vibronic states of benzene, obtained from the data presented in Fig. 6 and Fig. 7. It clearly shows that the EET is a function of the excess vibrational excitation, in a specific vibronic state. It is interesting to note that k_Q values are in the same range as those observed for the gas phase experiments [13]–[15], where a Bolzmann distribution of vibronic states is involved in the EET process.

It should be noted that our kinetic scheme does not include a dissociation channel for D^*-A. However the emission of biacetyl from the complex (Fig. 4) is blue shifted by about 2000 cm^{-1} with respect to that of the uncomplexed molecule (Fig. 5), indicative of emission from hot biacetyl that may have been formed by dissociation of D^*-A.

Table 1: Quenching constants for benzene quenched by biacetyl, as related to excess vibrational energy of the excited benzene, ΔE_{vib}.

Transition	Excitation Wavelength cm^{-1}	τ_f^D ns [35]	k_Q 10^6 $torr^{-1}$ sec^{-1}	ΔE_{vib} cm^{-1} [35]
6_0^1	38610	79	2.6	521
16_0^2	38565	89	2.2	486
6_1^2	38525	72	7.8	1042
$6_0^1\ 16_1^1$	38450	77	10.	767

Existing EET theories [1]–[4], [31], [32] are not directly applicable to the situation in hand. However, it is very probable that the requirement for spectral overlap, under the present conditions where EET takes place without prior vibrational relaxation, is met by a specific vibronic excitation. Our current and future studies will address this problem as well providing further data on the nature of the D–A complex and resolve some of the problems left open in this preliminary study [33], [34].

5. Conclusions

We have provided evidence for the possibility of electronic energy transfer between an excited donor to an acceptor in a supersonic jet expansion. Due to poor spectral overlap conditions are set for studying EET due mainly to short range exchange interaction in a D–A complex. The transfer is shown to depend on the specific vibronic excitation of the donor.

6. Acknowledgement

This work was supported by a grant from the VPR Technion Fund and the Frocht Fund for Electro Optics Research.

7. References

[1] Th. Förster, Ann. Phys. (Leipz) **2**, 55 (1948).
[2] Th. Förster, Disc. Faraday Soc. **27**, 7 (1959).
[3] Th. Förster in "Modern Quantum Chemistry" Istanbul lectures, O. Sinanoglu Ed. (Academic Press, N.Y. 1965).
[4] S. Speiser, J. Photochem., **22**, 195 (1983).
[5] D.L. Dexter, J. Chem. Phys. **21**, 836 (1953).
[6] D. Getz, A. Ron, M.B. Rubin, and S. Speiser, J. Phys. Chem., **84**, 768 (1980).
[7] S. Speiser and J. Katriel, Chem. Phys. Lett., **102**, 88 (1983).
[8] S. Hassoon, H. Lustig, M.B. Rubin, and S. Speiser, J. Phys. Chem., **88**, 6367 (1984).
[9] S. Hassoon, M.B. Rubin and S. Speiser, J. Phys. Chem., **90**, 912 (1986).
[10] M. Kaschke, N.P. Ernsting, B. Valeur, and J. Bourson, J. Am. Chem. Soc., **94**, 5757 (1990).
[11] H. Inakawa and W.A. Noyes, Jr., J. Chem. Phys. **37**, 583 (1962).
[12] W.A. Noyes, Jr. and I. Unger, Advan. Photochem. **4**, 49 (1966).
[13] E.K.C. Lee, M.W. Schmidt, R.G. Shortbridge, Jr., and G.A. Haninger, Jr., J. Phys. Chem. **73**, 1805 (1969).
[14] G.A. Haninger, Jr., E.K.C. Lee, J. Phys. Chem. **73**, 1815 (1969).
[15] G.L. Loper and E.K.C. Lee, J. Chem. Phys. **63**, 264 (1975).
[16] H.M. Lin, M. Seaver, K.Y. Tang, A.E.W. Knight, and C.S. Parmenter, J. Chem. Phys. **70**, 5442 (1979).
[17] J.T. Dubois and M. Cox, J. Chem. Phys., **38**, 2536 (1963).
[18] J.T. Dubois and F. Wilkinson, J. Chem. Phys., **38**, 2541 (1963).
[19] J.T. Dubois and R.L. Van Hemert, J. Chem. Phys., **40**, 923 (1964).
[20] O. Schnepp and M. Levy, J. Am. Chem. Soc. **84**, 172 (1962).
[21] D.E. Poeltl and J.K. McVey, J. Chem. Phys. **40**, 1801 (1984).
[22] Y. Tomioka, H. Abe, N. Mikami and M. Ito, J. Phys. Chem. **88**, 5186 (1984).
[23] F. Lahmani, C. Lardeux-Dedoner and A. Zchnacker Rentien, J. Chem. Phys. **92**, 4159 (1990).
[24] Li and Lubman, Rev. Sci. Instrum. **601**, 499 (1989).
[25] P.R.R. Langridge-Smith, D.V. Brumbaugh, C.A. Haynam and D.H. Levy, J. Phys. Chem. **85**, 3742 (1981).
[26] M. Gurnick, J. Chaiken, T. Benson, and J.D. McDonald, J. Chem. Phys. **74**, 99.
[27] S.M. Beck, D.E. Powers, J.B. Hopkins and R.E. Smalley, J. Chem. Phys. **73**, 2019 (1980).

[28] F.M. Behlen, D.B. McDonald, V. Sethuraman and S.A. Rice, J. Chem. Phys. **75**, 5685 (1981).

[29] S.M. Beck, J.B. Hopkins, P.E. Powers and R.E. Smalley, J. Chem. Phys. **74**, 43 (1981).

[30] W.R. Lambert, P.M. Felker, J.A. Syage and A.H. Zewail, J. Chem. Phys. **81**, 2195 (1984).

[31] S.H. Lin, Proc. R. Soc. Lond. A **335**, 51 (1973).

[32] S.H. Lin and H. Eyring, Proc. Nat. Acad. Sci. USA **72**, 4205 (1975).

[33] J. Bigman and S. Speiser (to be published).

[34] Y. Karni and S. Speiser (to be published).

[35] K.G. Spears and S.A. Rice, J. Chem. Phys., **55**, 5561 (1971).

Mode Specifity in Collisions of Aniline with Surfaces covered with Organized Organic Monolayers

Y. Paz and R. Naaman

Department of Chemical Physics
Weizmann Institute of Science
Rehovot, Israel

Abstract:

Aniline has been scattered from three organic substrates and a LiF single crystal surface. Applying multiphoton ionization and time of flight measurements the vibrational, rotational and translational energy distributions were measured simultaneously. The NH_2 "umbrella" like mode was found to be a very efficient accepting mode in the energy transfer process. The less rigid the surface, the greater the efficiency with which this mode is populated. The mode specificity does not exist for the rigid LiF surface. A model is presented, which explains all observations based on the collision time and kinematics.

Introduction:

Most studies to date on collisional energy transfer between gas phase molecules and surfaces have involved diatomics colliding with single crystal inorganic substrates. The advantage of working with a simple collider is obvious. However one does expect that if the scattering molecule is more complex, new information may emerge on the molecule-surface interaction. For example polyatomic molecules have vibrational modes that are characterized by very different frequencies and sites on the molecule. Hence it is expected that they will be affected differently by a collision with a surface.

In the present work, the energy transfer process between aniline and well defined inert organized organic amphiphilic surfaces were investigated. The results were compared to those obtained for scattering from LiF single crystal surface. The results indicate mode specificity in the translation to vibration energy transfer process. The extend of this specificity depends strongly on the surface.

Aniline was chosen for this study for two reasons. Firstly, its spectroscopy is well understood (1-4). Its vibrational modes can be classified as NH_2 related motion, ring motion and C-H related vibrations. Each type of mode is characterized by a

J. Jortner et al. (eds.), Mode Selective Chemistry, 429–441.

different frequency, and can be assigned to a well defined site in the molecule. Secondly, information is available on its energy transfer properties in the gas phase (5,6), which indicates mode specifity. The aniline-surface scattering process can therefore be compared with the known data on the gas-aniline energy transfer. The large range of vibrational frequencies makes the aniline a good probe for the gas-surface collision dynamics since it has the potential of probing processes occuring on various time scales. Here we applied this "probe" for the investigation of surfaces coated by organized organic monolayers (OOM). This type of substrates are of interest because of various reasons. They are a unique class of surfaces which may represent more realistically surfaces that affect the chemistry occurring in bulk and industrial processes. They have the advantage that they can be varied in a controlled manner. Thus, one can alter the interaction of the surface with the molecule by replacing the outer group of the chain. This varies the interaction potential between the molecule and the surface. The length of the chain can also be varied, while leaving the group facing the colliding molecule unchanged.
Recent work in our laboratory has investigated the scattering of NO from OOM. The translational (7), rotational (8) and electronic (9) energy distributions in the scattered molecules have been examined. Since NO serves as the "model system" for surface-gas scattering, the data obtained from OOM surfaces could be compared to that from inorganic single crystals. The results indicated that the nature of the scattering of NO from the OOM and inorganic surfaces was similar. It was found that the vibrational excitation of the scattered NO is extremely inefficient. It is expected that the situation will be very different for aniline (10).

The mechanisms for rotational and translational energy transfer between gas phase molecules and surfaces are well understood and experimentally documented. Much less is known about the mechanism for vibrational energy deposition. Several experiments have probed vibrational energy distribution in NO/surface scattering (11-20). Vibrational excitation upon collision with a surface has been shown to be inefficient for diatomic molecules (21). However, when NH_3 was scattered from Au(111), multiple quanta excitation of the umbrella mode were observed (22).

In the present study the energy deposition in a much larger polyatomic molecule scattered from surfaces is presented. The mechanism for vibrational excitation is deduced by simultaneously probing vibrational, rotational and translational energy transfer and compared with simple model calculations.

Experimental:

The experimental setup is similar to that used in previous experiments (8). A pulsed (nominal width of 50 μsec.) molecular beam of aniline seeded in argon, helium, or hydrogen is produced by a pulsed nozzle (Jordan Inc.). After being skimmed and collimated in two differentially pumped chambers, the molecular beam strikes the surface of a sample at an incident angle of 45 degrees. The sample temperature is controlled by Eurotherm 818 temperature controller.

The scattered aniline is detected in the same chamber using Resonance Enhanced Multi (two) Photon Ionization (REMPI). Tuneable UV radiation is obtained by doubling the output frequency of a Nd:Yag pumped dye laser using a WEX-1 (Quanta-Ray) unit. A mixture of Rhodamine B dye (80%) and Rhodamine 590 (20%) is used to ensure a flat gain curve of the dye in the region of interest. The laser beam is collimated by an iris and introduced to the probe region unfocussed. The laser power is reduced to about 500 $\mu J/cm^2$ per pulse, to avoid saturation effects. The ion signal is collected and amplified by a microchannel plate (MCP) detector. The laser energy is monitored continuously by a Laser Precision RJP 735 power meter. The REMPI signal and the laser intensity are processed by a boxcar integrator (Stanford Instruments) and stored in a PC-AT computer that also regulates the scanning of the dye laser.

Aniline at incident energies between 130 and 500 meV is scattered from n-octadecyltrichlorosilane (OTS) and perfluorodecanoic acid (PFDA) monolayers supported on silicone and mica slides respectively (23). The former exposes a 18 carbon saturated hydrocarbon chain, while the latter is a $CF_3(CF_2)_8COOH$ chain with the COOH group bound to the surface. The characteristics of OTS have been described previously, that of PFDA are similar to the PFAE used before (23). Previous scattering experiments indicated that the OTS monolayer is soft, or floppy, while the perfluorinated one is much more rigid. These surfaces because of their low energy remain clean even in the atmosphere. Contact angle measurements indicate that even after many months the properties of these surfaces do not vary. The results were reproducible with surfaces kept for months. Yet we have found that after few days of exposing the PFDA substrate to aniline the contact angle properties of this type of surface degraded. In correlation with the contact angle degradation, results obtained in the scattering experiments, on degraded surfaces, were not reproducible. This type of effect was not observed with the OTS substrate. The data presented obtained from proper surfaces where no sign of degradation could be recognized.

In addition to the surfaces covered with an OOM, a LiF single crystal surface was used. The LiF freshly cleaved and introduced immediately into the vacuum chamber where it was kept at a temperature of at least 400K. In the past He scattering was used to probe the cleanliness of the surface. It has been shown that the TOF signal of aniline can provide the same information. Figure 1 presents the TOF signal of aniline scattered from a clean LiF surface (A) and of aniline scattered from LiF that was left in the vacuum chamber for more than 24 hours (B). Data was taken at two scattering angles- one corresponds to specular scattering (upper curves) and the second to normal scattering (lower curve in each figure). As the surface gets dirty, the normal component increases relative to the specular component. Note the differences in the arrival time of the two components. We have found that the surface, as characterized by the TOF, does not vary during the experiment. Teflon surfaces were also used. They were prepared by cutting a Teflon block and exposing the freshly cut surface to the beam. No further characterization has been made in this case.

The laser intersects the scattered beam 2.0 cm above the surface. Time of flight (TOF) measurements are performed by varying the delay between the laser and the pulsed nozzle. The distance between the surface and the laser probe region, in these studies, is 4.6 cm. The flight time measurements are accurate to about 10 μsec.

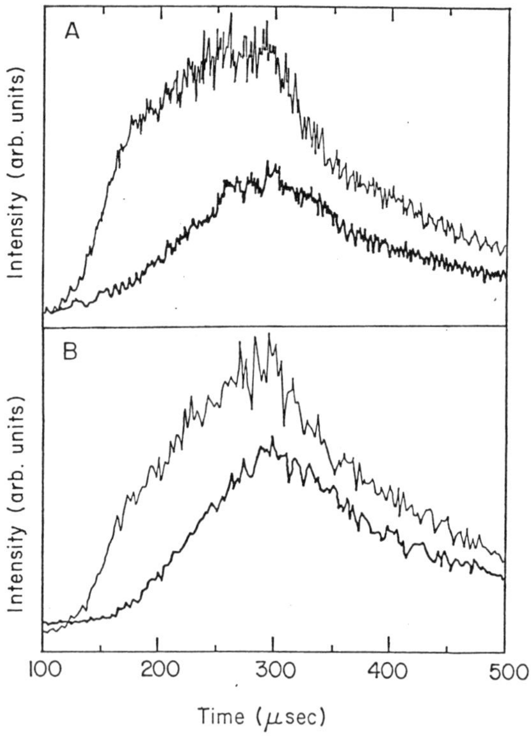

Figure 1: The time of flight signal obtained for aniline scattered from a clean LiF surface (A) monitored at specular angle (upper curve) and at normal angle (lower curve). In B the same measurements are presented from substrate maintained for 24 hours in the vacuum chamber. Notice the similar intensities for the two curves in B versus the large intensity difference in A.

Results:

Aniline was scattered from two organic surfaces and from Teflon and LiF. In figure 2 the resonance enhanced multiphoton ionization (REMPI) spectra of aniline are presented. Spectra shown are; aniline in a beam, produced by expanding 1.4 mbar aniline with 2000 mbar of helium (A), aniline scattered from OTS (B) and PFDA (C). In figure 3 similar spectra are shown for scattering from

Teflon (A) and LiF (B). The spectra result from the transition from the ground state (S_0) to the first singlet (S_1) state. They were recorded with a delay between the pulsed nozzle and the laser that provided the strongest signal. In figures 2B and 2C the spectra are dominated by a hot band that corresponds to the electronic excitation of molecules containing single quantum of the $-NH_2$ inversion mode. This vibration denoted as the I mode has a frequency of 40.9 cm^{-1} in the ground electronic state. The intensity ratio (R) between the I^1_1 transition and the origin is smaller in figure 2C compared to figure 2B. In figure 3 this ratio is even smaller. A clear pattern exists where R decreases along the series 2B,2C,3A and 3B. Since the Franck-Condon factor for this transition is about the same as that for the origin, the spectrum shown in figure 2B suggests population inversion. In addition, in figures 2B and 2C the two quanta excitation of this mode can be observed. Other vibrational modes of aniline scattered from the organic surfaces, were also found to be populated, but to a far lesser extent than the I mode. The T vibration is the out of plane bending mode of the aromatic ring, relative to the NH_2 group and has a frequency of 233 cm^{-1} in the ground state. The population in this mode is about ten percent of the ground state population. This is almost independent of the type of organic surface.

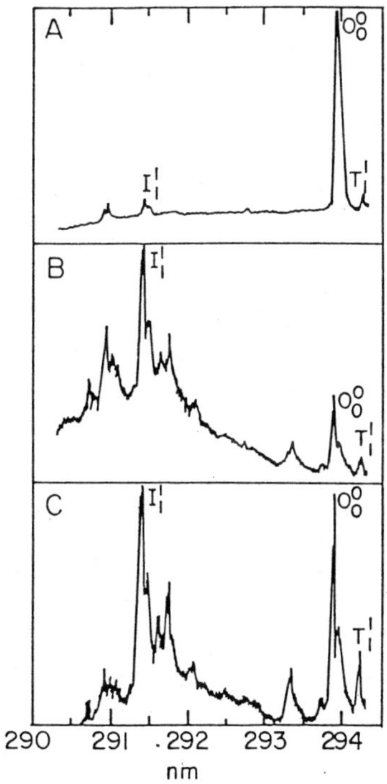

Figure 2: REMPI spectra of aniline in the molecular beam (A), scattered from OTS (B) and PFDA (C).

In the case of scattering from OTS (figure 3B) the population in two quanta of the I mode (corresponding to 421 cm^{-1}) is larger than that in the T mode, which is lower in energy by almost 200 cm^{-1} (2).

The special properties of the I mode are not apparent at all in the aniline scattered from the LiF surface. Here many modes are excited with no preference for any of them.

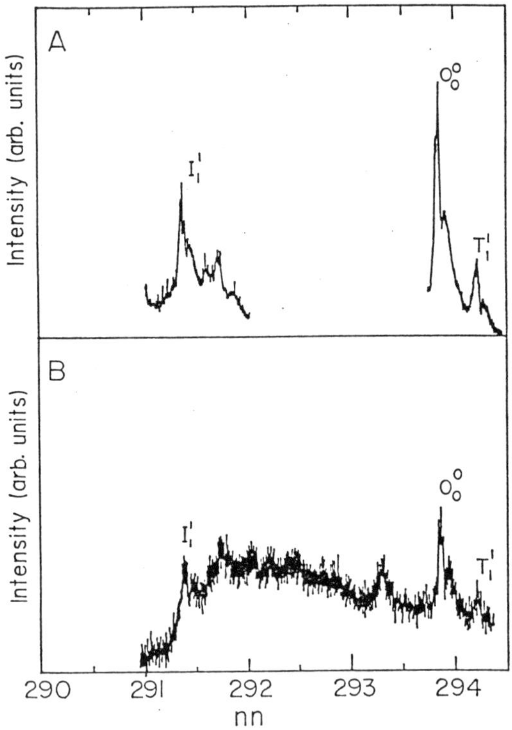

Figure 3: REMPI spectra of aniline scattered from Teflon surface (A) and from LiF single crystal surface (B).

In figure 4, the Time Of Flight signal (TOF) is shown for scattering from OTS, probing the origin transition (broken line) and the I^1_1 transition (solid line). A long tail is observed when the origin is monitored. It indicates long accommodation time up to hundreds of microsecond. For comparison, figure 5A presents the TOF monitored also at the origin and the I_1^1 wavelengths, but for aniline scattered from LiF. In 5B, the same data is shown as in 5A, but normalized. Here the TOF signal is independent of the vibrational energy of the scattered aniline. The tail at the long time regime does not exist for molecules scattered from LiF.

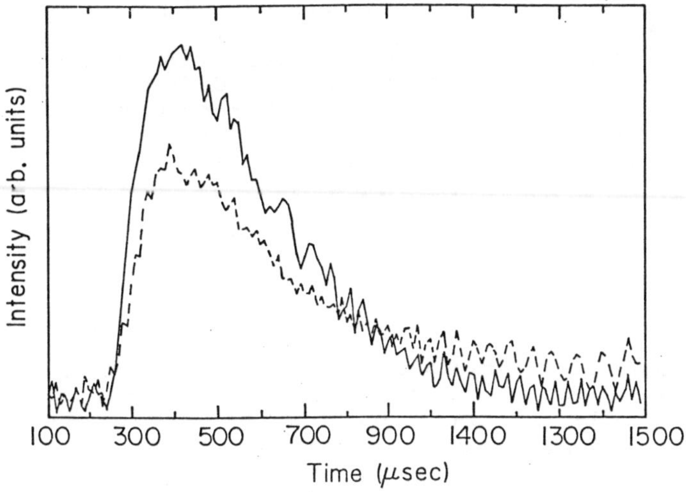

Figure 4: The time of flight signal of aniline scattered from OTS with no vibrational energy (broken line) and vibrationally excited to the I_1 state (solid line).

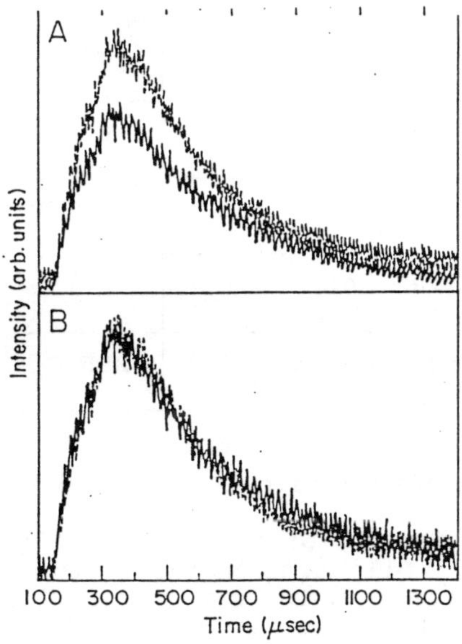

Figure 5: Time of flight signal of aniline scattered from LiF surface in the ground state (solid line) and vibrationally excited to the I mode (broken line). In A the data is presented as obtained, in B it is shown after normalizing the peak heights.

The collision energy has been varied from about 150 meV up to 630 meV. For OTS, for all collision energies below about 600 meV, the translational energy distribution is Boltzmann like. For aniline scattered from PFDA, thermal distributions were obtained for collision energies up to 460 meV. For higher collision energies the TOF signal was analyzed assuming that the intensity (F) observed can be characterized by the equation:

$$F=A\exp[-W(\frac{L}{tV_s}-1)^2]+\frac{B}{t^2}\exp[\frac{mL^2}{2kTt^2}] \quad (1)$$

Where V_s is the stream velocity, $W=\frac{\gamma M^2}{2}$ is a width factor, γ is the heat capacity ratio and M is the aerodynamic Mach number. L is the surface-laser distance and A and B are normalization factors, indicating the contribution of the stream versus Boltzmann component respectively. Residence time shorter than the experimental resolution (20 μsec) as been assumed. Hence in the analysis we ignored the tail observed at the long flight times for ground state molecules scattered from OTS.

Table 1: Translational Energy of Aniline Scattered from Various Surfaces

Surface	E_c(meV) ($\pm$2%)	V_s (m/sec)	W	T(K)	B/(A+B)	E_s(meV)
OTS	630	260$\pm$7	10$\pm$2	100$\pm$40	0.68 $\pm$0.15	32$\pm$2
PFDA	585	210$\pm$10	5$\pm$3.5	280$\pm$100	0.40 $\pm$0.15	21$\pm$3
PFDA	630	282$\pm$10	9.6$\pm$5	190$\pm$100	0.29 $\pm$0.14	39$\pm$2
LiF (295K)	585	850$\pm$40	3.2$\pm$0.2	255$\pm$7	0.21 $\pm$0.02	350$\pm$35

E_s is the scattering energy calculated based on V_s.

In table 1 the results of the above analysis are shown. Data presented only for collision energies for which non thermal distributions were obtained. The ratio B/(A+B) is a measure for the percentage of the distribution which is simulated by a Boltzmann type distribution. The results indicate that, for a given collision energy, the aniline is scattered from PFDA with more kinetic energy than when scattered from OTS. This is in agreement with results obtained before for various molecules and atoms that were scattered from the two type of OOM (7). Hence the notion that the perfluorinated monolayer is more rigid than the OTS is proved. For both OOM, only about 10% of the initial collision energy remains as a translational energy in the scatterer. When scattered from LiF the situation varies and about 60% of the collision energy remains as translation. Another important outcome from the kinetic energy measurements is that at low collision energies, where the vibrational energy distribution is highly non thermal, the translational energy distribution suggests that this degree of motion accommodates with the surface.

The rotational energy distribution was obtained using program simulating the rotational band contour to fit the experimental band contour. Due to the congestion of lines in the spectrum, such a fit could not be achieved for the I_1^1 line. For the origin, the rotational energy distribution could not be fitted to a simple one temperature Boltzmann distribution. For aniline scattered from OTS, the best fit was obtained assuming two temperatures, 50 and 300K.

Discussion:

The results presented above point to a mode specific behavior in the collision process between aniline molecules and low energy surfaces. To rationalize the observations, it is important to review the energy transfer results obtained for aniline colliding with He in the gas phase. In the work done by Liu et. al. (6) it was found that at low collision energies, the I_1 state is efficiently populated. On increasing the collision energy the population in this mode decreases, while the population in other low frequency vibrations increases. The gas phase data suggests that the excitation of the I_1 mode occurs for small impact parameter collisions. It is possible to relate the surface scattering results to the gas phase if we consider the collision time. When an aniline molecule collides with a "floppy surface", for which the collision time is long, it encounters a process similar to the low energy collision in the gas phase. Molecules colliding with a rigid surface have a short collision time, similar to the high energy collisions in the gas phase.

The "collision time" dependent process is different from the naive point of view on energy transfer between gas phase molecules and surface. Usually it is justified to assume that the surfaces do not "absorb" a large fraction of the collision energy. In this case, one expects that heavy (rigid) surfaces will be more efficient in promoting vibrational excitation in scattered molecules than light (floppier) surfaces. This notion arises from the conservation of energy requirement that if energy is absorbed by the surface, less energy will remain for distribution among the internal degrees of freedom of the scattered species.

From the data presented here, it is clear that in the present case, the simple description described above is not valid. Instead the gas phase like behavior seems to provide better insight on the mechanism. The data point to the fact that in the present study, the "softer" the surface, the larger proportion of molecules scattered at the peak of the TOF curve are in their vibrationally excited I_1 state.

Here a model is presented that rationalizes all the above observations.

The model:

The model has to describe two important features in the results, the special behavior of the I mode and its efficient population. Secondly it has to rationalize the variation of the population in this mode with different surfaces. It seems that a simple correlation exists between the collision time and the population in the I mode. The longer the encounter time between the aniline and the surface, the more efficiently the I mode is populated. The model must be able to rationalize also the gas phase observations.

In what follows a schematic model is suggested that can account qualitatively for all the observations. It assumes a rode with a mass M_A and moment of inertia I and a low frequency bending mode, that collides with a sphere with a mass M_b that is bound by a spring with a frequency ν to the surfaces M_c. Upon collision the rode can be "vibrationally excited" or rotationally excited. In addition energy can be transferred to the "surface spring". The ratio between the amount of energy in each channel is controlled by "adiabatic parameters"(24), one for the bending mode of the rode and the other one for the spring. In order to account for the gas phase behavior the sphere (M_b) is taken as a free sphere with small mass relative to M_A. The results of the rode-spring collision model can be summarized as follows:

a) Only molecules that collide with an impact parameter smaller than some value d_s, are reflected back on a single collision. The magnitude of d_s depends on the surface's mass and its vibrational frequencies.

b) Only collisions with an impact parameter smaller than some critical value (d_c) efficiently induce vibrational excitation in the rode.

The experimental observations can all be rationalized by the above model, assuming that those molecules not reflected back, upon their first encounter with the surface collide few more times. It is assumed that they finally are scattered with thermal energy distribution in all degrees of freedom.

The diameter of aniline is about 6Å while the chain occupies an area with a diameter of about 5Å. In most collisions, the aniline long axis is not parallel to the surface. Therefore it is justified to assume that almost for all encounters, the scattered molecule collides with one chain only. However, even if the collision of the molecule is with two chains simultaneously the discussion below is valid. For simplicity, it is helpful to describe the monolayers' chain as if it is a spring composed of two parts; a head group connected to the other parts of the chain with a spring. This picture is of course an oversimplification of the physical picture, since for example only a single frequency is associated with the surface. Still this simple model is shown to be very helpful for describing the physical effects.

The surface effective mass (M_{su}) depends on the mass of the functional group exposed to the vacuum (M_b), the mass of the other components in the chain (M_c) and the ratio between the "spring frequency" and the inverse collision time.

$M_{su}=M_b+\beta M_c$

M_{su} is also a measure for the rigidity of the surface. It contains both the masses and the dynamics that affect the kinematics of the collision.

$$\beta=\exp\left[-\frac{V_c}{2\pi\nu L}\right] \tag{2}$$

β is an adiabatic parameter (24) that introduces into the model the collision time. v_c is the collision velocity, ν is the "spring frequency" and L the characteristic interaction distance between the monolayer and the colliding molecule.
Using conservation laws, it is possible to derive the conditions for which the center of mass of the aniline molecule will reverse its direction of motion upon a **single** collision with the surface:

$$\frac{M_A}{M_{su}}+M_A\frac{d^2}{I}+2\frac{\Delta E_{sur}}{M_A V_A^2}<1 \tag{3}$$

Where I is the moment of inertia of aniline, d is the impact parameter, V_a is the collision velocity and ΔE_{sur} is the amount of vibrational energy transferred to the surface. The expression for ΔE_{sur} can be derived based on gas phase theory (25):

$$\Delta E_{sur}=\frac{2\hat{M}_A M_B M_C}{(\hat{M}_A+M_B)^2(M_B+M_C)}(\pi\nu L)\,csch^2\left(\frac{\pi\nu L}{V_A}\right) \tag{4}$$

Where the effective mass of the collider (26) is:

$$\hat{M}_A=\frac{M_A}{1+d^2\frac{M_A}{I}} \tag{5}$$

Hence, it is now possible to derive the expression for the maximum value of the impact parameter that will result in aniline reversing direction upon a single collision:

$$d_s=\sqrt{I\left(\frac{1}{M_A}-\frac{1}{M_{su}}-\frac{\Delta E_{sur}}{M_A E_c}\right)} \tag{6}$$

E_c is the collision energy.

When analyzing equation 4, one finds that for reasonable values of L, ν and V_A the term $\frac{\pi \nu L}{V_A}$ obtains values between 0.05 to 0.25. In this range of values, the csch function is proportional to V_A. Therefore the third term in equation 6, that is proportional to $csch^2/V_A^2$, is almost independent of the collision energy. This prediction of the model is consistent with the experimental observation where the ratio between the collision energy and the final translational energy of the scatterer is about constant. The above discussion also suggests that d_s increases with the increasing of the surface rigidity (M_{su}).

The relative population in the vibrational excited state depends on the ratio $(d_c/d_s)^2$. The more rigid the surface, the larger d_s and smaller this ratio will be. Therefore smaller fraction of the scattered molecules will be in the excited I mode. This result is consistent with the experimental observations.

The model suggests that the extensive population in the I mode is a result of a "filtering" process. Most of the molecules scattered back directly are necessarily those that collided with small impact parameter. More rigid surfaces (larger M_{su}) exhibit less selective filtering and the population in the I mode is relatively small. In the limit of very rigid surface (like LiF) one may assume that all the colliding molecules are scattered back after colliding once with the surface. Here no mode selectivity is predicted and other low frequency modes may compete efficiently with the I mode. With the less rigid surfaces, because the population is measured at a fixed time window in the time of flight distribution, population inversion seems to exist. In fact the population in the ground vibrational state is higher if one integrates over all the time of flight spectra and all the J states.

Conclusions:

The mode selective excitation in aniline scattered from floppy surfaces has been demonstrated. The results can be explained by a simple model that assumes that the time of the collision affects the extent of excitation to the low frequency I mode.

The population in the I mode can serve as an indication for the surface rigidity. The more rigid the surface, the less selective is the vibrational excitation process. The ability to probe several vibrational modes proved to be fruitful in obtaining information on the surface dynamics. Despite the "complexity" of the systems studied, a simple picture emerges from the results which can be rationalized completely based on the dynamics of the collision.

In recent experiments, applying the information obtained in this study, we were able to monitor structural changes as function of temperature, by monitoring the population in the I mode.

Acknowledgements:

This work was partially supported by the US-Israel Binational Science Foundation. We would like to thank J. Sagiv and R. Maoz for preparing the OOM and S. Cohen for critical reading of the manuscript.

References:

1. D.A. Chernoff and S.A. Rice, J. Chem. Phys. 70,2511 (1979).
2. M. Quack and M. Stockburger, J. Mol. Spectrosc. 43,87 (1972).
3. J.C.D. Brand, D.R. Williams and T.J. Cook, J. Mol. Spectrosc. 20,359 (1966).
4. J.M. Hollas and G.H. Kirby, J. Mol. Phys. 15,441 (1969)
5. D.A. Chernoff and S.A. Rice, J. Chem. Phys. 70,2521 (1979).
6. K. Liu, G. Hall, M.J. McAuliffe, C.F. Giese, and W.R. Gentry, J. Chem. Phys. 80,3494 (1984).
7. S.R. Cohen, R. Naaman, and J. Sagiv, Phys. Rev. Lett. 58,1208 (1987).
8. S.R. Cohen, R. Naaman, and J. Sagiv, J. Chem. Phys. 88,2757 (1988).
9. S.R. Cohen, R. Naaman and G.G. Balint-Kurti, Chem. Phys. Letts. 152,269 (1988); Chem. Phys. 134,119 (1989).
10. Y. Paz and R. Naaman, Chem. Phys. Letts. 172,120 (1990); J. Chem. Phys. 94,4921 (1991).
11. H. Vach, J. Hager, and H. Walther, J. Chem. Phys. 90,6701 (1989).
12. M. Asscher, W.L. Guthrie, T.H. Lin, and G.A. Samorjai, J. Chem. Phys. 78,6992 (1983).
13. J. Misewich and M.M.T. Loy, J. Chem. Phys. 84,1939 (1986).
14. J. Misewich, H. Zacharias, and M.M.T. Loy, Phys. Rev. Lett. 55,1919 (1985).
15. J. Misewich, P.A. Roland, and M.M.T. Loy, Surf. Sci. 171,483 (1986).
16. C.T. Rettner, F. Fabre, J. Kimman, and D.J. Auerbach, Phys. Rev. Lett. 55,1904 (1985).
17. J. Persson and R. Ryberg, Phys. Rev. Lett. 48,549 (1982).
18. M. Asscher, G.A. Samorjai, and Y. Zeiri, J. Chem. Phys. 81,1507 (1984).
19. E. Kolodney, D. Baugh, P.S. Powers, H. Reisler, and C. Wittig, Chem. Phys. Lett. 145,177 (1988).
20. F. Frenkel, J. Haeger, W. Krieger, H. Walther, G. Ertl, J. Segner, and W. Vielhaber, Chem. Phys. Lett. 90,225 (1982).
21. R.B. Gerber, Chem. Rev. 87,29 (1987).
22. B.D. Kay, T.D. Raymond, and M.E. Coltrin, Chem. Phys. Letts. 59,2792 (1987).
23. a) R. Maoz and J. Sagiv, J. Colloid. Interface Sci. 100,465 (1984). b) J. Gun, R. Iscovici, and J. Sagiv, ibid 101,201 (1984).
24. R.D. Levine and R.B. Bernstein, "Molecular Reaction Dynamics," (Oxford University, New York, 1974) Chap. 5.
25. B.H. Mahan, J. Chem. Phys. 52,5221 (1970).
26. J. Kimman, C.T. Rettner, D.J. Auerbach, J.A. Barker and J.C. Tully, Phys. Rev. Lett. 57,2053 (1986).

$NO(X^2\Pi)$ PRODUCT STATE DISTRIBUTIONS IN MOLECULE–SURFACE DISSOCIATIVE SCATTERING: $n, i - C_3F_7NO$ FROM MgO(100)

E. KOLODNEY,[a] P.S. POWERS, L. HODGSON,
H. REISLER and C. WITTIG
Department of Chemistry, University of Southern California,
Los Angeles, California, U.S.A.

ABSTRACT. Molecule-surface collision induced dissociation (CID) of $n - C_3F_7NO$ and $i - C_3F_7NO$ following direct inelastic scattering off single crystal MgO(100) has been studied at the hyperthermal kinetic energy range of 1-7 eV. The $NO(X^2\Pi)$ fragment was detected state-selectively using two-photon, two-frequency ionization. Product state distributions and angular distributions were integrated to give dissociation yields, which increased sharply with E_{incident}. Nascent NO rotational and spin-orbit state distributions are reported for several E_{incident} and surface temperatures. In most cases, rotational excitation could be described by separate temperatures for each spin-orbit state. The upper $^2\Pi_{3/2}$ state was underpopulated relative to statistical predictions. In addition, $NO(v'' = 1)$ was undetectable (e.g. $[v'' = 1]/[v'' = 0] \leq 2 \times 10^{-2}$ at $E_{\text{incident}} = 5.0$ eV and $T_s = 840$ K). The CID results are compared to NO state distributions derived from the photodissociation of the isolated, expansion cooled molecule with well defined excess energies ($E^\dagger$) above D_0. Possible CID mechanisms are discussed in light of the experimental findings and the similarities between the CID and photodissociation results. All of the results are compatible with direct inelastic scattering followed by unimolecular reaction on the S_0 and/or T_1 potential surface.

1. Introduction

The impulsive molecule-surface scattering event in the hyperthermal energy range (1-10 eV) is of fundamental scientific and practical importance. The basic collisional processes leading to the activation/dissociation of a polyatomic molecule scattered from a single crystal surface will also govern the entrance channel dynamics of gas-surface processes in which the molecule needs to overcome some energetic reaction barriers. In this paper we will describe an experimental work aimed at understanding the dynamical details of the mechanisms involved in a direct inelastic dissociative scattering of a polyatomic molecule ($n, i - C_3F_7NO$) off single crystal surface.

---

[a] Present address: Department of Chemistry, Technion – Israel Institute of Technology, Haifa 32000, Israel

J. Jortner et al. (eds.), Mode Selective Chemistry, 443–455.

Recently, several important direct processes have been observed for diatomic and polyatomic molecules scattered off stiff, chemically inert crystals. High CID efficiencies were reported using mass spectrometric detection [1-3]. Also a variety of molecular hyperthermal surface ionization and dissociative ionization processes were identified [4]. In our experiments, detection is laser based and the main experimental observables are the CID yields versus the incident kinetic energy ($E_{incident}$) and the ($NO(X^2\Pi)$) product state distributions (rotational and spin-orbit) [5]. Product state distributions are particularly useful in identifying statistical mechanisms and deviations from statistical behavior as evidenced successfully by numerous photodissociation experiments [6]. The aliphatic nitroso molecules, RNO, were chosen since (i) they all have low D_o values ($< 2\,eV$); (ii) the R species can be made heavy (e.g. perfluorination) to allow aerodynamic acceleration up to ~ 10 eV ($E_{incident} > D_o$); (iii) the NO fragment can be sensitively detected by photoionization and (iv) the gas-phase, $E^\dagger$ dependent unimolecular reactions of several isolated RNO molecules have been studied in detail, and were found to exhibit common features, which can be compared with molecule-surface CID results [6]. The use of two isomers was motivated, in part, by a desire to identify molecule-surface orientation effects in CID. We report fully resolved NO internal state distributions, as well as CID yields vs. $E_{incident}$. In a parallel series of experiments, we studied photoinitiated unimolecular reactions of the same molecules using expansion cooling and well-defined $E^\dagger$ values up to 4500 cm^{-1}. The possibility of using this detailed information as a comparative database is examined in a general discussion of possible CID mechanisms.

2. Experimental

A schematic diagram of the experimental set-up is shown in Fig. 1. A pulsed, supersonic molecular beam of n,i-heptafluoronitrosopropane (C_3F_7NO, Fluorochem) seeded in H_2 or He carrier gas was accelerated to kinetic energies of 1-7 eV. The beam was skimmed and collimated in a two-stage differentially pumped region before it entered the UHV chamber (base pressure 4×10^{-10} torr) where it was scattered from cleaved, single crystal MgO(100) surface. The surface was mounted on a manipulator and located 43 cm from the pulsed nozzle. Nascent NO fragments (CID products) were detected state selectively using two-photon two-frequency laser ionization via the $A^2\Sigma^+ \leftarrow X^2\Pi\,\gamma$ system. The output from a 10 Hz Nd:YAG-laser pumped dye laser (Quanta-Ray DCR2/PD1) was doubled to give approximately 5 mJ of pulsed, tunable radiation near 280 nm. Part of this output was converted to $\sim$ 226 nm using the second anti-Stokes line from a high pressure H_2 Raman scattering cell. Approximately 1 mJ of the 280 nm radiation was overlapped with the 226 nm radiation, providing two-frequency excitation. Both beams traveled similar path lengths between splitting and recombination, resulting in a temporal overlap better than $\pm0.3ns$. When using $\sim 4\mu$J per pulse (in a 1 mm^2 beam) of tunable radiation in the 226 nm region to excite γ(O–O) rotational transitions, and $\sim$ 1 mJ near 280 nm to pump the ionization transition, high S/N was achieved (detection sensitivity of 10^{-12} Torr per quantum state) without even partially saturating the γ band. The low 226 nm fluence also prevented dissociation of the parent molecule. The probe beams were perpendicular to the scattering plane, crossing it 10-15 mm from the surface, for an in-plane angular resolution of approximately 6° and full integration out of plane.

Ions were collected using a channeltron (Galileo)-electrode arrangement, whose output was amplified and sent to a transient recorder (Nicolet) and computer system (LSI 11/23). Data processing included gated integration with dynamic base line subtraction, signal averaging, and shot-to-shot normalization of the signals, using separate photodiodes to monitor the 226 and 280 nm energies. Rotational state distributions were measured by probing the γ band in the range 223.4-227.0 nm. Relative level densities $N_{J''}$ were extracted by dividing the normalized, integrated signal intensities by the appropriate line strengths (Hönl-London factors) [7]. Only well-resolved and unambiguously assigned lines were used.

All measurements and procedures for extracting rotational distributions were checked routinely in the scattering chamber with 300 K NO samples using the same configuration as when detecting scattered species. Typical pressures ranged from 5×10^{-10} to 1×10^{-8} Torr. The MgO crystal was x-ray analyzed in order to determine the plane and azimuthal directions prior to mounting. Once inside the UHV chamber, the crystal was annealed at 1100-1200 K for several hours. Fast Ar-atom bombardment was sometimes used for surface cleaning before thermal annealing. These surface preparation and cleaning procedures resulted in reliable He diffraction patterns, a sharp, background free Xe scattering angular distributions (16° FWHM at $E_{incident} = 2.1$ eV) and contaminants free Auger spectra. In order to ensure surface cleanliness (adsorbate-free and direct scattering conditions), the surface temperature Ts was maintained at 600 K during all experiments, except those in which Ts dependences were investigated.

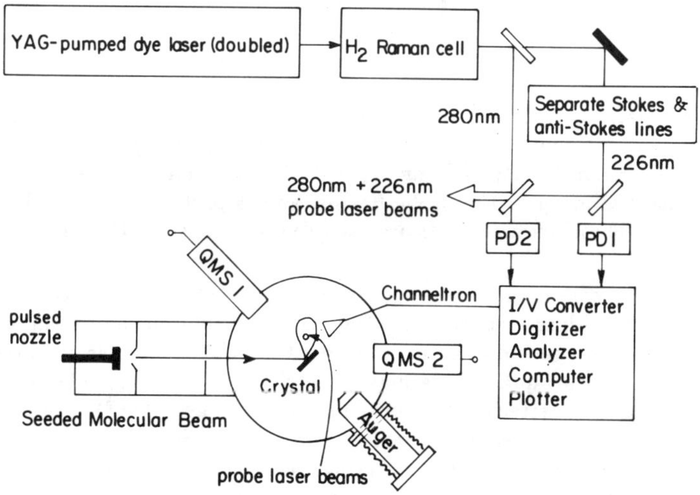

Figure 1. Schematic drawing of the overall experimental arrangement used in the molecule-surface CID studies.

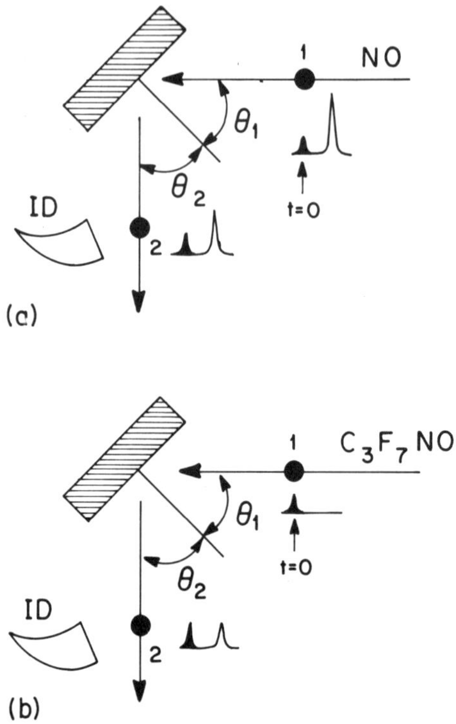

Figure 2. Schematic drawing showing probe/surface configurations for (a) detecting scattered NO from an incident beam of known composition and (b) detecting NO from CID of C_3F_7NO (ID = ion detector (channeltron)). The t=0 pulse (darkened) appears concomitantly with the firing of the probe beams, and is probably due to photons striking the channeltron. These arrangements are used to estimate the CID yield.

Throughout the experiments, two-photon, two-frequency laser ionization was used as the main probe, providing state-selective fragment distributions and beam diagnostics in the energy, time, and space domains (laser based TOF and angular scattering measurements). In addition, two quadrupole mass spectrometers (QMS's, UTI-100C) were used for measuring angular scattering distributions, time-of-flight (TOF) spectra, and relative intensities of the C_3F_7NO beams, as well as serving as residual gas analyzers. With the exception of the laser detection, the scattering/detection arrangement is that of fixed detectors and a rotating crystal. One QMS was oriented at 45° relative to the incident molecular beam, 26 cm from the surface, and was equipped with a collimator to give 1.1° resolution both in and out of the scattering plane. A second QMS was placed in-line with the incident molecular beam for diagnostics and $E_{incident}$ measurements.

Relative and absolute CID yields, as well as NO contamination, were measured using the configuration shown schematically in Fig. 2. Figure 2(a) shows an NO reference beam in two detection positions, before and after surface scattering. NO was detected at the $R_{11}+R_{21}$ bandhead ($0.5 \leq J" \leq 5.5$) with a 1.0 cm^{-1} laser bandwidth. Fig. 2(b) shows the same configuration for the seeded C_3F_7NO beam. The CID signal at position 2, relative to signals for other C_3F_7NO $E_{incident}$ values and the signal from the NO reference beam, enabled us to determine relative and absolute CID yields as a function of $E_{incident}$.

3. Results

3.1. COLLISION-INDUCED DISSOCIATION ON MgO(100)

3.1.1. *CID Kinetic energy dependence.* The $E_{incident}$ dependence of the CID yield for $n-$ and $i-C_3F_7NO$ is presented in Fig. 3 for Ts = 600 K. It is the NO fragment density, summed over internal states and scattering angles, relative to the corresponding signal from an NO reference beam, also summed over internal states and scattering angles. Product density/flux transformations were not done (due to insufficient information about the scattered velocities) and also we did not assess the fraction of scattered molecules that dissociate after the detection region. In estimating absolute CID yields, a reference NO beam with well-controlled incident flux was used for calibration. This flux was either equal

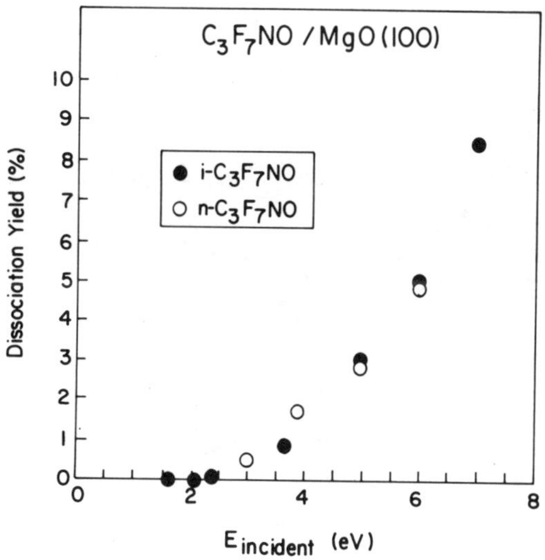

Figure 3. CID yields for $n-C_3F_7NO$ and $i-C_3F_7NO$ *vs.* $E_{incident}$ at $T_s = 600 \pm 20$ K. NO was monitored at the $Q_{11}+R_{21}$ bandhead. Uncertainty in absolute yields is $\pm$ 50%; relative values are $\pm$ 20%.

to the incoming C_3F_7NO flux or in a well-known ratio. The uncertainty in the relative CID yields is estimated to be ± 20%, and that of the absolute yield is ±50%. The important features emerging from Fig. 3 are: (a) Strong kinetic energy dependence of the CID yields, reflecting the direct inelastic nature of the scattering event. (b) The data points for both $n-$ and $i - C_3F_7NO$ fall on the same curve. This observation does not support a strongly orientation dependent excitation mechanism which is expected to show different $E_{incident}$ variation for the different geometries of the two isomers. (c) The apparent threshold for the CID of $i - C_3F_7NO$ at $T_s = 600K$ was measured at 2.35 ± 0.20 eV ($D_0 = 1.69$ eV for $n - C_3F_7NO \rightarrow n - C_3F_7 + NO$, and 1.64 eV for $i - C_3F_7NO \rightarrow i - C_3F_7 + NO$), and from Fig. 3, a similar value can be inferred for $n - C_3F_7NO$. This "threshold" depends on experimental sensitivity, and presently corresponds to a CID yield of 0.1%. In these experiments, θ_1 was $37 \pm 2°$, and the probe beams were near the specular angle.

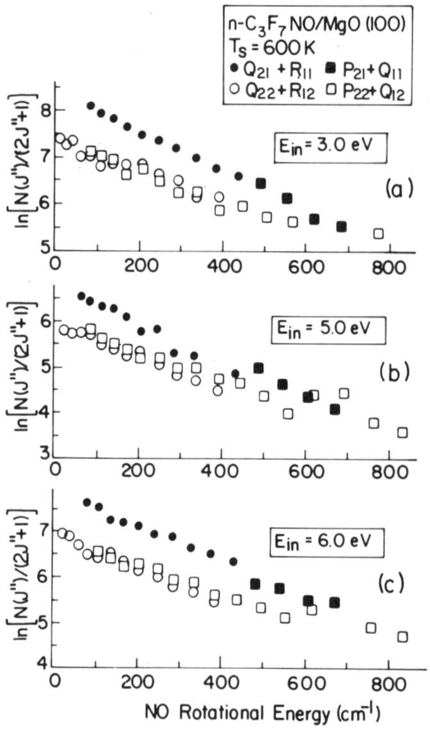

Figure 4. Boltzmann plots of NO rotational distributions from $n - C_3F_7NO$ CID at three $E_{incident}$ values and $T_s = 600 \pm 20$ K. Each distribution represents an average of several different measurements. The filled and empty symbols indicate rotational populations of $^2\Pi_{1/2}$ and $^2\Pi_{3/2}$ NO spin-orbit states, respectively.

3.1.2. *Rotational and Spin Orbit Excitations.* Figures 4 and 5 show Boltzmann plots { i.e., $\ln[N(J'')/2J''+1)] vs E_{rot}$} of NO rotational distributions from the CID of $n-$ and $i-C_3F_7NO$ at several $E_{incident}$ values, with $T_s = 600$ K and $\theta_1 = 40°$ (see Fig. 2). These distributions were obtained by averaging results from 4 to 5 spectra, taken at the specular scattering angle.

Several important features are immediately apparent from these figs. (a) NO spin orbit excitation from CID is low, $[^2\Pi_{3/2}/[^2\Pi_{1/2}] = 0.36$, corresponding to a temperature of 170 K, and this ratio changes little with $E_{incident}$ and surface temperature T_s. This ratio is much smaller than the corresponding ratios for scattered NO beam ($E_{incident} = 0.9$ eV) [12] and 300 K static NO samples all taken under the same conditions and detection configuration in the UHV chamber. (b) The rotational distributions for both NO spin-orbit states produced by CID of $n-C_3F_7NO$ appear Boltzmann-like, as shown in Fig. 4, although the upper $^2\Pi_{3/2}$ state has higher rotational excitation than $^2\Pi_{1/2}$ for all $E_{incident}$ values. The uncertainty in the rotational state populations is ± 6%.

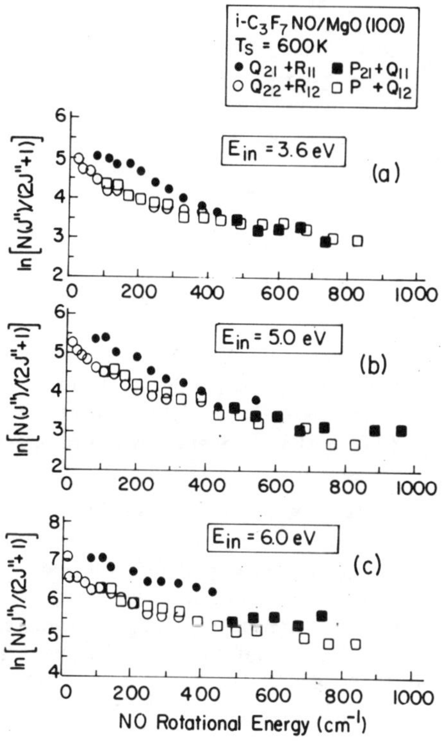

Figure 5. Boltzmann plots of NO rotational distributions from $i-C_3F_7NO$ CID at three $E_{incident}$ values and $T_s = 600 \pm 20$ K. Each distribution represents an average of several different measurements. The filled and empty symbols indicate rotational populations of $^2\Pi_{1/2}$ and $^2\Pi_{3/2}$ NO spin-orbit states, respectively.

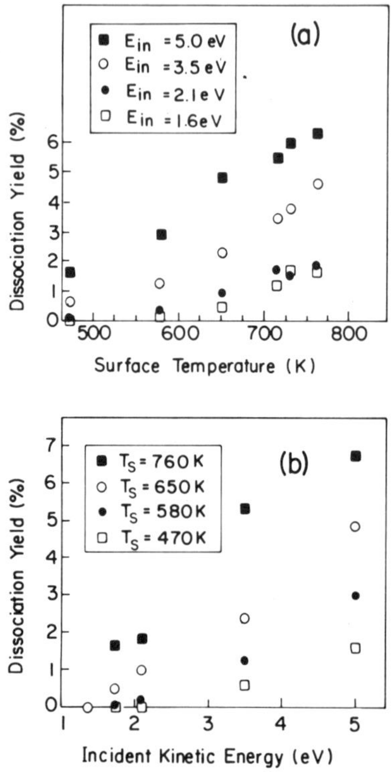

Figure 6. (a) Dissociation yield (±50%) of $i-C_3F_7NO$ scattered from MgO(100) versus surface temperature (±10K) for several E_{incident} values. A small correction factor was applied to the $T_s > 650$ K values. (b) Dissociation yield (±50%) of $i-C_3F_7NO$ scattered from MgO(100) versus E_{incident} (±0.25 eV) at several surface temperatures (±10 K).

In general, NO rotational excitations were modest, and when straight lines could be drawn (namely, temperatures assigned to the rotational distributions) $T_{\text{rot}} \sim 500$ K was usually obtained for $^2\Pi_{3/2}$, while T_{rot} in the range 350-420 K was obtained for $^2\Pi_{1/2}$. It should be emphasized that rotational temperatures are used only as a convenient, although arbitrary way to characterize the distributions using a single parameter, they cover limited ranges of E_{rot} and are not associated with equilibrium conditions at the surface.

Several attempts were made to detect NO ($v'' = 1$) from CID in the γ(1-1) region. However, no signal was observed with $E_{\text{incident}} = 5.0$ eV and $T_s = 600$ or 840 K. If we assume similar spin-orbit and rotational distributions in $v'' = 0$ and 1, and take into account the different Franck-Condon factors, an estimated upper limit of 2% can be placed on the $[v'' = 1]/[v'' = 0]$ ratio. However, more detailed measurements are necessary in order to fully quantify NO vibrational excitation. In the case of NO/metal collisions [8], it was proposed that short-lived negative ion resonances were the source of vibrational excitation. However, based on the electron affinities of NO (0.02 eV) [9] and C_3F_7 (2.3 eV)

[9], and recalling that MgO is a good insulator, the formation of transient NO negative ions in the present experiments seems unlikely.

3.1.3. *The Surface Temperature Effect.* We found surface temperature dependence only for the CID yields, (as shown in Fig. 6), while the internal NO excitations (rotational, vibrational, electronic) did not show any meaningful variation with T_s.

Figure 6 shows the dependence of the CID yield of $i-C_3F_7NO$ scattered from MgO(100) on surface temperature in the range of 470-760 K, for $E_{incident}$ in the range of 1.6-5.0 eV. The lower T_s limit was to ensure surface cleanliness and to prevent $i-C_3F_7NO$ adsorption on the surface during the course of an experiment. The CID yield is strongly increasing with both $E_{incident}$ and T_s. Moreover, the incident kinetic energy threshold for CID is a function of surface temperature. The experimental threshold for the CID of $i-C_3F_7NO$ is 2.35 ± 0.20 eV at $T_s = 580$ K, less than 1.6 eV at 650 K, and greater than 2.35 eV at 470 K. We attempted to see if a larger energy gap between D_0 and $E_{incident}$ (0.34 eV) could be bridged by the T_s contribution, but found no observable CID signal for $E_{incident} = 1.3$ eV and T_s up to 650 K.

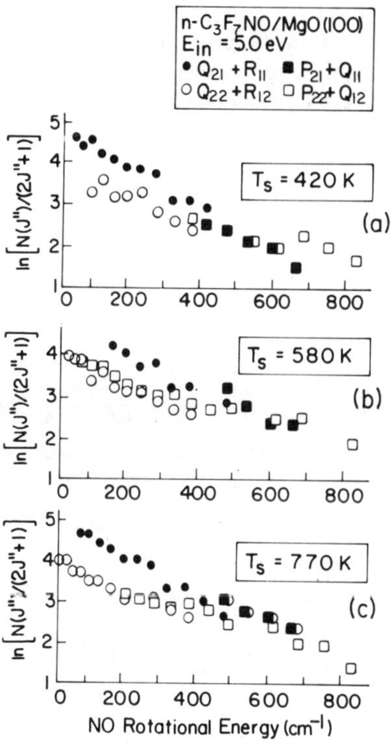

Figure 7. Boltzmann plots of NO rotational distributions from $n-C_3F_7NO$ CID at $E_{incident} = 5.0$ eV and three T_s values. Each distribution represents an average of several different measurements. The filled and empty symbols indicate rotational populations of $^2\Pi_{1/2}$ and $^2\Pi_{3/2}$ NO spin-orbit states, respectively.

As mentioned before, no pronounced or systematic dependence of the NO fragment internal excitation on either E_{incident} or T_s, was found. Only weak variation of the rotational distribution is shown in Fig. 7 for $n - C_3F_7NO$ at E_{incident} =5.0 eV for T_s in the range 400-800 K. The results for other combinations of E_{incident} and T_s were essentially the same.

We carefully checked for changes of the surface structure and stoichiometry upon heating, which could influence our measurement of the dissociation yield. He diffraction did not show any sign of surface reconstruction upon heating [10]. Furthermore, an Auger spectra of the MgO(100) surface in the range T_s = 300-1000 K showed chemical stability of the surface composition ([Mg]/[O] ratio) to within ± 3%.

More detailed description of the surface temperature effect can be found elsewhere [11].

3.2. THE C_3F_7NO PHOTODISSOCIATION EXPERIMENT AND RESULTS

The experimental arrangement used to study the collision-free one-photon photolysis of expansion cooled n, $i - C_3F_7NO$ has been described previously [13]. Briefly, photolysis was done using a YAG-pumped-dye system. A probe laser producing 226 nm, collinear and counterpropagating to the photolysis laser was used to detect the nascent NO photofragments via one photon LIF, at variable time delays following photolysis.

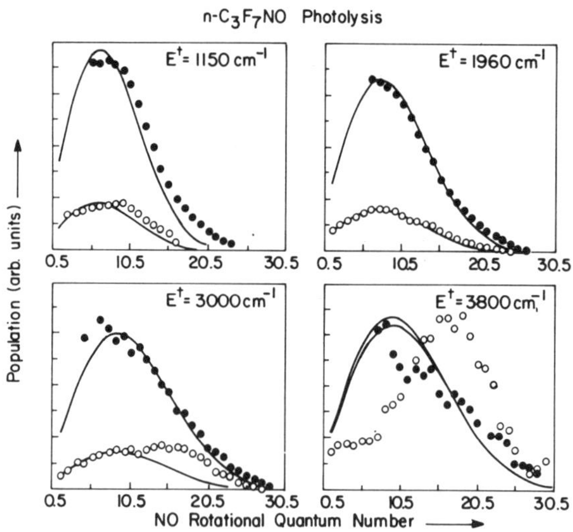

Figure 8. NO state distributions from $n - C_3F_7NO$ photolysis at various $E^{\dagger}$. The circles are data and the solid lines are statistical (prior) distributions. Rotational distributions are statistical at low $E^{\dagger}$ and show deviations at higher $E^{\dagger}$.

NO internal state distributions were obtained at several values of $E^\dagger$ for both $n-$ and $i-C_3F_7NO$, as shown in Figs. 8 and 9, respectively. $NO(X^2\Pi)$ is produced predominantly in $v'' = 0$ ($< 1\%$ in $v'' = 1$). An upper limit for D_0 was obtained from the highest observed J'' level in the long tail of the rotational distribution. The solid lines in Figs. 8 and 9 correspond to the calculated statistical distributions (priors) of product excitations at fixed $E^\dagger$ values [14]. The prior distributions for each spin-orbit state were separately normalized to the data and the experimental relative spin-orbit populations are as shown in the figures. We find, as is common with other nitroso molecules, that although the rotational populations of each of the spin-orbit states is statistical, the spin-orbit ratio is colder than statistical. The remarkably good fit, with the exception of the higher excess energies, is consistent with previous results, in which S_0 and/or T_1 were implicated as the reactive surfaces, following radiationless decay of S_1 [5,15]. Notice, however, that at high $E^\dagger$ deviations from statistical behavior occur. In particular, for $n - C_3F_7NO$ the $NO(^2\Pi_{3/2})$ rotational excitations at $E^\dagger >\sim 2000 cm^{-1}$ greatly exceed the statistical predictions.

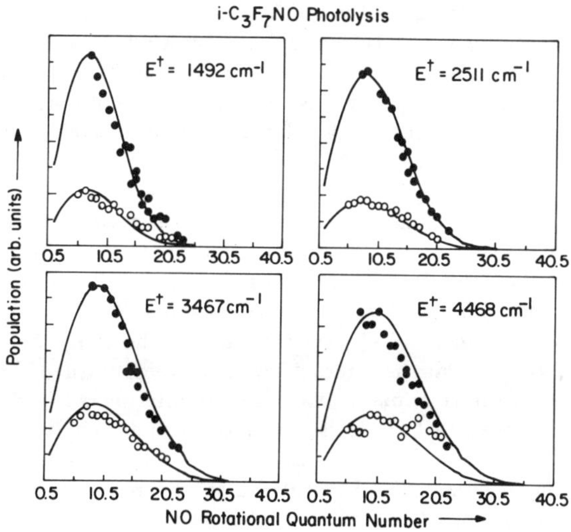

Figure 9. NO state distributions from $i - C_3F_7NO$ photolysis at various $E^\dagger$. The circles are data and the solid lines are statistical (prior) distributions. Rotational distributions are statistical at low $E^\dagger$ and show deviations at higher $E^\dagger$.

4. Discussion

We presented experimental results concerning the dissociation of a polyatomic molecule following an impulsive collision with an insulator single crystal surface. In order to rationalize these results in terms of the simplest mechanism which is consistent with and supported by the observations, three possible basic mechanisms are examined: (a) direct bond rupture caused by strong collisional distortion of the C–NO bond upon impact, a rapid process that could be efficient for certain approach geometries, (b) dissociation following molecule-surface charge transfer processes, as inferred in previous hyperthermal surface scattering experiments [2, 4], and treated theoretically in terms of surface harpooning and the formation of temperary induced ions [16], and (c) vibrational excitation within the S_0 and/or T_1 manifolds, followed by intramolecular vibrational redistribution and statistical unimolecular decomposition. With (a), the very similar variations of CID yield with E_{incident} for iso and normal isomers do not support a strongly orientation-dependent excitation mechanisms. So this is probably not a dominant dissociation mechanism. With regard to mechanism (b) we argue that entrance channel charge transfer interactions are less likely to be the dominant process on an excellent insulator like MgO. We have found [11] that the CID yields for $i - C_3F_7NO$ scattered from a semiconductor (GaAs(100)) and metallic (Ag(111)) surfaces are actually smaller than for MgO(100) (e.g. CID yield of 1.8% for MgO(100), 0.9% for GaAs(100) and 0.5% for Ag(111) at $E_{\text{incident}} = 4.0$ eV and $T_s = 580$ K). A dominant surface molecule charge transfer mechanism would imply greater yields for the silver surface. As a further check on the relative importance of the electronic mechanism [8] we measured the probability for direct vibrational excitation in collisions of NO with MgO(100) at high T_s. No NO ($v'' = 1$) was observed up to $E_{\text{incident}} = 0.90$ eV and $T_s = 775$ K, placing an upper limit of 0.1% on the $[v'' = 1]/[v'' = 0]$ ratio.

Thus, we conclude that mechanism (c) is favored, and the intramolecular motions, distortions and couplings that occur during the molecule-surface collision probably leads to statistical unimolecular dissociation. In what follows, we will examine the experimental evidence that support this choice. First, the NO state distributions obtained in the CID experiments for both $n-$ and $i - C_3F_7NO$ throughout the range $3.0 < E_{\text{incident}} < 7.0$ eV, as well as a several T_s, can be characterized by temperature-like distributions for each spin-orbit state. Despite some scatter and undulations, the distributions shown in Figs. 4, 5 and 7 appear rather statistical and depend weakly on E_{incident} and T_s. Second, the similarity between the photodissociation and CID results also suggests the possibility of common mechanisms. In contrast to the molecule-surface collision that generates a broad distribution of internal energies, in the photodissociation experiment, molecules are prepared with single $E^\dagger$ values and the measurements reflect the product state branching ratios. These can well be represented by prior distributions, implicating statistical unimolecular decomposition, except for the highest $E^\dagger$, where the onset of dynamical bias is observed (see Figs. 8, 9). Naturally, the relative contributions of different CID mechanisms, the efficiency of collision-induced excitation and the shape of E_{internal} distribution may depend also on the parent molecule. For example, we did not observe CID for CF_3NO at $E_{\text{incident}} = 3.6$ eV and $T_s = 600$ K, $D_0 = 1.73$ eV). Under the same conditions, a significant signal, corresponding to $\sim 1\%$ yield, was obtained with C_3F_7NO. This places an upper limit of 0.1% on the CF_3NO CID yield at $E_{\text{incident}} = 3.6$ eV, indicating less efficient CID than for C_3F_7NO. Similar dependence on the colliding molecule was observed in the CID of a series of alkyl iodides [2]. Clearly, in addition to the state-resolved measurements described here, more work with different molecules and surfaces, and with angular and TOF resolution is needed in order to help unravel mechanisms.

Acknowledgement

We thank Professors R.B. Gerber, A. Amirav, J. Pfab, P.D. Dapkus, and D. Baugh, for helpful suggestions and fruitful discussion, and Dr. G. Ziegler for Auger measurements. Research supported by AFOSR and ARO.

References

[1] (a) E. Kolodney, A. Amirav, R. Elber and R.B. Gerber, Chem. Phys. Lett. **111**, 366 (1984); (b) R.B. Gerber and A. Amirav, J. Phys. Chem. **90**, 4483 (1986).

[2] A. Danon, E. Kolodney and A. Amirav, Surf. Sci. **193**, 132 (1988).

[3] A. Danon, A. Amirav, J. Silberstein, Y. Salman and R.D. Levine, J. Phys. Chem. **93**, 49 (1989).

[4] A. Danon and A. Amirav, J. Phys. Chem. **93**, 5549 (1989).

[5] E. Kolodney, P.S. Powers, L. Hodgson, H. Reisler, and C. Wittig, J. Chem. Phys. **94**, 2330 (1991).

[6] H. Reisler, M. Noble and C. Wittig in *Molecular Photodissociation Dynamics*, edited by J. Baggott and M.N.R. Ashfold (Royal Society of Chemistry, London, 1987).

[7] L.T. Earles, Phys. Rev. **48**, 423 (1935).

[8] (a) C.T. Rettner, F. Fabre, J. Kimman, D.J. Auerbach, Phys. Rev. Lett. **55**, 1904 (1985); (b) C.T. Rettner, J. Kimman, F. Fabre, D.J. Auerbach, H. Morawitz, Surf. Sci. **192**, 107 (1987).

[9] B.M. Smirnov, Negative Ions (McGraw Hill, New York, 1982).

[10] E. Kolodney and A. Amirav, Surf. Sci. **155**, 715 (1985).

[11] P.S. Powers, E. Kolodney, L. Hodgson, G. Ziegler, H. Reisler, and C. Wittig, Submitted to J. Phys. Chem.

[12] E. Kolodney, D. Baugh, P.S. Powers, H. Reisler and C. Wittig, Chem. Phys. Lett. **145**, 177 (1988).

[13] Y.Y. Bai, A. Ogai, C.X.W. Qian, L. Iwata, G.A. Segal and H. Reisler, J. Chem. Phys. **90**, 3903 (1989).

[14] R.D. Levine and R.B. Bernstein, *Molecular Reaction Dynamics* (Oxford University Press, 1987).

[15] M. Noble, C.X.W. Qian, H. Reisler, and C. Wittig, J. Chem. Phys. **85**, 5763 (1986).

[16] (a) J.W. Gadzuk and S. Holloway, J. Chem. Phys. **84**, 3502 (1986); (b) J.W. Gadzuk and S. Holloway, Chem. Phys. Lett. **114**, 314 (1985); (c) S. Holloway and J.W. Gadzuk, Surf. Sci. **152**, 838 (1985).

IMPLICATIONS OF GIANT OPTICAL VIBRATIONS CREATED BY ULTRASHORT LASER PULSES

BENJAMIN FAIN
School of Chemistry
Tel Aviv University
Ramat Aviv
Tel Aviv 69978
Israel

1. Introduction

Interaction of the matter with ultrashort (femtosecond) electromagnetic pulses raise up some interesting problems. One of such problems is related to the question how the energy of the electromagnetic field in visible and u.v. region can be transformed into the phonon energy.

The "conventional" mechanism of such a tranformation can be described as follows[1-7]. Electrons in a metal or in a semiconductor are heated to a temperature T_e greater than the lattice temperature T_L. Then the electron-phonon interaction causes a relaxation of the electron temperature towards the equilibrium with $T_e=T_L$. This relaxation can be described by rate equations[6] with transitions rates being proportional to squared electron phonon matrix elements. The excitation of phonons in this mechanism has relatively large enertia with typical delay time of several picoseconds. On the other hand there is another almost inertialess mechanism of excitation of phonons via impulsive excitations (see reviews[8-9]). Usually this excitation is described in the Placzek approximation[8]. This approximation cannot be used in the case when the laser frequency is resonant to the interband energy differences. General quantum-mechanical analysis based on the density matrix formalism has been performed in the previous papers[10-12]. In these works general formulae for the density matrix jumps induced by laser pulses has been derived.

In this paper we will derive mean values of the phonon amplitudes resulted from the action of the ultrashort pulses. There is a crucial difference in the phonon excitation in two above mentioned mechanisms. The conventional mechanism[1-7] is a quasi-equilibrium one. The excited phonons can be characterized by the thermodynamic equilibrium distribution with the slowly varying temperature. The phonons excited by the ultrashort pulses can be named "giant" vibrations: a mean number of phonons in the one specific mode of vibrations is of the order of the magnitude of total number N of atoms in the crystal. In the contrast to the quasi-equilibrium mechanism the coherent phonons are excited by the ultrashort pulses. These properties of the excited phonons will be shown below.

Giant optical vibrations excited by the ultrashort pulses can cause some interesting phenomena in the solid. Among them there are parametric generation of the acoustic phonons and desorption via Raman phonon processes. These possibilities will be also analyzed in the present paper.

J. Jortner et al. (eds.), Mode Selective Chemistry, 457–466.

We will employ formulae for the density matrix jumps[12] resulted from the ultrashort laser pulses. The interaction energy with these pulses is represented in the form

$$V = [V(\Omega)e^{-i\Omega t} + V(-\Omega)e^{i\Omega t}]\, l(t) \tag{1.1}$$

where the dimensionless function $l(t) \leq 1$ (its maximum value is defined as unity) describes the pulse shape, while Ω is the laser frequency. The pulse duration is defined as

$$\int_{-\infty}^{\infty} l(t)\, dt = T \tag{1.2}$$

Resonant contributions to the density matrix jumps have the form[12]

$$\Delta\rho_{nv;nv'} = \frac{4\pi^2}{\hbar^2} \sum_{w} \rho_{ow;ow}\, V_{nv;ow}(\Omega)\, V_{ow;nv'}(-\Omega)\, \hat{l}(\omega_{nv;ow}-\Omega)\, \hat{l}(\Omega-\omega_{nv;ow})$$

$$(n \neq 0)$$

$$\Delta\rho^{res}_{ov;ov'} = -\frac{2\pi^2}{\hbar^2} \sum_{w} V_{ov;nw}(-\Omega)\, V_{nw;ov'}(\Omega)\, \hat{l}(\Omega-\omega_{nw;}\, \hat{l}(\omega_{nw;ov'}-\Omega) \tag{1.3}$$

$$(\rho_{ov;ov} + \rho_{ov';ov'}) \tag{1.4}$$

where $\rho_{ov;ov'}$ etc. are unperturbed density matrix elements. Indices n,o are electronic quantum numbers, while v, v′, w are vibrational ones. The rotating wave approximation is assumed

$$|\,\omega_{nv;ow} - \Omega\,| \lesssim T^{-1};\ \Omega T \gg 1 \tag{1.5}$$

and the frequencies $|\,\omega_{ov;ow}\,|$ of the vibrational motions may be neglected in comparison with Ω

$$\Omega \gg |\,\omega_{ov;ow}\,| \tag{1.6}$$

The ultrashort pulses are defined by the relation

$$T^{-1} \gtrsim |\,\omega_{ov;ow}\,| \tag{1.7}$$

The Furier transforms $\hat{l}(x)$ in formulae (1.3), (1.4) are defined as

$$\hat{l}(x) = \frac{1}{2\pi} \int_{-\infty}^{\infty} l(t)\, e^{ixt}\, dt \tag{1.8}$$

(There are misprints in formulae (3.11) and (3.15) of the Reference 12: sign minus has to be put before last sum of (3.11) and before the right hand side of (3.15)).

The nonresonant contribution to the density matrix jumps has the form[12]

$$\Delta\rho^{nr}_{ov;ov'} = \frac{i}{\hbar^2}(\rho_{ov;ov} - \rho_{ov';ov'})\sum_n <v|V^2_{on}(\Omega)|v'> \frac{2\omega_{no}}{\omega^2_{no} - \Omega^2}\,\Upsilon \tag{1.9}$$

where the pulse duration Υ (different from (1.2)) is defined as

$$\Upsilon = \int_{-\infty}^{\infty} l^2(t)dt \tag{1.10}$$

Formula (1.9) has been obtained in the Placzek's approximation and it is assumed that

$$|\Omega \pm \omega_{no}|T \gg 1 \tag{1.11}$$

In the next section we will use formulae (1.3), (1.4) and (1.9) in order to find the amplitudes of optical vibrations induced by the pulses satisfying conditions (1.5), (1.7) or (1.11).

2. Vibration Unduced by Non-Resonant Pulses

Ultrashort laser pulses (satisfying (1.17)) create coherent vibrational movement in the solid. In this and the next sections amplitudes of these vibrations will be derived. We will start from the non-resonant case (1.9). Formula (1.9) gives off-diagonal elements of the density matrix at some moment of time after action of the laser pulse.

The mean value of the destruction operator a of the optical mode (with the vector $k \approx 0$) can be found as

$$<a> = \sum_v \Delta\rho^{nr}_{o,v;o,v-1} a_{v-1,v} = \sum_v \Delta\rho^{nr}_{o;v;o,v-1}\sqrt{v}$$

$$= \frac{i}{\hbar^2}\sum_v (\rho_{ov;ov} - \rho_{o,v-1;o,v-1})\sqrt{v}\sum_n <v|V_{on}{}^2(\Omega)|v-1> \frac{2\omega_{no}}{\omega^2_{no} - \Omega^2}\,\Upsilon \tag{2.1}$$

In Placzek's approximation one gets

$$\frac{1}{\hbar}\sum_n \frac{2\omega_{no}}{\omega^2_{no} - \Omega^2}|V_{on}|^2 = \frac{1}{2}\alpha'|\vec{E}(\Omega)|^2 Q \tag{2.2}$$

Here $\alpha' = \frac{\partial\alpha}{\partial Q}$, and α is defined by the equation

$$\vec{P} = \alpha \vec{E} \tag{2.3}$$

where $\vec{P}$ is total dipole moment of the system, and $\vec{E}$ is the intensity of the electric field; Q is the amplitude of the opitcal mode vibration, it may be expressed through the creation and destruction operators

$$Q = \left[\frac{\hbar}{2\omega_v}\right]^{1/2} (a^+ + a) \tag{2.4}$$

while ω_v is the frequency of the optical vibration.

Substituting (2.2) and (2.4) into (2.1) we get

$$\langle a \rangle = -\frac{i}{\hbar} \alpha' |\vec{E}(\Omega)|^2 \left[\frac{\hbar}{2\omega_v}\right]^{1/2} \Upsilon \tag{2.5}$$

For the estimates it is convenient to introduce the polarizibility of one atom (molecule) of the solid: $\tilde{\alpha}$, and the coordinate R of the vibration between atoms of the solid

$$Q = \sqrt{NM}\ R \tag{2.6}$$

where M is some effective mass of the atoms. Then

$$\frac{\partial \alpha}{\partial Q} = \frac{1}{\sqrt{NM}} \frac{\partial \alpha}{\partial R} = \sqrt{\frac{N}{M}} \frac{\partial \tilde{\alpha}}{\partial R} \ ; \ \alpha = N\tilde{\alpha} \tag{2.7}$$

Substituting (2.7) into (2.5) one gets

$$\langle a \rangle = -i \sqrt{N} \frac{1}{\hbar} \frac{\partial \tilde{\alpha}}{\partial R} |\vec{E}(\Omega)|^2 \left[\frac{\hbar}{2M\omega_v}\right]^{1/2} \cdot \Upsilon \tag{2..8}$$

To estimate this amplitude we note that

$$\tilde{\alpha} \sim \frac{e^2 a_o^2}{\hbar \omega_{el}} \ ; \ \frac{\partial \tilde{\alpha}}{\partial R} \sim \frac{\tilde{\alpha}}{a_o} \ ; \ \left[\frac{\hbar}{2M\omega_v a_o^2}\right]^{1/2} \sim \left[\frac{\omega_v}{\omega_{el}}\right]^{1/2} \sim \left(\frac{m}{M}\right)^{1/4} \tag{2.9}$$

Here a_o is the interatomic distance ~ 10^{-8}cm; ω_{el} is typical electronic frequency ~ 10^{15}sec^{-1} and m is the mass of the electron. Using relations (2.9) we get

$$\langle a \rangle \sim \sqrt{N} \frac{e^2 a_o^2}{\hbar^2} |\vec{E}|^2 \Upsilon\, \omega_{el}^{-1} \left[\frac{\omega_v}{\omega_{el}}\right]^{1/2} \sim \sqrt{N}\ 10^{18} |\vec{E}|^2 \Upsilon\, \omega_{el}^{-1} \left[\frac{\omega_v}{\omega_{el}}\right]^{1/2} \tag{2.10}$$

where $ea_o \sim 10^{-18}$ e.s.u. For the fluence $3mJ/cm^2$ with 200 fs pulse (in the experiment of the desorption induced by femtosecond laser pulses[13]) one obtains

$$\langle a\rangle \sim 10^{-3}\sqrt{N} \tag{2.11}$$

It should be mentioned that relations (2.5), (2.8), (2.10), (2.11) give the amplitude of the vibration which remains active in the solid after the passage of the laser pulse

$$\langle a(t)\rangle = \langle a\rangle e^{i\omega_v t}e^{-t/\tau_v} \tag{2.12}$$

where τ_v is the vibrational relaxation time

$$\tau_v \gg \omega_v^{-1} \gtrsim T, \tilde{T} \tag{2.13}$$

3. Vibration Created by Resonant Pulse

In this section we will apply formulae (1.3) and (1.4) to the interband transitions in solids. In the microscopic theory matrix elements of $V_{nv;ow}$ type can be derived from the electron-phonon interaction. Eigenfunctions taking into account the electron-phonon interaction has the form

$$\Psi_{nv} = \Psi_n^o \Psi_v + \sum_{mw} \frac{H'_{mw;nv}}{E_{nv} - E_{mw}} \Psi_m^o \Psi_w^o \tag{3.1}$$

where $H'_{mw;nv}$ is a matrix element of the electron-phonon interacuon energy. Indices n,m are electronic quantum numbers, while v, w are phonon quantum numbers. Then matrix elements of the interaction with the electromagnetic field obtain the form

$$\langle n,v|V|n',v'\rangle = V_{nn'}\delta_{vv'} + \sum_m \left[V_{nm}\frac{H'_{mv;n'v'}}{E_{n'v'} - E_{mv}} + \frac{H'_{nv;mv'}}{E_{nv} - E_{mv}} V_{mn'} \right] \tag{3.2}$$

Employing formulae (1.3) and (1.4) we can now find the density matrix jumps resulted from the ultrashort laser pulses: $\Delta\rho_{n,\vec{k},v;n,\vec{k},v'}$ and $\Delta\rho_{o\vec{k}v;o\vec{k}v'}$. Here o and n are indices of the bands of the solid, and $\vec{k}$ are electron wave-vectors in the band. The electron-phonon interaction energy matrix element can be presented as

$$H'_{n\vec{k}v;n\vec{k}v'} = M_{n\vec{k}}^{-\vec{q}}\, \delta_{\vec{k},\vec{k}+\vec{q}}\, a_{\vec{q}} + \text{h.c.} \tag{3.3}$$

where $\vec{q}$ is the phonon wave-vector. The mean value of operator a_q has the form

$$\langle a\rangle_o = \sum_{\vec{k},v} \Delta\rho_{o,\vec{k},v;o,\vec{k},v-1} \sqrt{v} \; ; \; \langle a\rangle_n = \sum_{\vec{k},v} \Delta\rho_{n,\vec{k},v;n,\vec{k},v-1} \sqrt{v} \tag{3.4}$$

Using these formulae and neglecting the wave vector of the electromagnetic field we obtain expressions for <a> in zero and n-th band (details of the derivation will be presented elsewhere)

$$\langle a\rangle_o = \frac{4\pi}{\hbar^2} \sum_{\vec{k}} \rho_{o\vec{k};o\vec{k}} \left| V_{ok\,;\,nk}(\Omega) \right|^2 \frac{M^q_{n\vec{k}} - M^q_{o\vec{k}}}{\hbar} \hat{I}\,(\Omega-\omega_{n\vec{k},o\vec{k}})$$

$$\times \frac{\partial \hat{I}\,(\Omega-\omega_{n\vec{k},o\vec{k}})}{\partial\Omega} (2\bar{v}+1)\, \omega_v \tag{3.5}$$

$$\langle a\rangle_n = \frac{4\pi}{\hbar^2} \sum_{k} \rho_{o\vec{k};o\vec{k}} \left| V_{o\vec{k};\,n\vec{k}}(\Omega) \right|^2 \frac{M^q_{n\vec{k}} - M^q_{o\vec{k}}}{\hbar\omega_v} \hat{I}^2(\Omega-\omega_{n\vec{k};o\vec{k}}) \tag{3.6}$$

Here we will give estimates of these amplitudes. For this purpose we have to estimate electron-phonon matrix elements M^q_{nk}. Very comprehensive review of these and other quantities can be found in books[14,15]. For the interaction with optical phonons characterized by the deformation energy (not the most favorable case) M^q_{nk} has order of the magnitude

$$M^q_{nk} \sim \frac{1}{\sqrt{N}} \left[\frac{\hbar}{2M\omega_v a_o^2}\right]^{1/2} \frac{e^2}{a_o} \tag{3.7}$$

Taking into account that $\hat{I}(\Omega-\omega_{nk;ok}) \sim T$ we obtain the following estimates

$$\langle a\rangle_n \sim \frac{e^2 a_o^2 \vec{E}^2 T^2}{\hbar^2} \left[\frac{\omega_e}{\omega_v}\right]^{1/2} \sqrt{N} \tag{3.8}$$

$$\langle a\rangle_o \sim \langle a\rangle_n \frac{\partial \hat{I}(\Omega-\omega_{no})}{\partial\Omega}\, \omega_v \tag{3.9}$$

Using the numerical example given at the end of section 2 one gets

$$\langle a\rangle_o \lesssim \langle a\rangle_n \sim \sqrt{N} \left[\frac{\omega_e}{\omega_v}\right]^{1/2} \sim 30\, \sqrt{N} \tag{3.10}$$

Number of phonons in the coherent state described by formulae (3.5), (3.6) or by estimates (3.8), (3.10) is

$$n \sim \langle a \rangle^2 \gtrsim N \tag{3.11}$$

Thus the ultrashort laser pulse produces the giant vibration. In contrast to the quasi-equilibrium phonon distribution the giant vibration is characterized by phonons condensed in one (optic) vibrational mode. Number of phonons in this mode is of the order of the magnitude of total number of phonons modes.

4. Implications of Giant Optical Vibration

Giant optical vibration creates entirely non-equilibrium state of the solid. Distribution of phonon energies cannot be characterized by the temperature. Giant optical vibration is a coherent state. Relaxation of this state towards the equilibrium cannot be described by usual rate equations.

Particularly the giant optical vibration (before it decays) can cause parametric instability of acoustic phonons[16,17]. The threshold of such instability can be easily estimated. The phonon Hamiltonian can be written in the form

$$H = \sum_k \hbar\omega_{\vec{k}} a^+_{\vec{k}} a_{\vec{k}} + \hbar\omega_v a^+ a + \hbar F_{\vec{k}}(a+a^+)\,(a_{\vec{k}} a_{-\vec{k}} + a^+_{\vec{k}} a^+_{-\vec{k}}) \tag{4.1}$$

The third term is the third order anharmonic term containing interaction of the giant vibration $\underset{\sim}{a}$ with acoustic phonons satisfying the wave-vector conservation law

$$\vec{k}_1 + \vec{k}_2 = \vec{q} \approx 0 \tag{4.2}$$

The equations for the amplitudes of the acoustic waves $a^+_{\vec{k}}$ and $a^+_{-\vec{k}} = a^*_{\vec{k}}$ follow from the Hamiltonian (4.1) (in the rotating wave approximation)

$$\dot{a}_{\vec{k}} = -i\omega_{\vec{k}} a_{\vec{k}} - i\, F_k \langle a \rangle a^+_{-\vec{k}} - \gamma_k a_{\vec{k}}$$

$$\dot{a}^+_{-\vec{k}} = i\omega_{\vec{k}} a^+_{-\vec{k}} + i\, F^*_k \langle a^+ \rangle a_{\vec{k}} - \gamma_k a^+_{-\vec{k}} \tag{4.3}$$

These are equations for the mean values of the amplitudes of acoustic waves, and decay constants are introduced phenomenologically.

It is easy to show that for giant vibration with the frequency ω_v, the threshold condition of instability of acoustic phonons with

$$\omega_k \approx \frac{\omega_v}{2} \tag{4.4}$$

has the form

$$|F_k|^2 |\langle a \rangle|^2 > \gamma_k \tag{4.5}$$

One can get the following estimate for the anharmonic coefficient F_k

$$F_k \sim \left(\frac{\omega_v}{\omega_{el}}\right)^{1/2} \frac{\omega_v}{\sqrt{N}} \tag{4.6}$$

Substituting this estimate and estimate (3.8) into relation (4.5) one gets

$$\frac{e^2 a_o^2 \vec{E}^2 T^2 \omega_v}{\hbar^2} > \gamma_k \tag{4.7}$$

For the numerical example given at the end of Section 2 ($ea_o \sim 10^{-18}$ e.s.u., the fluence $3mJ/cm^2$, $T = 200fs$) we get

$$\omega_v > \gamma_k \tag{4.8}$$

which is obviously fulfilled.

Another implication of the existence of giant optical vibration may be the phonon induced desorption. Observation of the desorption induced by femtosecond laser pulses has been reported recently[13]. A possible explanation of this kind of phenomenon can be connected with the giant optical vibration induced by the ultrashort laser pulse.

The desorption can be described in terms of dynamics of states of the molecule-surface system[18-20]. The dynamics of the system are characterized by the probability $P_\alpha(t)$ that the molecule is in state $|\alpha\rangle$ (which could be bound or desorptive) at the time t. The rate of this quantity is given by the master equation

$$\frac{dP_\alpha(t)}{dt} = \sum_\beta \left[- W_{\alpha\beta}P_\alpha(t) + W_{\beta\alpha}P_\beta(t)\right] \tag{4.9}$$

This equation is a basis for the theory of thermal desorption. At finite temperatures there are transition rates from lower to upper bound levels. Ultimately this can lead to the transition to the continuum, desorptive levels, i.e., to the desorption. Transitions between state $|\alpha\rangle$ are characterized by the energy conservation laws

$$E_\alpha - E_\beta = \hbar\omega_q \tag{4.10}$$

where $\omega_{\vec{q}}$ is the phonon frequency. The giant optical vibration with frequency ω_v, in general, does not satisfy these conservation laws which require different frequencies for different transitions. Therefore the giant vibration can contribute to the transition rates through the two-phonon transitions characterized by the conservation laws

$$E_\alpha - E_\beta = \hbar\omega_v \pm \hbar\omega_q \tag{4.11}$$

The interaction energy used in desorption kinetics[18] can be written in the form

$$V = -\hat{B}_{\vec{k}} (a_{\vec{k}} + a_{\vec{k}}^+) \tag{4.12}$$

Here matrix elements of $B_{\vec{k}}$ may be written in the form

$$B_k = \frac{1}{\sqrt{N}} \left[\frac{\hbar}{2M\omega_k} \right]^{1/2} \frac{U_{\alpha\gamma}}{a_o} \tag{4.13}$$

where $U_{\alpha\gamma} \sim U_o$ is a characteristic interaction energy constant (which has order of the depth of the potential well of adsorbed molecule) and a_o is the interatomic distance. The two-phonon transitions may be called phonon Raman (Stokes and anti-Stokes) processes. The transition probabilities are characterized by the extra factor having the order of the magnitude of square of the ratio

$$\frac{V_{\alpha\gamma}}{\hbar\omega_v} \sim \frac{1}{\sqrt{N}} \left[\frac{\hbar}{2M\omega_v a_o^2} \right]^{1/2} \frac{U_o}{\hbar\omega_v} \langle a \rangle \tag{4.14}$$

assuming that $\hbar\omega_v$ is of the order of magnitude of characteristic difference $|E_\alpha - E_\gamma|$. Substituting estimate (3.8) into (4.14) we get

$$\frac{V_{\alpha\gamma}}{\hbar\omega_v} \sim \frac{U_o}{\hbar\omega_v} \frac{e^2 a_o^2 \vec{E}^2}{\hbar^2} T^2 \tag{4.15}$$

Again using numerical example of the end of section 2, we get

$$\frac{e^2 a_o^2}{\hbar^2} \vec{E}^2 T^2 \sim 1 \tag{4.16}$$

and

$$V_{\alpha\gamma}/\hbar\omega_v \sim \frac{U_o}{\hbar\omega_v} > 1 \tag{4.17}$$

In this case the perturbation theory on which the calculation of the transition rates is based, is not valid anymore. This implies that giant vibration's influence on the adsorbate-surface bond is not small, and the desorption rate will be essentially enhanced in comparison with thermal desorption.

Two above examples - parametric instability of acoustic phonons and the giant vibrations' induced desorption show that the giant vibration leads to essentially non-equilibrium processes and drastic changes in the system.

References

1. V.L. Ginzburg and V.P. Shabanskii, Dokl. Acad. Nauk SSSR, 100, 445 (1955).
2. M.I. Kaganov, I.M. Lifshitz and L.V. Tanatarov, Soviet Physics JETP, 4, 173 (1957).
3. S.I. Anisimov, B.L. Kapeliovich and T.L. Perel'man, Sov. Phys. JETP, 39, 375 (1974).
4. H.E. Elsayed-Ali, T.B. Norris, M.A. Pessot and G.A. Mourou, Phys. Rev. Lett., 58, 1212 (1987).
5. R.W. Schoenlein, W.Z. Lin and J.G. Fujimoto, Phys. Rev. Lett., 58, 1680 (1987).
6. P.B. Allen, Phys. Rev. Lett., 59, 1460 (1987).
7. P.B. Corkum, F. Brunel and N.K. Sherman, Phys. Rev. Lett., 61, 2886 (1988).
8. K.A. Nelson and E.P. Ippen, Adv. Chem. Phys. 75, 1 (1989).
9. S.H. Lin, B. Fain and N. Hamer, Adv. Chem. Phys. 79, 133 (1990).
10. B. Fain, S.H. Lin and N. Hamer, J. Chem. Phys. 91, 4485 (1989).
11. B. Fain, S.H. Lin and W.X. Wu., Phys. Rev. A40, 824 (1989).
12. B. Fain and S.H. Lin, J. Chem. Phys. 93, 6387 (1990).
13. J.A. Prybyla, T.F. Heinz, J.A. Misewich, M.M.T. Loy and J.H. Glownia, Phys. Rev. Lett., 64, 1537 (1990).
14. B.F. Gantmacher and J.B. Levinson, Scattering of Electrons in Metals and Semiconductors, ("Nauka" Moscow, 1984) (in Russian).
15. J.M. Ziman, Electrons and Phonons (Clarendon, Oxford, 1960).
16. R. Orbach, Phys. Rev. Lett., 16, 15 (1966).
17. V.M. (B). Fain, Sov. Phys. JETP Letters 4, 67 (1966).
18. H.J. Kreuzer and Z.W. Gorfel, Physisorption Kinetics, Vol. 1 of Springer Series in Surface Sciences, Eds. G. Ertl and R. Gomer (Springer, Berlin. 1986).
19. B. Fain, S.H. Lin and Z.W. Gortel, Surface Science 213, 531 (1989).
20. G. Korzeniewsky, E. Hood and H. Metiu, J. Vac. Sci. Technol., 20, 594 (1982).

FEMTO-SECOND PROCESSES AND ULTRAFAST BIOLOGICAL ELECTRON TRANSFER

S. H. LIN, R. G. ALDEN, C. K. TANG, Y. FUJIMURA AND
F. SUGAWARA
Department of Chemistry and
Center for the Study of Early Events in Photosynthesis
Arizona State University
Tempe, AZ 85287-1604 USA

ABSTRACT. In this paper, we present the excitonic-vibronic coupling model to treat ultrafast photo-induced electron transfer in photosynthetic reaction centers. The density matrix method is used. The dynamical behaviors of reaction centers due to photo-induced electron transfer are described by numerical calculations.

1. INTRODUCTION

The primary event in the conversion of solar energy into chemical energy during the process of photosynthesis is the separation of charge across the photosynthetic membrane. The charge separation takes place in the reaction center (RC), which is an integral membrane protein complex. The bacterial photosynthetic RC has been characterized in greater detail, both structurally and functionally, than any other integral membrane protein complex (for reviews, see [1-6]). From the three-dimensional structure of the RC [7-13], the identity and orientations of the redox active cofactors are known. Figure 1 shows the arrangement of the nine cofactors of the *Rb. sphaeroides* RC. Two bacteriochlorophyll *a* (Bchl) molecules together form an electronically coupled pair (P) situated near the periplasmic side of the membrane. Below P and to either side are two monomer Bchls (B_A and B_B), two bacteriopheophytin *a* (Bphe) molecules (H_A and H_B) and finally two ubiquinone molecules (Q_A and Q_B). Situated approximately between the two quinones is a nonheme iron. The two branches of this cofactor arrangement are related by an approximate two-fold rotational axis. The two-fold symmetry extends to the two RC core protein subunits, L and M. Each of these subunits has a molecular weight of nearly 30,000 and each contains five membrane spanning α helices [14-18].

Light energy, absorbed by the antenna of the photosynthetic membrane, is transferred, *via* a series of energy transfer reactions, to the coupled pair of Bchl molecules in the RC. Within the next 3 to 4 picoseconds (ps) electron transfer takes place from the excited singlet state of P(P*), ultimately forming the charge separated state $P^+H_A^-$.

J. Jortner et al. (eds.), Mode Selective Chemistry, 467–484.

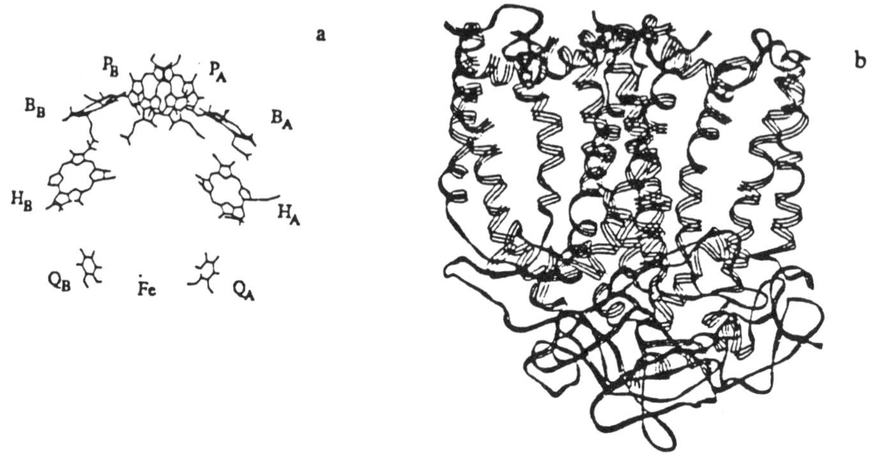

Figure 1. The structure of the reaction center from Rhodobacter sphaeroides R-26. The two-fold symmetry axis is approximately in the plane of the paper. a) The cofactors, with the phytyl and isoprenoid tails omitted for clarity (modified from Ref. [13]). b) The three protein subunits (from Ref. [13]). The L subunit is to the right, the M subunit is to the left, and the globular domain of the H subunit is at the bottom of the figure.

Electron transfer (ET) from the H_A anion (H_A^-) to the primary quinone, Q_A, proceeds with a time constant of about 200 ps. This is followed by transfer from Q_A to Q_B in hundreds of microseconds. The back reaction times are all at least an order of magnitude slower than the forward times; thus the quantum yield is near unity. There is apparently little or no ET from P to H_B. One of the outstanding questions of the RC structure-function relationship is what causes the strong functional asymmetry in this highly symmetric structure. A related and well-investigated question concerns the mechanism of the ET between P and H_A [19-32]. In particular, it is assumed that the states P*, $P^+B_A^-$, and $P^+H_A^-$ are pure states and are the only states important in the reaction. Additionally, it has been assumed that all states considered are at vibrational equilibrium before ET occurs.

Using ultrafast transient absorption experiments [30-31], it was discovered that one could not model the initial ET reaction as a single exponential decay of P* forming $P^+H_A^-$ at all wavelengths. It became clear that the room temperature decay kinetics measured were wavelength dependent. Different groups have interpreted the wavelength dependent kinetics differently. Kirmaier and Holten have proposed that the wavelength dependence results from real conformational heterogeneity in the RC population [30]. Zinth and coworkers [31] have coupled measurements of the wavelength dependent kinetics to a spectral analysis of the decay component amplitudes to support a model involving $P^+B_A^-$ as a true intermediate. Below 77 K, the variance in the rate constant for ET with wavelength apparently disappears (at least within the time resolution of the instruments used for the analysis) [30]. In addition, a structurally based analysis of changes in RC fluorescence with an applied electric field at 77 K have indicated that the direction of the difference dipole between P* and the initial charge separated state is not consistent with 100% transfer *via* $P^+B_A^-$ as a true intermediate [25].

In this paper we shall present the so-called excitonic-vibronic coupling model to treat the primary ET of RCs. Using this model, it is not necessary to assume the Markoff approximation and the approximation that vibrational dephasing and relaxation complete before ET takes place. This model can be used to calculate the steady state and time-resolved spectra. In the femto-second (fs) region, due to the large energy uncertainty, the fs time-resolved spectra consist of the contributions from the dynamics of population and coherence. This problem will be examined in this paper.

2. EXCITONIC-VIBRONIC COUPLING MODEL

Conventional ET theory utilizes the Markoff approximation and the adiabatic and Condon approximations [33-36]. Complete dephasing and vibrational relaxation are assumed to have occurred before ET. On the timescale of initial ET in the RC, a careful examination of these assumptions is warranted. In particular, we have developed an excitonic-vibronic coupling model for ET based on the structure and electronic states involved in the initial ET reaction of the *Rhodopseudomonas* (*Rps.*) *viridis* RC [37,38]. Instead of using just the three traditional electronic states thought to be involved in the initial ET (P*, P^+B^- and P^+H^-), this model utilizes a basis set of eight electronic states involving P, B_A and H_A, including such states as PB^+H^- and PB*H, and explicitly considers the two excitonic excited states of P. This model (see Fig. 2) does not assume fast dephasing and vibrational relaxation; six vibrational modes are included, as has been proposed by Warshel [39].

We have used the density matrix method to obtain the dynamic behavior of this system without assuming vibrational relaxation prior to ET. That is,

$$\frac{d\hat{\rho}}{dt} = -i\hat{L}_o\hat{\rho} - i\hat{L}'\hat{\rho} - \hat{\Gamma}\hat{\rho} \tag{2-1}$$

where $\hat{\rho}$ denotes the density matrix of the system and $\hat{L}_o$ and $\hat{L}'$ represent the Liouville operators corresponding to the zeroth order Hamiltonian and the Hamiltonian describing the interaction among the above-mentioned eight vibronic manifolds. $\hat{\Gamma}$ is the damping operator whose matrix elements describe the relaxation and dephasing properties of the system. The diagonal and off-diagonal matrix elements of $\hat{\rho}$ provide the information of the dynamics of population and coherence of the system, respectively.

The rate constants and populations of the intermediates predicted by this model will depend on the energy levels of the eight electronic states, the interaction energies, the coupling constants, and the magnitudes of the dephasing and vibrational relaxation constants. However, independent of the precise values one uses for these calculations, this model predicts that the observed ET rates should be sensitive to small changes in the energies and the coupling between the electronic states as well as the coupling between protein vibrational modes and the chromophores.

The ET takes place before the vibronic manifold {av} and the vibronic manifold {bv'}. The vibronic manifold {cv"} plays the role of "virtual states". Notice that from Eq. (2-1) we obtain

$$\frac{d\rho_{av,av}}{dt} = \frac{2}{\hbar} \sum_{v'} \text{Im}\,(H'_{av,bv'}\rho_{bv',av}) + \frac{2}{\hbar} \sum_{c}^{c\neq b} \sum_{v''} \text{Im}\,(H'_{av,cv''}\rho_{cv'',av}) - (\hat{\Gamma}\hat{\rho})_{av,av} \tag{2-2}$$

$$\frac{d\rho_{bv',av}}{dt} = -(i\omega_{bv',av} + \Gamma_{bv',av})\,\rho_{bv',av} + \frac{i}{\hbar} H'_{bv',av}(\rho_{b\overline{v}',bv'} - \rho_{av,av}) - \frac{i}{\hbar} \sum_{c}^{c\neq a,b}{}'' \sum_{v''} (H'_{bv',cv''}\rho_{cv'',av} - \rho_{bv',cv''}H'_{cv'',av}) \tag{2-3}$$

etc., where $\Gamma_{bv',av}$ represents the dephasing constant.

If the interaction matrix elements $H'_{av,bv'}$, $H'_{av,cv''}$, etc., are small, the perturbation method may be used. To the second-order approximation, we obtain

$$\frac{d\rho_{av,av}}{dt} = \frac{2}{\hbar} \sum_{v'} \text{Im}[H'_{av,bv'}\rho_{bv',av}(o)\, e^{-t(i\omega_{bv',av} + \Gamma_{bv',av})}] + \sum_{v'} \int_o^t d\tau\, M_{av,bv'}(t-\tau)\, [\rho_{bv',bv'}(\tau) - \rho_{av,av}(\tau)] + \sum_{v_{\wedge\wedge}}{}' \sum_{v''} \int_o^t d\tau\, M_{av,cv''}(t-\tau)[\rho_{cv'',cv''}(\tau) - \rho_{av,av}(\tau)] - (\Gamma\rho)_{av,av} \tag{2-4}$$

where for example

$$M_{av,bv'}(t) = \frac{2}{\hbar^2}|H'_{bv',av}|^2\, e^{-t\,\Gamma_{bv',av}} \cos \omega_{bv',av}t \tag{2-5}$$

represents the memory kernel for the vibronic transition $av \leftrightarrow bv'$.

In the Markoff approximation Eq. (2-5) becomes

$$\frac{d\rho_{av,av}}{dt} = \frac{2}{\hbar} \sum_{v'} \text{Im}[H'_{av,bv'} \rho_{bv',av}(o)\, e^{-t(i\omega_{bv',av} + \Gamma_{bv',av})}] + \sum_{v'} W_{av,bv'}(\rho_{bv',bv'} - \rho_{av,av}) + \sum_{c}' \sum_{v''} W_{av,cv''}(\rho_{cv'',cv''} - \rho_{av,av}) - (\hat{\hat{\Gamma}}\rho)_{av,av} \tag{2-6}$$

where for example, $W_{av,bv'}$ denotes the rate constant for the vibronic transition $av \leftrightarrow bv'$

$$W_{av,bv'} = \frac{2}{\hbar^2} |H'_{bv',av}|^2 \frac{\Gamma_{bv',av}}{\Gamma^2_{bv',av} + \omega^2_{bv',av}} \tag{2-7}$$

This is equivalent to applying the steady state approximation to $\rho_{bv',av}(t)$ i.e.,

$$\rho_{bv',av}(t) = \frac{\frac{i}{\hbar} H'_{bv',av}}{i\omega_{bv',av} + \Gamma_{bv',av}} (\rho_{bv',bv'} - \rho_{av,av}) \tag{2-8}$$

In this case, the coherence term involving $\rho_{bv',av}(o)$ in Eq. (2-6) disappears.

Next, we consider the term $(\hat{\hat{\Gamma}}\rho)_{av,av}$. Notice that

$$(\hat{\hat{\Gamma}}\rho)_{av,av} = \sum_{v_1} (\Gamma^{av_1,av_1}_{av,av} \rho_{av_1,av_1} - \Gamma^{av,av}_{av_1,av_1} \rho_{av,av}) + \sum_{c}'^{c\neq a} \sum_{v''} (\Gamma^{cv'',cv''}_{av,av} \rho_{cv'',cv''} - \Gamma^{av,av}_{cv'',cv''} \rho_{av,av}) \tag{2-9}$$

The first term describes the vibrational relaxation, while the second term is due to electronic relaxation. For the case in which vibrational relaxation is much faster than electronic processes, we can assume that vibrational equilibrium is established before electronic processes take place. In this case, we may set

$$\rho_{av,av}(t) = \rho_{aa}(t)\, P_{av} \tag{2-10}$$

where P_{av} represents the Boltzmann distribution and $\sum_{v} P_{av} = 1$, to obtain

$$\frac{d\rho_{aa}}{dt} = (W_{b\to a}\rho_{bb} - W_{a\to b}\rho_{aa}) + \sum_{c}'(W_{c\to a}\rho_{cc} - W_{a\to c}\rho_{aa}) - \sum_{c} (\Gamma^{cc}_{aa}\rho_{cc} - \Gamma^{aa}_{cc}\rho_{aa}) \tag{2-11}$$

where for example

$$W_{a\to b} = \sum_{v}\sum_{v'} P_{av} W_{av,bv'} \tag{2-12}$$

and

$$\Gamma_{aa}^{bb} = \sum_{v}\sum_{v'} P_{bv'} \Gamma_{av,av}^{bv',bv'} \tag{2-13}$$

Equation (2-12) represents the conventional expression of ET in the Fermi Golden Rule form; it describes the direct ET.

The ET by the super-exchange mechanism can also be obtained from the generalized master equations given by Eqs. (2-2) and (2-3). In this case, the perturbation calculation has been carried out to the fourth-order approximation. A complete analysis has been carried out. For example, for the case in which the direct ET is negligible, in the Markoff approximation we find

$$\frac{d\rho_{av,av}^{(4)}}{dt} = -\frac{2}{\hbar^4}\sum_{v'} \frac{1}{\Gamma_{bv',av}^2 + \omega_{bv',av}^2} \left| \sum_{c}\sum_{v''} \frac{H'_{av,cv''}H'_{cv'',bv'}}{\omega_{av,cv''}} \right|^2 \times \rho_{av,av} \tag{2-14}$$

where $\rho_{av,av}^{(4)}$ represents the fourth-order approximation of $\rho_{av,av}$. The conventional expression can be obtained when we introduce the Condon approximation and make the additional approximation of replacing $\omega_{av,cv''}$ by ω_{ac}, i.e.,

$$\frac{d\rho_{av,av}^{(4)}}{dt} = -\frac{2}{\hbar^2}\sum_{v'} \frac{\Gamma_{bv',av}}{\Gamma_{bv',av}^2 + \omega_{bv',av}^2} \left| \sum_{c} \frac{H'_{ac}H'_{cb}}{\hbar\omega_{ac}} \right|^2 \times |\langle\Theta_{av}|\Theta_{bv'}\rangle|^2 \rho_{av,av} \tag{2-15}$$

where H'_{ac} and H'_{cb} represent the electronic matrix elements and $|\langle\Theta_{av}|\Theta_{bv'}\rangle|^2$ denotes the Franck-Condon factor.

3. ULTRAFAST TIME-RESOLVED ABSORPTION

In this section, we shall discuss the application of the generalized linear response theory to the fs time-resolved absorption spectra. For the case in which the pump and probe pulses do not overlap, we can solve the Liouville equation twice, first for the Liouville equation with the pumping laser and then for the Liouville equation with the probe laser. Notice that the total Hamiltonian H is written as

$$\hat{H} = \hat{H}_o + \hat{V}(t) \; ; \quad \hat{V}(t) = -\vec{\mu}\cdot\vec{E}(t) \tag{3-1}$$

where $\hat{V}(t)$ represents the interaction between the system and radiation field. Here $\vec{\mu}$ denotes the dipole operator in the dipole transition and $\vec{E}(t)$ is the optical electric field. In this case, the rate of energy absorption by the molecular system is given by

$$Q = \langle\frac{d\hat{H}}{dt}\rangle = -\langle\vec{\mu}\rangle\cdot\frac{d\vec{E}}{dt} = -\vec{P}\cdot\frac{d\vec{E}}{dt} \tag{3-2}$$

where $\langle\hat{H}\rangle$ represents the average value of $\hat{H}$.

To show the application of Eq. (3-2), we notice that

$$\vec{P}(t) = \vec{P}(\omega)e^{-it\omega} + \vec{P}(-\omega)e^{it\omega} \; ; \quad \vec{E} = \vec{E}(\omega)e^{it\omega} + \vec{E}(-\omega)e^{it\omega} \tag{3-3}$$

and for a randomly oriented system

$$\vec{P}(\omega) = \chi(\omega)\,\vec{E}(\omega) \; ; \qquad \vec{P}(-\omega) = \chi(-\omega)\,\vec{E}(-\omega) \tag{3-4}$$

where $\chi(\omega)$ represents the susceptibility. Substituting Eqs. (3-3)-(3-4) into Eq. (3-2) and using the rotating wave approximation (RWA) we obtain

$$Q = 2\omega\,\chi''(\omega)\,|\vec{E}(\omega)|^2 \tag{3-5}$$

where $\chi''(\omega)$ represents the imaginary part of $\chi(\omega)$. Eq. (3-2) will be used in this work to treat pump-probe experiments.

It should be noted that the time average over the optical oscillation period is often applied to Eq. (3-2), but in the RWA, the oscillating terms disappear.

To treat the pump-probe experiment with non-overlapping pulses, we start with the stochastic Liouville equation

$$\frac{d\hat{\rho}}{dt} = -i\hat{L}_o\hat{\rho} - \frac{i}{\hbar}[\hat{V}_s,\hat{\rho}] - \hat{\Gamma}\hat{\rho} = -i\hat{L}_o\hat{\rho} - i\hat{L}_1\hat{\rho} - \hat{\Gamma}\hat{\rho} \tag{3-6}$$

where $\hat{V}_s$ represents the optical interaction due to the probe pulse, $\hat{V}_s = -\vec{\mu}\cdot\vec{E}_s(t)$, and

$$\vec{E}_s = L_s(t-t_p)[\vec{E}_s(\omega_s)e^{-it\omega_s} + \vec{E}_s(-\omega_s)\,e^{it\omega_s}] \tag{3-7}$$

Here $L_s(t-t_p)$ denotes the pulse shape function and t_p represents the probing time. For simplicity, we shall omit the subscript "s" (scanning) in the following derivation.

Making the substitution

$$\hat{\rho} = e^{-it\hat{L}'_o}\hat{\sigma} = \hat{G}(t)\hat{\sigma} \tag{3-8}$$

where $\hat{L}'_o = \hat{L}_o - i\Gamma$, Eq. (3-6) can be integrated as

$$\hat{\sigma}(t) = \hat{\sigma}_i + \Delta\hat{\sigma}(t) \tag{3-9}$$

where

$$\Delta\hat{\sigma}(t) = -i \int_{t_i}^{t} d\tau\ \overline{L}_1(\tau)\hat{\sigma}(\tau) \tag{3-10}$$

and $\hat{\sigma}_i$ represents $\hat{\sigma}$ at $t = t_i$, t_i representing the time when the pumping laser is switched off. It should be noted that $\hat{\sigma}_i$ should be determined by solving the Liouville equation with the presence of the pumping laser. The polarization $\vec{P}$ can now be calculated

$$\vec{P} = Tr[\hat{\rho}(t)\vec{\mu}] = Tr[\hat{G}(t)\ \Delta\hat{\sigma}(t)\vec{\mu}] \tag{3-11}$$

It follows that

$$\vec{P}(\omega) = \frac{i}{\hbar} \sum_{nmm'} [\hat{G}(\Delta t)\hat{\rho}_i]_{nm} \int_0^{\Delta t} d\tau\ L(\Delta t'-\tau)$$

$$\times\ [\vec{\mu}_{mm'}(\vec{\mu}_{m'n}\cdot\vec{E}(\omega))\ e^{i\tau(\omega+\omega_{nm'})}$$

$$-\ \vec{\mu}_{m'n}(\vec{\mu}_{mm'}\cdot\vec{E}(\omega))\ e^{i\tau(\omega+\omega_{m'm})}] \tag{3-12}$$

where $\Delta t' = t - t_p$ and

$$Q = i\omega\ [\vec{P}(-\omega)\cdot\vec{E}(\omega) - \vec{P}(\omega)\cdot\vec{E}(-\omega)]\ L(t-t_p) \tag{3-13}$$

for the case $\omega \gg 1/T$, T being the pulse duration of probing light at $t = t_p$, we obtain

$$Q = \frac{\omega}{\hbar} \sum_{nmm'} [\hat{G}(\Delta t)\hat{\rho}_i]_{nm}[(\vec{\mu}_{mm'}\cdot\vec{E}(-\omega))(\vec{\mu}_{m'n}\cdot\vec{E}(\omega))\ \times$$

$$(J(\omega+\omega_{nm'}) + J(\omega_{m'm}-\omega)) - (\vec{\mu}_{m'n}\cdot\vec{E}(-\omega))(\vec{\mu}_{mm'}\cdot\vec{E}(\omega))\ \times$$

$$(J(\omega+\omega_{m'm}) + J(\omega_{nm'}-\omega))] \tag{3-14}$$

where for example

$$J(\omega+\omega_{nm'}) = \int_0^{\Delta t} d\tau L(\tau)e^{i\tau(\omega+\omega_{nm'})} \tag{3-15}$$

If we use

$$L(t-t_p) = \exp[-\frac{2}{T}\ |t - t_p|] \tag{3-16}$$

then Eq. (3-15) becomes

$$J(\omega+\omega_{nm'}) = \frac{e^{\Delta t(-\frac{2}{T} + i(\omega+\omega_{nm'}))} - 1}{-\frac{2}{T} + i(\omega+\omega_{nm'})} \tag{3-17}$$

For a system of randomly oriented molecules, the expression for Q is simplified as

$$Q = |\vec{E}(\omega)|^2 \sum_{nm} [\hat{G}(\Delta t)\hat{\rho}_i]_{nm} F_{nm}(\omega) \tag{3-18}$$

where $F_{nm}(\omega)$ represents the band shape function

$$F_{nm}(\omega) = \frac{\omega}{3\hbar} \sum_{m'} (\vec{\mu}_{mm'} \cdot \vec{\mu}_{m'n})[J(\omega-\omega_{m'n}) + J(-\omega+\omega_{m'm}) - \{J(\omega+\omega_{m'm}) + J(-\omega-\omega_{m'n})\}] \tag{3-19}$$

Eq. (3-18) shows that Q is related to the optical intensity, band shape function and the dynamics of the system described by $[\hat{G}(\Delta t)\rho_i]_{nm}$.

From the above derivation, we can see that Q in Eq. (3-18) provides the time-resolved spectroscopy for describing the dynamics of the system. To determine $[\hat{G}(\Delta t)\hat{\rho}_i]_{nm}$, it is necessary to solve the master equation for describing the dynamical process under consideration (see Eq. (2-1)).

For a molecular system, the adiabatic approximation can be used in the lower excited electronic states. We shall assume that the fs dynamics under consideration is taking place in the electronic state a and the probe laser induces the electronic transition a→b. In the adiabatic approximation, from Eq. (3-18) we find

$$\chi''(\omega) = \frac{1}{6\hbar} \sum_{vv'u} [\hat{G}(\Delta t)\hat{\rho}_i]_{av,av'} (\vec{\mu}_{av',bu} \cdot \vec{\mu}_{bu,av})[J(\omega-\omega_{bu,av}) + J(-\omega+\omega_{bu,av'}) - \{J(\omega+\omega_{bu,av'}) + J(-\omega-\omega_{bu,av})\}] \tag{3-20}$$

where (bu) represents the vibronic states of the b electronic manifold, while av and av' denote the vibronic states of the a electronic manifold. It should be noted that the absorption coefficient is proportional to $\omega\chi''(\omega)$.

For an allowed transition, Eq. (3-20) becomes

$$\chi''(\omega) = \frac{1}{6\hbar} |\vec{\mu}_{ab}|^2 \sum_{vv'u} [\hat{G}(\Delta t)\hat{\rho}_i]_{av,av'} \langle\Theta_{av'}|\Theta_{bu}\rangle\langle\Theta_{bu}|\Theta_{av}\rangle \times [J(\omega-\omega_{bu,av}) + J(-\omega+\omega_{bu,av'})] \tag{3-21}$$

where $\langle\Theta_{bu}|\Theta_{av}\rangle$ and $\langle\Theta_{av'}|\Theta_{bu}\rangle$ represent the vibrational overlap integrals. Here for simplicity the anti-resonance terms given by the last two terms in Eq. (3-20) have been neglected. We shall introduce the normalized band-shape function $f_{av,av'}(\omega)$ to rewrite Eq. (3-21) as follows

$$\chi''(\omega) = \frac{\pi}{3\hbar} |\vec{\mu}_{ab}|^2 \sum_{vv'} [\hat{G}(\Delta t)\hat{\rho}_i]_{av,av'} \alpha_{av,av'}(\omega) \tag{3-22}$$

where

$$\alpha_{av,av'}(\omega) = \frac{1}{2\pi} \sum_u \langle\Theta_{av'}|\Theta_{bu}\rangle\langle\Theta_{bu}|\Theta_{av}\rangle [J(\omega-\omega_{bu,av}) + J(-\omega+\omega_{bu,av'})] \tag{3-23}$$

for $v \neq v'$, and

$$\alpha_{av,av}(\omega) = \frac{1}{\pi} \sum_u |\langle\Theta_{av}|\Theta_{bu}\rangle|^2 \frac{(\frac{2}{T})}{(\frac{2}{T})^2 + (\omega-\omega_{bu,av})^2} \tag{3-24}$$

Equation (3-22) describes the fs time-resolved absorption spectroscopy; it shows that the fs time-resolved absorption spectra consist of the coherence contribution with the band-shape function $\alpha_{av,av'}(\omega)$ and the population contribution with the band-shape function $\alpha_{av,av}(\omega)$.

4. DISCUSSION

Using the excitonic-vibronic model for fast ET, one can predict experimentally observable effects of excitation wavelength on the spectral and kinetic features of fast transient absorption measurements. As an example, the results from this model are presented, using the relative state energies shown in Figure 2, the coupling constants and vibrational frequencies given in Table 1, and the bath-induced dephasing constants given in Ref. [37], for excitation into the lower exciton band of P (Fig. 3) or the monomer Bchls (Fig. 4) of *Rps. viridis* with a vibrational lifetime (τ_{vib}) of 0.1 ps. Similar results are expected for *Rb. sphaeroides* RCs. The dephasing constants are in the 20 to 80 cm^{-1} range; the results are insensitive to changes in the pure dephasing constant between 5 and 150 cm^{-1}. Notice that the two forms of P^+B^- (states 3 and 7 of Fig. 2) have very different energies due to the large energy difference between the two excitonic states of P. Similarly, the energy difference is large for the two forms of P^+H^- (states 5 and 8 of Fig. 2). These energy differences have a profound effect on the wavelength dependence of the decay kinetics. Figure 3 shows the results predicted for excitation into the lowest exciton band of P. The model predicts that the overall decay of P* (*i.e.*, the decay of the stimulated emission) is nearly exponential. However, none of the remaining curves show a simple single exponential behavior. Consider how the kinetic behavior of a similar mixture of states would manifest itself in the time-dependent absorbance changes in the *Rb. sphaeroides* RC spectrum. Most of the populated states absorb in the spectral region between 770 and 810 nm in this species. This spectral region would

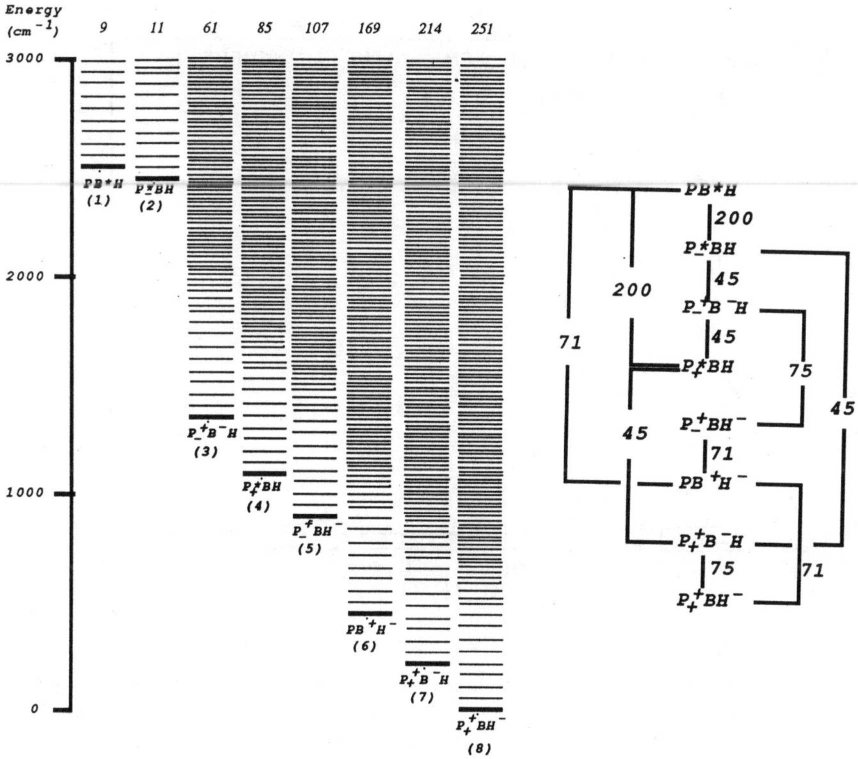

Figure 2. a) Relative energy level diagram for the reaction center of Rps. viridis. Eight vibronic states of the pigments are taken into account: (1) PB*H, (2) P_-^*BH, (3) $P_-^+B^-H$, (4) P_+^*BH, (5) $P_-^+BH^-$, (6) PB^+H^-, (7) $P_+^+B^-H$ and (8) $P_+^+BH^-$, in their order of electronic energy (omitting the subscript A of B and H). b) Proposed kinetic scheme. The number shown along the lines connecting the states are the electronic coupling constants.

TABLE 1. Vibrational modes and coupling constants

Transitions	Vibrational Modes (cm^{-1})					
	100	224	750	1200	1400	1520
	Coupling Constants					
$P \rightarrow P^*$	0.06	0.03	0.02	0.03	0.02	0.04
$P \rightarrow P^+$	0.3	0.1	0.07	0.01	0.09	0.1
$H \rightarrow H^-$	0.4	0.25	0.12	0.08	0.18	0.12

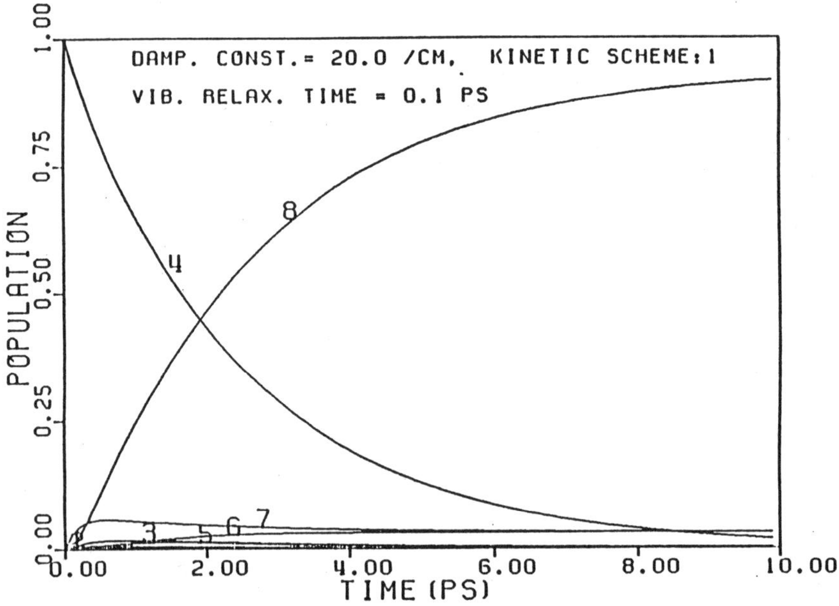

Figure 3. Model calculations of the time evolution of the population of the vibronic states of the pigments in the reaction center of Rps. viridis with a vibrational relaxation time of 0.1 ps (from Ref. [37]). Photon absorption to the band of P_+^*BH is used as the initial condition with τ_{vib} = 0.1 ps. The numbers 1-8 refer to the eight electronic states shown in Figure 2.

contain contributions from B^+ and B^- (bleaching near 800 nm), contributions from the cationic form of the upper exciton state of P, P_-^+ (probably around 810 nm), contributions from electrochromic shifts due to P^+ formation (positive blue of 800, negative red of 800), and contributions due to absorbance from both the upper and lower excited exciton states of P. The basic motif of simple decay kinetics of the stimulated emission, but more complex kinetics in the 770 to 810 nm region is qualitatively similar to the observed experimental behavior.

Figure 4 shows predicted kinetic traces for RCs excited directly into the monomer Bchls. Three new states are observed under these conditions. Two of these are excited states that decay rapidly, the excited state of B and the excited upper exciton state of P. Indeed, excitation into B has been observed to give rise to short-lived excited states [19,21].

Figure 5 shows the calculated time-resolved absorption data for the P band of Rps. viridis. This region was chosen, because only the stimulated emission from P* and the bleaching of P are thought to be the spectra. The simulations are very similar to the femtosecond time-resolved absorption data recently reported for Rps. viridis as well as Rb. sphaeroides [30,31]. These analyses are currently being extended to

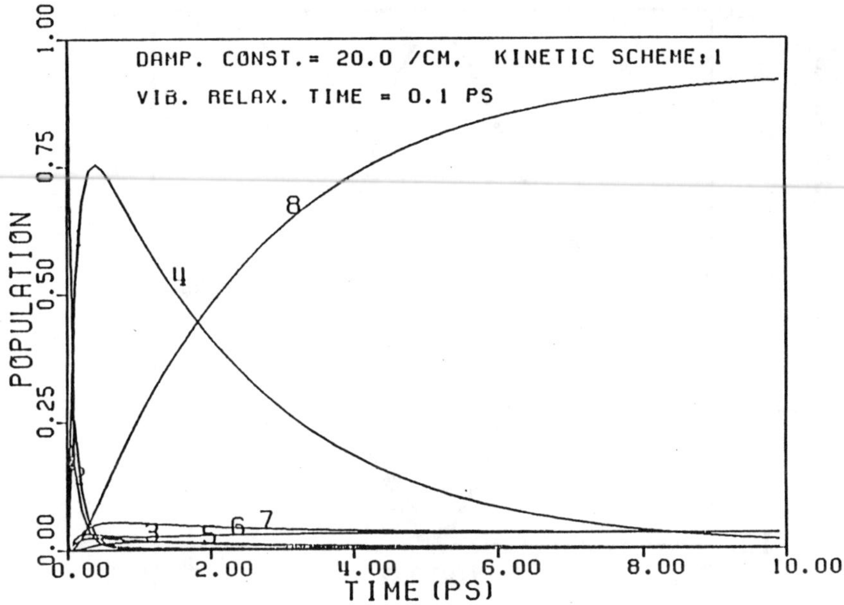

Figure 4. Time evolution of the population of the vibronic states of the pigments after photon absorption to the band of PB*H with a vibrational relaxation time of 0.1 ps with τ_{vib} = 0.1 ps (from Ref. [37]).

the spectroscopically more complex regions of the spectru, such as 760-810 nm in Rb. sphaeroides.

The above discussion represents an example of how the excitonic-vibronic coupling model we have developed can be used to evaluate excitation wavelength dependent changes in the spectral and kinetic parameters determined from the time resolved absorption work. The order and relative energies of the states in Figure 2 is a reasonable guess based on previous work, but other energetic schemes will be considered and tested experimentally or by molecular orbital calculations. By both modeling and measuring the picosecond timescale kinetics and spectra at many excitation wavelengths, we should be able to begin to elucidate some of the mechanistic details of this system.

In the version of our excitonic-vibronic model for ET described above, we have considered the system to preexist in a certain excited vibronic state, such as the excited lower exciton state of P, at time = 0. It is possible to build into the model the interaction between the system and the excitation light explicitly. This has a significant advantage for the analysis of ultrafast absorbance changes vs. time since it explicitly considers the effects of a coherent excitation source on the time evolution of the system. From the uncertainty principle, $\Delta E \Delta t \sim \hbar/2$, we can see that in analyzing the femtosecond

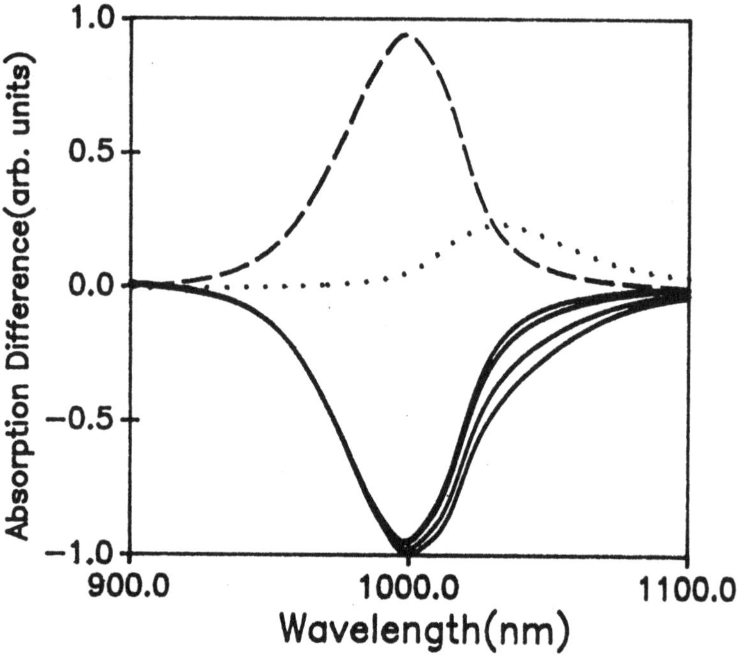

Figure 5. Calculated time-resolved absorption spectra of the P band region of Rps. viridis at 0.0, 1.0, 4.0, and 10.0 ps based upon the decays shown in Figure 5. The absorption spectrum (- - -) and stimulated emission (••••) are calculated using the following parameter; ω_1 = 140 cm^{-1}, ω_2 = 39 cm^{-1}, S_1 = 1.2, S_2 = .87, and Γ = 180 cm^{-1} with the displaced oscillator.

transient absorption spectra, both the dynamics of population and coherence (or phase) should be taken into account. Recently, we have developed theoretical treatments for femtosecond time-resolved experiments like absorption [40,41], fluorescence [42,43] and resonance Raman scattering [44,46].

The work in progress is to extend the excitonic-vibronic coupling model by including additional states in the system. In particular, we plan to include the effects of states involving the B branch cofactors. We also propose to consider anharmonic effects on the model. Such effects become particularly pronounced when the free energy gap between states involved in the ET reaction is large. To date, this effect has not been examined for ET in RCs. Additionally, we intend to incorporate the MO information from ENDOR experiments and the spin-spin coupling determined from the ^{3}P yield, in order to calculate more accurately the energy levels, state interactions and coupling constants that are required to apply the excitonic-vibronic coupling model to the mutant RCs.

Another rich source of information regarding the electronic states relevant to ET in mutant RCs is the steady state absorbance spectrum. A variety of models employing basis sets of electronic states that are mixed to describe the observed ground state transitions have been developed. Friesner and coworkers [47] have proposed a model assuming that the six excited states P_A*, P_B* (the two monomer excited states of P), B_A*, B_B*, H_A* and H_B* are coupled. In other words, they propose the mixing of the excited states of Bchl and Bphe in the A and B branches. A similar exciton model has also been considered by Fischer and coworkers [48,49]. So far we have only employed our excitonic-vibronic coupling model to study the dynamics of RCs; it can also be employed to calculate the steady state absorption spectra of RCs. In principle, a complete description of the time evolution of the absorbance changes can be produced from this model by calculating the rate of energy absorption which is related to the polarization $\vec{P} = Tr(\hat{\rho}\vec{\mu})$, where $\vec{\mu}$ is the dipole operator and $\hat{\rho}$ is determined from solving the Liouville equation (for detailed treatment, see Refs. [40] and [41]).

5. REFERENCES

[1] Parson, W. W. and Ke, B. (1982) 'Photosynthesis: Energy Conservation by Plants and Bacteria', Govindjee (ed.), Academic Press, New York, pp. 331.

[2] Kirmaier, C. and Holten, D. (1987) 'Primary photochemistry of reaction centers from the photosynthetic purple bacteria', Photosynth. Res. 13, 225-260.

[3] Feher, G., Allen, J. P., Okamura, M. Y. and Rees, D. C. (1989) 'Structure and function of bacterial photosynthetic reaction centers', Nature 339, 111-116.

[4] Rees, D. C., Komiya, H., Yeates, T. O., Allen J. P. and Feher, G. (1989) 'The bacterial photosynthetic reaction center as a model for membrane proteins', Ann. Rev. Biochem. 58, 607-633

[5] Budil, D. E., Gast, P., Change, C.-H., Schiffer, M. and Norris, J. R. (1987) '3-Dimensional x-ray crystallography of membrane proteins: Insights into electron transfer', Ann. Rev. Phys. Chem. 38, 561-583.

[6] Deisenhofer, J. and Michel, H. (1989) 'The photosynthetic reaction center from the purple bacterium Rhodopseudomonas viridis', Science 245, 1463-1473.

[7] Deisenhofer, J., Epp, O., Miki, K., Huber, R. and Michel, H. (1984) 'X-Ray structure analysis of a membrane-protein complex. Electron density map of 3 Å resolution and a model of the chromophores of the photosynthetic reaction center from Rhodopseudomonas viridis', J. Molec. Biol. 180, 385-398.

[8] Deisenhofer, J., Epp, O., Miki, K., Huber, R. and Michel, H. (1985) 'Structure of the protein subunits in the photosynthetic reaction

center of Rhodopseudomonas viridis at 3 Å resolution', Nature 318, 618-624.

[9] Michel, H., Epp, O. and Deisenhofer, J. (1986) 'Pigment protein interactions in the photosynthetic reaction center for Rhodopseudomonas viridis', EMBO J. 5, 2445-2451.

[10] Chang, C. H., Tiede, D., Tang, J., Smith, U., Norris, J. and Schiffer, M. (1986) 'Structure of Rhodobacter sphaeroides R-26 reaction center', FEBS Lett. 205, 82-86.

[11] Yeates, T. O., Komiya, H., Chirino, A., Rees, D. C., Allen, J. P. and Feher, G. (1988) 'Structure of the reaction centers from Rhodobacter sphaeroides R-26 and 2.4.1 protein cofactor (bacteriochlorophyll, bacteriopheophytin and carotenoid) interaction 4', Proc. Natl. Acad. Sci. USA 85, 7993-7997.

[12] Allen, J. P., Feher, G., Yeates, T. O., Komiya, H. and Rees, D. C. (1988) 'Structure of the reaction center from Rhodobacter sphaeroides R-26. Protein cofactor (quinones and Fe^{2+}) interactions 5', Proc. Natl. Acad. Sci. USA 85, 8487-8491.

[13] Allen, J. P. (1988) 'Light-energy Transduction in Photosynthesis: Higher Plant and Bacterial Models', S. E. Stevens, Jr. and D. A. Bryant (eds.), The American Society of Plant Physiologists, pp. 155.

[14] Williams, J. C., Steiner, L. A., Feher, G. and Simon, M. I. (1984) 'Primary structure of the L-subunit of the reaction center from hodopseudomonas sphaeroides', Proc. Natl. Acad. Sci. USA 81, 7303-7307.

[15] Youvan, D. C., Bylina, E. J., Alberti, M., Begusch, H. and Hearst, J. E. (1984) 'Nucleotide and deduced polypeptide sequences of the photosynthetic reaction center B870 antenna and flanking polypeptides from R. capsulata', Cell 47, 949-957.

[16] Belanger, G., Berard, J., Corriveau, P. and Gingras, G. (1988) 'The structural genes-coding for the L-subunit and M-subunit of Rhodospirillum rubrum photoreaction center', J. Biol. Chem. 263, 7632-7638.

[17] Trebst, A. (1987) 'The 3-dimensional structure of the herbicide binding niche on the reaction center polypeptides of Photosystem II', Z. Naturforsch C42, 742-750

[18] Michel, H. and Deisenhofer, J. (1988) 'Relevance of the photosynthetic reaction center from purple bacteria to the structure of Photosystem II', Biochemistry 27, 1-7.

[19] Woodbury, N. W., Becker, M., Middendorf, D. and Parson, W. W. (1985) 'Picosecond kinetics of the initial photochemical electron transfer reaction in bacterial photosynthetic reaction centers', Biochemistry 24, 7516-7521.

[20] Wasielewski, M. R. and Tiede, D. M. (1986) 'Subpicosecond measurements of primary electron transfer in Rhodopseudomonas viridis reaction centers using near infrared excitation', FEBS Lett. 204, 368-372.

[21] Breton, J., Martin, J. L., Migus, A., Antonetti, A. and Orszag, A. (1986) 'Femtosecond spectroscopy of excitation energy transfer and initial charge separation in the reaction center of the photosynthetic bacterium Rhodopseudomonas viridis', Proc. Natl.

Acad. Sci. USA 83, 5121-5125.

[22] Martin, J. L., Breton, J., Hoff, A. J., Migus, A. and Antonetti, A. (1986) 'Femtosecond spectroscopy of electron transfer in the reaction center of the photosynthetic bacterium Rhodopseudomonas sphaeroides R-26: Direct electron transfer from the dimeric bacteriochlorophyll primary donor to the bacteriopheophytin acceptor with a time constant of 2.8 ± 0.2 ps', Proc. Natl. Acad. Sci. USA 83, 957-961.

[23] Bixon, M., Jortner, J., Michel-Beyerle, M. E., Ogrodnik, A. and Lersch, W. (1987) 'The role of the accessory bacteriochlorophyll in reaction centers of the photosynthetic bacteria-intermediate acceptor in the primary electron transfer', Chem. Phys. Lett. 140, 626-630.

[24] Marcus, R. A. (1987) 'Superexchange versus an intermediate BChl mechanism in reaction centers of photosynthetic bacteria', Chem. Phys. Lett. 133, 471-477.

[25] Lockhart, D. J., Goldstein, R. F. and Boxer, S. G. (1988) 'Structure based analysis of the initial electron transfer step in bacterial photosynthesis-electric field induced fluorescence anisotropy', J. Chem. Phys. 90, 1408-1415.

[26] Holzapfel, W., Finkele, U., Kaiser, W., Oesterhelt, D., Scheer, H., Stilz, H. U. and Zinth, W. (1989) 'Observation of a bacteriochlorophyll anion radical during the charge separation in a reaction center', Chem. Phys. Lett. 160, 1-7.

[27] Warshel, A., Chu, Z. T. and Parson, W. W. (1989) 'Dispersed polaron simulations of electron transfer in photosynthetic reaction centers', Science 246, 112-116.

[28] Scherer, P. O. J. and Fischer, S. F. (1989) 'Quantum treatment of the optical spectra and the initial electron transfer process with the reaction center of Rhodopseudomonas viridis', Chem. Phys. 131, 115-127.

[29] Friesner, R. A. and Won, Y. (1989) 'Spectroscopy and electron transfer dynamics of the bacterial reaction center', Biochim. Biophys. Acta 977, 99-122.

[30] Kirmaier, C. and Holten, D. (1990) 'An inhomogeneous distribution of bacterial reaction centers underlies the observed temperature and wavelength dependence of the rates of the primary electron transfer reactions', Proc. Natl. Acad. Sci. USA 87, 3552.

[31] Holzapfel, W., Finkele, U., Kaiser, W., Oesterhelt, D., Scheer, H., Stilz, H. U. and Zinth, W. (1990) 'Initial electron transfer in the reaction center from Rhodobacter sphaeroides', Proc. Natl. Acad. Sci. USA 87, 5168-5172.

[32] Lockhart, D. J., Kirmaier, C., Holten, D. and Boxer, S. G. (1990) 'Electric field effects on the initial electron-transfer kinetics in bacterial photosynthetic reaction centers', J. Phys. Chem. 94, 6987-6995.

[33] Marcus, R. A. (1956) 'Theory of oxidation-reduction reactions involving electron transfer 1', J. Chem. Phys. 24, 966-978.

[34] Hopfield, J. J. (1977) 'Photoinduced charge transfer-critical test of mechanism and range of biological electron transfer processes', Biophys. J. 18, 311-321.

[35] Chu, Z. T., Lin, Y. Y. and Lin, S. H. (1987) 'A theoretical investigation of the mechanism of electron transfer in condensed media', J. Mol. Sci. 5, 127-138.
[36] Lin, S. H. (1989) 'Theory of photoinduced electron transfer in condensed media', J. Chem. Phys. 90, 7103-7113.
[37] Chu, Z. T. (1988) Ph.D. Thesis, Arizona State University; part of this work was presented at the Biophysical Society Meeting (1988) Biophys. J. 53, 66a; Alden, R. G., Chu, Z. T., Lin, S. H., Islampour, R., Allen, J. P. and Williams, J. C. (1991) Adv. Multiphoton Process. Spectry. 7, 201.
[38] Sugawara, M., Fujimura, Y., Yeh, C. Y. and Lin, S. H. (1990) 'Application of the density matrix method to the primary electron transfer in photosynthetic reaction centers', J. Photochem. Photobiol. A54, 321-331.
[39] Warshel, A. (1980) 'Role of the chlorophyll dimer in bacterial photosynthesis', Proc. Natl. Acad. Sci. USA 77, 3105-3109.
[40] Fain, B., Lin, S. H. and Hamer, N. (1989) 'Two-dimensional spectroscopy: Theory of nonstationary, time-dependent absorption and its application to femtosecond processes', J. Chem. Phys. 91, 4485-4494.
[41] Lin, S. H. and Fain, B. (1989) 'Application of the theory of two-dimensional spectroscopy to the real-time femtosecond transition state spectroscopy', Chem. Phys. Lett. 155, 216-220.
[42] Lin, S. H., Fain, B. and Yeh, C. Y. (1990) 'Ultrafast time-resolved fluorescence spectroscopy', Phys. Rev. A41, 2718-2729.
[43] Villaeys, A. A., Vallet, J. C. and Lin, S. H. (1990) 'Generalized susceptibility approach to transient nonlinear optical processes', Phys. Rev. A41, 2796-2807.
[44] For review, see Lin, S. H., Fain, B. and Hamer, N. (1990) 'Ultrafast processes and transition-state spectroscopy', Adv. Chem. Phys. 79, 133-267.
[45] Fain, B. and Lin, S. H. (1990) 'Space-time coherence induced by ultrashort electromagnetic pulses' J. Chem. Phys. 93, 6387-6397.
[46] Lin, S. H., Alden, R., Fain, B. and Villaeys, A. A. (1991) 'Femtosecond time-resolved resonance Raman scattering', J. Mol. Structure (in press).
[47] Won, Y. and Friesner, R. A. (1988) 'Simulation of optical spectra from the reaction center of Rhodopseudomonas viridis', J. Phys. Chem. 92, 2208-2214.
[48] Knapp, E. W., Scherer, P. O. and Fischer, S. F. (1986) 'Model studies of low-temperature optical transitions of photosynthetic reaction centers A-spectra, LD-spectra, CD-spectra, ADMR-spectra and LD-ADMR spectra for Rhodopseudomonas viridis', Biochim. Biophys. Acta 852, 295-305.
[49] Knapp, E. W., Fischer, S. F., Zinth, W., Sander, M., Kaiser, W., Deisenhofer, J. and Michel, H. (1986) 'Analysis of optical spectra from single crystals of Rhodopseudomonas viridis reaction centers', Proc. Natl. Acad. Sci. USA 82, 8463-8467.

ACTIVE CONTROL OF SELECTIVITY OF PRODUCT FORMATION IN A CHEMICAL REACTION: WHAT'S NEW?

STUART A. RICE

Department of Chemistry and The James Franck Institute,
The University of Chicago, Chicago, IL 60637

ABSTRACT. This paper summarizes the results of three recent analyses relevant to the theory of active control of selectivity of product formation in a chemical reaction. Two of the analyses concern fundamental aspects of the quantum dynamics of unbounded systems, while the third is concerned with an example, namely, the calculation of the optimal field for the control of the evolution of a particular molecular process (photodissociation of a diatomic molecule) and the quality of approximations to that field.

The fundamental aspects of the theory of active control of selectivity of product formation which we consider are both related to the possible influence of chaotic dynamics on the level of control that is achievable. We first report an analysis of of the Kolmogorov-Sinai entropy of an unbounded quantum system. Our analysis makes explicit use of the notion that, although successive measurements of the state of the system can extend indefinitely in time, any individual measurement is necessarily of finite duration. It is supposed that successive measurements are made on individual replica systems drawn in succession from an ensemble of identical systems. The nonzero uncertainty in energy associated with a finite duration measurement permits partitioning of Hilbert space into domains on which one can define the probability for finding the system in a particular state. The joint probability for finding the system traversing a particular succession of states can then be defined on the tensor product space formed from the Hilbert spaces associated with the system at the times of each of the measurements. The joint probability provides the essential ingredient for the definition of an information entropy associated with the evolution of the system. Given this definition of the measurement process, it is shown that the Kolmogorov-Sinai entropy of an unbounded quantum system is zero, hence the quantum dynamics cannot be chaotic. With this result in hand we discuss the achievability of optimal control of the evolution of an unbounded quantum system by the action of external fields. Previous work has established that the evolution of a quantum system with a discrete and bounded spectrum of states can, in principle, be fully controlled, i.e., that the system wave function can be guided by external fields to approach arbitrarily closely a selected target state wave function. The optimal control of the evolution of a quantum system with a discrete and bounded spectrum of states has also been studied, using a method of analysis that depends on the localized character of bound state wave functions and the fact that the spectrum of states is bounded. We show that

J. Jortner et al. (eds.), Mode Selective Chemistry, 485–525.

optimal control of the evolution of an unbounded quantum system is possible, in the sense that it is possible to minimize the difference between the product function formed at time t_0 from a localizing function and a continuum wave function and a similarly defined target function at time t_f. The result obtained does not establish that such optimal control is equivalent to full control, i.e. that the difference between the initial and target functions can be made arbitrarily small, which would require showing that the set of control functions is complete with respect to the function space of the system. The analysis establishes the existence of external fields that can optimally guide the unbounded system evolution in the absence of other constraints than the Schroedinger equation of motion, but does not provide an algorithm for the construction of such fields.

The third analysis reviewed treats the effects of pulse shaping in the Tannor-Rice scheme for selective photochemistry, as applied to the photodissociation of a diatomic molecule. Optimal control theory is used to generate the pulse shape which provides a defined enhanced yield of products. It is then shown that achievable approximations to the optimum pulse shape can provide a significant fraction of the enhancement of the product yield achieved with the optimum pulse.

I. Introduction

This paper reviews three recent developments in the theory of active control of selectivity of product formation in a chemical reaction. In order, we examine the argument that the Kolmogorov-Sinai entropy of an unbounded quantum system is zero[1], the complementary demonstration that the evolution of such a system can be controlled[2], and a demonstration that the optimal control field for a particular dynamical process (the photodissociation of a diatomic molecule) can be well approximated by a simple superposition of fields[3].

The achievement of selective control of product formation in a reaction has been sought persistently throughout the evolution of chemistry. To date, such control has only been accomplished at the macroscopic level, by manipulation of external variables such as temperature, pressure, solvent character, etc. The manipulation of macroscopic control parameters generates only passive control in the sense that the molecular dynamics which defines the reaction is influenced only indirectly through the statistical effects arising from interactions with other molecules in the system. Recent theoretical work has shown that it is, in principle, possible to influence the selectivity of product formation in a chemical reaction by use of external fields which modify the molecular dynamics. The approaches proposed, which are forms of active control, fall into two categories emphasizing, respectively, the use of time independent and time dependent control fields.

Brumer and Shapiro[4] have shown that control of product selectivity can be achieved if two degenerate exit channel wave functions can be made to interfere constructively or destructively, which interference will occur if there is simultaneous generation of two or more coherent excitation routes to the degenerate final states. This control scheme focuses attention on the properties of eigenstates of the system; its implementation requires controlling the phase difference between the two cw laser sources driving the different excitation routes.

Tannor and Rice[5], and Tannor, Kosloff and Rice[6], have shown that controlling the duration of propagation of a wave packet on an excited state electronic potential energy surface, by controlling the time delay between pump and dump pulses, can be used to influence the selectivity of products formed on the ground state potential energy surface. Rabitz and co-workers[7] have shown that it is possible to design optimally shaped pulses that will guide the evolution of a system from a chosen initial state to a selected final state in a specified time interval. Their work, which has been applied only to reactions on one potential energy surface, has been extended by Kosloff, Rice, Gaspard, Tersigni and Tannor[8] to the design of optimally shaped pulses that utilize modulation of wave packet evolution on an excited state potential energy surface to influence the selectivity of product formation on the ground state potential energy surface. This control scheme focuses attention on the temporal and spatial evolution of wavepackets on one or more potential energy surfaces. In the simplest situation, in which the shaping of the control field pulses is forsaken, implementation of this scheme requires the development of techniques to control both the temporal separation and the relative phases of the pump and dump pulses. The achievement of optimal control of the selectivity of product formation, which will typically involve generation of a pulse with complicated spectral content and time signature, requires further development of femtosecond pulse shaping technology, as exemplified by the work of Heritage et al[9] and Warren and co-workers[10].

The development of the theory of active control of selectivity of product formation in a chemical reaction has, as in most new areas of research, proceeded under the assumption that there is no fundamental limitation to what can be accomplished. To what extent is it possible to alter the selectivity of product formation in a chemical reaction by intervention at the level of individual molecule dynamics? One difficulty in answering the question posed arises from the extreme complexity of polyatomic molecule dynamics. Consider, first, the classical mechanical description of a polyatomic molecule. It is now widely appreciated that the classical dynamics of a typical polyatomic molecule is, in different (possibly interleaved) energy regimes, both quasiperiodic and chaotic[11]. In the chaotic regime the trajectory representing the evolution of the state of the system in phase space is extremely sensitive to the initial conditions. Specifically, in the chaotic regime two trajectories started with infinitesimally different initial conditions will diverge such that the distance between them grows exponentially with the time[12]. In contrast, in the quasiperiodic (or regular) regime two trajectories started with infinitesimally different initial conditions will diverge such that the distance between them grows only linearly with the time[12]. The measure of chaotic motion in a system described by classical mechanics is the Kolmogorov-Sinai (KS) entropy[13,14,15], which can be interpreted as the rate of growth in the information necessary to describe the evolution of the system. The KS entropy of a system in the chaotic regime is necessarily nonzero positive.

The characteristic features of chaotic dynamics lead to several nested questions: Is it possible to exert any control over the evolution of a system which exhibits chaotic dynamics? If partial control of the evolution of such a dynamical system is possible, does the extent of controllability have a natural limit, and is that limit a function of the KS entropy of the system? The existing theory of optimal control[16,17] does not directly address whether or not dynamical chaos influences the extent of controllability of the evolution of a system. Nevertheless, the demonstration that it is possible to find external fields which optimally control the evolution of the broad category of systems whose time dependence is describable by an equation of motion of the form

$$\frac{d}{dt}\mathbf{x}(t) = \mathbf{H}_0\mathbf{x}(t) + \sum_{l=0}^{r} u_l(t)\mathbf{H}_l\mathbf{x}(t)\,, \tag{1.1}$$

where $\mathbf{x}(t)$ represents the state of the system, $\mathbf{H}_0$ is the generator of the uncontrolled "bare" system dynamics, the $\{\mathbf{H}_l\}$ are operators in the system phase space and the $u_l(t)$ are time dependent scalar control functions, implies that the controlled system dynamics is regular even if the uncontrolled bare system dynamics is chaotic. We note that optimal control of the evolution of the system dynamics generated by $\mathbf{H}_0$ is not the same as full control; by analogy with the variational principle of quantum mechanics, the latter requires that the set of control functions $u_l(t)$ be complete in the sense that the function space defined by the bare system Hamiltonian is appropriately spanned. When the evolution of a system is optimally controlled the difference between the initial and target states is made as small as possible for the set of control functions used; when the evolution of a system is fully controlled that difference is made zero.

The generic structure of the quantum dynamics of a typical polyatomic molecule has not been worked out. With reference to the discussion of the last paragraphs, the relationship between the quantum dynamics and the classical dynamics of a system for which the classical dynamics is chaotic is poorly understood. Kosloff and Rice[18,19] have shown that in a quantum system which has only a discrete and bounded spectrum, the KS entropy is necessarily zero, implying that the system dynamics cannot be chaotic. Given this result it is not surprising that it can be shown that the evolution of such a system can be fully controlled[20]. In very recent work Zhao and Rice[1] have extended the Kosloff-Rice analysis to show that the KS entropy of an unbounded quantum system is zero provided that all of the sequential measurements made to determine the evolution of the system have finite duration in time. And, in a related analysis, Zhao and Rice[2] have shown that it is, in principle, possible to find a set of external fields with which one can optimally direct the evolution of an unbounded quantum system in the sense that a particular initial state is guided to have minimal difference from a specified final state in a defined finite time. This analysis does not establish the possibility of full control of the evolution of an unbounded quantum system, since it does not include a proof that there exists a set of control functions that appropriately spans the function space of the dynamical system. In technical terms this means that in the variational formulation that defines the optimal control functions one can only establish the existence of a minimum, and not prove that the minimum achieves the value zero. However, the demonstration that the KS entropy of an unbounded quantum system is zero implies that, in principle, full control of the evolution of such a system is possible.

We also discuss an example of control of evolution of a realistic quantum system using the Tannor-Rice scheme in the generalized form developed by Kosloff, Rice, Gaspard, Tersigni and Tannor. This analysis[3] explicitly acknowledges that, with present day technology, the most convenient way to excite multiple molecular transitions, simultaneously and coherently, is to use ultrashort, transform limited, optical pulses. The wave packets so generated have a particularly simple form in the limit that the excitation is of infinitesimal duration; in that limiting case the spectral content of the wave packet spans all states of the system connected to the initial state with nonvanishing transition moments, and the amplitudes of the states contributing to the wave packet do not depend on the nuclear dynamics. However, infinitesimally short laser pulses cannot be generated, and pulses shorter than 50 fs are difficult to generate reliably. A 50 fs Gaussian pulse has a spectral bandwidth of only 300 cm^{-1}, which is not large enough to excite a great number of the states of most molecules. Moreover, the amplitudes of the components of the typical wave packet generated by a 50 fs pulse are influenced by coupling to nuclear motions with comparable periods, hence so is the evolution of the wave packet. We must then expect that with fixed shape 50 fs pump and dump pulses we will achieve only a lower limit to the control of selectivity of product formation.

We report progress in answering three questions raised by the above considerations:

(i) What are the requirements for an experimental demonstration of the Tannor-Rice scheme when applied to a real reaction?

(ii) What improvement in selectivity of product formation is generated when optimally shaped pulses replace the fixed shape pump and dump pulses used in (i)?

(iii) Is there a simple approximation to the optimal pulse shape found in (ii), one which can be generated with present day technology?

Our approach to problem (i) is to examine the modulation of the product yield in the photodissociation of a diatomic molecule as a function of the separation of pump and dump pulses. Although controlling the amount of product formed in the photodissociation of a diatomic molecule does not demonstrate the enhancement of the formation of a particular reaction product selected from a set of possible reaction products, it does depend on the same features of the wave packet dynamics and wave packet interference as does enhancement of the yield of one of several reaction products. The real reactions we have studied are the photodissociation of the van der Waals molecule HgAr and the photodissociation of I_2. Only the photodissociation of Hg Ar is reviewed herein. HgAr is very nearly an ideal vehicle to study the simplest version of the Tannor-Rice scheme in that the level spacings are small, so that even 80 fs pump and dump pulses can generate marked modulation of the photofragment population as a function of pump-dump pulse delay. We also show that rotational dephasing does not significantly degrade the modulation of product yield generated by controlling the pump-dump pulse delay.

Our approach to problem (ii) is to carry out an optimization of the pulse shape using the method of Kosloff, Rice, Gaspard, Tersigni and Tannor[8]. It is found that the use of optimally shaped pulses in place of fixed shape pump and dump pulses increases the photofragmentation yield per pulse by many orders of magnitude. The calculated pulse shapes are not overwhelmingly complicated, but they are also not simple. Were attention to be focussed solely on using the optimally shaped pulse to control the photofragmentation yield, we would generate the classical conundrum of the best being the enemy of the good. Aside from verifying our theoretical analysis, it serves little purpose to predict that a photofragmentation reaction can be efficiently driven by a specially shaped pulse if that shaped pulse is too complicated to produce in the laboratory. Accordingly, we argue that the use to which we put the calculated optimal pulse shape is as important as is its calculation.

Our approach to problem (iii) is to calculate the Fourier transforms of the optimally shaped pulses found in (ii) so as to identify the spectral content and durations of the principal components of the pulses. We then construct simple approximations to the optimally shaped pulses by superposing a few Gaussian pulses selected to have spectral content and widths which are close to those of the major Fourier components of the optimally shaped pulses. We find that these approximations to the optimally shaped pulses suffice to generate photofragmentation yields within a factor of two of those generated by the optimally shaped pulses.

As of this date there has not yet been an experimental demonstration of the control of selectivity of product formation in a reaction involving a polyatomic molecule using either the Brumer-Shapiro or the Tannor-Rice schemes. However, the basic principles of both schemes have been demonstrated in experiments which modulate the production of specific "products" in processes in diatomic molecules. In particular, Gordon and co-workers21 have shown that the photoionization of HCl can be controlled by varying the relative phase of coherent one photon and three photon excitation routes, and Scherer and co-workers22 have shown that the fluorescence of I_2 can be controlled by varying the time delay between phase locked femtosecond pump and dump pulses. These experimental results suggest that, with the to be expected development of laser technology, active control of selectivity in a chemical reaction will become a reality.

II. The Kolmogorov-Sinai Entropy of an Unbounded Quantum System

A. BACKGROUND

It is now well established that the classical dynamics of the typical nonintegrable mechanical system can exhibit deterministic chaos[11]. Amongst the clearest manifestations of deterministic chaos are an exponential growth with time of the separation of system trajectories which have infinitesimally different initial conditions, and the associated exponential growth with time in the amount of information required to predict the properties of a particular trajectory. The latter can be used to generate a quantitative measure of deterministic chaos, namely the KS entropy[12,13,14,15], h_{KS}, with which we shall be concerned throughout this section. For the present it suffices to note that h_{KS} is zero if the system dynamics is quasiperiodic (i.e. regular), h_{KS} is infinite if the system dynamics is truly random, and h_{KS} is a constant larger than zero if the system displays deterministic chaos.

In contrast with the situation in classical dynamics, our present knowledge of the role which underlying classical deterministic chaos plays in the quantum dynamics of a system is very limited[23]. Our physical intuition concerning the behavior of mechanical systems is so strongly rooted in classical dynamics that there is a strong expectation that quantum systems must also support deterministic chaos. Yet, for the case that a quantum system has (only) a discrete spectrum of states it can be shown that the quantum KS-entropy is zero even if the corresponding classical KS-entropy is nonzero[18,19], which implies that such a system does not exhibit deterministic chaos.

How can we define a KS-entropy for an unbounded quantum system? How is that KS-entropy related to deterministic chaos in the corresponding classical system?

Zhao and Rice[1] base their analysis of the KS-entropy of an unbounded quantum system on an extension of the work of Kosloff and Rice[18]. As background to this analysis it is convenient to provide a simple definition of the KS-entropy of a system described by classical dynamics. To do so we quote (with some minor omissions) the following from a recent book by Schuster[24].

"Consider the trajectory $\mathbf{x}(t)$, and suppose that the d-dimensional phase space is partitioned into boxes of size ℓ^d. The state of the system is now measured at intervals of time τ. Let $p(i_0,...,i_n)$ be the joint probability that $\mathbf{x}(t = 0)$ is in box i_0, that $\mathbf{x}(t = \tau)$ is in box i_1, ... , and that $\mathbf{x}(t = n\tau)$ is in box i_n. The quantity

$$h(n) = -\sum_{i_0}\sum_{i_1}\cdots\sum_{i_n} p(i_0,\cdots,i_n)\log[p(i_0,\cdots,i_n)]$$

is proportional to the information needed to locate the system on a special trajectory $i_1^*,...,i_n^*$ with precision ℓ (if one knows, a priori, only the probabilities $p(i_0,...,i_n)$). Therefore $h(n + 1) - h(n)$ is the additional information needed to predict in which cell i_{n+1}^* the system will be if we know that it was previously in $i_1^*,...,i_n^*$. This means that $h(n + 1) - h(n)$ measures our loss of information about the system from time $n\tau$ to time $(n + 1)\tau$.

The KS-entropy is defined as the average rate of loss of information:

$$h_{KS} = \lim_{\tau\to 0}\lim_{\ell\to 0}\lim_{N\to\infty}(N\tau)^{-1}\sum_{n=0}^{N-1}[h(n + 1) - h(n)]$$

$$= -\lim_{\tau\to 0}\lim_{\ell\to 0}\lim_{N\to\infty}(N\tau)^{-1}\sum_{i_0}\sum_{i_1}\cdots\sum_{i_N} p(i_0,\cdots,i_N)\log[p(i_0,\cdots,i_N)].$$

The limit $\ell \to 0$ (which must be taken after $N \to \infty$) makes h_{KS} independent of the particular partition."

The process which is used to define h_{KS} can be thought of as a determination of the overlaps of successive partitions of the system phase space, with the successive partitions defined at particular times by the natural evolution of the system. The determination of these successive overlaps can, in turn, be thought of as effected by successive measurements carried out on the evolving system. Kosloff and Rice[18] used the latter observation to define the KS-entropy of a bounded quantum system with a discrete spectrum, and we adopt the same notion to define the KS-entropy of an unbounded quantum system. We assume that observations are made on replica systems drawn successively from an ensemble, and that these measurements may extend to $t \to \infty$. However, we also assume that individual measurements are of finite duration, each with a corresponding uncertainty in energy. We show that when the measurements made on the system have these characteristics, the KS-entropy of an unbounded quantum system is zero.

B. GENERAL REMARKS

To establish the notation we shall use we consider some of the generic properties of quantum systems[25,26].

The quantum mechanical description of a system is based on a Hilbert space $\mathbb{H}$, a positive density operator $\hat{\rho}$, and a set of operators $\{ \hat{A}_1, \hat{A}_2, \hat{A}_3, ... \}$ which represent all possible physical observables. The expectation values of the operators corresponding to observables are reals.

In general, an unbounded quantum system will have both discrete and continuum states. The discrete states of the system are defined by

$$\hat{H} | m > = E_m | m >, \quad m = 1, 2, 3, ... , \tag{2.1}$$

and the continuum states by

$$\hat{H} | \lambda > = \lambda | \lambda >, \quad \lambda \in \Omega . \tag{2.2}$$

The domain $\Omega \in \mathbf{R}$ supports the continuum part of the spectrum of the system. The set which is the union of the discrete and continuum eigenvectors,

$$\{ | m > \ | \ m = 1, 2, 3, ...; \ | \lambda > | \ \lambda \in \Omega . \} , \tag{2.3}$$

generates a complete state space $\mathbb{V} \in \mathbb{H}$. A complete set of orthogonal projection operators can be constructed from the basis states of $\mathbb{V}$,

$$\{ \hat{P}_i \,|\, i = 1, 2, 3... ; \ \hat{Q}(\lambda) \,|\, \lambda \in \Omega \}. \tag{2.4}$$

The operator which projects out the discrete state $| m >$ is defined by

$$\hat{P}_m = | m >< m |, \tag{2.5}$$

while the projection operator for the entire continuum is defined by

$$\hat{Q} = \int_\Omega | \lambda > d\lambda < \lambda | = \int_\Omega \hat{Q}(d\lambda) . \tag{2.6}$$

We now divide the domain Ω into subdomains $\{\lambda, \lambda+\delta\lambda\}$, denoted E_λ. Each of these subdomains defines a Borel set which contains physically significant states, hence each includes a set of continuous reals which are the physical observables $\lambda \in \Omega$. We define the (differential) projection operator associated with the subdomain E_λ as

$$\delta\hat{Q} = \int_\lambda^{\lambda + d\lambda} | \lambda' > d\lambda' < \lambda' | = \int_{E_\lambda} \hat{Q}(d\lambda) = \hat{Q}(E_\lambda) \tag{2.7}$$

on the subspace spanned by the kets of the interval $(\lambda, \lambda+\delta\lambda)$.

The projection operators satisfy the standard relations

$$\hat{P}_m\hat{P}_n = \hat{P}_m\delta_{mn}, \quad \hat{Q}(E_\lambda)\hat{Q}(E_{\lambda'}) = \hat{Q}(E_\lambda)\delta_{E_\lambda E_{\lambda'}}, \tag{2.8}$$

$$\hat{P}_m\hat{Q}(E_\lambda) = \hat{Q}(E_\lambda)\hat{P}_m = 0, \quad \text{for all m and } E_\lambda \in \Omega , \tag{2.9}$$

and the closure condition can be written

$$\hat{1} = \sum_m \hat{P}_m + \int_\Omega \hat{Q}(d\lambda)$$

$$= \sum_m \hat{P}_m + \sum_{E_\lambda} \hat{Q}(E_\lambda) . \tag{2.10}$$

The probability for finding state $| m >$ in the system with density operator $\hat{\rho}$ is

$$\text{prob}[\hat{\rho}, m] = \text{trace}[\hat{\rho}\hat{P}_m] \tag{2.11}$$

and the probability for the system to have an energy in the domain E_λ is

$$\text{prob}[\hat{\rho}, E_\lambda] = \text{trace}[\hat{\rho}\hat{Q}(E_\lambda)], \ E_\lambda \in \Omega . \tag{2.12}$$

In general, a system cannot be represented as being in a pure state. In that case the system wave function, $|\Psi>$, is in an incoherent mixture of pure states, and can be represented in the form

$$|\Psi> = \sum_m a_m |m> + \int_\Omega d\lambda\, a(\lambda)\, |\lambda>, \tag{2.13}$$

with the corresponding density operator

$$\begin{aligned}\hat{\rho} &= |\Psi><\Psi| \\ &= \sum_{a,\, a'} |a> \omega_{aa'} < a'|.\end{aligned} \tag{2.14}$$

The sum in (2.14) formally includes the necessary integrals over the continuum. The time evolution of the density operator is a strongly continuous one parameter group of positive linear maps T on the state space $\mathbb{V}$ of $\mathbb{H}$, such that $T : \mathbb{V} \rightarrow \mathbb{V}$ with[27,28,29]

$$\text{trace}[\, T_t(\hat{\rho})\,] = \text{trace}[\,\hat{\rho}\,] = 1\,. \tag{2.15}$$

Thus all the $\hat{\rho} \in \mathbb{V}$ belong to the trace class with unit trace. At time t, after the mapping T, the density operator is related to the density matrix at time zero by

$$\hat{\rho}_t = T_t(\hat{\rho}) = \hat{U}_H(t)\, \hat{\rho}_0\, \hat{U}_H^{+}(t) \tag{2.16}$$

with

$$\hat{U}_H(t) = e^{-i\hat{H}t}, \tag{2.17}$$

Let the Hamiltonian of the system be represented as

$$\hat{H} = \hat{H}_0 + \varepsilon \hat{H}_I\,, \tag{2.18}$$

where $\hat{H}_0$ is diagonal, $\hat{H}_I$ is the off diagonal part of the Hamiltonian and ε is a pure number. If we define the operators

$$\hat{Z}_0\, \hat{\rho} = -i\, [\, \hat{H}_0, \hat{\rho}\,]\;, \tag{2.19}$$

$$\hat{Z}_I\, \hat{\rho} = -i\, [\, \hat{H}_I, \hat{\rho}\,]\,, \tag{2.20}$$

$$\hat{Z} = \hat{Z}_0 + \varepsilon \hat{Z}_I\,, \tag{2.21}$$

and if

$$[\hat{H}, \hat{Z}] = 0, \tag{2.22}$$

then the density operator at time t can be written[25,26]

$$\hat{\rho}_t = e^{\hat{Z}t}\hat{\rho}_0 = \hat{U}_t\hat{\rho}_0 \tag{2.23}$$

with the generator $\hat{Z}$.

C. THE QUANTUM KS-ENTROPY: A DEFINITION

By the spectral decomposition theorem[28], (i) an Hermitian operator $\hat{A}$ has a corresponding unique spectral family of projection operators, (ii) the operator $\hat{A}$ admits a spectral resolution into the complete operator space constructed from the family of orthogonal projection operators

$$\{\hat{P}_i \mid i = 1, 2, 3\ldots;\ \hat{Q}(\lambda) \mid \lambda \in \Omega\}, \tag{2.24}$$

and (iii) we can write

$$\hat{A} = \sum_m a_m\hat{P}_m + \int_\Omega \lambda\, \hat{Q}(d\lambda)$$

$$= \sum_m a_m\hat{P}_m + \sum_{E_\lambda} \lambda\, \hat{Q}(E_\lambda), \tag{2.25}$$

where a_1, a_2, a_3, ... and $\lambda \in \Omega$ are the expectation values of $\hat{A}$.

To define the average information entropy and the KS-entropy for an unbounded quantum system we proceed as follows.

Imagine that an observable $\hat{A}$ is measured at successive times $\{ t_i \mid i = 0,1,\ldots,n \}$ on replicas selected from an ensemble of identical systems with the density operator $\hat{\rho}$. We assume that each measurement is of nonzero duration Δ, and that the measurement time is small compared to the interval between measurements. This measurement procedure, for example, defines the energy of the system with an uncertainty h/Δ. At time t_j the density operator is $\hat{U}_t^j\hat{\rho}$, where $\hat{U}_t$ is the time evolution operator defined in (2.23). The mapping on the Hilbert space generated by the system dynamics is $\hat{U}_t^j : \mathbb{V}^{(j)} \rightarrow \mathbb{V}^{(j)}$, where $\mathbb{V}^{(j)}$ is the Hilbert space of the system at time t_j. For a sequence of measurements at $t_0 = 0$, $t_1 = \tau$,... , $t_n = n\tau$ the joint probability measure is defined on the tensor product Hilbert space

$$\mathcal{K} = \mathbb{V}^{(0)} \otimes \mathbb{V}^{(1)} \otimes \mathbb{V}^{(2)} \otimes \cdots\cdots \otimes \mathbb{V}^{(n)}. \tag{2.26}$$

The combined density operator for the correlated state of the system is defined to be

$$\hat{\rho}_n = \hat{\rho} \otimes \hat{U}\hat{\rho} \otimes \hat{U}^2\hat{\rho} \otimes \cdots\cdots \otimes \hat{U}^n\hat{\rho}\,. \tag{2.27}$$

The superoperator for the observable $\hat{A}$ on the tensor product space is then

$$\hat{A}_n = \hat{A}^{(0)} \otimes \hat{A}^{(1)} \otimes \hat{A}^{(2)} \otimes \cdots\cdots \otimes \hat{A}^{(n)}, \tag{2.28}$$

and the corresponding superprojection operator for the outcomes is

$$\hat{P}_{j_0 j_1 \cdots j_n} = \hat{P}_n = \hat{P}^{(0)}_{j_0} \otimes \hat{P}^{(1)}_{j_1} \otimes \hat{P}^{(2)}_{j_2} \otimes \cdots\cdots \otimes \hat{P}^{(n)}_{j_n} \tag{2.29}$$

where $n = j_0 j_1 j_2 \cdots j_n = 1, 2, \cdots$ is a collective index and runs over all possible correlated states. The projection operator $\hat{P}^{(i)}$ at $t = t_j$ in equation (2.29) can be either for a discrete eigenstate or for a subdomain of the continuum states.

Using the definitions introduced above, the joint probability for the set of measurements to have the outcomes defined in (2.29) is

$$p_{j_0 j_1 \cdots j_n}(\hat{A}_n) = \mathrm{trace}[\, \hat{P}_n \hat{\rho}_n \,]\,. \tag{2.30}$$

The average information entropy is then defined to be

$$h(\hat{U},\hat{A},\hat{\rho}) = \lim_{n \to \infty} \frac{1}{n} h(\hat{\rho}_n,\hat{A}_n), \tag{2.31}$$

and we define the quantum mechanical KS-entropy by

$$h_{KS}(\hat{U}) = \sup_{\hat{A} \in \{\hat{A}_i\}} h(\hat{U},\hat{A},\hat{\rho})\,, \tag{2.32}$$

where the joint information entropy is

$$\begin{aligned} h(\hat{\rho}_n, \hat{A}_n) &= -\sum_j p_j(\hat{A}_n) \ln[p_j(\hat{A}_n)] \\ &= -\sum_{j_0} \sum_{j_1} \cdots \sum_{j_n} p_{j_0 j_1 \cdots j_n}(\hat{A}) \ln\left(p_{j_0 j_1 \cdots j_n}(\hat{A})\right) \end{aligned} \tag{2.33}$$

and the sum formally includes the continuum.

D. BOUNDED QUANTUM SYSTEMS REVISITED

The KS-entropy is designed to represent only the deterministic chaos associated with the motion generated by the evolution operator $\hat{U}_t$. The Kosloff-Rice[18] argument concerning the KS-entropy was based on the observation that for a quantum system with a discrete spectrum the evolution operator has a discrete spectrum, hence it is possible to find a number n such that the operator $\hat{U}_t^n$ is as close to the identity operator as one desires, which implies that the KS-entropy of such a system is zero.

We now rederive the Kosloff-Rice result from a different point of view.

The joint probability (2.30) contains information about all the overlaps of the states in the set of measurements made as the system evolves. Clearly, the outcome of a measurement at time t_n is conditional on the outcomes of all prior measurements. Suppose the observable $\hat{A}$ is measured at $t = 0$. If the outcome of the measurement is a_0, the corresponding state is $| a_0 >$ and the probability to have the outcome a_0 is p_{a_0}. The outcome of a measurement of the same observable at $t = \tau$ is dependent on the outcome of the measurement at $t = 0$. In effect, the measurement at $t = \tau$ determines only the the states generated from $| a_0 >$ by the natural system evolution in the interval from $t = 0$ to $t = \tau$. The same argument holds for each time interval between measurements.

In a bounded quantum system the evolution is entirely determined by phase factor changes. At $t = 0$ the probability that a measurement will have outcome a_0 is

$$p_{a_0} = \mathrm{trace}[\hat{P}_{a_0}\hat{\rho}] = \omega_{a_0 a_0} \tag{2.34}$$

where the density operator is

$$\hat{\rho} = \sum_{a,\, a'} | a > \omega_{aa'} < a' | \tag{2.35}$$

and the projection operator is

$$\hat{P}_{a_0} = | a_0 >< a_0 | \tag{2.36}$$

At time $t = \tau$ the conditional probability for a measurement having outcome a_1, given the outcome of the measurement at $t = 0$, is

$$p_{a_1} = \mathrm{trace}[\hat{P}_{a_1}\hat{\rho}] = \delta_{a_0 a_1} \tag{2.37}$$

where the time dependent density operator is

$$\hat{\rho}(t_1) = \sum_{a, a'} e^{i(\Delta_{a'} - \Delta_a)t_1} | a > \omega_{aa'} < a' | \tag{2.38}$$

and the projection operator is

$$\begin{aligned} \hat{P}_{a_1} &= | a_1 >< a_1 | / \omega_{a_0 a_0} \\ &= | a_1^{(0)} >< a_1^{(0)} | . \end{aligned} \tag{2.39}$$

The exponential factor in (2.38) describes the time evolution and the projection operator in (2.39) has been scaled to display the condition imposed by the prior measurement at $t = 0$. Such scaling is valid because at $t = \tau$ the scaled states, just like the original unscaled states, form a complete state space which is a subspace of the tensor product space defined in (2.26). Proceeding in this fashion we find for the joint probability that measurements at $t = 0, t = \tau, \ldots, t = n\tau$ will yield the outcomes $a_0, a_1, \ldots, a_n$

$$\begin{aligned} p_{a_0 a_1 \cdots a_n}(\hat{A}_n) &= \mathrm{trace}[\, \hat{P}_n \hat{\rho}_n \,] \\ &= p_{a_0} \delta_{a_0 a_1} \delta_{a_1 a_2} \cdots \delta_{a_{n-1} a_n}. \end{aligned} \tag{2.40}$$

Substituting (2.40) into equation (2.33) for the joint information entropy yields

$$\begin{aligned} h(\hat{\rho}_n, \hat{A}_n) &= -\sum_{j_0} \sum_{j_1} \cdots \sum_{j_n} p_{j_0 j_1 \cdots j_n}(\hat{A}) \ln\left(p_{j_0 j_1 \cdots j_n}(\hat{A})\right) \\ &= -\sum_{j_0} p_{j_0} \ln(p_{j_0}) = h_0 \quad , \end{aligned} \tag{2.41}$$

which is finite at $t = 0$.

Inserting (2.41) into (2.31), for a given evolution operator $\hat{U}$ and a density operator $\hat{\rho}$, leads to the result that the average information entropy is zero,

$$\begin{aligned} h(\hat{U}, \hat{A}, \hat{\rho}) &= \lim_{n \to \infty} \frac{1}{n} h(\hat{\rho}_n, \hat{A}_n) \\ &= \lim_{n \to \infty} \frac{1}{n} h_0 = 0 \end{aligned} \tag{2.42}$$

and so is, from (2.32), the KS-entropy of the bounded quantum system.

E. THE KS-ENTROPY OF AN UNBOUNDED QUANTUM SYSTEM

We now consider the information entropy and the KS-entropy of a quantum system whose spectrum includes both discrete and continuum states. The density operator, formally including the continuum, is

$$\hat{\rho} = \sum_{a, a'} | a > \omega_{aa'} < a' | \quad , \tag{2.43}$$

which can be decomposed into

$$\hat{\rho} = \hat{\rho}_B + \hat{\rho}_{BC} + \hat{\rho}_C , \tag{2.44}$$

where the individual density operators are for the bound, continuum and bound-continuum states, respectively. We have

$$\hat{\rho}_B = \sum_m \sum_{m'} | m > \omega_{mm'} < m' | , \quad m, m' = 1, 2, \cdots \tag{2.45}$$

$$\hat{\rho}_C = \int_\Omega d\lambda \int_\Omega d\lambda' \, | \lambda > \omega_{\lambda\lambda'} < \lambda' | \tag{2.46}$$

and

$$\hat{\rho}_{BC} = \sum_m \int_\Omega d\lambda \, | \lambda > \omega_{\lambda m} < m | + \sum_m \int_\Omega d\lambda \, | m > \omega_{m\lambda} < \lambda | . \tag{2.47}$$

Suppose that at $t = 0$ a measurement made on the system yields the value a_0 within the Borel set $(a_0, a_0 + \delta a)$ in the continuous spectrum. This measurement is associated with the projection operator

$$\hat{Q}_{a_0} = \int_{a_0}^{a_0+\delta a} d\lambda \, | \lambda >< \lambda | = \int_{E_{a_0}} d\lambda \, | \lambda >< \lambda | \tag{2.48}$$

The probability for finding a_0 in the Borel set $(a_0, a_0 + \delta a)$ is

$$p(a_0) = \mathrm{trace}[\,\hat{\rho}\hat{Q}_{a_0}\,]$$

$$= \mathrm{trace}[\,(\,\hat{\rho}_B + \hat{\rho}_{BC} + \hat{\rho}_C\,)\,\hat{Q}_{a_0}\,] = \mathrm{trace}[\,\hat{\rho}_C\hat{Q}_{a_0}\,]\,. \tag{2.49}$$

Substituting (2.45), (2.46), (2.47) and (2.48) into (2.49) one obtains

$$\hat{\rho}_C\hat{Q}_{a_0} = \int_\Omega d\lambda_1 \int_\Omega d\lambda_2\, |\,\lambda_1 > \omega_{\lambda_1\lambda_2} < \lambda_2\,| \int_{a_0}^{a_0+\delta a} d\lambda\, |\,\lambda >< \lambda\,|$$

$$= \int_\Omega d\lambda_1 \int_{a_0}^{a_0+\delta a} d\lambda\, |\,\lambda_1 > \omega_{\lambda_1\lambda} < \lambda\,| \tag{2.50}$$

and

$$p(a_0) = \mathrm{trace}[\,\hat{\rho}_C\hat{Q}_{a_0}\,]$$

$$= \int_\Omega d\lambda' < \lambda'\,| \left(\int_\Omega d\lambda_1 \int_{a_0}^{a_0+\delta a} d\lambda\, |\,\lambda_1 > \omega_{\lambda_1\lambda} < \lambda\,| \right) |\,\lambda' >$$

$$= \int_{a_0}^{a_0+\delta a} \omega_{\lambda\,\lambda}\, d\lambda \tag{2.51}$$

As before, the outcome of a measurement at $t = \tau$ is conditioned by the outcome of the measurement at $t = 0$. The projection operator associated with a_1 being the measured outcome is

$$\hat{Q}^{(0)}_{a_1} = \frac{1}{p(a_0)} \int_{a_1}^{a_1+\delta a} d\lambda\, |\,\lambda >< \lambda\,| \tag{2.52}$$

Note that, just as in our discussion of the quantum system with a discrete spectrum, the projection operator defined in (2.52) is scaled to account for the condition imposed by the prior measurement.

The probability for finding the measurement outcome a_1 is

$$p(a_1^{(0)}) = \mathrm{trace}[\,\hat{\rho}_C\hat{Q}^{(0)}_{a_1}\,]. \tag{2.53}$$

The density operator now is time-dependent, and it can be written in the form

$$\hat{\rho}_c(t) = \int_\Omega d\lambda \int_\Omega d\lambda' \, | \lambda > \omega_{\lambda\lambda'}(t) < \lambda' | \tag{2.54}$$

with the time dependent coefficient

$$\omega_{\lambda\lambda'}(t) = \omega_{\lambda\lambda'}(0) \exp\{ I(\lambda'-\lambda)t\} . \tag{2.55}$$

The argument of the right hand side of (2.53) is now

$$\hat{\rho}_c \hat{Q}^{(0)}_{a_1} = \frac{1}{p(a_0)} \int_{a_1}^{a_1+\delta a} d\lambda \, | \lambda >< \lambda | \int_{a_0}^{a_0+\delta a} d\lambda_1$$

$$\times \left(\int_\Omega d\lambda_2 \, | \lambda_1 > \omega_{\lambda_1\lambda_2} < \lambda_2 | + \int_{\Omega,\, \lambda_2 \neq a_0} d\lambda_2 \, | \lambda_2 > \omega_{\lambda_2\lambda_1} < \lambda_1 | \right)$$

$$= \frac{1}{p(a_0)} \int_{a_0}^{a_0+\delta a} d\lambda \int_\Omega d\lambda_2 \, | \lambda > \omega_{\lambda\lambda_2} < \lambda_2 | \, \delta_{a_0a_1} \tag{2.56}$$

and (2.53) assumes the form

$$p(a_1^{(0)}) = \mathrm{trace}[\, \hat{\rho}_c \hat{Q}^{(0)}_{a_1} \,]$$

$$= \frac{1}{p(a_0)} \int_\Omega d\lambda' < \lambda' | \left(\int_{a_0}^{a_0+\delta a} d\lambda \int_\Omega d\lambda_2 \, | \lambda > \omega_{\lambda\lambda_2} < \lambda_2 | \right) | \lambda' > \delta_{a_0a_1}$$

$$= \frac{1}{p(a_0)} \int_{a_0}^{a_0+\delta a} W_{\lambda\lambda} \, d\lambda \, \delta_{a_0a_1} = \delta_{a_0a_1} . \tag{2.57}$$

Notice that in (2.56) and (2.57) we have used

$$\int_{a_0}^{a_0+\delta a} d\lambda_0 \int_{a_1}^{a_1+\delta a} d\lambda_1 \; F(\lambda_0, \lambda_1) < \lambda_0 \, | \lambda_1 >$$

$$= \int_{a_0}^{a_0+\delta a} d\lambda_0 \int_{a_1}^{a_1+\delta a} d\lambda_1 \; F(\lambda_0, \lambda_1)\, \delta(\lambda_0 - \lambda_1)$$

$$= \begin{pmatrix} \int_{a_0}^{a_0+\delta a} d\lambda_0 F(\lambda_0, \lambda_0), & \text{if } a_0 = a_1 \\ 0, & \text{otherwise} \end{pmatrix} = \int_{a_0}^{a_0+\delta a} d\lambda_0 F(\lambda_0, \lambda_0)\delta_{a_0 a_1} \tag{2.58}$$

Similarly, in an n-time measurement the joint probability for finding the outcomes $a_0, a_1, \ldots, a_n$ is

$$p(a_0, a_1, \cdots, a_n) = \text{trace}[\, \hat{Q}_n \hat{\rho}_n \,]$$
$$= p(a_0)\delta_{a_0 a_1}\delta_{a_1 a_2} \cdots \delta_{a_{n-1} a_n}. \tag{2.59}$$

Substituting (2.59) into (2.33), the joint information entropy is found to be

$$h(\hat{\rho}_n, \hat{A}_n) = -\sum_{j_0} \sum_{j_1} \cdots \sum_{j_n} p(j_0 j_1 \cdots j_n) \ln(p(j_0 j_1 \cdots j_n))$$
$$= -\sum_{j_0} p(j_0) \ln(p(j_0)) = h_0 \tag{2.60}$$

where h_0, the information entropy at t = 0, is finite. Then the average information entropy is zero,

$$h(\hat{U}, \hat{A}, \hat{\rho}) = \lim_{n \to \infty} \frac{1}{n} h(\hat{\rho}_n, \hat{A}_n) = \lim_{n \to \infty} \frac{1}{n} h_0 = 0 \tag{2.61}$$

and, using (2.32), we conclude that the KS-entropy for a quantum system with a continuum spectrum of states is zero:

$$h_{KS}(\hat{U}) = \sup_{\hat{A} \in \{\hat{A}_i\}} h(\hat{U}, \hat{A}, \hat{\rho}) = 0. \tag{2.62}$$

The preceding argument started with the assumption that the measurements made correspond to successive projections of subsets of the continuum states of a system whose spectrum has both discrete and continuum states. However, equations (2.44) and (2.58) imply that the density operator defined in equation (2.47) does not contribute to the information entropy, so the system average information entropy is the sum of the contributions corresponding to the discrete and continuum parts of the spectrum. Since we have shown that the quantum KS-entropies corresponding to dynamics in the discrete part of the spectrum and to dynamics in the continuum part of the spectrum are both zero, (2.61) holds for the general case when the measurements of the system involve both discrete and continuum states.

III. Optimum Control of Transformations in an Unbounded Quantum System

A. BACKGROUND

There have been only a few studies of the limitations of controllability of a quantum system.

Huang, Tarn and Clark[20] consider a quantum system which has a discrete and bounded spectrum of states. They show, without consideration of the influence of external constraints, that it is possible to find control fields which will guide an initial state of the system ($t = 0$) to a particular final state of the system at the specified time $t = t_f$. In the case considered the system is fully controlled in the sense that all of the population can be transferred from the initial state to the selected final state. The arguments used in developing this result are valid only for a quantum system with a discrete and bounded spectrum.

Peirce, Dahleh and Rabitz[30] have developed the formalism for the determination of the external fields which will optimally control the evolution of a bounded quantum system; they do include, generically, the influence of external constraints on the system. However, their proof of the existence of control fields relies on the properties of L^2 functions in Hilbert space, so cannot be directly applied to the case of unbounded quantum systems.

The spectrum of an unbounded quantum system has, in general, both bound states and continuum states. It is the lack of localization of the continuum state wave functions which poses the difficulty in developing an existence theorem for the optimal control fields for such a system. Zhao and Rice[2] have shown that the evolution of a modified continuum wave function, generated by the product of a localizing function and a continuum wave function, can be optimally controlled by the application of external fields. In this section we review the Zhao-Rice analysis and a few applications.

B. GENERAL REMARKS

We consider a general quantum mechanical system with Hamiltonian

$$\mathbf{H}_0 = \mathbf{T} + \mathbf{V}, \tag{3.1}$$

where $\mathbf{T}$ is the kinetic energy operator and $\mathbf{V}$ is the potential energy operator, assumed time independent for the present. The system wave function is governed by the time-dependent Schroedinger equation

$$i\frac{\partial}{\partial t}\Psi(\mathbf{x}, t) = \mathbf{H}_0\Psi(\mathbf{x}, t) \tag{3.2}$$

$\left(\frac{h}{2\pi} = 1\right)$ subject to the initial condition $\Psi(\mathbf{x}, t_0) = \Psi_0(\mathbf{x})$, which has the formal solution

$$\Psi(\mathbf{x}, t) = \mathbf{U}(t, t_0)\Psi(\mathbf{x}, t_0). \tag{3.3}$$

As usual, $\mathbf{U}(t, t_0)$ is the evolution operator

$$\mathbf{U}(t, t_0) = \exp\{-i\mathbf{H}_0(t - t_0)\}, \tag{3.4}$$

t is the time and $\mathbf{x}$ represents all of the coordinates. The time evolution operator exhibits the group property, and the set of all time evolution operators forms a semi-group for the positive time domain.

To control the evolution of a system it must be subjected to external fields which, in effect, defines a new system with Hamiltonian

$$\mathbf{H} = \mathbf{H}_0 + \mathbf{H}'(u), \tag{3.5}$$

where $\mathbf{H}'$ is the operator representing the coupling between the "bare" system with Hamiltonian $\mathbf{H}_0$ and the external field. The strength of the external field is adjusted by varying the parametric function u(t), which must be invariant under all possible transformations of the system coordinates. The wave function of the controlled system satisfies

$$i\frac{\partial}{\partial t}\Psi(\mathbf{x}, t) = [\mathbf{H}_0 + \mathbf{H}']\Psi(\mathbf{x}, t), \tag{3.6}$$

subject to the same initial condition as in (3.2), $\Psi(\mathbf{x}, t_0) = \Psi(\mathbf{x})$. Eq. (3.6) defines the time evolution of the quantum mechanical system in the admissible domain $t \in [0, t_f]$, with the final time t_f determined by the properties of the specific system being controlled.

The control problem to be addressed can be formulated as the following question: For a given initial state $\Psi(\mathbf{x}, t_0)$ of $\mathbf{H}_0$ at t_0, and a given target state $\widehat{\Psi}(\mathbf{x}, t_f)$ of $\mathbf{H}_0$ at t_f, can one guide the evolution of the state of the "bare" system in the time domain $t \in [0, t_f]$ by adjusting the external fields represented in $\mathbf{H}$ so as to force the system wave function $\Psi(\mathbf{x}, t_f)$ to approach the target wave function $\widehat{\Psi}(\mathbf{x}, t_f)$ as closely as permitted within the function space defined by the $u_l(t)$?

To answer this question, Zhao and Rice adopt the following **definition**:

A quantum mechanical system is said to be controllable if, for a given target wave function $\widehat{\Psi}(\mathbf{x}, t_f)$, the wave function defined by Eq. (3.6) satisfies the condition

$$\| \Psi(\mathbf{x}, t_f) - \widehat{\Psi}(\mathbf{x}) \| \leq \varepsilon \tag{3.7}$$

for any given small $\varepsilon > 0$, with the norm defined as $\| f \| = < f \mid f >^{1/2}$ on a suitable Hilbert space.

C. OPTIMAL CONTROL OF THE EVOLUTION OF A QUANTUM MECHANICAL SYSTEM

The Peirce-Dahleh-Rabitz analysis requires that the wave function of the system to be controlled be square integrable, which is not the case when the system is in a continuum state such as characterizes the products of a fragmentation reaction. The analysis which follows addresses this problem by separating the Hilbert space of the system into the part associated with the discrete spectrum and its complement associated with the continuous spectrum. We carry out this separation implicitly by defining a new function $\zeta(\mathbf{x},t)$, related to the system wave function by

$$\zeta(\mathbf{x},t) = \mathbf{P}(\mathbf{x})\Psi(\mathbf{x},t) \, . \tag{3.8}$$

This new function is an element of the Hilbert space $\mathbb{H} = L^2(\mathbf{R}^d)$, which has inner product $< f \mid g >$ and norm $\| f \| = < f \mid f >^{1/2}$ for the time domain $t \in [\,0, t_f\,]$. The function $\mathbf{P}(\mathbf{x})$ is a localization operator which transforms the delocalized wave function $\Psi(\mathbf{x}, t)$ to a localized function $\zeta(\mathbf{x},t)$ on the defined Hilbert space. The operator $\mathbf{P}(\mathbf{x})$ is necessarily continuous and bounded. The norm $\| \zeta(\mathbf{x},t) \|$ is bounded as

$$\| \zeta(\mathbf{x},t) \| \leq N, \ \in \ L^2(\mathbf{R}^d), \text{ and } t \in [\,0, t_f\,], \tag{3.9}$$

where N is a constant.

The control operator $\mathbf{H}'$ is chosen to be a linear integral operator with the kernel $u(\mathbf{x}, \mathbf{x}', t)$,

$$\mathbf{H}'\Psi(\mathbf{x}, t) = \int_{\mathbf{R}^d} u(\mathbf{x}, \mathbf{x}', t)\Psi(\mathbf{x}', t)\, d\mathbf{x}' \qquad t \in [\,0, t_f\,]. \tag{3.10}$$

The form (3.10) is capable of representing a wide class of external control fields. $\mathbf{H}'$ is assumed to be a bounded perturbation to the bare system Hamiltonian $\mathbf{H}_o$, i.e.

$$\| \mathbf{H}' \|_{op} \leq M_1, \tag{3.11}$$

and the kernel operator $u(\mathbf{x}, \mathbf{x}', t)$ is required to have the norm

$$|| u(\mathbf{x},\mathbf{x}',t) || \le M_2 \in L^2(\mathbf{R}^d) \tag{3.12}$$

for $t \in [0, t_f]$, where M_1 and M_2 are constants. Since $u(\mathbf{x},\mathbf{x}', t)$ is a bounded operator it shares a common domain with the function $\zeta(\mathbf{x},t)$ on the defined Hilbert space $\mathbb{H}$.

We now consider the functional

$$J(u) = < \zeta(\mathbf{x},t_f) - \widehat{\zeta}(\mathbf{x}) \mid \zeta(\mathbf{x},t_f) - \widehat{\zeta}(\mathbf{x}) >(u) \,. \tag{3.13}$$

The new target function $\widehat{\zeta}(\mathbf{x})$ is related to the target wave function $\widehat{\Psi}(\mathbf{x})$ as before, i.e.

$$\widehat{\zeta}(\mathbf{x}) = \mathbf{P}(\mathbf{x})\widehat{\Psi}(\mathbf{x}) \,. \tag{3.14}$$

We take the set of all admissible control functions $u(\mathbf{x},\mathbf{x}', t)$ to define a control function space $\mathbb{U}$. Using the definition of the infimum, there exists a sequence of control functions

$$\{u_n(\mathbf{x}, \mathbf{x}', t) \mid n=1,2,3......\} \in \mathbb{U} \tag{3.15}$$

in the control function space with the property

$$\lim_{n \to \infty} J(u_n) = \inf_{u \in \{u_n\} \in \mathbb{U}} J(u) \,. \tag{3.16}$$

The sequence $\{u_n\}$ is bounded and each element has a finite norm, as noted in (3.12). Then, the monotone sequence convergence theorem implies that

$$\lim_{n \to \infty} u_n = \overline{u} \tag{3.17}$$

and that the sequence is weakly convergent. For each control function u_n in the sequence $\{u_n\}$ we can obtain a wave function $\Psi_n(\mathbf{x},t)$ by solving the Schroedinger equation. These wave functions, which are in one-to-one correspondence with the control functions, define a sequence

$$\{\Psi_n(\mathbf{x},t) \mid n=1,2,3......\}, \;\; \notin / \in \; L^2(\mathbf{R}^d) \tag{3.18}$$

for $t \in [0, t_f]$, where the symbol "$\in$" is appropriate only when the system is bounded and the wave function is localized. Correspondingly, a sequence of the localized functions, $\zeta_n(\mathbf{x},t)$, is obtained by applying the localization operator $\mathbf{P}(\mathbf{x})$ to (3.11):

$$\{\zeta_n(\mathbf{x},t) \mid n=1,2,3......\}, \in L^2(\mathbf{R}^d), \tag{3.19}$$

for $t \in [\,0, t_f\,]$. These functions are elements of the Hilbert space $\mathbb{H}$. For every element of the sequence (3.19), the scalar product of Eq. (3.13) defines a bilinear functional in the Hilbert space $\mathbb{H}$. Using the same argument as above, the sequence (3.19) is weakly convergent and defines a limit

$$\lim_{n \to \infty} \zeta_n = \zeta(\overline{u}) = \overline{\zeta}. \tag{3.20}$$

With the knowledge that the sequences defined in (3.15) and (3.19) are at least weakly convergent, we examine the limiting behavior of the functional defined in (3.13). We shall show that the functional defined in (3.13) can be minimized, and that the input control function $u(\mathbf{x},\mathbf{x}',t)$ is the minimizing function. This result follows from

$$\begin{aligned} J(\overline{u}) &= J(\lim_{n \to \infty} u_n) \\ &= \langle \lim_{n \to \infty} \zeta_n(\mathbf{x},t) - \hat{\zeta}(\mathbf{x}) \mid \lim_{n \to \infty} \zeta_n(\mathbf{x},t) - \hat{\zeta}(\mathbf{x}) \rangle = \langle \overline{\zeta} - \hat{\zeta} \mid \overline{\zeta} - \hat{\zeta} \rangle \\ &= \langle \overline{\zeta} \mid \overline{\zeta} \rangle - \langle \overline{\zeta} \mid \hat{\zeta} \rangle - \langle \hat{\zeta} \mid \overline{\zeta} \rangle + \langle \hat{\zeta} \mid \hat{\zeta} \rangle \end{aligned} \tag{3.21}$$

and

$$\begin{aligned} \lim_{n \to \infty} J(u_n) &= \inf_{u \in \{u_n\}} J(u) \\ &= \lim_{n \to \infty} \inf \langle \zeta_n \mid \zeta_n \rangle - \lim_{n \to \infty} \left(\langle \zeta_n \mid \hat{\zeta} \rangle + \langle \hat{\zeta} \mid \zeta_n \rangle \right) + \langle \hat{\zeta} \mid \hat{\zeta} \rangle. \end{aligned} \tag{3.22}$$

A comparison of Eqs. (3.13) and (3.14) shows that

$$J(\overline{u}) \le \inf_{u \in \{u_n\}} J(u) \tag{3.23}$$

so $J(\overline{u})$ takes the value of the lower limit of the functional (3.13) and $\overline{u}$ is the minimizing function, which demonstrates that there exists a solution $\overline{u}$, and a corresponding solution for the control problem defined by Eq. (3.6), which will minimize the functional in equation (3.13). If the value of the minimum attained is zero the system is fully controllable; if not, the optimal control solution defines the maximum attainable control of the evolution of the system for the given set of control functions.

D. TWO EXAMPLES

We now consider, as a special case, a quantum system with both discrete and continuum components of the spectrum, and we seek control of the evolution of the system in the subset of discrete states. This situation is not covered by the work of Peirce, Dahleh and Rabitz.

The eigenstates of the bare system with Hamiltonian $\mathbf{H}_0$ define a complete state space which we formally represent in the form

$$\hat{P} + \hat{Q} = 1 , \tag{3.24}$$

where

$$\hat{P} = \sum_m | m >< m |, \quad < m | m' > = \delta_{mm'} \tag{3.25a}$$

$$\mathbf{H}_0 | m > = E_m | m >, \qquad m = 1, 2, 3.... \tag{3.25b}$$

and

$$\hat{Q} = \sum_n \int_{\varepsilon_n}^{\infty} d\varepsilon \, | n, \varepsilon >< n, \varepsilon |, \qquad < n, \varepsilon | n', \varepsilon' > = \delta_{nn'} \, \delta(\varepsilon - \varepsilon') \tag{3.26a}$$

$$\mathbf{H}_0 | n, \varepsilon > = (\varepsilon_n + \varepsilon) | n, \varepsilon >, \quad n = 1, 2, 3... \tag{3.26b}$$

with

$$< m | n, \varepsilon > = 0, \text{ for all m, n and } \varepsilon. \tag{3.27}$$

$\hat{P}$ and $\hat{Q}$ are projection operators that separate the spectrum into the set of discrete states and the set of bound-continuum states, respectively. The projection operators satisfy the standard relations

$$\hat{P}^2 = \hat{P}, \; \hat{Q}^2 = \hat{Q}, \; \hat{P}\hat{Q} = \hat{Q}\hat{P} = 0. \tag{3.28}$$

In (3.26a) each integral runs over one segment of the continuum from its threshold ε_n as defined in (3.26b) to infinity.

A system described by Eq. (3.6) can be decoupled by applying the projectors

$\hat{P}$ and $\hat{Q}$ to the wave function and then invoking the completeness relation[31,32]. This procedure yields

$$\begin{aligned} \Psi(\mathbf{x},t) &= \hat{P}\Psi(\mathbf{x},t) + \hat{Q}\Psi(\mathbf{x},t) \\ &= \Psi_B(\mathbf{x},t) + \Psi_C(\mathbf{x},t), \end{aligned} \tag{3.29}$$

where $\Psi_B(\mathbf{x},t) = \hat{P}\Psi(\mathbf{x},t)$ is the bounded part of the wave function and $\Psi_C(\mathbf{x},t) = \hat{Q}\Psi(\mathbf{x},t)$ is the bound-continuous part of the wave function. In general, a wave function can be expanded in the form

$$| \Psi(t) > = \sum_m a_m(t) \,| m > + \sum_n \int_{\varepsilon_n}^{\infty} d\varepsilon \; b_n(\varepsilon, t) \,| n, \varepsilon > . \tag{3.30}$$

Applying both of the projectors in Equation (3.24) to the system described by (3.6) yields the following coupled equations for the wave functions:

$$\begin{aligned} i\frac{\partial}{\partial t}\Psi_B &= \mathbf{H}_{BB}\,\Psi_B + \mathbf{H}_{BC}\,\Psi_C \\ i\frac{\partial}{\partial t}\Psi_C &= \mathbf{H}_{CB}\,\Psi_B + \mathbf{H}_{CC}\,\Psi_C . \end{aligned} \tag{3.31}$$

We take the initial condition to be $\Psi(\mathbf{x},0) = \Psi_B(\mathbf{x},0) + \Psi_C(\mathbf{x},0)$, where the matrix elements of the Hamiltonian are

$$\begin{aligned} \mathbf{H}_{BB} &= \hat{P}\mathbf{H}\hat{P}, \quad \mathbf{H}_{BC} = \hat{P}\mathbf{H}\hat{Q} \\ \mathbf{H}_{CB} &= \hat{Q}\mathbf{H}\hat{P}, \quad \mathbf{H}_{CC} = \hat{Q}\mathbf{H}\hat{Q} . \end{aligned} \tag{3.32}$$

The bounded part and the bound-continuous part of the wave function can be formally separated by solving Eq. (3.31). From the second line of Eq. (3.31) one obtains

$$\begin{aligned} \Psi_C(t) = -i \int_0^t ds \, \exp\left\{-i \int_s^t \mathbf{H}_{CC}(t')\, dt'\right\} \mathbf{H}_{CB}(s)\Psi_B(s) \\ + \exp\left\{-i \int_0^t \mathbf{H}_{CC}(t')\, dt'\right\} \Psi_C(0) . \end{aligned} \tag{3.33}$$

Substituting this expression into the first line of Eq. (3.31) yields the time evolution of the localized part of the wave function in the form

$$i\frac{\partial}{\partial t}\Psi_B = \mathbf{H}_{BB}\Psi_B - i\,\mathbf{H}_{BC}\int_0^t ds\,\exp\left\{-i\int_s^t \mathbf{H}_{CC}(t')\,dt'\right\}\mathbf{H}_{CB}(s)\Psi_B(s) \tag{3.34}$$

where we have assumed that the initial condition is $\Psi_C(0) = 0$. In this case the localized part of the wave function is formally separated from the bound-continuous part of the wave function, and the time evolution of the localized wave function is formally independent of that of the bound-continuous wave function.

To determine if it is possible to optimally control the evolution of the localized part of the wave function we choose $\mathbf{P}(\mathbf{x}) = \hat{P}$, and the minimization functional to be

$$\begin{aligned} J(u) &= <\Psi(\mathbf{x},t_f) - \hat{\Psi}(\mathbf{x}) \mid \hat{P} \mid \Psi(\mathbf{x},t_f) - \hat{\Psi}(\mathbf{x})>(u) \\ &= <\Psi_B(\mathbf{x},t_f) - \hat{\Psi}_B(\mathbf{x}) \mid \Psi_B(\mathbf{x},t_f) - \hat{\Psi}_B(\mathbf{x})>(u). \end{aligned} \tag{3.35}$$

Note that we have used the identity $\hat{P}^2 = \hat{P}$ and $\hat{\Psi}_B(\mathbf{x}) = \hat{P}\hat{\Psi}(\mathbf{x})$. As before, the arguments that lead to (3.16) also show that (3.35) is minimizable. Therefore, in any quantum system which supports both discrete and continuum states, the evolution in the subspace of the discrete states can be optimally controlled.

A more interesting case arises when we consider the control of evolution of a system described by resonances embedded in the continuum. It is conventional to represent the energy of a resonant state as a complex number, the imaginary part of which defines the lifetime of the resonance. It is now established that the properties of a wide class of resonances can be calculated by the complex scaling (rotation-dilation) method[33,34,35]. This method is based on the analytic continuation of the unitary group of scaling defined by

$$\mathbf{U}(\eta)\Psi(\mathbf{x}) = \eta^{3(N-1)/2}\Psi(\eta\mathbf{x}), \tag{3.36}$$

where $\eta = |\eta|\exp(i\theta)$ is a complex number. The corresponding transformed Hamiltonian is

$$\mathbf{H}(\eta) = \mathbf{U}(\eta)\mathbf{H}\mathbf{U}(\eta)^{-1}. \tag{3.37}$$

If the potential energy operator allows for an analytic continuation into the complex plane, then under a complex scaling of coordinates the branch cuts extending from each threshold of the continuum rotate about their respective thresholds by an angle -2θ, and the rotated continua will, in general, expose resonances whose complex eigenvalues are stable under variations of θ, while the spectrum of bound states is unaffected.

In somewhat more detail, the transformation based on the complex scaling parameter defined above scales the radial coordinate r as

$$r \rightarrow \eta\, r = |\eta|\, r\, e^{i\theta} \tag{3.38}$$

and also defines the dilatation analytic Hamiltonian family of operators displayed in (3.37). The transformation operator $\mathbf{U}(\eta)$ satisfies

$$\mathbf{U}^{+}(\eta^{*}) = \mathbf{U}^{-1}(\eta). \tag{3.39}$$

Since the dilatation transformation of the Hamiltonian makes the complex scaled wave function $\Psi(\eta)$ an L^2 function, we are led to redefine the scalar product as

$$< \Psi(\eta^{*}) \mid \Psi(\eta) > \tag{3.40}$$

The projection operator for state k corresponding to this new definition of the scalar product is, in the coordinate representation,

$$\hat{\rho}_{kk} = \mid \Psi_k(\eta) >< \Psi_k(\eta^{*}) \mid \tag{3.41}$$

which satisfies $\hat{\rho}^2 = \hat{\rho}$ even though it is not self-adjoint.

In the representation which uses the complex scaling coordinates, the bare system Hamiltonian $\mathbf{H}_0(\eta)$ defines the eigenvalue problem

$$\mathbf{H}_0(\eta)\, \Psi(\eta) = E\, \Psi(\eta). \tag{3.42}$$

The wave function which satisfies Eq. (3.42) belongs to L^2 and can be expended in an L^2 basis set. Diagonalizing the Hamiltonian matrix

$$< \Psi(\eta^{*}) \mid \mathbf{H}_0(\eta) \mid \Psi(\eta) > \tag{3.43}$$

leads to the eigenvalues and the eigenvectors which correspond to the bound and quasi-bound states of the bare system. These eigenvectors are used to construct the projection operators for the all the bound and quasi-bound states. The projection operator for the bound states can be written

$$\hat{P} = \sum_{m} \mid m >< m \mid , \tag{3.44}$$

and the projection operator for the quasi-bound states embedded in the continuum is

$$\hat{Q} = \sum_j \int_{c_j} dz_j \mid z_j >< z_j \mid \tag{3.45}$$

with

$$\mathbf{H}_0(\eta) \mid z_j > = z_j \mid z_j > , \tag{3.46}$$

where | m > is the eigenvector of a bound state with a real eigenvalue E_m and z_j is a complex eigenvalue of $\mathbf{H}_0(\eta)$ with eigenvector $| z_j >$. The states $| z_j >$ represent the continuum part of the spectrum of the system. The contour c_j is constructed by distorting the path of integration for the jth segment of the continuous spectrum (see (3.4a)), and the summation in (3.45) is over all the contributions to the continuous spectrum.

For present purposes we need to apply the complex scaling method to the analysis of the evolution of a system. Under the transformation $\mathbf{U}(\eta)\Psi = \Psi(\eta)$ the time dependent Schroedinger equation takes the form

$$i \frac{\partial}{\partial t} \Psi(\eta, t) = \mathbf{H}(\eta) \Psi(\eta, t) \tag{3.47}$$

with the initial condition

$$\Psi(\eta, t_0) = \Psi_0(\eta) . \tag{3.48}$$

For our controlled system the complex scaled Hamiltonian operator is, of course, $\mathbf{H}(\eta) = \mathbf{H}_0(\eta) + \mathbf{H}'(\eta)$. The wave function which satisfies (3.47) is localized, square integrable, and has the norm defined by (3.40). Clearly, we can now take the localization operator required for the optimization process to be identical to the projection operators defined in Eqs. (3.44) and (3.45), i. e.

$$\mathbf{P}(\mathbf{x}) = \hat{P} + \hat{Q} . \tag{3.49}$$

To demonstrate the feasibility of optimal control of the evolution of the system (3.47), we define the functional

$$J(u) = < \Psi(\eta, t_f) - \widehat{\Psi}(\eta) \mid \mathbf{P}(\mathbf{x}) \mid \Psi(\eta, t_f) - \widehat{\Psi}(\eta) > (u) \tag{3.50}$$

where the target function is scaled as

$$\widehat{\Psi}(\eta) = \mathbf{U}(\eta) \widehat{\Psi} . \tag{3.51}$$

The argument which leads to Eq. (3.16) again shows that (3.51) is minimizable, which implies that the system as defined can be optimally controlled.

IV. Controlled Photodissociation of a Diatomic Molecule

We now examine, as an example, the use of external fields to control the photodissociation of a diatomic molecule. Since the details of the general theory are given elsewhere[3,8], we shall only outline the optimal control theory approach to optical pulse shaping. For the simple reaction studied in this Section we need consider only two electronic potential energy surfaces; one of these surfaces is used to mediate a chemical reaction on the other surface. Adopting a semiclassical representation of the electromagnetic field and its interaction with the molecule, and the Born-Oppenheimer approximation for the electronic states of the molecule, the equation of motion for this system can be represented in the form

$$\begin{pmatrix} \mathbf{H}_u & \mathbf{V}_{gu} \\ \mathbf{V}_{ug} & \mathbf{H}_g \end{pmatrix}\begin{pmatrix} \psi_u \\ \psi_g \end{pmatrix} = \begin{pmatrix} \mathbf{H}_u & \mu.\boldsymbol{E}(t) \\ \mu.\boldsymbol{E}(t) & \mathbf{H}_g \end{pmatrix}\begin{pmatrix} \psi_u \\ \psi_g \end{pmatrix} \tag{4.1}$$

with $\boldsymbol{\mu}$ the negative of the transition dipole operator and $\boldsymbol{E}(t)$ the amplitude of the classical electric field. The first step in the optimal control theory approach is to specify what is wanted, namely, the sought for final state of the system. As we are interested in the ground state wavefunction at some final time t_f, we define the objective functional as a projection of the final state of the system,

$$I = \langle \Psi_f | \hat{P} | \Psi_f \rangle, \tag{4.2}$$

where $\hat{P}$ is a projection operator for the selected product and Ψ_f is the (two component) wavefunction at time t_f.

The optimization procedure is subject to several important physical constraints; the particular constraints appropriate to an experiment are determined, in part, by the technical details of that experiment. We choose as one constraint that the energy of the electric field is to be conserved in the optimization process:

$$\int_{t_o}^{t_f} \boldsymbol{E}^2(t)dt = E. \tag{4.3}$$

A second necessary constraint is that the system always satisfy the Schroedinger equation. These two constraints introduce two Lagrange multipliers λ and Φ, respectively. In analogy with Ψ, Φ is a two component function defined by

$$\Phi = \begin{pmatrix} \phi_u \\ \phi_g \end{pmatrix}, \tag{4.4}$$

so that the objective functional may be written as

$$\bar{I} = \langle \Psi_f | \widehat{P} | \Psi_f \rangle + i \int_{t_o}^{t_f} \left(\langle \Phi | i\hbar\partial_t - \mathbf{H} | \Psi \rangle - \text{c.c.} \right) dt + \lambda \left[\int_{t_o}^{t_f} \boldsymbol{E}^2(t) dt - E \right] \tag{4.5}$$

where c.c. stands for complex conjugate. It can be shown that (4.5) leads to the following coupled sets of equations to be solved:

$$i\hbar\partial_t \Psi(x, t) = \mathbf{H}\Psi(x, t), \tag{4.6}$$

$$i\hbar\partial_t \Phi = \mathbf{H}\Phi, \tag{4.7}$$

$$|\Psi(t_o)\rangle = \begin{pmatrix} 0 \\ \psi_g(t_o) \end{pmatrix}, \tag{4.8}$$

$$|\Phi(t_f)\rangle = \widehat{P} \, | \Psi(t_f)\rangle, \tag{4.9}$$

which are further linked by the equation for the optimal field $\boldsymbol{E}$,

$$\boldsymbol{E}(t) = O(t)/\lambda, \tag{4.10}$$

where

$$O(t) = \mathrm{Im}\left(\langle \psi_u | \mu | \phi_g \rangle + \langle \psi_g | \mu | \phi_u \rangle \right). \tag{4.11}$$

The Lagrange multiplier λ is determined by substituting (4.10) into (4.3), giving

$$\lambda = \pm \left(\frac{1}{E} \int_{t_o}^{t_f} O^2(t) dt \right)^{1/2}. \tag{4.12}$$

In the calculations reported by Amstrup et al, the potential energy surface for each state is a one-dimensional Morse function taken from the literature (see e.g. references 36 and 37 for a review of efficient and accurate methods to calculate Morse wavefunctions). We assume that at t = 0 the wavefunction is an eigenfunction of a specific vibrational state.

The "target" wavefunctions are chosen to be complex Gaussian wavepackets of the form

$$\chi(R) = \left(\frac{2}{\pi\delta_R}\right)^{1/4} \exp\left(-(R - R_{cent})^2/\delta_R + ip_0(R - R_{cent})\right). \tag{4.13}$$

The classical coordinate R_{cent} and momentum p_0 give the central location of $\chi(R)$ in phase space, while δ_R gives the spatial spread of the wavepacket and R_{cent} is chosen such that it is close to the end of the grid while not giving any "waste" of wavefunction passing the grid at late times in the forward time propagation. The parameter δ_R is chosen such that the width of the wavepacket is of the same order as suggested by preliminary calculations with Gaussian laser pulses. The final time is also chosen using considerations derived from preliminary calculations. The parameter p_0 is chosen such that the kinetic energy of the wavepacket is of the same order as the dissociation energy.

The initial guess for the electric field is the sum of two Gaussian pulses, each of the form

$$\boldsymbol{E}_j(t) = A_j \exp\left(-\alpha_j(t - t_j)^2\right) \cos\left(\omega_j t + \varphi_j\right). \tag{4.14}$$

The values of these parameters were selected by analysis of separate Wigner swarm calculations. The pulse durations were chosen to correspond to those readily achievable experimentally.

Amstrup et al take μ to be independent of the internuclear distance.

We now examine the dissociation of the van der Waals molecule HgAr in its excited electronic A state, using the higher excited C state as the mediating state, as illustrated in Fig. 1. For this system we describe some of a results obtained when the full optimization scheme is used, and also the results when the restricted electric field, including only two Gaussian pulses, is used.

The initial guess for the electric field is shown in Fig. 2(A), while the field after one iteration is shown in Fig. 2(B), and the (partially optimized) field after 14 iterations in Fig. 2(C). Fig. 3 shows the initial and target wavefunctions, while Fig. 4 shows the wavefunctions at the final time t_f corresponding to the electric fields in Fig. 2(A) - (C). Power spectra corresponding to the electric fields in Fig. 2(A) - (C) are shown in Fig. 2(a) - (c). The power spectra are the Fourier transforms of the electric fields; note that the units have been changed to wavenumbers.

We also consider a simple approximation to the electric field shown in Fig. 2(B). Three Gaussian pulses are used. The central times and widths of these pulses are taken from Fig. 2(B), while the central frequencies are taken from Fig. 2(b). The resulting wavefunction with this electric field is shown in Fig. 5 along with the results using only the first and either one of the last two of these Gaussian pulses. The total energy of the electric fields in all these calculations is the same. The values of our objective functional, $\bar{I}$, and the dissociation yield are tabulated with a few other intermediate calculations in Table 2. The dissociation yield is defined as the amount of the final wavefunction that has

Table 1. Parameters for HgAr.

	HgAr
Time parameters:	
Initial time (t_0)	0
Final time (t_f)	36000 au (871.2 fs)
Number of time steps	80001
Grid parameters[a]:	
R_{min}	3.5 a_0 (1.9Å)
R_{max}	20.0 a_0 (10.6Å)
Number of points	512
Target parameters:	
R_{cent}	13.0 a_0 (6.88Å)
δ_R	0.5 au (0.14Å^2)
p_0	26.36 au ($5.18 \cdot 10^{-18}$g cm/s)
Total Energy [b] :	0.001 au
Total Energy [c] :	1.9 mJ/cm^2
Electronic trans. moment	3.34 D

[a] $1\ a_0 = 4\pi\epsilon_0\,\hbar^2/(\mathrm{m}\,e^2) = 0.529 \cdot 10^{-10}$m is the Bohr radius
[b] It is actually $\int_{t_0}^{t_f}(\mu\mathcal{E}(t))^2 dt$, that is given here
[c] Given the value above, and the electronic transition moments below

Table 2. Objective function values ($\langle\Psi_f|\widehat{P}|\Psi_f\rangle$) and dissociation yield ($\int_{10}^{20}|\Psi_f(R)|^2 dR$) for HgAr dissociation. Simul1 is from the calculation with 3 Gaussian pulses; Simul2 with the first two, and Simul3 with the first and last Gaussian pulse from the simulated field.

Iteration number	$\langle\Psi_f\vert\widehat{P}\vert\Psi_f\rangle$	Dissociation yield	$\langle\Psi_f\vert\Psi_f\rangle$
0	$3.27 \cdot 10^{-6}$	$8.96 \cdot 10^{-5}$	1.000
1	$6.52 \cdot 10^{-2}$	$8.18 \cdot 10^{-2}$	1.000
5	$9.39 \cdot 10^{-2}$	$1.63 \cdot 10^{-1}$	1.000
9	$9.90 \cdot 10^{-2}$	$1.19 \cdot 10^{-1}$	1.000
14	$1.34 \cdot 10^{-1}$	$1.68 \cdot 10^{-1}$	1.000
Simul1	$4.06 \cdot 10^{-2}$	$7.02 \cdot 10^{-2}$	1.000
Simul2	$4.54 \cdot 10^{-2}$	$6.04 \cdot 10^{-2}$	1.000
Simul3	$4.71 \cdot 10^{-3}$	$4.15 \cdot 10^{-2}$	1.000
Gaussian[a]	$8.63 \cdot 10^{-2}$	$1.51 \cdot 10^{-1}$	1.000

[a] From the 47th iteration with the electric field restricted to 2 Gaussian pulses

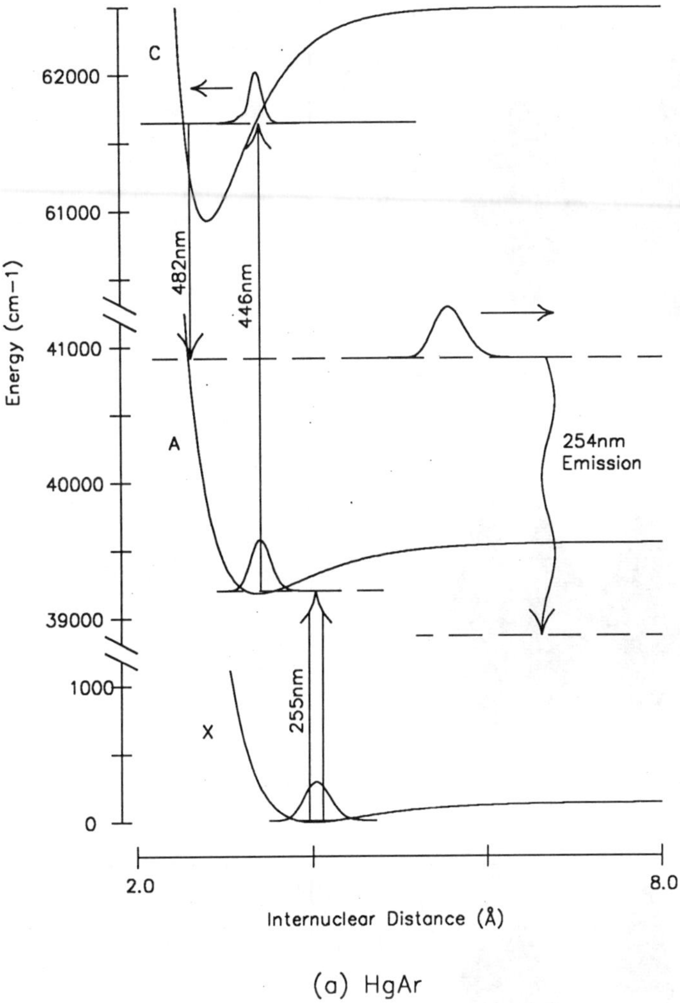

Figure 1. Potential energy surfaces for HgAr. The double arrows correspond to resonant continuous-wave excitation.

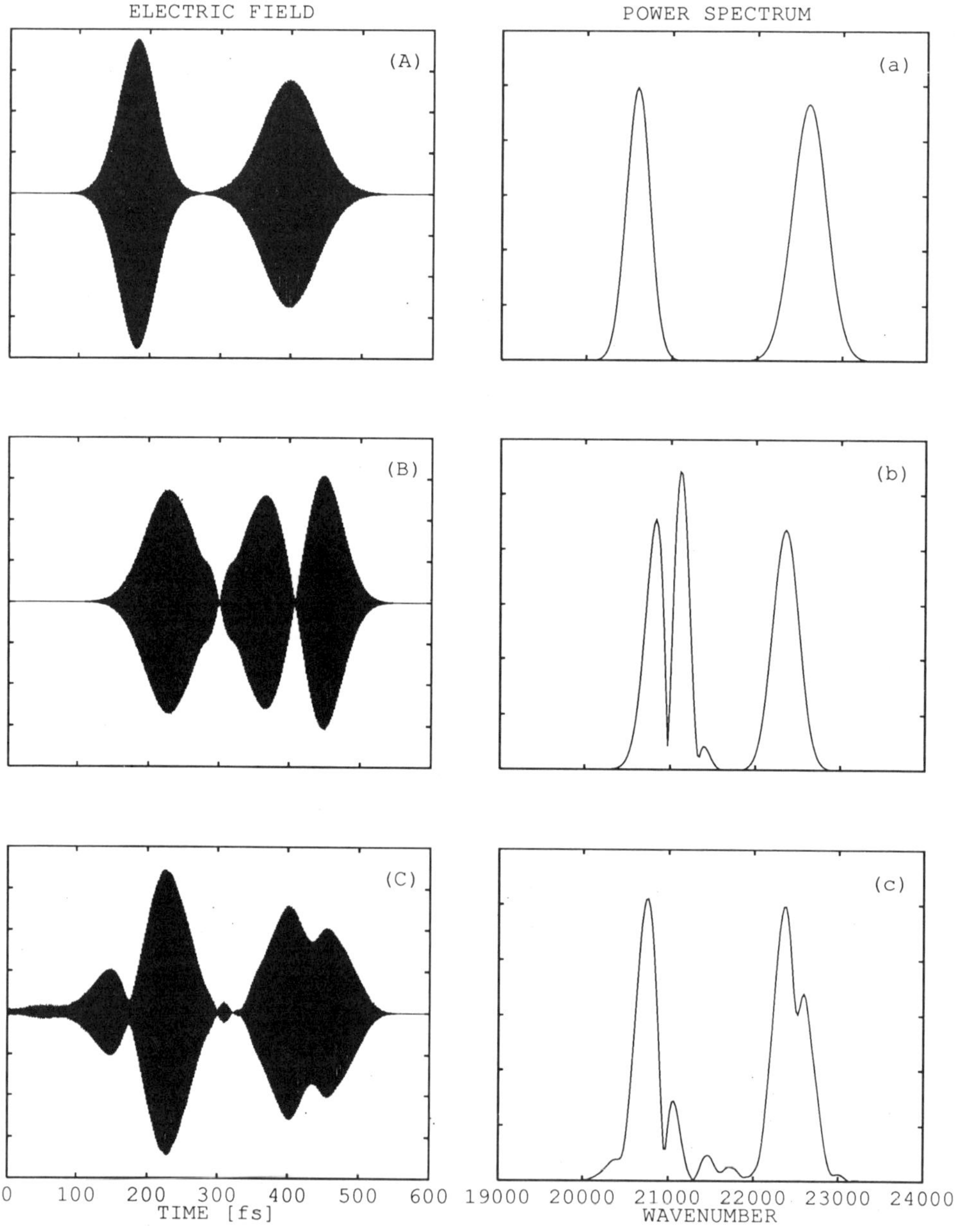

Figure 2. Electric fields for HgAr dissociation: (A) initial guess and (B) after one iteration and (C) after 14 iterations. (a) - (c) show the corresponding power spectra (absolute values).

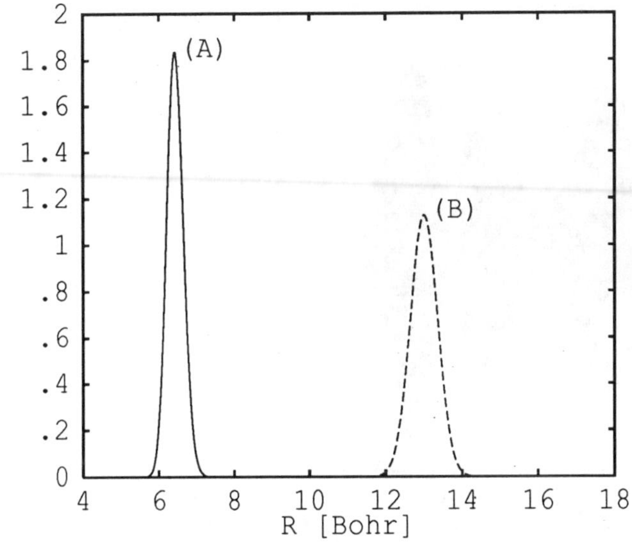

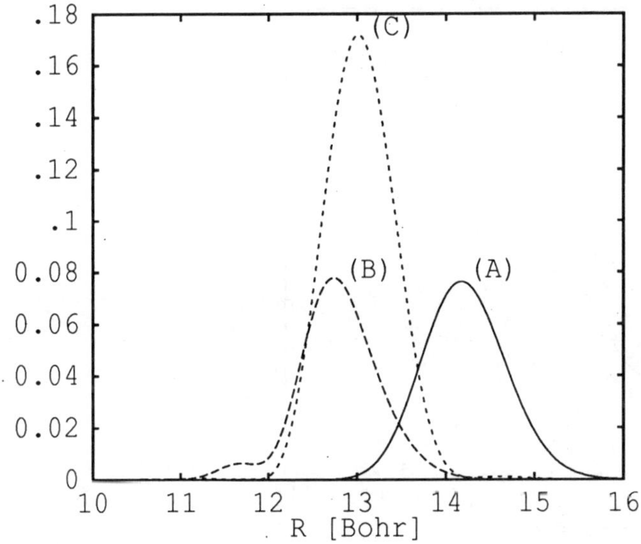

Figure 3. (A) Initial wave function (v = 0), and (B) target wave function. Both are numerically squared.

Figure 4. (A) Wave function (numerically squared and x1000) at final time t_f given the initial electric field. (B) Wave function (num. squar.) at t_f given the electric field after first iteration. (C) Wave function (num. squar.) at t_f given the electric field after 14 iterations.

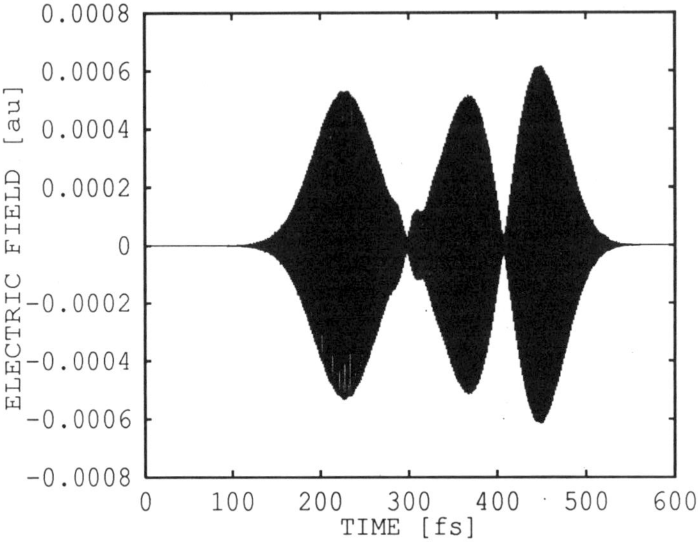

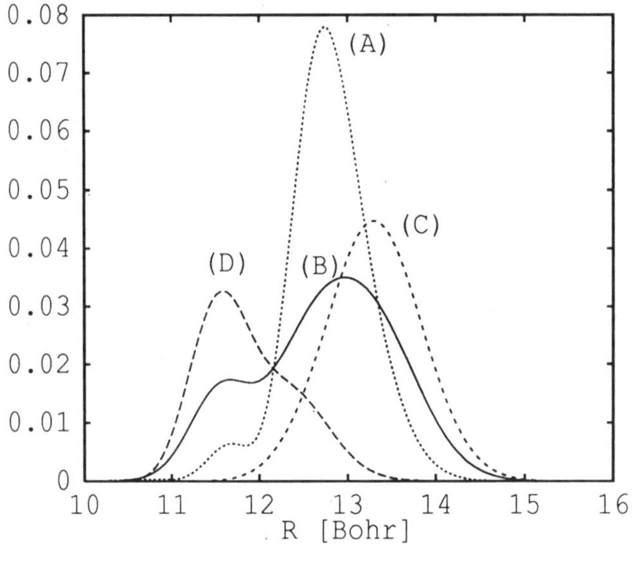

Figure 5. Simulation of the electric field from Fig. 2(A) by use of 3 Gaussian pulses centered as in Fig. 2(A) and with frequencies from Fig. 2(B)

Figure 6. Wave functions (numerically squared) at final time, given the electric fields: (A) after one iteration (same wave function as in Fig. 4(B); (B) the simulated field with 3 Gaussian pulses (shown in Fig. 5); (C) using only the first two Gaussian pulses from the simulated field; and (D) using only the first and last Gaussian pulse from the simulated electric field.

passed $R = 10\ a_0$:

$$Y = \int_{10}^{\infty} |\Psi_f(R)|^2 dR. \tag{4.15}$$

Included in the table are values of $\langle\Psi_f|\Psi_f\rangle$ to show conservation of the norm of the wavefunction. From this table and the figures we see that there is a very large enhancement of the objective functional as well as of the dissociation yield with the first iteration, while further iterations only give modest enhancements. Some iterations even decrease the amount of dissociation, though the value of the objective functional increases. We also see that the approximation to the partially optimized pulse which uses three Gaussian pulses, as well as the simpler approximation which includes only the first and last of these pulses, gives results that are comparable to those obtained with the partially optimized pulse.

We now consider the results of calculations where the electric field is generated by a sum of two Gaussian fields, as in (4.14). There are 10 parameters to optimize: the amplitudes (A_j), the frequencies (ω_j), the phases (φ_j), the widths (α_j), and the central times (t_j) of each pulse. The result after 47 iterations is given in the final line of Table 2. By comparison with the other results in this table, we see that further optimization of the parameters in the two/three pulse Gaussian sequence introduced will enhance the value of the objective functional, indeed almost as much as for later iterations in the full optimization scheme.

The calculations presented in the previous paragraphs have ignored the influence of molecular rotation on the induced dissociation yield. This approximation is intuitively plausible for the time scale over which the pump and dump pulses, or the optimally shaped pulses, act. That time scale is, for HgAr, about 500 fs, whereas the characteristic time associated with a rotational transition, defined by the rotational constant of the molecule, is of the order of 50,000 fs. Nevertheless, it is worthwhile examining the influence of molecular rotation on the control of selectivity of product formation. Amstrup et al report the results of such a calculation, using second order perturbation theory. By virtue of the rotational selection rule $\Delta J = \pm 1$, transitions from a particular rovibrational component of the wavepacket on the A surface of HgAr will be connected to the two corresponding (J + 1, J - 1) rovibrational components of the wavepacket on the C surface, thereby creating a coherent superposition of these levels. The full wavepacket on the C surface is constructed by superposing the pairs of rovibrational levels generated by the transitions from all the rovibrational components of the wavepacket on the A surface. Let the initial state be the appropriate thermal distribution of rotational levels. The dissociation yield generated by a pulse sequence can be calculated individually for each of the rotational levels in the initial state. Since the initial rotational sublevels are not coherently related to each other no interference between them can occur; the total dissociation yield is an incoherent sum of the yields for each initial J. Amstrup et al find that there is a small but progressive degradation of the photofragment yield as J increases, as is expected since the energy differences between the levels generated by $J \rightarrow J + 1$ and $J \rightarrow J - 1$ increase as J increases; the consequence of this energy increase is a beat frequency in the wavepacket evolution that increases as J increases, thereby also contributing to an increase in the unphasing rate. Nevertheless, for the range of J appropriate to HgAr the degradation of selectivity generated by the rotational motion is negligibly small for all practical purposes.

V. Discussion

That the KS-entropy of an unbounded quantum system is zero is not intuitively obvious. An examination of the arguments used shows that the definition of the measurement process underlies all else. The crucial assumption in the Zhao-Rice analysis is that the duration of a measurement must be finite, which implies that the uncertainty in the expectation value of the energy of the system cannot be reduced to zero. The subdomain of Hilbert space projected out by each observation made on the system then has nonzero measure. Since the measure of the nth subdomain does not approach zero as $t_n \rightarrow \infty$, it follows that the KS-entropy of the system must be zero.

We can make this result more plausible by exploiting some qualitative arguments related to an approximation often used in the analysis of unbounded quantum systems. As a concrete example, consider a polyatomic molecule confined in a large box. At low energy the states of the system correspond to the translational, rotational and vibrational motions of the intact molecule. Because the system is confined to a box the spectrum of states, including translational states, is discrete. At some higher energy the states of the system correspond to, say, the translational, rotational and vibrational motions of molecular fragments; as at low energy, the spectrum is discrete by virtue of the confinement of the system imposed by the box. One of the approximations often used to describe the overall behavior of an unbounded quantum system is to first consider the states of that system confined to a box, and then to examine how those states are changed as the size of the box becomes indefinitely large. Of course, the spectrum of states remains discrete as long as the box is of finite size, but the density of states increases very rapidly as the box size increases. As long as the size of the box is finite, the KS-entropy of the bounded quantum system is zero. It is intuitively plausible that by choosing an appropriately sized box the properties of the "box-quantized" system can be made to resemble closely the properties of the corresponding unbounded quantum system. Assuming that is so, it is reasonable to infer that the KS-entropy of an unbounded quantum is zero.

The qualitative arguments just cited are inadequate in the limit of infinite volume of the confining box, i.e. when the quantum system is unbounded. These arguments also fail to give any information concerning the existence of a correspondence limit for the dynamical behavior. The same difficulty exists with respect to the approach to the correspondence limit for the KS-entropy of a system with a discrete spectrum, e.g. the behavior of a point particle in a stadium with fixed size in the limit as the mass of the particle becomes infinite; as long as the spectrum is discrete the quantum Kolmogorov entropy is zero even though the classical KS-entropy is nonzero. These unresolved difficulties are associated with the nonanalytic nature of the approach to the correspondence limit.

Given that the KS entropy of an unbounded quantum system is zero we expect that, In principle, the evolution of such a system should be controllable. Zhao and Rice have shown that there exist external fields with which it is possible to optimally control the time evolution of an unbounded quantum system. The sense in which we define optimal control is that for a given class of external fields it is possible to minimize the difference between the projection of a continuum wave function, generated at time t_0 by the product of a localizing function and a continuum wave function of the system of interest, and a similarly defined target function at time t_f. It can be shown that for at least one class of choices of the localizing function the optimized projected continuum wave function can be inverted to yield the corresponding continuum wave function of the system of interest.

It is important to note that the Zhao-Rice analysis refers to the case that that are no constraints acting on the system, unlike the real world situation. The extent of control of a quantum system will depend on the active constraints and, in general, must be expected to be less than for an unconstrained quantum system.

The results of the calculations of the laser pulse shape that optimizes the photodissociation of HgAr provide some justification for the belief that it is possible to design fields which will optimize the yield of particular products of a chemical reaction. Of course, much remains to be done. In particular, although the general formalism for the calculation of optimal control fields is applicable to systems with any number of degrees of freedom, visualization of the wavepacket dynamics is very difficult when the system has more than two degrees of freedom. One consequence of that difficulty is inhibition of the use of physical intuition to make simple physical approximations. It seems likely to be worthwhile to develop approximate dynamical descriptions of systems with many degrees of freedom (e.g. reaction path representations) which will permit, in turn, development of useful approximations to the optimal control fields that generate particular products.

VI. Acknowledgements

The work reviewed in this paper represents the contributions of many co-workers whom I wish to thank, namely, D. Tannor, R. Kosloff, P. Gaspard. S. Tersigni, M. Zhao, R. J. Carlson and B. Amstrup. The research program has been supported by the National Science Foundation.

References

1. M. Zhao and S. A. Rice, submitted to J. Chem. Phys.

2. M. Zhao and S. A. Rice, submitted to J. Chem. Phys.
3. B. Amstrup, R. J. Carlson, A. Matro and S. A. Rice, J. Phys. Chem. in press
4. P. Brumer and M. Shapiro, Chem. Phys. Lett. **82**, 177 (1986); C. Asaro, P. Brumer and M. Shapiro, Phys. Rev. Lett. **60**, 1634 (1988); M.Shapiro, J.W. Hepburn and P. Brumer, Chem. Phys. Lett. **149**, 451 (1988); T.Seideman, M. Shapiro and P. Brumer, J. Chem. Phys. **90**, 7132 (1989).
5. D. J. Tannor and S. A. Rice, J. Chem. Phys. **83**, 5013 (1985).
6. D. J. Tannor, R. Kosloff and S. A. Rice, J. Chem. Phys. **85**, 5805 (1986); For additional details on the numerical methods employed see D. Kosloff and R. Kosloff, J. Comp. Phys. **52**, 35 (1983).
7. S. Shi and H. Rabitz, Chem. Phys. **139**, 185 (1989): S. Shi, A. Woody and H. Rabitz, J. Chem. Phys. **88**, 6870 (1988); **92**, 2927 (1990).
8. R. Kosloff, S. A. Rice, P. Gaspard, S. Tersigni and D. J. Tannor, Chem. Phys. **139**, 201 (1989).
9. A. M. Weiner, J. P. Heritage and R. N. Thurston, Opt. Lett. **11**, 153(1986); R. N. Thurston, J. P. Heritage, A. M. Weiner and W. J. Tomlinson, EEE J. Quant. Elect., **QE-22**, 682 (1986); A. M. Weiner and J. P. Heritage, Revue Phys. Appl. **22**, 1619 (1987); A. M. Weiner, D. E. Leaird, G. P. Wiederecht and K. A. Nelson, Science **247**, 1317 (1990).
10. F. Spano, M. Haner and W. S. Warren, Chem. Phys. Lett. **135**, 97 (1987); Ultrafast Phenomena **V**, Springer-Verlag, Berlin (1986), eds. G. R. Fleming and A. Siegman, p. 514.
11. See, for example, A. Lichtenberg and M. A. Lieberman, *Regular and Stochastic Motion*, Springer-Verlag, New York (1983).
12. Ya. B. Pesin, Math. Surveys **32**, 55 (1977).
13. A. Kolmogrov, Dokl. Akad. Nauk. SSSR **124**, 774 (1959).
14. V. A. Rokhlin, Math. Surveys **22**, 1 (1967).
15. S. N. Rasband, *Chaotic Dynamics of Nonlinear Systems,* John Wiley & Sons, New York (1990).
16. A. E. Bryson and Y. Ho, *Applied Optimal Control*, Hemisphere, Washington, D. C. (1975).
17. D. G. Luenberger, *Introduction to Dynamic Systems*, Wiley, New York (1979).
18. R. Kosloff and S. A. Rice, J. Chem. Phys. **72**, 1340 (1981); S. A. Rice, and R. Kosloff, J. Phys. Chem. **86**, 2153 (1982).
19. J. Manz, J. Chem. Phys. **91**, 2190 (1989).
20. G. H. Huang, T. J. Tarn and J. W. Clark, J. Math. Phys. **24**, 2608 (1983).
21. S. M. Park, S-P. Lu and R. J. Gordon, submitted to J. Chem Phys.
22. N. F. Scherer, A. Ruggiero, M. Du and G. R. Fleming, J. Chem. Phys. **93**, 856 (1990); N. F. Scherer, R. J. Carlson, A. Matro, M. Du. A. J. Ruggiero, V. Romero-Rochin, J. A. Cina. G. R. Fleming and S. A. Rice, submitted to J. Chem. Phys.
23. See, for example, M. C. Gutzwiller, *Chaos in Classical and Quantum Mechanics*, Springer-Verlag, New York (1990).
24. H. G. Schuster, *Deterministic Chaos*, Physik-Verlag, Weinheim, FRG (1984).
25. A. Messiah, *Quantum Mechanics*, Vol. I, North-Holland Publishing Co., Amsterdam (1961).

26. E. Merzbacher, *Quantum Mechanics*, 2nd Ed., John Wiley & Sons, New York (1970).
27. E. B. Davies, *Quantum Theory of Open Systems*, Academic Press, New York (1976).
28. T. F. Jordan, *Linear Operators for Quantum Mechanics*, John Wiley & Sons, New York (1969).
29. J. von Neumann, *Mathematical Foundations of Quantum Mechanics*, Princeton University Press (1955).
30. A. P. Peirce, M. A. Dahleh and H. Rabitz, Phys. Rev. A **37**, 4950 (1988).
31. B. H. Bransden, " *Atomic Collision Theory* ", The Benjamin/Cummings Publ. Co, Inc. (1983).
32. S. Nordholm and S. A. Rice, J Chem. Phys. **62**, 157 (1975).
33. W. P. Reinhart, in " *Mathematical Frontiers in Computational Chemical Physics* ", D. G. Truhlar, ed., Springer-Verlag, New York (1988).
34. S. Chu, in " *Resonances in Electron-Molecule Scattering, Van der Waals Complexes, and Reactive Chemical Dynamics* ", D. G. Truhlar, ed., American Chemical Society, Washington, D. C. (1984).
35. C. A. Chatyidinmitriou-Dreismann, Adv. Chem. Phys. in press.
36. J. P. Dahl and M. Springborg, J. Chem. Phys, **88**, 4535 (1988).
37. H. Kobeissi, J. Comp. Phys. **61**, 351 (1985).

THE PROSPECTS FOR IMPULSIVELY DRIVEN, MODE-SELECTIVE CHEMISTRY IN CONDENSED PHASES

Keith A. Nelson
Department of Chemistry
Massachusetts Institute of Technology
Cambridge, MA 02139, USA

ABSTRACT. A discussion is presented on the possibility of inducing mode-selective chemistry in ordered crystalline solids excitation and observing its occurrence in real time through impulsive stimulated Raman scattering. Localized reactions of individual molecules or molecule pairs and collective chemical reactions associated which structural phase transitions are discussed.

1. Introduction

Most if not all examples of "mode-selective chemistry" have involved small or medium-sized molecules, or clusters, in the gas phase. The reports in this volume provide an indication of the rapidly improving ability to demonstrate mode-selective chemistry in these systems. Here we suggest that in fact there may be a rich variety of mode-selective chemistry to be carried out in condensed phases. The focus of this discussion is on crystalline solids, but examples in other phases such as liquid crystals or ordered films might also be found. The use of ultrashort pulses to induce and monitor chemical change is also emphasized, but other possibilities are mentioned.

Mode-selective chemistry in solids is not expected to provide a general tool for chemical synthesis or material fabrication, although some specialized possibilities exist. As in the gas phase, the main motivation for pursuing this goal is improved understanding of the roles of energy localization and flow, relaxation, and interactions between reactive and other modes in mediating reaction dynamics, branching ratios, and yields.

In this article a rather free-ranging discussion of some of the possibilities for achieving mode-selective chemistry in crystalline solids is presented. Two possible paths are emphasized. The first involves extension of impulsive stimulated Raman scattering (ISRS) techniques such that femtosecond light pulses could be used to drive large-amplitude displacements of selected lattice vibrational modes whose motions can lead to chemical reaction. The second involves selection of samples in which collective chemical change in the form of structural phase transitions facilitates manipulation of the free energy surface governing reaction.

2. Impulsive stimulated Raman scattering

It has been shown that a sufficiently short light pulse can "impulsively" excite Raman-active vibrational modes of crystals or molecules [1-3]. The only requirement is that the pulse duration be short compared to the vibrational period so that a sudden ("impulse") force is exerted. Following excitation, the coherent vibrational oscillations and decay of the excited modes can

J. Jortner et al. (eds.), Mode Selective Chemistry, 527–533.

be monitored in real time with variably delayed probe pulses whose duration is also short compared to the vibrational period. Impulsive stimulated Raman scattering (ISRS) data showing time-dependent oscillations of molecular vibrations, optic phonons, and other excitations have been presented.

In most ISRS experiments to date, the coherent vibrational amplitude produced is at most a few thousandths of an angstrom. This is smaller than mean thermal displacements or even quantum uncertainties. Yet as early as 1985, calculations showed that coherent vibrational displacements in the 0.1-1 A range could be achieved with modest Raman-scattering differential polarizabilities and readily obtainable femtosecond pulse intensities [4]. This could have important consequences in chemically reactive materials. In some organic molecular crystals, for example, neighboring molecules could be "pushed" toward each other along a Raman-active optic phonon coordinate such that thermal or photochemical bimolecular reaction or polymerization were assisted [5]. Structural phase transitions and other solid-state rearrangements might also be induced. Unfortunately, coherent vibrational amplitudes approaching 0.1 A have not been reached in any controlled experiment to date. Unwanted nonlinear optical processes such as multiphoton ionization lead to optical damage well before such amplitudes are achieved in most or all crystalline samples.

The challenge in this endeavor is therefore to drive large-amplitude phonon displacements through ISRS excitation while circumventing or reducing unwanted processes, especially those that lead to material damage. We discuss several experimental approaches which can be used to circumvent optical damage or minimize its consequences while coherent optic phonon-driven chemistry in molecular crystals is induced and monitored. The combination of these approaches may lead to successful execution of mode-selective chemistry in a variety of crystalline solids.

2.1. MULTIPLE-PULSE ISRS EXCITATION

The first method is designed to avert damage due to the high peak intensity of focussed femtosecond pulses. The idea is simple: instead of driving large amplitudes with one intense ultrashort pulse, one could drive a selected vibrational mode repetively in a manner analogous to repetitively pushing a child on a swing. An intense excitation pulse can be split into a sequence of smaller pulses timed to drive the vibrational mode once each period, gradually building up a large vibrational amplitude. This approach has now been demonstrated through the use of femtosecond "pulse-shaping" techniques developed by researchers at Bell Communications Research [6]. In an a-perylene crystal, a pulse sequence was used to amplify a selected mode by about a factor of ten [7]. Ideally in this method, sample damage could be avoided by adjusting the intensity of each pulse in the sequence to be just below the damage threshold for a single ultrashort pulse. In that case the total optical energy that could be used would be equal to the maximum single-pulse energy multiplied by the number of pulses. This would be the case if there were no cumulative damage mechanisms. In the α-perylene crystal, the total energy in the pulse sequence was greater than that that could be used in a single pulse, but only by about half as much as the optimum. The cumulative damage mechanism is not known, but presumably electronic excited states produced by pulses in the beginning of the sequence absorb additional photons from later pulses and undergo ionization or some other type of breakdown. To add to the complexity, there appears to be some dependence of the damage threshold on the repetition rate of the experiment. The original experiments were done at an 8-KHz repetition rate; more recent work has indicated that the damage threshold is higher when the repetition rate is about 400 Hz. This suggests that some excitation has a lifetime on the millisecond scale. The electronic excited state normally produced in a-perylene has a lifetime of about 100 ns [8]. It is possible that with low probability an electron may be freed from its parent molecule and may become trapped elsewhere in the lattice for long times. This is consistent with the observation of weak photoconductivity in α-perylene crystals [9].

Absorption of light from a subsequent pulse by the trapped electron or the parent ion could then initiate crystal damage.

Experiments thus far indicate that multiple-pulse excitation can be expected to yield about tenfold amplification of selected phonon modes. The extent of amplification is limited ultimately by phonon dephasing and cumulative damage mechanisms.

2.2. RESONANCE ENHANCEMENT

Another approach to increasing the coherent vibrational amplitude produced through ISRS is to tune the excitation wavelength to near an electronic absorption band. In this case the Raman-scattering differential polarizability, to which the coherent phonon amplitude is linearly proportional, may show resonance enhancement due to vibronic coupling involving the electronic excited state. It is not uncommon to see resonance enhancement of two to three orders of magnitude at a strong absorption peak. However, excitation at such a wavelength would produce a high density of electronic excited states whose dynamics would be observed in competition with the ground-state vibrational dynamics initiated by ISRS. In addition, the optical damage threshold of nearly any sample is greatly reduced for resonant excitation. The best strategy is probably to use excitation wavelengths near the red edge of an absorption band. In this manner one can hope for a significant increase -- perhaps tenfold -- before the density of excited states becomes excessive.

ISRS excitation on or near electronic absorption bands has been carried out in dye molecules [10-12] and in molecular iodine [13]. In the latter, evidence for some degree of anharmonicity corresponding to excitation of significant amplitude was noted.

2.3. DAMAGE CONTROL

There are several rather mundane but potentially important experimental approaches that can be used to generate and observe large-amplitude phonon responses even if some sample damage cannot be avoided. One is to continually move the sample so that fresh regions are always irradiated. To the extent that slow damage processes can occur as discussed above, this may avert damage entirely by providing many seconds for any region to recover following irradiation. If some damage cannot be avoided, this approach still allows data to be collected before the response of interest is influenced significantly by deterioration of the sample. Obviously this approach is best suited to crystals which can be grown in large sizes of good optical quality. For example, it has been used in spectroscopic characterization of polymer sheets which undergo irreversible photochemical change [14].

An additional possibility is to use multichannel detection techniques so that data can be recorded at many times following each excitation event. The most straightforward way to do this would be to measure transient absorption with a single excitation pulse and many probe pulses, each delayed by a different amount and each monitored by its own detector. Some more elegant approaches exist. For example, the excitation pulse and a single, cylindrically focussed probe pulse could be crossed at a large angle such as 90 degrees. The transmitted probe light would reach a multichannel detector, and different parts of the beam profile would correspond to different delay times following excitation. In some cases, it may be possible to use a chirped probe pulse (crossed at a small angle with the excitation pulse) so that different probe wavelengths reach the sample at different delay times. The transmitted probe light would then be dispersed onto a multichannel detector [15]. In any of these approaches, as much temporal information as possible is extracted from each repetition of the experiment. This minimizes the total number of repetitions necessary for any sample or sample region.

2.4. OUTLOOK

Clearly, ISRS excitation of coherent vibrational amplitudes in the 0.1-1 A range is not a trivial task. However, through judicious combination of several methods designed to avoid or minimize the effects of electronic excitation and optical damage, there should be good prospects for realization of this goal in some materials. The possibility of mode-selective chemistry which could be induced under microscopically controlled conditions of molecular (wavepacket) position and momentum and observed in real time provides substantial motivation for efforts toward this end.

3. Structural Phase Transitions and Cooperative Chemical Reactions

Another avenue for mode-selective chemistry in crystalline solids arises from the special nature of those crystals in which structural phase transitions involve change which is clearly chemical in nature. In some dramatic cases, covalent bonding undergoes change and the phase transition is thought of as a "cooperative chemical reaction". The phase transition in methyl *p*-dimethylaminobenzenesulfonate, $(CH_3)_2$-N-C_6H_4-SO_3-CH_3, in which a methyl group jumps from the sulfonate group on each molecule to the amine group of each neighbor along a chain to form the zwitterionic *p*-trimethylammoniumbenzenesulfonate crystal, is an excellent example [16,17]. Another impressive example is offered by the large family of electron donor-acceptor mixed-stack aromatic crystals, in which a "neutral monomer" phase of alternating, evenly spaced donors and acceptors undergoes a transition into an "ionic dimer" phase in which neighboring donors and acceptors are paired and substantial charge transfer has occurred [18,19]. In fact, essentially every kind of chemical reaction known -- those involving changes in van der Waals interactions, hydrogen bonding, ionic bonding, covalent bonding, or molecular electron density -- is represented in some crystalline solids undergoing structural phase transitions. In a restricted but real sense any structural phase transition offers prospects for conducting and observing mode-selective chemistry.

There are several factors that make structural phase transitions an excellent proving ground for solid-state mode-selective chemistry. First, the motions of crystalline constituents which take place in a phase transition are usually describable mainly in terms of one or a very small number of lattice vibrational coordinates [20,21]. In other words, one or two lattice phonons usually provide the "collective reaction coordinate" along which motion must take place for the phase transition (i.e. for "cooperative chemical reaction") to occur. Some experimental control over motion along lattice phonon directions may be exerted through various techniques such as ISRS, irradiation at far-IR frequencies, application of uniaxial stress or electric fields, or even ultrasonic excitation in cases where acoustic phonon displacements are involved.

A second factor of considerable interest is that in phase transitions it is a free energy surface that controls the motions of interest [20], rather than a potential energy surface as in the case of localized chemical reactions. The forms of the free energy surfaces for continuous (second-order) and discontinuous (first-order) structural phase transitions at different temperatures are shown in Fig. 1. Note that cooperative chemical reactions with or without barriers can occur. Of special importance is that fact that the form of the free energy surface can be manipulated extensively just by changing the sample temperature. This is in contrast to the potential energy surfaces mediating localized chemical change, and offers considerable experimental control over cooperative reaction conditions and dynamics. For example, near a second-order phase transition temperature a very small force in the direction of the collective reaction coordinate produces a response much larger than usual since the curvature of the free energy surface is very low.

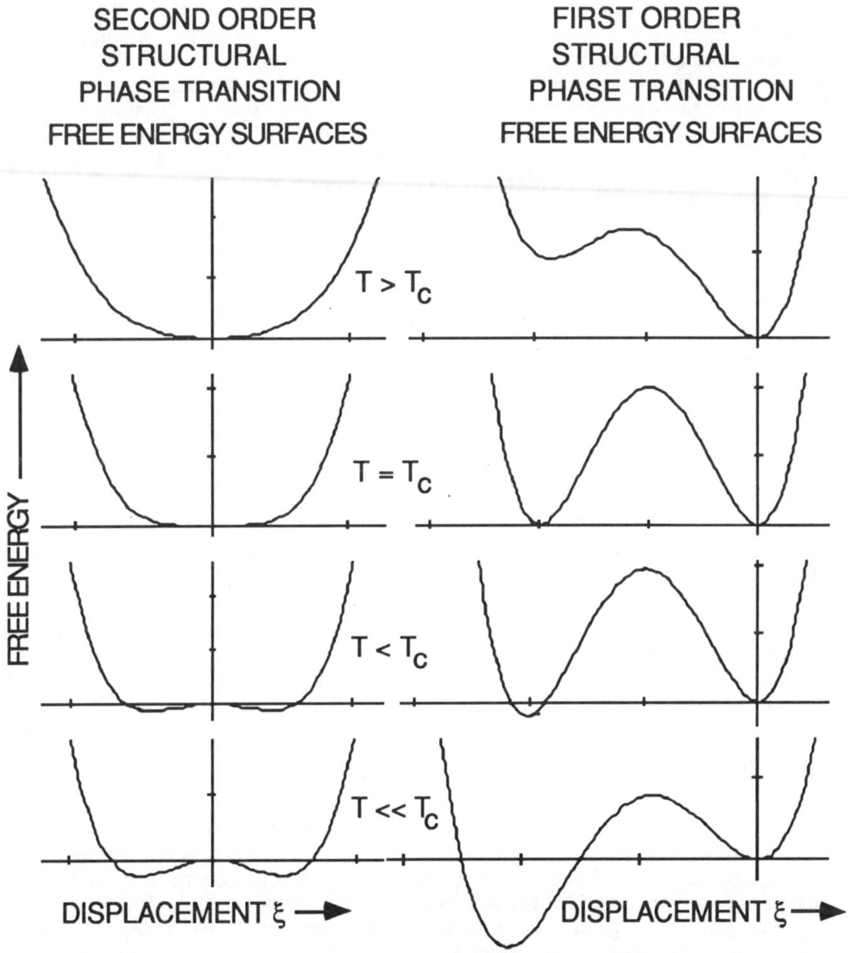

Figure 1. Illustration of the temperature-dependent free energy surfaces for first and second-order structural phase transitions. In continuous (second-order) transitions, there is only one stable phase at any temperature. The two minima below the phase transition temperature Tc correspond to different possible domains of the same structure with displacements in opposite directions. The equilibrium structure changes continuously at Tc and below. In discontinuous (first-order) transitions, a free energy barrier separates different phases (corresponding to inequivalent local free energy minima) which may coexist. The equilibrium structure changes discontinuously at Tc.

A final factor of practical importance is the ease with which cooperative chemical change, which influences the macroscopic optical, electronic, mechanical and other properties of the sample, can be detected experimentally when it is induced.

Impulsive excitation of acoustic and optic phonon modes involved in structural phase transitions has been carried out in a number of inorganic and organic crystalline solids [3,22]. Motion along the collective reaction coordinate has been induced and monitored in systems with and without barriers. This has permitted new information to be extracted about the statistical mechanics of some phase transitions and about the microscopic mechanisms of others. In a certain sense "mode-selective chemistry" has already been carried out in these experiments: a very small amount of energy has been imparted to a selected lattice vibrational coordinate which coincides with or couples to the collective reaction coordinate, and motion along that coordinate has occurred in a phase-coherent way which could not be driven by the same amount of thermalized energy. However, it should be noted that in these experiments the magnitudes of coherent displacements away from the free energy minimum have always been small (less than mean thermal amplitudes) and the displacements have always been followed by relaxation back toward the initial state. Permanent or semipermanent crystalline rearrangement has not been achieved. For example, in no case has a crystal in a metastable (i.e. supercooled or superheated) structure been driven over a free-energy barrier, through a first-order phase transition, and into the stable phase through impulsive excitation. As in the case of localized chemistry, achieving this would require coherent phonon amplitudes exceeding mean thermal amplitudes.

4. Prospects

There are a large number of crystalline solids in which localized or collective chemical change involving mainly a small number of well defined lattice vibrational degrees of freedom occurs. In principle, any such system offers the possibility of mode-selective chemistry induced by a nonthermal distribution of energy in the relevant lattice modes. Of main concern in this discussion has been coherent excitation of lattice vibrations through impulsive stimulated Raman scattering. For localized chemical reactions, large-amplitude ISRS excitation of coherent optic phonons represents a possible method for mode-selective control and real-time observation. Reaching sufficient vibrational amplitudes remains a difficult task. Several experimental approaches for progress in this direction have been outlined.

Structural phase transitions have also been discussed in relation to mode-selective chemical reactions. In this case, the chemistry occurs collectively and a free energy surface rather than a potential energy surface controls the reaction dynamics. Extensive manipulation of this surface is therefore possible through control of the sample temperature. A modest degree of reversible collective chemical change can be driven through ISRS even with small coherent vibrational amplitudes, and can be observed in real time. Long-lived rearrangements from metastable to stable phases could be driven by large-amplitude coherent phonon displacements.

5. References

1. S. De Silvestri, J.G. Fujimoto, E.P. Ippen, E.B. Gamble, Jr., L.R. Williams, and K.A. Nelson, Chem. Phys. Lett. 116, 146 (1985).

2. S. Ruhman, A.G. Joly, and K.A. Nelson, J. Chem. Phys. 86, 6563 (1987).

3. Y.-X. Yan, L.-T. Cheng, and K.A. Nelson, in Advances in Nonlinear Spectroscopy, (Adv. in Spectrosc. series, v. 16), ed. by R.J.H. Clark and R.E. Hester (Wiley, Chichester, 1988), p. 299.

4. Y.-X. Yan, E.B. Gamble, Jr., and K.A. Nelson, J. Chem. Phys. 83, 3591 (1985).

5. A.G. Joly, S. Ruhman, B. Kohler, and K.A. Nelson, in Ultrafast Phenomena VI, ed. by T. Yajima, K. Yoshihara, C.B. Harris, and S. Shionoya (Springer-Verlag, Berlin, 1988), pp. 506-510.

6. A.M. Weiner, J.P. Heritage, E.M. Kirschner, J. Opt. Soc. Am. B 5, 1563 (1988).

7. A.M. Weiner, D.E. Leaird, G.P. Wiederrecht, and K.A. Nelson, Science 247, 1317-1319; J. Opt. Soc. Am. B, in press.

8. B. Walker, H. Port, and H.C. Wolf, Chem. Phys. 92, 177 (1985).

9. W. Warta, R. Stehle, and N. Karl, Appl. Phys. A 36, 163 (1985).

10. M.J. Rosker, F.W. Wise, and C.L. Tang, Phys. Rev. Lett. 57, 321 (1986); K.A. Nelson and L.R. Williams, Phys. Rev. Lett. 58, 745 (1987).

11. A. Mokhtari and J. Chesnoy, Phys. Rev. A 38, 3566 (1988).

12. H.L. Fragnito, J.-Y. Bigot, P.C. Becker, and C.V. Shank, Chem. Phys. Lett. 160, 101 (1989).

13. N. Scherer, L. Ziegler, and G.R. Fleming, to be published.

14. D.N. Rao, R. Burzynski, X. Mi, and P.N. Prasad, Appl. Phys. Lett. 48, 388 (1986).

15. R.J.D. Miller, personal communication.

16. C.N. Sukenik et al., J. Am. Chem. Soc. 99, 851 (1977).

17. P.N. Prasad and J. Swiatkiewicz, Mol. Cryst. Liq. Cryst. 93, 25 (1983).

18. J. B. Torrance et al., Phys. Rev. Lett. 47, 1747 (1981).

19. Y. Iwasa et al., Phys. Rev. B 42, 2374 (1990).

20. A.D. Bruce and R.A. Cowley, Structural Phase Transitions (Taylor & Francis, 1981).

21. Light Scattering Near Phase Transitions, H.Z. Cummins and A.P.Levanyuk, eds. (North-Holland, 1983).

22. L.-T. Cheng and K.A. Nelson, Phys. Rev. B. 37, 3603 (1988); L.-T. Cheng and K. A. Nelson, Phys. Rev. B 39, 9437 (1989); T.P. Dougherty, G.P. Wiederrecht, and K.A. Nelson, Ferroelectrics, in press; Phys. Rev. B, submitted.

MODE SELECTIVE CHEMISTRY

R.D. Levine
The Fritz Haber Research Center for Molecular Dynamics
The Hebrew University of Jerusalem
Jerusalem 91904, Israel

and

Joshua Jortner
School of Chemistry
Tel Aviv University
Ramat Aviv, Tel Aviv 69978, Israel

ABSTRACT. Selectivity of product formation in activated chemical reactions can be accomplished by the control of energy acquisition, storage and disposal in the molecular system. This goal can be achieved by 'passive control' via energy acquisition, i.e., selecting the initial conditions and letting the system evolve under its own Hamiltonian, and by the 'active control' of energy storage and disposal, i.e., modifying the equations of motion by imposing external, additional terms in the Hamiltonian. We discuss the possibility and conditions for non-statistical reaction dynamics in molecules, van der Waals molecules, clusters, surfaces, condensed phases and biophysical systems, which result in selective chemistry.

1. Introduction

The achievement of selective control of product formation in a chemical reaction constitutes a primary goal, which has been pursued throughout the evolution of chemistry. Most chemical reactions require an energy of activation. In many reactions there will be more than one set of products. The problem is whether the energy of activation can be provided or manipulated in a manner designed to preferentially favor a particular product set. Traditionally, control of the pathways of chemical reactions has been accomplished on the macroscopic level by the modification of external variables, such as temperature

J. Jortner et al. (eds.), Mode Selective Chemistry, 535–571.

or pressure. On the microscopic level, selectivity was shown possible by the choice of solvents or catalysts, which modify the activation required for distinct reaction paths. The current science and art of modern photochemistry, which utilizes the merging of laser sources and supersonic beams, focused on a variety of schemes for the realization of photoselective chemistry (Jortner and Levine 1981). Perhaps the most successful such schemes were associated with the exploitation of laser monochromaticity to achieve selective absorption and thereby induce the reaction of one component in a mixture. The most familiar example is laser isotope separation via isotope-selective atomic ionization and the fragmentation of D_2H^+ to form either H^++D_2 or D^++DH (Carrington and McNaib 1989, Hartke *et al.* 1989) via isotope-selective excitations of resonances coupled to different exit channels. Isotope-selective high-order multiphoton molecular vibrational excitation, which utilizes the high intensities of the laser fields (Letokhov 1989) is another important illustration of this idea.

From the points of view of intramolecular and intermolecular dynamics, selectivity pertains to the issue to what extent it is possible to control the evolution of a molecular quantum system? Stated alternatively we inquire what are the ways to control energy acquisition, storage and disposal in molecular systems, resulting in the alternation of product formation. There are two ways to accomplish the desired goal. The first is to control energy acquisition by selecting the initial conditions and then let the system evolve under its own Hamiltonians. This is sometimes known as 'passive control' since there is no intervention during the time evolution towards the products. Much of what we have to say is concerned with passive control. The problem is the rapid (or not so rapid?) erasure of initial details in the necessarily nonlinear mechanical systems that are highly excited olyatomic molecules.

The second way is 'active control'. Here one controls energy storage and disposal by the modification of the equations of motion. This intervention is accomplished by imposing, from the outside, additional terms in the Hamiltonian. Active control is a topic of considerable current interest (J.A. Beswick, P. Brumer, D.J. Tannor, R. Kosloff, and S.A. Rice in this volume, Brumer and Shapiro 1986, Shapiro and Brumer 1989, Mukamel and Shan 1985, Scherer *et al.* 1991, Park *et al.* 1991). Here the additional perturbation is due to a laser field. A different version of active control has however long been practiced by chemists: selectivity can be readily achieved by a suitable choice of a solvent or of a catalyst, etc. In this manner, energy storage and disposal for distinct reaction paths can be varied.

Unimolecular reactions are reactions of energy rich polyatomic molecules. For a **uni-**molecular process to be possible it is necessary that the chemical reaction rate be slow as compared to the rate by which the molecule is energized (Lindemann 1922). The early Hinshelwood (1927) approach to collisional activation assumed that due to the averaging over all manners of collision the energized molecules are an ensemble characterized only by its total energy. However, 'in a molecule of moderate but not too great complexity it is not impossible that there may be several distinct modes of activation corresponding to particular divisions of the energy among a limited number of vibrational (or rotational) degrees of freedom' (Hinshelwood and Fletcher 1933). 'To a first approximation each of these modes may be associated with a separate probability of transformation, because internal redistribution may be difficult without collisions'.

The idea of a mode selective unimolecular reaction is thus quite old. What is new is that progress in laser light sources and supersonic jet spectroscopy is providing hard experimental evidence (Crim 1990, Manz and Parmenter 1989). The one new conceptual component is our improved understanding (Rice 1981, Brumer 1981) of the dynamics of nonlinear systems (and our molecules are nonlinear since they can dissociate or isomerise). This provides a detailed understanding how, by the time a molecule dissociates, the memory of any initially selective excitation can be erased. The question becomes then one of degree. How far does the intramolecular vibrational energy redistribution (IVR) compete against dissociation. The simplest assumption (Rice-Ramsperger 1927, Kassel 1928) is that the dissociating molecule is characterized only by its total energy (and other good quantum numbers such as the total angular momentum, Marcus-Rice 1951). Experiments in the 1960's on chemically–activated molecules (Rabinovich and Setser 1964) and with hot atoms (Rowland 1970) indicate (Oref and Rabinovitch 1979) an essentially complete erasure of initial conditions. An example showing the considerable fluxionality (Zare *et al.* 1990) of the energy rich molecule is the dissociation of spiropentane (II), Figure 1. This has been prepared (von Doering *et al.* 1968) by chemical activation but using CD_2. The CD_2 group is found in any of the possible positions in the product.

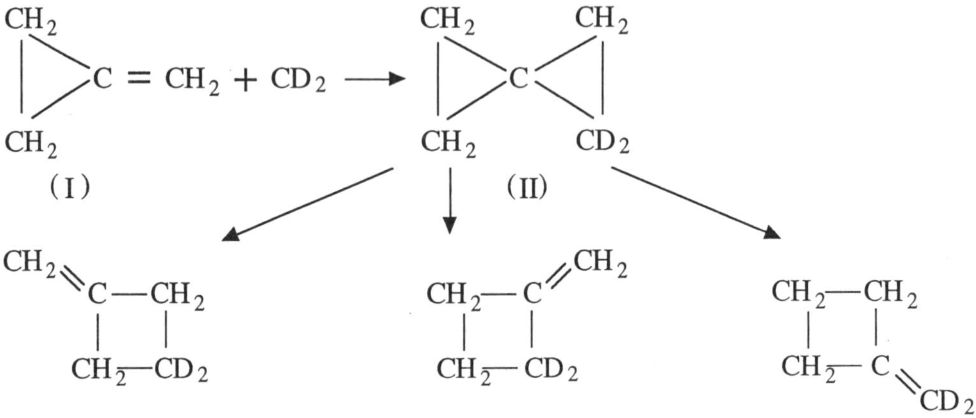

Figure 1. The experimental evidence of von Doering *et al.* (1968) for the considerable fluxionality of chemically activated spiropentane.

In the 1970's and early 80's, infrared multiphoton pumping provided additional experimental evidence (Lupo and Quack 1987) in support of rapid IVR. This was complemented by theoretical and computational studies of IVR in two coupled anharmonic oscillators (Oxtoby and Rice 1976). It led to the point of view that mode selective chemistry, if at all possible, is the exception that proves the rule (Marcus 1983).

At the same time that selectivity was ruled an exception in unimolecular reactions, clear and sometimes spectacular selectivity was being demonstrated for direct bimolecular collisions (Polanyi 1971,Bauer 1978, Levine and Bernstein 1987, Whitehead 1988, Rubahn and Bergmann 1990). Of course, in these direct reactions the switchover from reactants to products occurs within a few vibrational periods. There is not the time for extensive IVR, showing again that it is a matter of time scales. One of our purposes here is to discuss the relevant times.

The dissociation of van der Waals adducts (Levy 1981) was an early example (Beswick and Jortner 1981, Jortner and Levine 1981) of a non-statistical unimolecular process. The relevant parameter is the frequency mismatch between the chemical and the van der

Waals bonds. For this reason, questions of IVR involving modes of different frequencies (K.K. Lehman, P.R. Fleming *et al.*, M. Quack, J.C Light, D.H. Levy, G.D. Barrio and D. Nesbit, in this volume) and the dynamics of dissociation of van der Waals molecules continue to be of interest. The overtone excitation (F.F. Crim, in this volume) and single photon excitation (S. Rosenwaks, in this volume) of HOD leading to selective dissociation is one experimental example of a poor (HO *vs.* OD) mistmatch in a system of purely chemical bonding. The dynamics of HOOH (P.R. Fleming, in this volume) and in acetylene and its derivatives (K.K. Lehmann, in this volume) is another.

The spectroscopic interrogation of IVR dynamics (Lehman *et al.* 1990, K.K. Lehman in this volume, Pate *et al.* 1991) provides some of the lower limits for IVR, where in substituted acetylenes the IVR times fall in the range 0.2-6.0 nsec. Thus IVR at low levels of excitations can be surprisingly low, preserving the identity of the initial state as time scales which considerably exceed chemical processes, e.g. isomerization (D.L. Thompson in this volume). On the other hand, modifying the rigidity of the molecular framework can considerably enhance the rate of IVR (C.S. Parmenter and Z.Q. Zhao, in this volume).

Two possible extensions of the selectivity demonstrated in van der Waals adducts are laser induced desorption of large physisorbed molecules (Zare and Levine 1987) and reaction dynamics in rare gas solvents (Wilson and Levine 1988, Levine 1988, Li and Wilson 1990). Closely related is mode selectivity in surface scattering (R. Naaman and E. Kolodney, in this volume), and in clusters (N. Ben-Horin *et al.*, J. Jellinek, A. Heidenreich and J. Jortner, F.G. Amar and R.B. Gerber *et al.*, in this volume).

Passive control of selectivity can be traced to the inherent vibrational level structure of many-atom systems (i.e., frequency mismatches), as well as to the nature of the experimental conditions, which govern the ultrafast processes of energy acquisition and storage. Coulomb explosions of polyatomic molecules (E.W. Schlag in this volume, Müller-Dethlefs *et al.* 1984) and of surface states (Stockbaum and Madey 1983) involve ultrafast (10^{-15}–10^{-14}sec) site-selective energy acquisition in conjunction with bond specific energy disposal (Jortner and Levine 1990). Reactive energy disposal on the femtosecond (and subfemtosecond) time scale may be excited for high-energy cluster surface collisions, which demonstrated the occurrence of nuclear D-D fusion (Beuhler *et al.* 1989). The utilization of ultrafast "chemical clocks" of Coulomb explosion and possibly high-energy cluster collisions show potential applications for selective chemistry.

Proceeding from passive control to active control of selectivity by laser fields (P. Brumer, D.J. Tannor and S.A. Rice in this volume, Jortner and Kommandeur 1978, Mukamel and Sham 1985, Tannor and Rice 1985, Tannor *et al.* 1986, Kosloff *et al.* 1989, Brumer and Shapiro 1986, Shapiro and Brumer 1989, Siedeman *et al.* 1989) we note that the field of laser chemistry is being extended to encompass the utilization of phase coherence, pulse sequencing and pulse shaping (K.A. Nelson in this volume). The two major schemes were advanced for active radiative control of selectivity, which involve the exploitation of coherent radiative interference effects (Brumer and Shapiro 1986, Seideman *et al.* 1989, P. Brumer in this volume) and a pump-dump scheme (Tannor and Rice 1985, D.J. Tannor in this volume). A central message emerging from our discussion of active control is that active radiative control schemes necessarily also incorporate the elements of passive control, i.e., IVR and spreading of nuclear wavepackets have to be avoided (or minimized) in order to render active control schemes feasible in real life.

The active control of selectivity in photobiology is a topic of central interest. The pursuit of the task of life involves the conversion of solar energy into photochemical energy in photosynthetic reaction centers of bacteria and of plants (Deisenhofer and Michel 1989, Michel-Beyerle 1985, 1990). This conversion process (S.H. Lin *et al.* in this volume, Bixon *et al.* 1991) proceeds via a sequence of well organized, highly efficient, directional and specific electron transfer steps across the photosynthetic membrane. The primary charge separation processes are highly selective, proceeding across a single branch of the two-branched reaction center. Nature utilizes incoherent solar radiation for photosynthesis and does not have lasers at its disposal to initiate radiative active control. Rather, nature imposes active control by a suitable construction of the protein medium of the reaction center, exerting symmetry breaking effects (Michel-Beyerle *et al.* 1988, Bixon *et al.* 1991), which accomplish remarkable selectivity in the primary charge separation in photosynthesis.

2. Resonances - an overview

Our purpose is to estimate the extent of IVR in polyatomic molecules energized to chemically significant levels. A point we shall return to is that unlike the case of two coupled oscillators, a multimode molecule can contain energy above an isomerization or a dissocication limit and yet the energy per mode can be quite low (Schlag and Levine 1989).

An important yet not settled question is whether for realistic molecular potentials the extent of IVR scales with the total energy or with the mean energy per mode. While we shall here argue for the latter, the former point of view has its supporters. It may also be that the difference is a difference between relatively rigid and relatively fluxional molecules and has no all-embracing unique answer.

Towards our purpose we shall use all the available evidence, theoretical, computational and experimental. As in our earlier discussion (Jortner and Levine 1990) the conceptual tool we shall employ is that of a mechanical resonance.

At the energies of interest the vibrational modes of molecules are anharmonic. Of course, they are not only anharmonic but also coupled to one another but let us start with a zeroth order description when such coupling is absent. Even then the frequency of any particular mode will depend on the distribution of energy amongst the available modes. In classical mechanics the zeroth order description corresponds to the energy being a function only of the actions (action $\equiv\hbar$ quantum number), $\mathbf{I} = I_1, I_2, \cdots$ of the different modes,

$$E = H_0(\mathbf{I}) \tag{1}$$

The frequency $\omega_j(\mathbf{I})$ of the j'th vibrational mode is given by

$$\omega_i(\mathbf{I}) = \partial H_0(\mathbf{I})/\partial I_j \quad . \tag{2}$$

In quantum mechanics where the $I_j/\hbar$'s are integers, equation (1) is the Dunham expansion of the energy in terms of the vibrational quantum numbers. $H_0(\mathbf{I})$ is thus available from experiment whenever overtone spectra have been fitted to a Dunham expansion.

IVR is facile when two or more modes are 'in resonance', i.e. if there is a set of (not all positive) integers $n_1, n_2, \cdots$ such that

$$\mathbf{n}\cdot\omega \equiv \sum_j n_j \omega_j(\mathbf{I}) \simeq 0 \quad . \tag{3}$$

We sometimes refer (Weeks and Levine 1991) to the set of integers $\mathbf{n}$ as a Fermi vector. The most familiar example is the 1,-2,0 resonance in CO_2 involving the symmetric stretch and the bend modes.

As the notation in (3) and (2) emphasizes, the frequencies ω_j depend on the actions of the different modes. At the same total energy E, there can be a whole range of sets of actions $\mathbf{I} = I_1, I_2, ...$ which satisfy equation (1). The different partitions of the total energy correspond to different frequencies for each one of the modes. At the same total energy there can be different resonances depending on the partitioning of the total energy.

The coupling between the modes, which has thus far been neglected, can efficiently induce energy exchange only amongst modes which are nearly in resonance. Otherwise, it requires 'dynamical tunneling' and the efficiency is exponentially small. The strength of the coupling determines the 'width' of the resonance i.e. how far off-resonance does the influence of the resonance extend.

An isolated resonance is manifested as a perturbation in the spectrum. The 'width' is measured by the range of states affected by the perturbation. A loss of selectivity due to IVR requires that resonances 'overlap' i.e. that the widths of nearby resonances have a common area so that an excitation initially localized on a particular mode can migrate to the mode resonantly coupled to it and then to the third mode coupled via the second (and overlapping) resonance, and so on. A key question is whether, at the energy of interest, a connected region of overlapping resonances exists throughout the constant energy plane. If it does, however selective the initial excitation, the molecule will rapidly sample a representative region of the available phase space and selectivity will be lost.

Two factors come therefore into play, the density of resonances and their width. A problem that very much deserves a solution is that we do not know enough about either factor, for multimode systems. In discussing this problem it is important to recognize that, in part, the language of resonances is so appealing because while we are using classical mechanics there is an immediate transcript into quantum mechanics. In particular, a resonance with a given Fermi vector $\mathbf{n}$ corresponds to n_j quanta being added (or removed if n_j is negative) to the mode j. There are therefore at least two reasons why very high order ($\mathbf{n}\cdot\mathbf{n} \gg 1$) resonances are not as important as the lower order ones: (i) At the energies of interest to us, a medium sized polyatomic molecule has on the average, very few (one, two) vibrational quanta per mode. (ii) For realistic molecular intermode coupling, the width of the resonance diminishes exponentially as its order increases.

For realistic molecular Hamiltonians $H_0(\mathbf{I})$, where the vibrational spacings become narrower at higher excitations one can show that the density of resonances will be large enough, however weak the coupling, only near the dissociation energy of all modes, (figure 2). This is, of course, a very high energy. The onset of facile IVR for any realistic molecule, with finite coupling strength will surely be at a far lower energy. Even so, the result does point out to the energy per mode rather than to the total energy as being the relevant parameter.

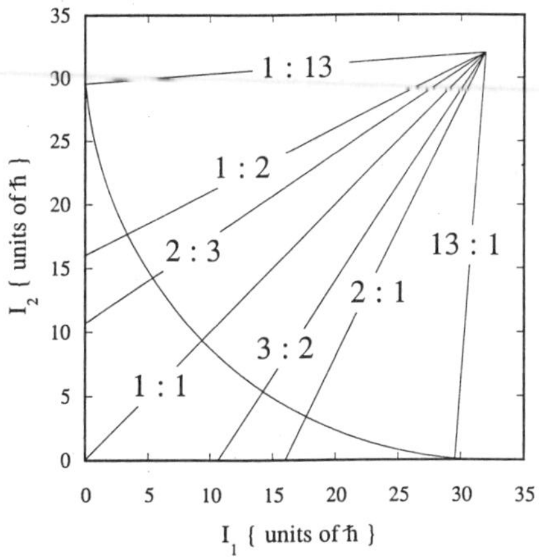

Figure 2. The resonance lines (along which equation 3 is satisfied the label is $n_1;n_2$), for two equivalent anharmonic oscillators. The hyperbola is the line where the total energy equals te bond dissociation energy. The resonances converge at the energy where both bonds are broken. Figure by D.E. Weeks.

The strong coupling criterion for IVR is $\overline{V}\rho > 1$ where $\overline{V}$ is the average coupling strength and ρ is the density is familiar from the theory of radiationless transitions (Bixon and Jortner 1968). It is still an open question if, and if so, under what conditions, the increase in the number of resonances as their order increases (Stuchebrukhov *et al.* 1989) can compensate for their decreasing width. For any system of more than two vibrational modes there is another, and far less discussed, route to IVR (Engel and Levine 1989). For a system of s modes, $s > 2$, resonances can coexist. If these have a finite width then at the point of their intersection they can interact. The result is that the constant E surface is covered by a web of crossing resonances, figure 3. It is sufficient for this web to be dense enough. Here too, the question of the density of crossings vs. the coupling strength is the determining factor.

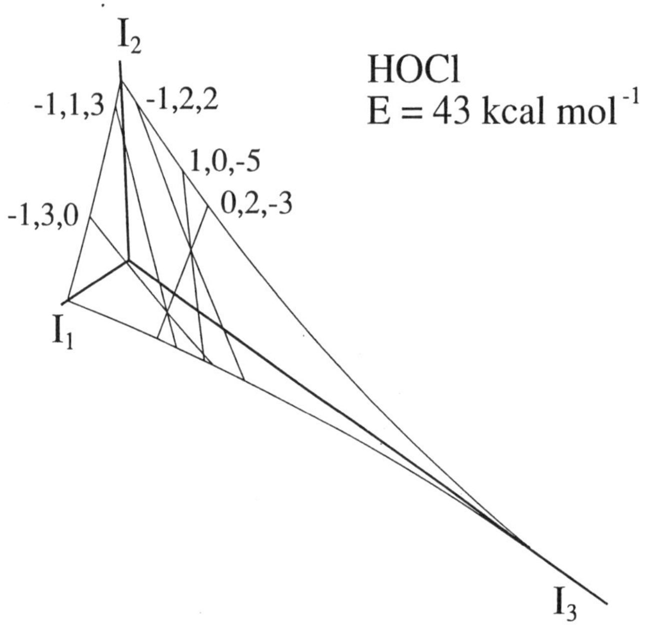

Figure 3. The intersection of lower order resonance lines with the surface of constant energy (E = 43 kcalmol^{-1}) of HOCl. Figure by D.W. Weeks. The Fermi vectors are identified, with modes 1,2,3 being the OH stretch, bend and OCl stretch, respectively. Note that for three or more modes, resonances can coexist.

Up to this point we have been concerned with intrastate IVR. The isomorphism between the two major classes of intramolecular radiationless tansitions, i.e., interstate electronic relaxation and intrastate IVR, is well established (Jortner and Levine 1981). Ladder diagrams (Bixon and Jortner 1968, 1969, Jortner and Mukamel 1973) are useful for the description of both processes. Intersystem crossing in large organic molecules (Amirav *et al.* 1988) allows for the exploration of the role of resonances in inducing mode selectivity. The intramolecular dynamics of $S_1 \rightarrow \{T_1\}$ intersystem crossing (ISC) from a low lying vibronic S_1 state (the doorway state) to the first dense triplet $\{T_1\}$ manifold for anthracene and its derivatives in supersonic jets (Amirav *et al.* 1988) is dominated by the mediated coupling $S_1 - \{T_x\} - \{T_1\}$ where $\{T_x\}$ is a sparse manifold of a higher triplet state. The role of the mediating $\{T_x\}$ manifold for electronic relaxation is analogous to mediated IVR, where a sparse manifold of vibrational states with a small number of quanta per mode effectively couples to the doorway states. Another class of analogies where the dissipative manifold corresponds to a genuine dissociative continuum involves accidental predissociation (Jortner and Levine 1981), accidental vibrational predissociation (Jortner and Levine 1981) and mediated dissociation from high stretch vibrational overtone excitations (P.R. Fleming *et al.* in this volume). The mediated ISC is expected to depend strongly on the energy matching of the doorway S_1 state with the $\{T_x\}$ resonances. This is quite dramatically demonstrated (Amirav *et al.* 1988) in the extreme and irregular variations of the ISC rates from photoselected low-lying vibronic levels of the S_1 state of 9,10 dibromoanthracene (Figure 4). These mediated ISC rates

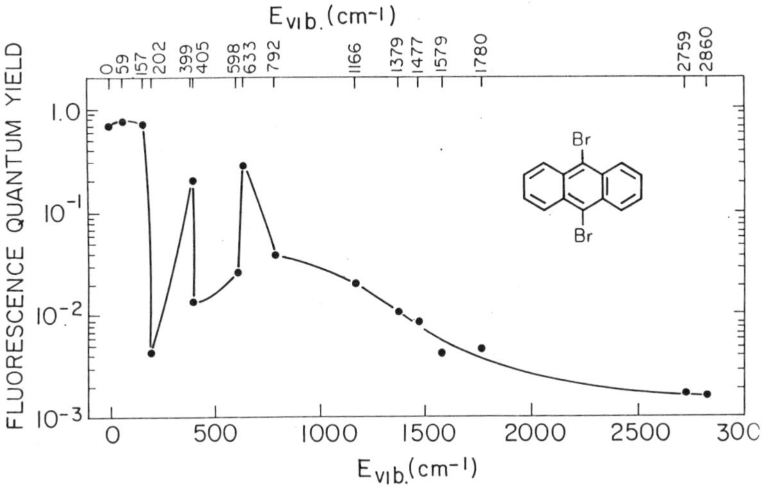

Figure 4. Absolute fluorescence quantum yield from photoselected vibronic levels of the S_1 manifold of jet-cooled 9,10 dibromoanthracene (Amirav *et al.* 1988). The numbers on the upper horizontal scale represent the energies of the vibronic levels above the electronic origin of S_1. Note the resonance effects in the range E_{vib}=0–800 cm^{-1}, which manifests mode selectivity. IVR within the S_1 manifold sets in at $\gtrsim$800cm^{-1} eroding mode selectivity at higher energies.

can also be drastically modified by level shifts (~ 40 cm^{-1}) induced by complexing of the organic molecule with Ar (Amirav *et al.* 1988). The interstate dynamics of the low-lying S_1 vibronic states provides a good example for mode selectivity induced by resonance effects. Of course, for higher-energy vibronic S_1 states ($\gtrsim$ 800 cm^{-1} for anthracene and its derivatives) IVR and the S_1 manifold erodes the signature of the initial state and selectivity is lost (Fig. 4). Spectroscopic evidence for intermolecular energy transfer in van der Waals molecules was recently presented by Dayton *et al.* (1991).

3. The classical power spectrum

The classical power spectrum (Noid *et al.* 1981) is a direct representation of the distribution of the frequencies of the motions executed during the time evolution of the system. A particular spectrum is generated by running a classical trajectory and computing the time autocorrelation function of, say, a bond length. The power spectrum is the Fourier transform of the autocorrelation function. One can convincingly argue (Heller 1981) that the optical spectrum can be well approximated by initiating several classical trajectories with initial conditions that representatively sample the Franck Condon region and obtaining an averaged spectrum. Here, however, we consider just a single trajectory.

If the actual Hamiltonian is well approximated by the zeroth order, separable, Hamiltonian $H_0(\mathbf{I})$, the spectrum consists of sharp lines of the frequencies $\omega_j(\mathbf{I})$. The reason is that in the zeroth order the actions $\mathbf{I}$ are constants of the motion. Each mode oscillates

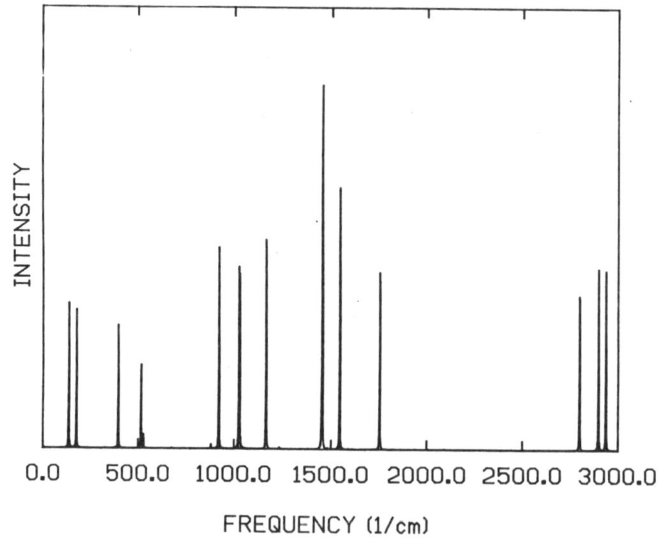

Figure 5. The classical power spectrum of CH_3ONO at a very low total energy, adapted from Sewell *et al.* (1991). In this limit all modes have a well defined frequency.

with its given frequency (which can depend on the initial distribution of energy amongst the modes). This is the strict Slater limit, generalized to allow 'self' and 'cross' anharmonicity.

For most polyatomics this separable approximation is strictly valid only at the very lowest energies. A classical power spectrum of methyl nitrite (CH_3ONO) from the computational work of Sewell *et al.* (1991) is shown in figure 5 to a total energy less than the quantum zero point energy. The sharp peaks correspondto the primary modes of CH_3ONO as determined by $H_0(\mathbf{I})$. The power spectrum shown in figure 6 is computed at the quantum zero point energy of CH_3ONO (28.87 kcal mol^{-1}) and indicates that although the primary peaks have broadened and congested regions have appeared around the base of such modes (e.g., the CH stretches) that are nearly resonant, the essential features of the zero order Hamiltonian $H_0(\mathbf{I})$ are still clearly identifiable. Even when the CH mode is excited to $v = 6$ (71.67 kcal mol^{-1}), which corresponds to an energy above the lowest threshold for dissociation, the primary peaks can still be identified, figure 7. Only when the energy is above the threshold for dissociating any of the CH_3ONO bonds is the identity of the primary modes erased.

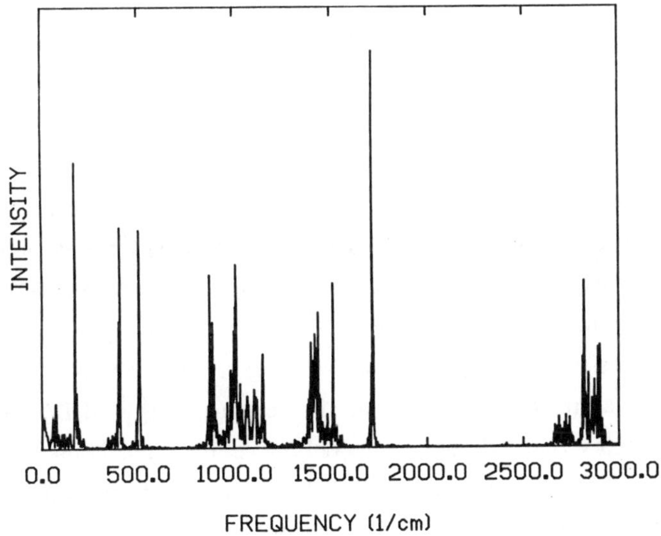

Figure 6. The classical power spectrum of CH_3ONO at an energy equal to the quantal zero point energy of the molecule (Sewell *at al.*, 1991).

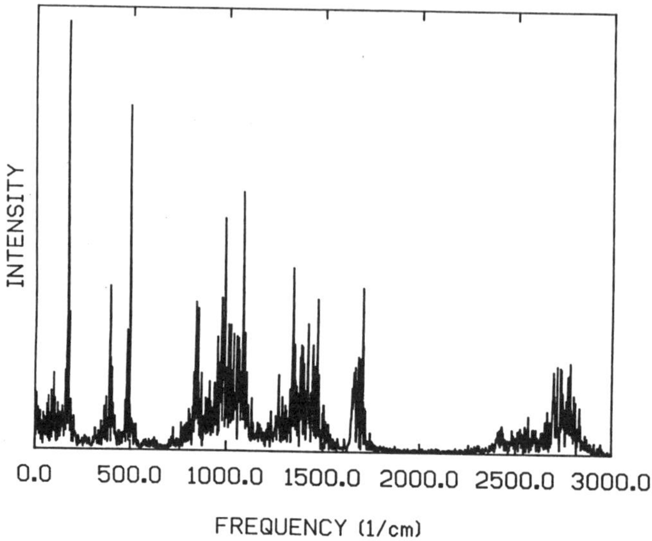

Figure 7. Same as figures 5,6 but at a total energy (71.67 kcalmol^{-1}) which is above the lowest dissociation threshold (and also above the threshold for isomerization. Sewell *et al.* 1991).

Figures 5-7 also exhibit the onset of 'fluctuations' in the intensities of the spectra (Levine 1987). At the lowest energy, a few lines carry most of the intensity and most other lines are barely visible. As the energy increases the line intensities become comparable but they vary erratically as a function of frequency. One expects these fluctuations to tend to a universal limit (and not to smooth out) as the energy is increased. Such fluctuations are also expected in the magnitude of individual rate constants of unimolecular dissociation (Levine 1988, Polik *et al.* 1988). The RRKM expression is then interpreted as determining the mean rate constant *vs.* energy. The mean is computed by giving equal weight to all regions of phase space at the same energy. If IVR is incomplete then the RRKM theory is strictly applicable only when the preparation process is itself non-selective as has been assumed in the Hinshelwood theory of collisional activation.

Experimentally, high resolution infrared spectroscopy (Lehman, in this volume, Parmenter and Zhao, in this volume, Pate *et al.* 1991, Brummel *et al.* 1991, Go *et al.* 1990, McIlroy and Nesbit 1990) is beginning to provide detailed evidence on the selectivity of mode-mode coupling. It remains however to show that such selectivity can maintain some memory of the initial state on time scales of chemical interest.

4. The rate of exploration of phase space

For a selective preparation process we need to know if the rate of dissociation can compete with the rate of IVR. In other words, can many molecules dissociate before a significant fraction of phase space has been sampled.

It is already evident from figures 6,7 and will be further discussed in section 5, that the process of IVR is itself selective. Subsets of modes exchange energy among themselves before larger subsets of modes become coupled. When such is the case, we are also concerned with the representability of the sampling of phase space.

The exploration of phase space can be conveniently discussed in terms of time correlation functions. The simplest is the autocorrelation function of the initially excited region. This can be obtained from the experimental spectrum by taking its Fourier transform. Dynamical computations can, of course, directly yield the correlation functions. Obtaining the information from experiment is however of obvious interest.

To obtain the most representative distribution $\rho(t)$ in phase space which is consistent with a given spectrum, one can proceed as follows (Remacle and Levine 1991): Let Q be a projection on the Franck-Condon region. Then the autocorrelation function $C(t)$ is given by $|C(t)|^2 = Tr(Q\rho(t))$. We now seek that density $\rho(t)$ which reproduces the t-dependence of $|C(t)|^2$, as determined by the spectrum and is otherwise of maximal entropy.

It is worthwhile to emphasize that this procedure determines the most delocalized (Levine 1981) $\rho(t)$ which is consistent with the optical spectrum. Any additional conditions which one may want to impose on $\rho(t)$ can only serve to reduce its entropy and thereby make $\rho(t)$ more localized.

The entropy, $S(t)$, of $\rho(t)$ is a measure of the (maximal) volume of phase space which can be sampled up to the time t. Towards understanding the t dependence it is useful to bear in mind a property of Fourier transforms (Heller 1980) which we often refer to as the time energy uncertainty relation: The t-dependence of $C(t)$ for earlier times, say $0 < t < T$ is primarily determined by the variation of the optical spectrum at lower resolution, i.e. on the frequency scale of $\omega > \hbar/T$. In other words, if we coarse grain the optical spectrum using a window function of spectral width $\hbar/T$, we hardly affect $C(t)$ for $t < T$. Since $S(t)$ is a function of $C(t)$, the same applies to it: The sampling of phase space at the earlier times is determined by the gross features of the optical spectrum. High resolution spectral details correspond to longer time evolution.

At high levels of excitation one often notes a 'nested' structure of the spectrum: At the poorest of all resolutions there is the overall Franck Condon envelope. Its width corresponds to the first excursion out of the localized initially excited region. Often this is so rapid ($\sim 10^{-14}\ s$) that one may fail to cover a wide enough frequency range in the experimental spectra. This rapid initial delocalization of the Franck Condon limited wavepacket is possible, of course, in a perfectly harmonic molecule and is not indicative of IVR.

Within the Franck Condon envelope one can find an inherent and quite broad structure due to the strong coupling of equivalent modes. An example (Pique *et al.* 1987) is the two CH vibrations in C_2H_2. Figures 6,7 provide another illustration, viz. the 3 CH modes of CH_3 which are evidently strongly mixed.

At higher resolution each one of the inherent ('giant') (Pique *et al.* 1988) spectral features is found to exhibit inherent detail. In C_2H_2 (Holme and Levine 1988) these are the celebrated clumps (Hamilton *et al.* 1986) and such clumps have been seen since for other molecules (Luo and Rizzo 1991). The density of clumps in C_2H_2 is significantly below the density of states. Indeed, at even higher spectral resolution the clumps can be seen to be a set of closely spaced lines.

The nested spectral structure implies that there are several time regimes in $C(t)$ and hence in $S(t)$, figure 8. The sampling of phase space, as judged from high overtone

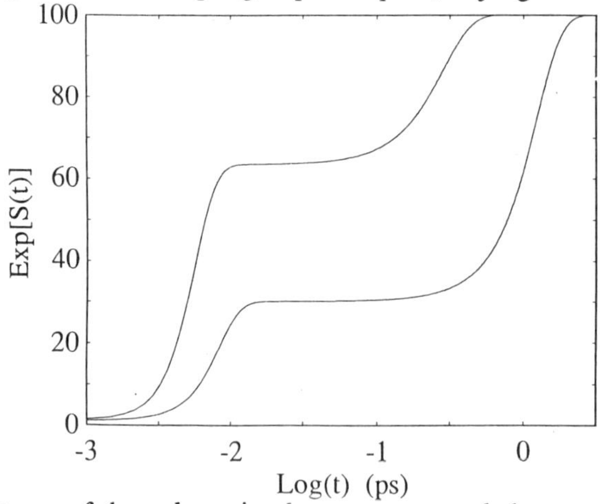

Figure 8. The entropy of the volume in phase space sampled up to the time t. (Adapted from Remacle and Levine 1991).

spectra, is sequential. A given stage is over at a time $T = \hbar/\Delta\omega$ where $\Delta\omega$ is the inherent resolution of a given spectrum. The sampling is *de jure* complete when $\Delta\omega$ is the lowest separation of states in the spectrum, i.e. of the order of the mean spacing of states, $\rho^* = 1/\Delta\omega$ where ρ^* is the density of states. In general, a given stage in the sampling of phase space is about over by the time $\hbar$ (density of spectral features). Figure 8. The entropy of the volume in phase space sampled up to the time t. (Adapted from Remacle and Levine 1991).

For polyatomic molecules of real interest, the density of states is so high that the ultimate sampling of phase space is far too slow. Strict RRKM behavior is only possible for a non-selective preparation. Yet, on practical grounds, two qualifications come in. The first is that sampling of a subspace, which is typically very much faster, may be all that is required for loss of selectivity. Such will be the case if the reaction coordinates of either one of two dissociation paths are coupled to the same subset of modes. Turned the other way round, to achieve selectivity, one needs dissociation paths which originate from distinct sets of modes. The less resonance coupling between the two sets - the better.

The other practical point is the selectivity of the preparation process. It is only under special circumstances (e.g., stimulated emission pumping) that one can excite a pure (in the quantum mechanical sense) initial state. More typically, due to hot bands etc., one excites a mixture. This is certainly the case in infrared multiphoton dissociation from a room temperature ensemble of molecules. The essential point is that even a very moderate incoherence in the initially prepared state has a significant effect (Remacle *et al.* 1991). The most direct argument invokes the 'fluctuations' in the magnitudes of rate constants in a narrow energy range. From the distribution function for such fluctuations (Levine 1988) one readily concludes that incoherent averaging over even, say, ten initial states already yields just about the average value. In entropic terms, for $S(t)$, to be well below its long time value, $\rho(t)$ needs to be a very localized density. If $\rho(t)$ is a mixture to begin with, its entropy starts at a finite value. Only if the number of components of the initial mixture is exponentially small compared to, say, the number of states of the transition state, can dissociation compete with IVR.

5. The separation of time scales

The experimental observation of different inherent spectra at different spectral resolutions leads to the conclusion that the sampling of phase space is sequential. Each stage has its own characteristic time scale. At each consecutive stage an increasing range of states is coupled. In terms of nested subspaces, one concludes that a subspace is representatively sampled before there is any extensive sampling of a larger subspace.

Two quite distinct excitation processes (e.g., direct overtone pumping *vs.* stimulated emission pumping) which can access different regions in phase space can therefore lead to quite distinct histories in terms of the sampling of phase space. Only eventually will the memory of the initial excitation be fully erased.

It follows from the sequential nature of the exploration of phase space that mode selectivity should be possible.

To design a selective excitation one needs to better understand the origin of the different time scales. One possible interpretation is that they reflect the order of magnitude differences in the coupling strengths as the resonance order is increased. Thus, in acetylene the 1:1 resonance between the two CH stretch modes leads to a rapid energy exchange while the 2:3 CH stretch - CC stretch resonance is manifested later (Holme and Levine 1988) and the stretch-bend resonances (Engel and Levine 1989) appear only much later. In general, high frequency stretch modes remain isolated for significant periods. So much so that in hydrides pumped to high overtones we see (Li *et al.* 1991) quite incomplete IVR prior to dissociation unless collisions with the other molecules around provide for another, external, mechanism for IVR.

6. Ultrafast chemical clocks

At this stage of our discussion it is apparent that it could be misleading to think about reactions in many-atom systems along the conventional lines which imply complete IVR preceeding unimolecular decay of energy-rich polyatomic molecules, as there will be interesting exceptions to such behavior. Passive control of selectivity can be accomplished by the inherent structure of and interactions within the many-atom systems (i.e., frequency mismatches) as well as by the nature of the experimental conditions pertaining to some unique aspects of energy acquisition and disposal. An interesting situation in the latter category involves the Coulomb explosion of charged polyatomic molecules (Schlag, this volume, Eberhardt *et al.* 1983, Müller-Dethlefs *et al.* 1984) and of surface states (Knotek and Feibelman 1987, Stockbauer and Madey 1983). Coulomb explosion occurring on the fsec time scale can beat IVR (Jortner and Levine 1990). The sequence of events leading to chemical selectivity is: (i) a site-specific inner core ionization or excitation to a high orbital; (ii) two-center Auger process; (iii) bond-selective Coulomb explosion. Selectivity originates from the site-selective energy acquisition in conjunction with the bond-specific energy disposal. Statistical computations of fragmentation patterns which do incorporate a site-selective excitation (Salman *et al.*, 1991) can qualitatively account for the differences in the observed (Habenicht *et al.* 1991) patterns for excitation at distinct sites.

The time scale for Coulomb explosion resulting in a bond rupture can be determined from a simple mechanical model (Jortner and Levine 1990). This characteristic time, τ_c, for the separation of the two positive ions is identified with the extra time required for dissociation because the ions started their descent on the repulsove potential curve when at rest instead of at their terminal velocity. A straightforward calculation (Jortner and Levine 1990) gives

$$\tau_c = \frac{l_c}{2v_c} \ln \left[\frac{4\langle R \rangle}{l_c} \right] , \quad (4)$$

where $v_c = (2E/\mu)^{1/2}$ is the terminal speed, $l_c = (e^2/E)$ is the terminal length, E is the terminal kinetic energy, $\mu$is the reduced mass and $\langle R \rangle$ is some characteristic length ($> l_c$). This simple argument results in the ultrashort Coulomb explosion time of 10^{-14}–10^{-15} s. The time scale for a Coulomb explosion is shorter by about one order of magnitude than the corresponding time scale for direct molecular dissociation (Bersohn and Zewail 1988, Bernstein and Zewail 1989, Beswick and Jortner, 1990).

The first example of site-specific inner-core excitation followed by selective dissociation involves photon-stimulated desorption of ions from surfaces, which is induced by the absorption of high-energy (20-100eV) photons (Knotek and Feibelman 1978, Stockbaum and Madey 1983). Ion desorption is due to an interatomic, two-center Auger decay, which susbequently results in Coulomb explosion. Such a process was reported for the maximal valence compound TiO_2. The excitation involves the formation of a 3p core hole in the Ti^{4+} ion. In maximal valence compounds, this hole is filled by an electron from the oxygen, accompanied by the ionization of one or two electrons of the oxygen. The resulting O^+ ion finds itself in close proximity to Ti^{4+}, with a subsequent Coulomb explosion occurring, which expels the O^+ ion from the surface. This physical picture results in two predictions (Stockbaum and Madey 1983): First, the site-specific primary ionization is invariant with respect to the mode, i.e., electron or photon impact of the site-selective excitation. Second, ion desorption from nonmaximal valency ionic compounds, e.g. Ti_2O_3, should be greatly reduced due to the occurrence of a one-center Auger process. These predictions are borne out by experiment. The excitation selectivity in this case is rather trivial, resulting from the ionization of a specific ion. A more refined site core excitation is encountered in molecular systems.

Chemical shifts of the K edge for different carbon atoms in an organic molecule bring up the realization of selectie site excitation (Schlag, this volume, Eberhardt *et al.* 1983, Müller-Dethlefs 1984). For an organic molecule ($C_{(1)}$–$C_{(2)}$–X, with the $C_{(1)}$ and $C_{(2)}$ carbon atoms being located in a different chemical environment, site-specific excitation of $C_{(1)}$ or of $C_{(2)}$ can lead to a two-center Auger process (Fig. 9), which results in a

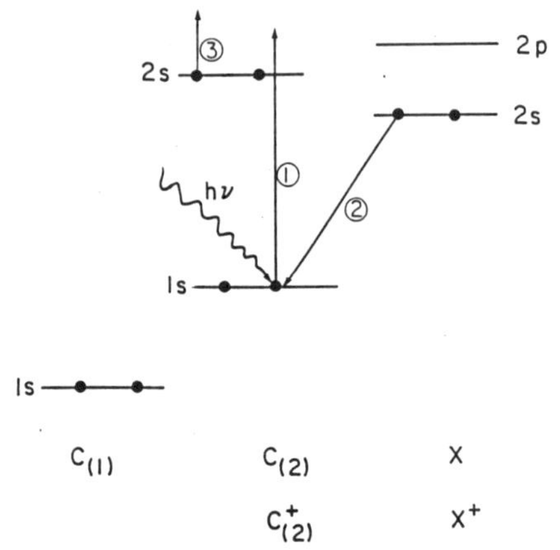

Figure 9. Schematic drawing of the levels involved in a Coulomb explosion following synchrotron excitation of a carbon 1s electron.

bond-selective Coulomb explosion of $(CH_3)_2CO$. The total ion yield (Fig. 10) reveals two thresholds for different carbon atoms, whose selective K-shell ionization produces either O^+ or CH_3^+. The molecule breaks locally around the site of the carbon atom where selective ionization occurs. The utilization of the ultrafast (10^{-14}–10^{-15}) "chemical clock" of Coulomb explosion shows potential applications for selective chemistry.

Reactive energy disposal on the femtosecond time scale, which precludes IVR, may also be exhibited for direct photodissociation of polyatomics (Zewail and Bernstein 1989). The time scale of direct photodissociation to neutral fragments is -10^{-13}, e.g.,

105 ± 30 fs for the dissociation of ICN (Zewail and Bernstein 1989). This time scale is

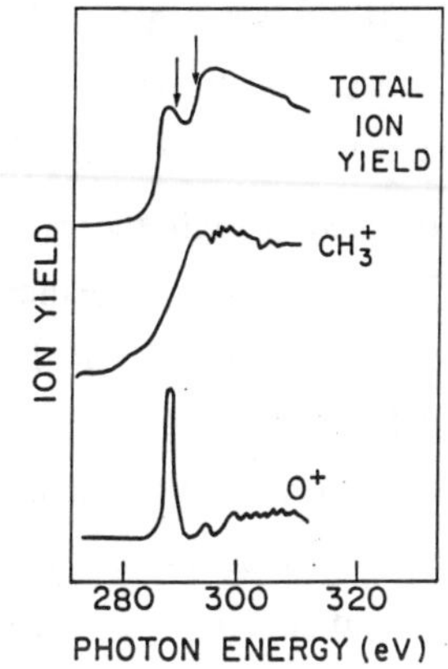

Figure 10. The yields of O^+ and of CH_3^+ following site-selective excitation of an inner core electron in $(CH_3)_2CO$. (Adapted from Eberhardt *et al.* 1983).

by about one order of magnitude larger than that for Coulomb explosion, and there is still an open question regarding the competition between (neutral) fragmentation and IVR on this 10^{-13} time scale. Ionic molecular fragmentation by Coulomb explosion constitutes one of the fastest reactive proceses, which freezes out IVR.

Proceeding from these ultrafast molecular processes towards clusters, we note that the Coulomb explosion was documented in molecular clusters. Doubly charged rare-gas clusters exhibit a remarkably well-characterized minimal "critical" size for the appearance of a stable cluster, e.g., the smallest Ar_n^{2+} is experimentally observed for n = 93. Molecular dynamics studies (Echt 1987, Gay and Berne 1982) on Xe_n^{2+} clusters have indicated that the n = 51 cluster dissociates after ~ 50 ps, while the n = 55 cluster is stable on the time scale of 300 ps. The fission of doubly charged rare-gas clusters is complicated by hole trapping effects (Schwentner *et al.* 1985, Saenz *et al.* 1984), which are considered for a single hole in a cluster but not elucidated in the context of Coulomb explosions. The time scale (5×10^{-10} s) of Coulomb explosion of large (n = 51) rare-gas clusters (Gay and Berne 1982) seems to be sufficiently long to allow for intracluster vibrational energy redistribution. The interesting issue of selective cluster dissociation due to Coulomb explosion is still open.

A novel fascinating aspect of ultrafast processes in clusters involves cluster-surface collisions (Beuhler *et al.* 1989). High energy collisions of charged molecular clusters with surfaces are expected to occur on a $< 10^{-15}$ sec time scale, the lower time limit is sufficiently fast to compete with intracluster and intramolecular IVR. A very intersting experiment in this category pertains to cluster impact fusion (Beuhler *et al.* 1989), where D-D fusion was accomplished by the collision of high-energy (200-300 keV) $(D_2O)_n^+$ (n = 25-1300) clusters with surfaces of TiD. The intramolecular, intracluster, cluster surface and surface compression processes compete effectively with intramolecular and intracluster IVR processes, where the collision energy seems to be selectively deposited in the cluster and in the surface impact domain. The understanding of these phenomena is in embryonic stags. Less esoteric cluser-surface and cluster-cluster high-energy collisions inducing a variety of nonreactive and reactive processes will be of considerable interest in the broad area of selectie chemistry.

7. Active radiative control

Up to this point we have been concerned with passive control of the selectivity of chemical reactions, which is limited to external intervention during the energy acquisition process. After preparation of the initial state (by optical or collisional excitation or by core ionization) the molecular system evolves unperturbed; its time evolution being governed by the free-molecule Hamiltonian and determined by the initial state. Passive control essentially pertains to the establishment of the condition for the survival of the characteristic of the initial doorway state on the time scale for energy disposal. Recent work focussed on active control of photochemical reactions, with the evolution of the system being controlled by both the free molecule Hamiltonian and by external means (Jortner and Kommandeur 1978, Mukamel and Shan 1985, Brumer and Shapiro 1986, Shapiro and Brumer 1989, Brumer and Shapiro 1989, Seideman *et al.* 1989, Chan *et al.* 1991, Tannor and Rice 1985, Tannor *et al.* 1986, Kosloff *et al.* 1989). The various schemes for radiative active control, which were recently discussed, introduce novel methods into the arsenal of laser chemistry, utilizing not only monochromaticity and high intensity of lasers but also phase coherence, short pulse durations, pulse sequencing and pulse shaping. Early studies (Jortner and Kommandeur 1978) of coherent optical effects in isolated large molecules focused on coherent driving by a sequence $\pi/2$ pulses of a doorway state, which is coupled to a background manifold of optically inactive levels. It was demonstrated that electronic relaxation or IVR for this model system can be retarded by a sequence of two delayed $\pi/2$ pulses, with the retardation being determined by the pulse's delay time relative to the temporal decay time $\hbar/2\pi\bar{V}^2\rho$ of the doorway state. Retardation of IVR by coherent driving in intense radiation fields was proposed (Mukamel and Shan 1985) and on the basis of model calculations it was inferred that when the Rabi frequency Ω exceeds the frequency spread of an isolated group of molecular eignestates, i.e., when $\Omega \gg 2\pi\bar{V}^2\rho/\hbar$, IVR processes are suppressed. These conclusions were validated and extended by model calculations (Tannor *et al.* 1986) on the retardation of the spread of a molecular wavepacket on an anharmonic excited state potential surface.

Two major schemes were introduced for active radiative control of chemical reactions. The first class encompasses coherent control in CW excitation, via pulse sequencing (Brumer and Shapiro 1986, Seideman *et al.* 1989, Chan *et al.* 1991). The second scheme involves a pump-dump process induced by a pair of fsec pulses, with control being accomplished by spacing and shaping the pulses. The coherent radiative control scheme of Brumer and Shapiro rests on the implications of quantum interference effects in the intermediate state(s) which affect the branching ratios for product formation in a two-photon (or multiphoton) process. Interference effects are well known in atomic and molecular multiphoton ionization (Park and Diebold 1990) where third harmonic generation can either suppress (Miller and Compton 1982, Li *et al.* 1988, Orlando *et al.* 1986) or enhance (Proctor *et al.* 1987) the ionization process. Here, the third harmonic one-photon (ω_3) absorption interferes with three photon (ω_1) absorption to an intermediate state. While these effects bear a close similarity to the problem at hand, they cannot be readily utilized to control chemical reactions and it is preferable to supply the two coherent ω_1 and ω_3 laser beams externally. A nice recent scenario for active control (Chan *et al.* 1991) involves a $|G\rangle \rightarrow |I\rangle$ excitation by three photon (ω_1) along the first path and by one-photon (ω_3) along a second path. The population of the intermediate $|I\rangle$ state is of the form (Chan *et al.* 1991)

$$\varepsilon_3^2|\langle G|\mu|I\rangle|^2 + \varepsilon_1^6|KG|t|I\rangle|^2 - 2\varepsilon_3\varepsilon_1^3\cos(\Delta\theta + \Delta_{13})|f_{13}| \tag{5}$$

where ε_3 and ε_1 are the electric fields of the ω_3 and ω_1 coherent pulses, respectively. $\Delta\theta$ is their phase difference, μ and t the one-photon transition moment and the three-photon transition operator, respectively, while $f_{13}\exp(i\Delta_{13})$ is the cross term transition matrix element. The interference term in Equation (5) can either retard or enhance dissociation or ionization from |I>. It was shown that such interference effects will result in a considerable degree of control over electronic branching ratios for dissociation of IBr (Chan *et al.* 1991). A closely related problem involves coherent control via pulse sequencing (Seideman *et al.* 1989), which exploits interference effects in a two-pulse sequence, with the first pulse exciting a superposition of two bound electronically excited intermediate states |1> and |2>, and the second pulse induces transitions to continuum states of the ground electronic configuration. The population probability of each continuum exit channel can be expressed (Seideman *et al.* 1991) as a sum of two direct terms involving the contribution of intermediate states and an interference term, which can either constructively enhance or distinctly reduce photodissociation (or ionization) into a given channel. The quantum interference effects in the intermediate states which affect the yield and/or the branching ratios of products can be realized for a coherent short-pulse one-photon excitation of a pair (or a small number) of levels, or alternatively for the CW coherent excitation of a single level (or a bunch of degenerate levels) via distinct excitation routes. This scheme may not be applicable for complex, congested level structure of the intermediate states. This intrinsic limitation will restrict the applicability of this interesting approach to small molecular systems.

The pump-dump scheme (Tannor and Rice 1985, Tannor *et al.* 1986, Kosloff *et al.* 1989) utilizes an excited electronic state to mediate a reaction on the ground electronic state of a triatomic molecule. The ground state potential surface has a stable equilibrium geometry and at least two exit channels. This model system is excited by a sequence of two (Gaussian) pulses. The first pulse pumps the system from its ground vibronic level to an electronically excited state producing a nonstationary nuclear wavepacket on the excited state potential surface. Subsequently, the dumping pulse brings the system back to the electronic ground state followed by nuclear dynamics on the ground state potential surface towards the exit channel(s). The problem assessed and solved (Tannor and Rice 1985, Tannor *et al.* 1986, Kosloff *et al.* 1989) was the determination of the temporal sequence and shape of the pump and dump pulses that optimize the desired product selectivity for a variety of triatomic model systems. The application of the Tannor-Rice-Kosloff scheme to a real-life situation raises some serious problems. The potential surfaces of polyatomic molecules are anharmonic. Consequently, spreading of the nuclear wavepackets will occur between and after the application of the two sequential pulses which will decrease the selectivity of product formation and possibly erode any selectivity in a pump-dump experiment on real molecules. This (and other) active control schemes, which utilize pulse sequencing, must also incorporate elements of passive control, which have to be operative during the long "time windows" between the pulses. Thus, the central issues of the erosion of the signature of the initially excited state (i.e., wavepacket) by IVR has to be also incorporated within the framework of a realistic active control scheme.

What has been accomplished up to date on the experimental front? Two experiments (Scherer *et al.* 1991 and Park *et al.* 1991) were recently reported beautifully demonstrating the manifestation of excitation of a diatomic molecule by two coherent light sources. Scherer *et al.* (1991) used a two-pulse sequence of femtosecond phase locked laser pulses to resonantly excite vibronic transitions of the B state of I_2. This two-pulse experiment involves the preparation and quantum interference of two nuclear wavepacket amplitudes in an excited state of a molecule. This experiment provides a novel approach to control the optically prepared probability amplitude on an excited state potential surface, which may be of considerable interest for selective intramolecular dynamics. Park *et al.* (1991) have accomplished coherent control of the ionization of HCl achieved by simultaneous 3+1 and 1+1 multiphoton excitation via the intermediate $j\ {}^3\Sigma(\Omega = 0^+)$ electronic state. Coherent phase control in accord with Eq. (5), was accomplished by varying the phases of the two laser beams. This remarkable result, which was accomplished for a single channel molecular process, provides a promising avenue for the control of branching ratios in multi-channel energy disposal.

8. Active control of charge separation in photosynthesis

We now proceed from molecular dynamics to dynamic processes in large biomolecules. The primary charge separation in photosynthesis constitutes a central energy conversion process in biology. Extensive structural information on the structure of reaction centers of photosynthetic bacteria (Deisenhofer and Michel 1989) allows to establish central structure-function relations for these systems. The X-ray structure of photosynthetic bacterial reaction clusters shows a quasisymmetric arrangement of the prosthetic groups (i.e., the bacteriochlorophyll dimer, the two accessory bacterioclorophylls, the two bacteriopheophytons and two quionoes) along the two protein subunits M (medium weight) and L (light weight) forming the L and M branches of the reaction center (Deisenhofer and Michel 1989). The two branches exhibit (approximately) two-fold rotation symmetry, with the spatial and orientational ordering of the pigments being almost identical to the two branches (Fig. 11). One outstanding problem is the undirectionality of the electron transfer process along the L branch of the reaction center (Michel-Beyerle *et al.*

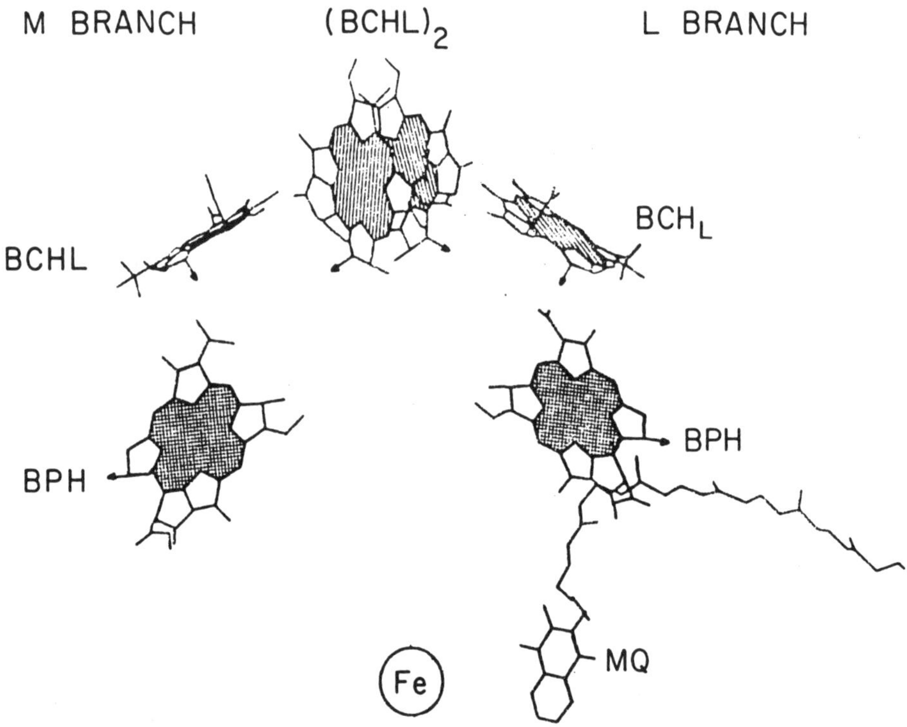

Figure 11. An overview of the structure of the prosthetic groups in the reaction center of Rps. viridis (Diesenhofer and Michel 1989).

1988), with the branching ratio for charge separation along the two branches being > 25 (at 90 K) in favour of the L branch (Bixon *et al.* 1991). Undirectionality is governed by symmetry breaking effects induced by the protein medium (Michel-Beyerle *et al.* 1988, Bixon *et al.* 1991) which originate from distinct electronic and nuclear contributions to the elementary electron transfer rates between the prosthetic groups favoring the electron transfer process along the L branch. These electron transfer rates are described within the framework of nonadiabatic multiphonon theory (Jortner 1976), being determined by intermolecular electronic coupling terms and nuclear Franck Condon factors. Small structural differences for the spatial ordering of the pigments across the two branches to a far way in modifying the electronic coupling terms in favor of those for the L branch. Concurrently, the energetics of the ion pair states is modified by different microscopic environment on the two branches, resulting in different nuclear Franck Condon factors for electron transfer between pigments on the L and M branches, favoring again the L branch (Bixon *et al.* 1991). The phenomenon of symmetry breaking originates from cumulative contributions of electronic and nuclear factors, inducing the unidirectionality of charge separation. Active control exerted by the protein medium accomplishes this remarkable selectivity in the primary charge separation in photosynthesis.

Acknowledgment

We thank Dr. David Weeks for his critical reading of the manuscript. This research was supported in part by the Binational Israeli-German James Franck Research Program for laser-matter interaction.

References

Amar, F.G. (1991) 'Cluster photoexcitation and evaporation: two regimes', in this volume.

Amirav, A., Horwitz, C. and Jortner, J. (1988) 'Optical selection studies of electronic relaxation from the S_1 state of jet-cooled anthracene derivatives', J. Chem. Phys. 88, 3092.

Barrio, G.D. (1991) 'Photofragmentation of rare gas-interhalogen van der Waals complexes: theoretical studies versus experimental data, in this volume.

Bauer, S.H. (1978) 'How energy accumulation and disposal affect the rates of reactions', Chem. Rev. 78, 147.

Bernstein, R.B. and Zewail, A.H. (1989) 'Femtosecond real-time probing of reactions. III. Inversion to the potential from femtosecond transition-state spectroscopy experiments', J. Chem Phys. 90, 829.

Berry, R.S. (1991) 'How we and molecules explore molecular landscapes', in this volume.

Bersohn, R. (1991) 'Temperatures of fragments of unimolecular dissociations', in this volume.

Bersohn, R. and Zewail, A.H. (1988) 'Time dependent absorption of fragments during photodissociation', Ber. Bunsenges. Phys. Chem. 92, 373.

Beswick, J.A. and Jortner, J. (1981) 'Intramolecular dynamics of van der Waals molecules', Adv. Chem. Phys. 47, 363.

Beswick, J.A. and Jortner, J. (1990) 'Time scales for molecular photodissociaton', Chem. Phys. Lett. 168, 246.

Beswick, J.A. (1991) 'Coherent excitation effects in photofragmentation', in this volume.

Beuhler, R.J., Friedlander, G. and Friedman, L. (1989) 'Cluster-impact fusion', Phys. Rev. Lett. 63, 1292.

Bixon, M. and Jortner, J. (1968) 'Intramolecular Radiationless Transitions', J. Chem. Phys. 48, 715.

Bixon, M. and Jortner, J. (1969) 'Comments on electronic relaxation processes', Molecular Crystals 213, 237.

Bixon, M., Jortner, J., and Michel Beyerle, M.E. (1991) 'On the mechanism of the primary charge separation in bacterial photosynthesis', Biochim. Biophys. Acta 1056, 301.

Brauman, J.I., Zare, R.N. and Levine, R.D. (1990) 'Fragment isotope distribution as a signature of molecular fluxionality', Chem. Phys. Lett. 172, 233.

Brumer, P. (1981) 'Intramolecular energy transfer: theories for the onset of statistical behavior', Adv. Chem. Phys. 47, 201.

Brumer, P. (1991) 'Developments in coherent radiative control of chemical reactions', in this volume.

Brumer, P. and Shapiro, M. (1986) 'Control of unimolecular reactions using coherent light', Chem. Phys. Lett. 126, 541.

Brumer, P. and Shapiro, M. (1989) 'One-photon mode selective control of reactions by rapid or shaped laser pulses: an emperor without clothes?', Chem. Phys. 139, 221.

Brummel, C.L., Mork, S.W. and Philips, L.A. (1991) 'Evidence of virational mode coupling in 2-fluroethanol via high-resolution infrared spectroscopy in a molecular beam', JACS 113, 4342.

Chan, C.K., Brumer, P. and Shapiro, M. J. Chem. Phys., in press.

Chesnovsky, O. (1991) 'Dynamics of triplet states in isolated molecules', in this volume.

Crim, F.F. (1990) 'State- and bond-selected unimolecular reactions', Science, 249, 1387.

Crim, F.F. (1991) 'Controlling the pathways in chemical reactions: bond-selected photodissociation and bimolecular research', in this volume.

Dayton, D.C., Block, P.A. and Miller, R.E. (1991) 'Spectroscopic evidence for near-resonant intermolecular energy transfer in the vibrational predissociation of C_2H_{2-HX} and C_2H_2–DX (X=Cl, Br and I) Complexes, J. Phys. Chem. 95, 2881.

Deisenhofer, J. and Michel, H. (1989) 'The photosynthetic reaction centre from the purple bacterium *Rhodopseudomonas viridis*', EMBO J. 8, 2149.

Eberhardt, W., Sham, T.K., Carr, R., Krummacher, S., Strongin, M., Weng, S.L. and Wesner, D. (1983) 'Site-specific fragmentation of small molecules following soft-X-ray excitation', Phys. Rev. Lett. 50, 1038.

Echt, O. (1987) in Elemental and Molecular Clusters, G. Benedek, T.P. Martin, G. Pacchioni (eds.), Springer-Verlag, Berlin, p. 263.

Fain, B. (1991) 'Implication of giant optical vibrations created by ultrashort laser pulses', in this volume.

Fleming, P.R., Luo, X. and Rizzo, T.R. (1991) 'Multiple laser probes of intramolecular dynamics at chemically significant energies', in this volume.

Gay, J.G. and Berne, B.J. (1982) 'Computer simulation of Coulomb explosions in doubly charged Xe microclusters', Phys. Rev. Lett. 49, 194.

Gerber, R.B., Alimi, R., Garcia-Vela, A. and Hurwitz, A. (1991) 'Dynamics of photoinduced reactions in clusters', in this volume.

Engel, Y.M. and Levine, R.D. (1989) 'Vibration-vibration resonance conditions in intramolecular classical dynamics of triatomic and larger molecules', Chem. Phys. Lett. 164, 270.

Even, U. and Ben-Horin, N. (1991) 'Dynamics of heteroclusters', in this volume.

Go, J., Bethardy, G.A. and Perry, D.S. (1990) 'Rotationally mediated intramolecular vibrational redistribution in jet-cooled *trans*-ethanol at 2990 cm^{-1}', J. Phys. Chem. 94, 6153.

Habenicht. W., Baiter, H., Müller-Dethlefs, K. and Schlag, E.W. (1991) '"Memory effects" in molecular fragmentation induced by site-specific core excitation using a reflection time-of-flight mass spectrometer', J. Phys. Chem.

Hamilton, C.E., Kinsey, J.L. and Field, R.W. (1986) 'Stimulated emission pumping: new methods in spectroscopy and molecular dynamics', Ann. Rev. Phys. Chem. 37, 493.

Heidenreich, A. and Jortner, J. (1991) 'Search for mode selective cluster isomerization', in this volume.

Heller, E.J. (1980) 'Quantum intramolecular dynamics: criteria for stochastic and nonstochastic flow', J. Chem. Phys. 72, 1337.

Heller, E.J. (1981) 'The semiclassical way to molecular spectroscopy', Acc. Chem. Res. 14, 368.

Hinshelwood, C.N. (1927) 'On the theory of unimolecular reactions', Proc. Roy. Soc. (London) A113, 230.

Hinshelwood, C.N. and Fletcher, C.J.M. (1933) 'Kinetics of the decomposition of molecules of intermediate complexity', Nature 131, 24.

Holme, T.A. and Levine, R.D. (1989) 'Theoretical and computational studies of highly vibrationally excited acetylene', Chem. Phys. 131, 169.

Jellinek, J. (1991) 'State selectivity in clusters - molecular interactions', in this volume.

Jortner, J. (1976) 'Temperature dependent activation energy for electron transfer between biological molecules', J. Chem. Phys. 64, 4860.

Jortner, J. and Kommandeur, J. (1978) 'Some features of coherent optical effects in large molecules', Chem. Phys. 28, 273.

Jortner, J. and Levine, R.D. (1981) 'Photoselective chemistry', Adv. Chem. Phys. 47, 1.

Jortner, J. and Levine, R.D. (1990) 'Selective chemistry', Is. J. Chem. 30, 207.

Jortner, J. and Mukamel, S. (1973) 'Preparation and decay of excited molecular states' in 'The World of Quantum Chemistry' R. Daudel and B. Pullman (eds.) Reidel, Dordrecht, p. 145.

Kassel, L.S. (1928) 'Studies in homogeneous gas reactions', J. Phys. Chem. 32, 225.

Klafter, J. and Braiman, Y. (1991) 'Energy transfer in molecular chains', in this volume.

Knotek, M.L. and Feibelman, P.J. (1978) 'Ion Desorption by core-hole Auger decay', Phys. Rev. Lett. 40, 71.

Kolodney, E. (1991) 'Product state distributions in molecule surfaces dissociative scattering', in this volume.

Kosloff, R. and Baer, R. (1991) 'Impulsive iterative inversion', in this volume.

Kosloff, R., Rice, S.A., Gaspard, P., Tersingi, S. and Tannor, D.J. (1989) 'Wavepacket dancing: Achieving chemical selectivity by shaping light pulses', Chem. Phys. 139, 201.

Lehmann, K.K. (1991) 'Statistical IVR and its frustrations: sub-Doppler spectra of substituted acetylene compounds', in this volume.

Lehmann, K.K., Pate B.H. and Scoles, G. (1990) 'Statistical intramolecular vibrational relaxation and its hindrance: The fundamentals of $(CH_3)_3C{-}C{\equiv}CH$ and $(CH_3)_3Si{-}C{\equiv}CH$', J. Chem. Phys. 93, 2152.

Levine, R.D. (1981) 'The information theoretic approach to intramolecular dynamics', Adv. Chem. Phys. 47, 239.

Levine, R.D. (1987) 'Fluctuations in spectral intensities and transition rates', Adv. Chem. Phys. 70, 53.

Levine, R.D. (1988) 'Quantal fluctuations in unimolecular rate constants', Ber. Bunsenges. Phys. Chem. 92, 222.

Levine, R.D. (1989) 'Intramolecular Dynamics', in J. Jortner and B. Pullman (eds.), Perspectives in Quantum Chemistry, Kluwer, pp. 41.

Levine, R.D. and Bernstein, R.B. (1987) Molecular Reaction Dynamics and Chemical Reactivity, Oxford University Press, Oxford.

Levy, D.H. (1981) 'van der Waals molecules', Adv. Chem. Phys. 47, 323.

Levy, D.H. (1991) 'Mode selected van der Waals chemistry: complexes of amino tetrazine', in this volume.

Letokhov, V.S. (1989) (ed.) 'Lasers in Atomic, Molecular and Nuclear Physics', World Scientific, Singapore.

Li, L., Wu, M. and Johnson, P.M. (1987) 'Photodissociation, third harmonic interferences, polarization effects and the MPI spectroscopy of chlorine, J. Chem. Phys. 86,

Li, L., Wu, M. and Johnson, P.M. (1988) 'One- and two-color multiphoton ionization of argon', J. Chem. Phys. 89, 3470.

Li, Y.S. and Wilson, K.R. (1990) 'What can gas phase reactions tell us about solution reactions?', J. Chem. Phys. 93, 8821.

Li, Y.S., Whitnell, R.M., Wilson, K.R. and Levine, R.D. (1991) 'Overtone induced dissociated of small molecules in rare gas solution,

Light, J.C. (1991) 'Vibrational level structures and resonanes in triatomic systems', in this volume.

Lin, S.H. (1991) 'Femtosecond processes and ultrafast biological electron transfer', in this volume.

Lindemann, F.A. (1922) 'The radiation theory of chemical action', Trans. Faraday Soc. 17, 598.

Lorquet, J.C., Engel, Y.M. and Levine, R.D. (1990) 'On the separation of time scales in the exploration of phase space of an isolated molecule', Chem. Phys. Letts. 175, 461.

Lupo, D.W. and Quack, M. 'IR-Laser photochemistry' (1987) Chem. Rev. 87, 181.

J. Manz and C.S. Parmenter, (1989) 'Mode selective unimolecular reactions', Chem. Phys. 139, 1.

Marcus, R.A. and Rice, O.K. (1951) 'The kinetics of the recombination of methyl radicals and iodine atoms', J. Phys. & Colloid Chem. 55, 894.

Marcus, R.A. (1983) 'On the theory of intramolecular energy transfer', Faraday Discuss. Chem. Soc. 75, 103.

McIlroy, A. and Nesbitt, D.J. (1990) 'Vibrational mode mixing in terminal acetylenes: High-resolution infrared laser study of isolated *J* states', J. Chem. Phys. 92, 2229.

Michel-Beyerle, M.E. (1985) (ed.) 'Antennas and reaction centers of photosynthetic bacteria', Springer-Verlag, Berlin.

Michel-Beyerle, M.E. (1990) (ed.) 'Reaction centers of photosynthetic bacteria', Springer-Verlag, Berlin'.

Michel-Beyerle, M.E., Plato, M., Deisenhofer, J., Michel, H., Bixon, M. and Jortner, J. (1988) 'Unidirectionality of charge separation in reaction centers of photosynthetic bacteria', Biochim. Biophys. Acta 93, 52.

Miller, J.C. and Compton, R.N. (1982) 'Third harmonic generation of multiphoton ionization in rare gases', Phys. Rev. A25, 2056.

Mukamel, S. and Shan, K. (1985) 'On the selective elimination of intramolecular vibrational redistribution using strong resonant laser fields', Chem. Phys. Letts. 117, 489.

Müller-Dethlefs, K., Sander, M., Chewter, L.A. and Schlag, E.W. (1984) 'Site-specific excitation in molecules at very high energies: changes in ionization patterns of CF_3CH_3', J. Phys. Chem. 88, 6098.

Naaman, R. (1991) 'Mode selectivity in energy distribution in aniline scattered from organic surfaces', in this volume.

Nelson, K.A. (1991) 'Ultrafast spectroscopy of chemically reactive systems with femtosecond pulses and pulse sequences', in this volume.

Nesbit, D. (1991) 'IR Laser studies of mode selective energy transfer dynamics in fragile molecules', in this volume.

Nikitin, E.E. (1991) 'The Landau method of estimating transition probabilities: collision-induced and radiation-induced events', in this volume.

Noid, D.W., Koszykowski, M.L. and Marcus, R.A. (1981) 'Quasiperiodic and stochastic behavior in molecules', Ann. Rev. Phys. Chem. 32, 267.

Oref, I. (1991) 'Random walk model for energy transfer at high temperatures', in this volume.

Oref, I. and Rabinovitch, B.S. (1979) 'Do highly excited reactive polyatomic molecules behave ergodically?', Acc. Chem. Res. 12, 166.

Orlando, T.M., Li, L., Anderson, S.L. and White, M.G. (1986) 'Third harmonic interference effects in the MPI spectrum of acetylene', Chem. Phys. Lett. 129, 31.

Park, S.M. and Diebold, G.J. (1990) 'Selective ionization of Ba and Sr isotopes based on a two-photon interference effect', Phys. Rev. A42, 417.

Park, S.M., Lu, S.P. and Gordon, J. (1991) J. Chem. Phys., in press.

Pate, B.H., Lehman, K.K. and Scoles, G. (1991) J. Chem. Phys. in press.

Pique, J.-P., Chen, Y., Field, R.W. and Kinsey, J.L. (1987) 'Chaos and dynamics on 0.5-300-ps time scales in vibrationally excited acetylene: Fourier transform of stimulated-emission pumping spectrum, Phys. Rev. Letts. 58, 475.

Pique, J.-P., Engel, Y.M., Levine, R.D., Chen, Y., Field, R.W. and Kinsey, J.L. (1988) 'Broad spectral features in the stimulated emission pumping spectrum of acetylene', J. Chem. Phys. 88, 5972.

Polik, W.F., Moore, B.C. and Miller, W.H. (1988) 'Quantum interference among competing unimolecular decay channels: asymmetric S_0D_2CO decay profiles', J. Chem. Phys. 89, 3584.

Proctor, M.J., Stockdale, J.A.D., Efthimiopoulos, T. and Fotakis, C. (1987) 'Third-harmonic generation and ionization proceses in Kr', Chem. Phys. Lett. 137, 223.

Quack, M. (1991) 'Mode selective vibrational redistribution and unimolecular reactions during and after IR laser excitation', in this volume.

Remacle, F. and Levine, R.D. (1991) 'The maximum entropy of an optical spectrum and the redistribution in phase space', Chem. Phys. Lett. 181, 307.

Remacle, F. and Levine, R.D. (1991) 'Does a dissociating molecule sample the available phase space'

Rice, S.A. (1981) 'An overview of the dynamics of intramolecular transfer of vibrational energy', Adv. Chem. Phys. 47, 117.

Rice, S.A. (1991) 'Active control of selectivity of product formation in a chemical reaction: what's new?', in this volume.

Rice, O.K. and Rampsperger, H.C. (1927) 'Theories of unimolecular gas reactions at low pressures', J. Am. Chem. Soc. 49, 1617.

Rosenwaks, S. (1991) 'Mode selective photodissociation of deuterated molecules', in this volume.

Rubahn, H.-G. and Bergmann K. (1990) 'The effect of laser-induced virational bond stretching in atom-molecule collisions', Ann. Rev. Phys. Chem. 41, 735.

Saenz, J.J., Soler, J.M., Garcia, M. and Echt, O. (1984) Chem. Phys. Lett. 109, 71.

Salman, I., Silberstein, J. and Levine, R.D. (1991) 'Non-statistical charged fragments distribution in a Coulomb explosion following a site-selective ionization, J. Phys. Chem., in press.

Scherer, N.F., Carlson, R.J., Matro, A., Du, M., Ruggeiro, A.J., Romero-Rochin, V., Cina, J.A., Fleming, G.R. and Rice, S.A. J. Chem. Phys., in press.

Schlag, E.W. (1991) 'Site-selective memory effects in molecular and ZEKE-spectroscopy', in this volume.

Schlag, E.W. and Levine, R.D. (1989) 'On the unimolecular dissociation of large molecules', Chem. Phys. Letts. 163, 523.

Schwentner, N., Koch, E.E. and Jortner, J. (1985) Electronic Excitations in Condensed Rare Gases, Springer-Verlag, Berlin.

Seideman, T., Shapiro, M. and Brumer, P. (1989) 'Coherent radiative control of unimolecular reactions: Selective bond breaking with picosecond pulses', J. Chem. Phys. 90, 7132.

Sewell, T.D., Thompson, D.L. and Levine, R.D. (1991), to be published.

Shapiro, M. (1991) 'Theory of chemical reactions and bond breaking with ultra short laser pulses', in this volume.

Shapiro M. and Brumer, P. (1989) 'Laser control of unimolecular decay yields in the presence of collisions', J. Chem. Phys. 90, 6179.

Soler, J.M., Saenz, J.J. and Garcia, N. (1984) 'The effect of ionization on magic numbers of rare-gas clusters', Chem. Phys. Lett. 109, 71.

Speiser, S. (1991) 'Electronic energy transfer between benzene and biacetyl in a supersonic jet expansion', in this volume.

Stockbauer, R. and Madey, T.E. (1983) 'Desorption of ions from surfaces: mechanisms of photon stimulated desorption', Ann. Isr. Phys. Soc. 6, 483.

Stuchebrukhov, A., Ionov, S. and Letokhov, V. (1989) 'IR spectra of highly vibrationally excited large polyatomic molecues and intramolecular relaxation', J. Phys. Chem. 93, 5357.

Tannor, D.J. and Jin, Y. (1991) 'Design of femtosecond pulse sequences to control photochemical products', in this volume.

Tannor, D.J., Kosloff, R. and Rice, S.A. (1986) 'Coherent pulse sequence induced control of selectivity of reactions: Exact quantum mechanical calculations', J. Chem. Phys. 85, 5805.

Tannor, D.J. and Rice, S.A. (1985) 'Control of selectivity of chemical reaction via control of wave packet evolution', J. Chem. Phys. 83, 5013.

Thompson, D.L. (1991) 'Mode specificity in unimolecular reactions', in this volume.

Troe, J. (1991) 'Adiabatic and nonadiabatic dynamics in unimolecular reactions', in this volume.

Weeks, D.E. and Levine, R.D. (1991) 'Sub-resonant distortion of phase space torii'

Whitehead, J.C. (1988) (ed.) Selectivity in Chemical Reactions, Reidel, Dordrecht.

Wilson, K.R. and Levine, R.D. (1988) 'Activated chemical reactions driven by accepted fluctuations', Chem. Phys. 152, 435.

Zhao, Z.-Q. and Parmenter, C.S. (1991) 'Observations of level mixing associated with the methyl rotor acceleration of intramolecular vibrational redistribution. p-fluorotoluene, in this volume.

INDEX